Biodiversity, Ecology and Conservation of Rivers and Streams of North East India

Biodiversity, Ecology and Conservation of Rivers and Streams of North East India

Edited by

Laishram Kosygin

AKANSHA PUBLISHING HOUSE

NEW DELHI • GUWAHATI • VISAKHAPATNAM

AKANSHA PUBLISHING HOUSE
4649-B/21, Ansari Road
Darya Ganj, New Delhi - 110 002
Ph.: 23263193 / 9811582579 / 09435116718
Fax: 011-23263193
Email: ektabooks@yahoo.com
www.akanshapublishinghouse.com

Biodiversity Ecology and Conservation of Rivers and Streams of North East India

First Published 2011
ISBN 978-81-8370-276-8

PRINTED IN INDIA

Published by Akansha Publishing House, New Delhi and Printed at Harmain Offset Press, Delhi.

Contents

Foreword

Research in the vast field of fish ecology, limnology and eco-conservation has shown rapid progress during the last few decades. Despite the recent developing trend of wetland studies there has been very little informations on the limnology, biodiversity and bio-ecology of the freshwater lentic water bodies of the North Eastern region of India. Interestingly the N.E. region although comprised of only 7.5% of the total geographical area of India, its drainage system involves the tributaries of three major river systems of South-east Asia: the Barak-Brahmaputra system, the Chindwin- Irrawadi system and Kaladan river system. Further, this N.E. region of India, being the core zone of Indo-Burma biodiversity Hot Spots, and tri-junction of three distinct biogeographical regions: extensions of Eastern Himalayas, Malayan Archipelago and Oriental region (India), indicate a high degree of endemism playing an important role in the speciation and evolution of species. Moreover, the South East Asia is a region of great diversity in respect of fish fauna. However, the biodiversity and bio-ecology of this region still remained unexplored. Apart from certain occasional reports on limnology and aquatic biology on the lotic water bodies of N.E. India there has not been any up to date reference work on this aspect.

It has been a pleasure for me to go through the contents of the book entitled "Biodiversity, Ecology and Conservation of Rivers and Streams of North East India" edited by Laishran Kosygin (Scientist, Zoological Survey of India). It embodies a compilation of 25 selected research articles from various authors based on various aspects of biodiversity, ecology, limnology, hydrology and eco-conservation of the lotic water bodies in the N.E. region of India, which is really a significant contribution to the subject. I am very hopeful that this book will be a highly useful document for a variety of scholars, students, teachers, researchers, scientists, and all workers in the field of lotic freshwater biology in general and of N.E. India in particular.

Dr. Kosygin deserves special appreciation for his painstaking and enthusiastic approach to complete the strenuous task. I would like to congratulate Dr. Kosygin over his remarkable contribution to highlight the available important works relating to the biodiversity, ecology and conservation of the lotic water bodies of North East India.

Prof H. Tombi Singh

The North Eastern
Pradesh, Assam,
is endowed with
important and
Churachandpur
the Barak
tributaries are
hydrologic region
flowing through
drainage
Malaysian
is unique
high level of
the region. Howev
discovery state.
exploitation and
quality
accused
a compilation
aspects of biodiversity
close water bodies
My
have enriched the
of India for his
I wish to express
of Botanic
Vice-Chancellor
inspire
Sciences, Manipur
Ecology Wing, Govt.
and Dr. K.K. Rajeshwari
suggestions and
Sciences
Research management
Scientist 'F' of NEH

Preface

The North Eastern Region of India, comprising of eight states namely Arunachal Pradesh, Assam, Manipur, Meghalaya, Mizoram, Nagaland, Sikkim and Tripura, is endowed with bountiful water resources, which are drained three internationally important and entirely separated drainages namely, Barak-Brahmaputra, Chindwin-Irrawaddy and Kaladan. The Brahmaputra in the northern part and the Barak flowing through the southern margins along with their numerous tributaries are the two main rivers that have created a very dynamic and powerful hydrologic regime in the region. On the other hand, the Chindwin-Irrawaddy flowing through Myanmar drains the easternmost flanks of the region. The Kaladan drainage drains some part of the southern Mizoram. At the confluence of the Indo-Malayan, Indo-Chinese and Indian biogeographical realms, the NE region of India is unique in providing a profusion of habitats, which features diverse biota with a high level of endemism. In the recent years, many new species are described from the region. However, biodiversity and ecology of the region are still under the discovery state. Most of the rivers of the region are under pressure due to over exploitation and other anthropogenic activities leading to deterioration of water quality and degradation of biodiversity etc. Therefore, it is necessary to obtain accurate information regarding the biodiversity and ecology of these rivers and streams in order to conserve and manage them holistically. The aim of the book is to provide baseline information on the lotic water bodies of this region for their conservation and future investigation.

My special thanks and appreciation go to the scientists whose contributions have enriched the book. I express my deep sense of gratitude to Zoological Survey of India for his valuable suggestions and encouragement to bring out this volume. I wish to express my sincere gratitude to Dr. S. Kaul, Director (Wetlands) of Ministry of Environment and Forests, Government of India and Prof. H. Tombi Singh, former Vice-Chancellor of Manipur University, who have been the source of constant inspiration. I am also thankful to Prof. Waikhom Vishwanath, Department of Life Sciences, Manipur University, Dr. M. Homeshor Singh, Head of Environment and Ecology Wing, Govt. of Manipur, Dr. H. Birkumar Singh of NEIST (CSIR), Imphal and Dr. R.K. Rajeshwari Devi, Reader of Oriental College, Imphal for their valuable suggestions and help. I owe my special thanks to Prof. S. P. Biswas, Head of Life Sciences Department, Dibrugarh University; Dr. Sabitry Ch. Bordoloi, Head of Resource management and Environment division, IASST, Guwahati; Dr. P. Kotoky, Scientist 'F' of NEIST, Jorhat; Dr. L. Arunkumar, Sr. Lecturer of Mayai Lambi

College, Manipur for various courtesies. I am also thankful to all well-wishers who helped and extended their cooperation in this endeavour.

I also express my deep sense of gratitude to my mother Smt S. Pramodini Devi and my wife Smt. R.K. Gyaneshwari Devi for their valuable encouragements and moral supports given to me.

Laishram Kosygin

Biodiversity and Conservation Strategies of Fishes of the Tlawng River, Mizoram, India with Some New Records

—*Laishram Kosygin**

[This paper presents a checklist of the fish fauna of Tlawng river, Mizoram, based on present collection and those reported by the earlier workers. In total, sixty-four species of fishes representing thirty-four genera and fifteen families have been recorded from the river. *Garra manipurensis* Vishwanath and Sarojnalini and *Tor barakae* Arunkumar and Basudha are reported here for the first time from the Mizoram state and *Glyptothorax telchitta* (Hamilton) and *Garra kempi* Hora, for the first time from the Tlawng river. Based on the IUCN status, the fish fauna of the river includes one endangered, three vulnerable and two near threatened species. Highest species composition was observed in the cyprinidae family (53.1%) followed by cobitidae (7.8%) and sisoridae (7.8%). Fish fauna of the river is a mixture of endemic hill stream, Assamese and some widely distributed forms. Strategies for conservation of threatened fishes are suggested.

Key words: Biodiversity, conservation, new records, Tlawng river, Mizoram.]

Introduction

The Tlawng river, which is also known as Dhaleswari river is the longest and one of the most important rivers of Mizoram, India. The river originates at an elevation of about 1070 m from the southern part of Myanmar and flows from the south to north direction passing through the *Aizawl District* and eventually joins the Barak river in Cachar of Assam. The length of the river is about 185 km in Mizoram and it traverses with a moderate slope.

The river has rich biodiversity and fishing activity in the river is considered as an exciting experience. However, much of the potential of the river systems and aquatic bioresources in Mizoram remains largely unexploited. There are few reports on the fishes of Tlawng river of Mizoram. Kar *et al.* (2000) reported nine fish species from the river. Later, Karmakar and Das (2007) recorded sixteen species of which *Devario naganensis, Garra annandalei, Garra nasuta, Lepidocephalichthys*

*Zoological Survey of India, Hilltop, Gopalpur-on-Sea 761 002, Ganjam District, Orissa
Email: lkzsi5@yahoo.com

berdmorei, Eutropiichthys vacha and *Conta conta* were reported for the first time from the Mizoram state. While reporting fishes of Barak drainages of North Eastern India, Kar and Sen (2007) included thirty-five species from the Tlawng river in Mizoram. However, there is no comprehensive list of fishes of the Tlawng river and information on their status. In the present study a systematic checklist of fish fauna of the river is prepared based on the present collection and those reported by the earlier workers. The status of the fishes and conservation strategies of the threatened species are given.

Materials and Methods

Fishes were collected from the Tlawng river, near Sairang of Mizoram and identified following Jayaram (1999), Talwar and Jhingran (1991) and by comparing with the type and others of specimens in the Zoological Survey of India (ZSI), Kolkata and Manipur University Museum of Fishes (MUMF). Species, which were not collected in the present survey but reported earlier from the river, are also included in this paper. The families have been arranged phylogenetically, following the classification of Nelson (1984) and species under a genus followed alphabetic sequence. The correct Zoological name with author citation, distribution and status (IUCN, 2010) are shown against each species.

Results and Discussion

A checklist of the fish fauna of the Tlawng river of Mizoram is furnished in Table 1.1. In the present collection a total of twenty-five species, which belongs to fourteen genera and six families were collected from the river. The cyprinid fishes *Garra manipurensis* Vishwanath and Sarojnalini, 1998 (Figure 1.1, 1.2) and *Tor barakae* Arunkumar and Basudha, 2003 are reported here for the first time from the Mizoram State and *Glyptothorax telchitta* (Hamilton, 1822) and *Garra kempi* Hora, 1921, for the first time from the Tlawng river.

Garra manipurensis was originally described from Manipur by Vishwanath and Sarojnalini (1988). However, Nebeshwar *et al.* (2009) remarked that the type locality as given in the original description might be due to wrong labeling as several subsequent collection from the Manipur river did not represent the species. Vishwanath and Kosygin (2002) and Nebershwar Sharma (2002) reported the fish from Barak river (Brahmaputra basin) in Manipur, India. In the present study identity of the specimens collected from the Mizoram were confirmed after comparing with the holotype in the Manipur University Museum of Fishes (MUMF 0130) of *G. manipuensis* and distribution of the fish is extended to the Mizoram state. Menon (1964) remarked that gradual shifting of vent forward from the base of anal fin and development of proboscis on snout as specialized adaptations of *Garra* to rapid running waters, which are also of great systematic and evolutionary significance within the genus. While reporting fishes of *Garra* from Manipur, Vishwanath (1993) remarked that in *lissorhynchus* and *yunnanensis* complexes of the genus, which do not possess a proboscis, the vent has shifted far forward of the anal fin, whereas in *gotyla* complex, where a proboscis is present, the vent is not far forward from the anal fin origin. However, in *G. manipurensis* both the specialized characters are lacking. Therefore, the fish may be considered as a primitive form of the genus.

Figure 1.1a and b. Comparative views of Tlawng river at Sairang, Mizoram showing the seasonal variations in water colour. A: clear water during pre-monsoon (April, 2007); B: Turbid water due to heavy inflow of eroded silts from the catchment area during monsoon/Post-Monsoon (September, 2007).

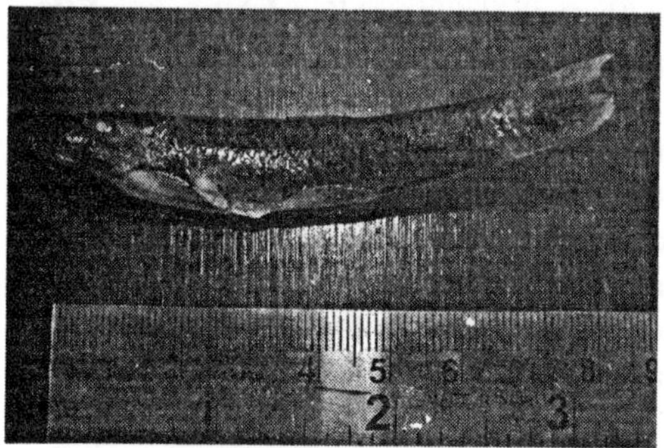

Figure 1.2. *Garra manipurensis*

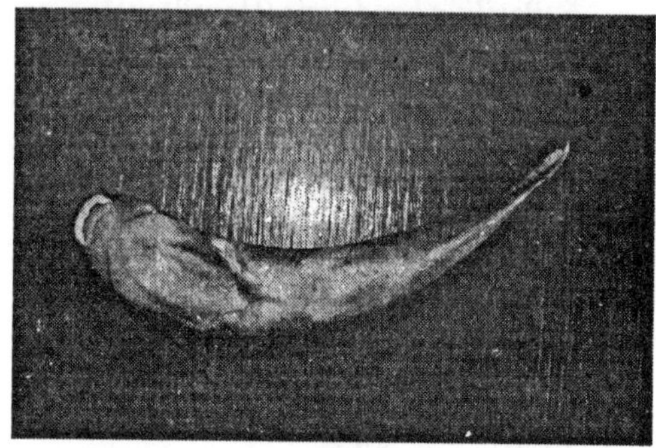

Figure 1.3. Ventral view of *Garra manipurensis.*

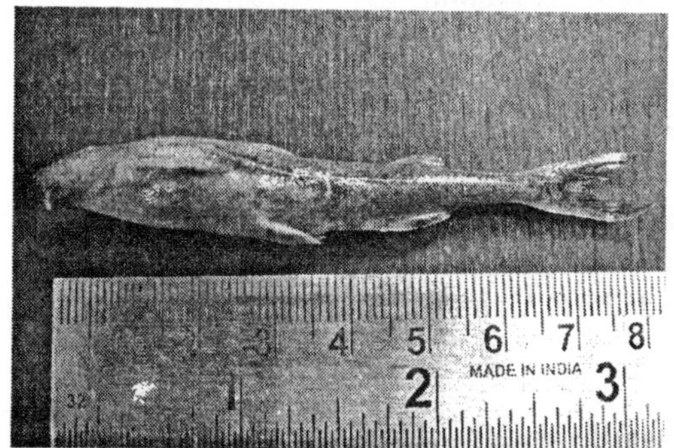

Figure 1.4. *Glyptothorax telchitta*

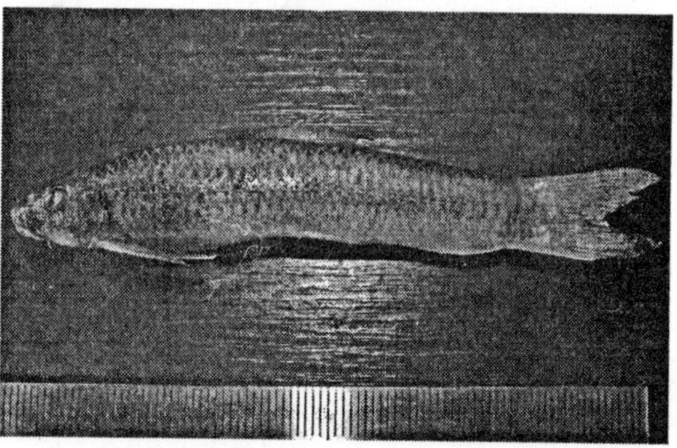

Figure 1.5. *Garra kempi.*

Table 1.1
Systematic list, distribution and IUCN status of the fishes of Tlawng river, Mizoram

Scientific Name	*Distribution*	*Status*
Order CYPRINIFORMES		
Family CYPRINIDAE		
Subfamily Danioninae (= Rasborinae)		
1. *Securicula gora* (Hamilton) $^{\alpha\delta}$	India (North India up to Orissa and Mizoram), Bangladesh, Nepal, Pakistan.	LC
2. *Salmophasia bacaila* (Hamilton)* $^{\alpha}$	India (Brahmaputra, Ganga, Mahanadi river systems), Bangladesh, Nepal, Pakistan.	LC
3. *Salmophasia phulo* (Hamilton) $^{\alpha}$	India (Brahmaputra, Ganga river systems), Bangladesh.	LC
4. *Barilius barna* (Hamilton) $^{\alpha\beta}$	India (Brahmaputra, Ganga, Mahanadi river systems), Bangladesh, Nepal.	LC
5. *Barilius bendelisis* (Hamilton)* $^{\alpha\beta}$	India (through out except Kerala), Bangladesh, Nepal, Pakistan, Sri Lanka.	LC
6. *Barilius shacra* (Hamilton) $^{\beta}$	India (Brahmaputra, Ganga, Yamuna rivers), Bangladesh, Nepal.	LC
7. *Barilius tileo* (Hamilton) $^{\alpha\delta}$	India (Assam, Manipur, West Bengal, Mizoram), Bangladesh, Nepal.	NE
8. *Barilius vagra* (Hamilton)* $^{\alpha\delta}$	India (Brahmaputra, Ganga, Indus, Yamuna river systems), Bangladesh, Nepal, Pakistan, Sri Lanka.	LC
9. *Barilius* sp1.*#	India (Mizoram)	NE
10. *Barilius* sp2.*#	India (Mizoram)	NE
11. *Esomus danricus* (Hamilton) $^{\alpha}$	India (throughout north India), Bangladesh, Myanmar, Nepal, Pakistan, Sri Lanka.	LC
12. *Devario aequipinnatus* (McClelland)*	India (Andhra Pradesh, Manipur, Mizoram, Tamil Nadu,), Bangladesh, Myanmar, Nepal, Sri Lanka, Tailand.	LC
13. *Devario naganensis* (Chaudhuri) $^{\alpha\beta}$	India (Manipur, Nagaland, Mizoram), Bangladesh, Nepal, Pakistan.	VU
14. *Ampblypharyngodon mola* (Hamilton)*	India (throughout except Kerala), Bangladesh, Myanmar, Nepal, Pakistan.	LC
Subfamily Cyprininae		
15. *Tor putitora* (Hamilton) $^{\beta}$	India (all along the Himalaya), Bangladesh, Nepal, Pakistan.	EN

(Contd...)

Scientific Name	*Distribution*	*Status*
16. *Tor tor* (Hamilton) $^{\alpha}$	India (foothills of Himalaya, Assam, Bihar, Madhya Pradesh, N. Bengal, Manipur, Mizoram, Nagaland), Bangladesh, Pakistan.	NT
17. *Tor barakae* Arunkumar &Basudha	India (Manipur, Mizoram)	DD
18. *Puntius conchonius* (Hamilton) $^{\alpha}$	India (Brahmaputra, Ganga, Mahanadi, Cauvery river systems), Bangladesh, Nepal, Pakistan.	LC
19. *Puntius ticto* (Hamilton) $^{\alpha\beta}$	India (throughout), Bangladesh, Myanmar, Nepal, Pakistan, Sri Lanka, Thailand.	LC
20. *Cirrhinus* sp.		NE
21. *Cirrhinus mrigala* (Hamilton) $^{\alpha\delta}$	India (throughout north India), Bangladesh, Myanmar, Nepal, Pakistan.	LC
22. *Cirrhinus reba* (Hamilton)* $^{\delta}$	India (throughout India), Bangladesh, Nepal, Pakistan.	NE
23. *Labeo calbasu* (Hamilton) $^{\delta}$	India (throughout except Kerala), Bangladesh, Myanmar, Nepal, Pakistan, Thailand.	LC
24. *Labeo rohita* (Hamilton) $^{\delta}$	India (throughout India), Bangladesh, Myanmar, Nepal, Pakistan, Sri Lanka.	LC
25. *Labeo* sp.*#	India (Mizoram)	NE
Sufamily Garrinae		
26. *Grossocheilus latius* (Hamilton) $^{\alpha\beta}$	India (Brahmaputra, Ganga, Mahanadi river systems, western ghats), Bangladesh, Nepal, Pakistan.	LC
27. *Garra annandalei* Hora $^{\beta}$	India (Darjeeling Himalaya, Mizoram), Bangladesh, Nepal.	LC
28. *Garra gotyla* (Gray) $^{\alpha}$	India (Brahmaputra, Ganga basin) Bangladesh, Nepal, Myanmar, Pakistan.	LC
29. *Garra kempi* Hora*	India (Abor hills, Arunachal Pradesh, Manipur, Mizoram, Nagaland), Tibet, Myanmar	LC
30. *Garra lissorhynchus* (McClelland)*	India (Brahmaputra river systems, Assam Himalaya, Manipur, Nagaland, Mizoram).	LC
31. *Garra manipurensis* Vishwanath & Sarojnalini*	India (tributaries of Barak river in Manipur and Mizoram)	VU
32. *Garra naganensis* Hora*	India (Brahmaputra basin in Manipur, Mizoram and Nagaland).	LC

(Contd...)

Scientific Name	Distribution	Status
33. *Garra nasuta* (McClelland)* β	India (Brahmaputra basin in Assam, Manipur, Mizoram and Nagaland), Myanmar, south China, Vietnam.	LC
34. *Garra* sp.*#	India (Mizoram)	NE
Family PSILORHYNCHIDAE		
35. *Psilorhynchus balitora* (Hamilton) α	India (Yamuna river, Gomti, river, Assam, N. Bengal), Bangladesh, Myanmar.	LC
Family BALITORIDAE		
Subfamily Balitorinae		
36. *Balitora brucei* (Gray)*	India (Assam, Darjeeling, Manipur, Meghalaya, Mizoram), Bangladesh, Myanmar, Nepal.	NT
Subfamily Nemacheilinae		
37. *Acanthocobitis botia* (Hamilton) αβ	India (Brahmaputra, Ganga river systems), Pakistan, Thailand, China.	LC
38. *Schistura* sp.*#	India (Mizoram)	NE
Family CODITIDAE		
Subfamily Botiinae		
39. *Botia dario* (Hamilton) α	India (Bihar, Cachar, Meghalaya, Manipur, Mizoram, N. Bengal, Punjab, UP), Bangladesh.	LC
40. *Botia rostrrata* Günther* α	India (Meghalaya, Mizoram, W. Bengal), Bangladesh.	VU
Subfamily Cobitinae		
41. *Pangio pangia* (Hamilton) α	India (north eastern part of Bengal, Assam, Manipur, Mizoram), Myanmar.	LC
42. *Lepidocephalichthys annandalei* Chaudhuri α	India (Teesta river system, N. Bengal, Meghalaya), Bangladesh.	LC
43. *Lepidocephalichthys berdmorei* (Blyth) β	India (Manipur, Mizoram,), Bangladesh, Myanmar, Thailand, Malay Peninsula, China.	LC
Order SILURIFORMES		
Family BAGRIDAE		
Subfamily Bagrinae		
44. *Sperata seenghala* (Sykes) α	India (Cauvery, Ganga, Godavari, Krishna, Yamuna river systems, Mizoram), Bangladesh, China, Pakistan.	LC

(Contd...)

Scientific Name	*Distribution*	*Status*
Family SCHILBEIDAE		
Subfamily Schilbeinae		
45. *Clupisoma garua* (Hamilton) $^{\alpha\delta}$	India (throughout north India up to Mahanadi river), Bangladesh, Nepal, Pakistan.	LC
46. *Eutropichthys vacha* (Hamilton) $^{\beta}$	India (north India up to Mahanadi river), Bangladesh, Myanmar, Nepal, Pakistan, Thailand.	LC
Family: AMBLYCEPIDAE		
47. *Amblyceps mangois* (Hamilton) $^{\alpha}$	India (all along the foot hills of Himalaya), Bangladesh, Myanmar, Nepal, Thailand.	LC
Family: SISORIDAE		
Subfamily Glyptosterninae		
48. *Glyptothorax conirostrae conirostrae* (Steindachner) $^{\alpha}$	India (Himachal Pradesh, J. & K., Siliguri, N. Bengal, Mizoram), China.	DD
49. *Glyptothorax telchitta* (Hamilton)*	India (Bihar, N. Bengal, U.P., Manipur, Mizoram), Bangladesh, Nepal, Pakistan.	LC
50. *Glyptothorax* sp1.*#	India (Mizoram)	NE
51. *Glyptothorax* sp2.*#	India (Mizoram)	NE
52. *Glyptothorax* sp3.*#	India (Mizoram)	NE
Family ERETHISTIDAE		
53. *Conta conta* (Hamilton) $^{\beta}$#	India (Mahanadi river, N. Bengal, U.P., Assam, Manipur), Bangladesh.	DD
54. *Pseudolaguvia shawi* Hora $^{\alpha}$	India (Brahmaputra drainage), Bangladesh.	LC
55. *Pseudolaguvia tuberculata* (Prashad and Mukherji) $^{\beta}$	India (Mizoram), Myanmar.	DD
Order Beloniformes		
Family Belonidae		
56. *Xenentodon cancila* (Hamilton) $^{\beta}$	India (widely distributed), Bangladesh, Myanmar, Pakistan, Sri Lanka, Thailand.	LC
Order SYNBRANCHIFORMES		
Family MASTACEMBELIDAE		
57. *Macrognathus pancalus* Hamilton $^{\alpha}$	India (widely distributed), Bangladesh, Pakistan.	LC
58. *Mastacembelus armatus* (Lacèpéde)* $^{\alpha\beta}$	India (throughout India), Bangladesh, Myanmar, Pakistan, China, Hainan	LC

(Contd...)

Scientific Name	*Distribution*	*Status*
	Island, Java, Malaya, Nepal, Sri Lanka, Sumatra, Thailand, Tonkin, Vietnam.	
Order PERCIFORMES		
Family CHANDIDAE		
59. *Chanda nama* Hamilton $^{\alpha}$	India (Ganga, Krishna, Mahanadi, Brahmaputra basins), Bangladesh, Nepal, Pakistan.	LC
60. *Parambassis ranga* (Hamilton) $^{\alpha}$	India (widely distributed), Bangladesh, Myanmar, Pakistan, Thailand, Malaysia.	LC
61. *Parambassis tenasserimensis* Roberts $^{\alpha}$	India (Mizoram), Myanmar.	NE
Family NANDIDAE		
Subfamily Badinae		
62. *Badis badis* (Hamilton) $^{\alpha}$	India (Ganga basin in Himachal Pradesh to Brahmaputra basin in Assam and Mahanadi basin), Nepal, Bangladesh.	LC
Family GOBIIDAE		
Subfamily Gobiinae		
63. *Glossogobius giuris* (Hamilton)* $^{\alpha\delta}$	India (throughout India), Bangladesh, Myanmar, Pakistan, Nepal, Sri Lanka, East coast of Africa, Japan, Australia, South Pacific.	NE
Family CHANNIDAE		
64. *Channa* sp.#	India (Mizoram)	NE

Note: * = Present collection; α = Reported by Kar and Sen (2007); β = Reported by Karmarkar and Das (2007); δ = Reported by Kar *et al.* (2000); # = Yet to be confirmed.

(EN = endangered; VU = vulnerable; NT = near threatened; LC = least concern; DD = data deficient; NE = not evaluated).

On the other hand, the position of vent in *Garra kempi* is shifted far forward and almost in the middle of pelvic and anal fins origins. The fish may be considered as a specialized form of the *yunnanensis* complex. *Garra kempi* is reported from Arunachal Pradesh (Hora, 1921), Manipur (Vishwanath *et al.* 1987), Meghalaya (Sen, 1995), Mizoram (Kar and Sen, 2007), Nagaland (Hora and Mukerji, 1935; Kosygin and Vishwanath, 1998) in India; Tibet (Yunfei, 1987) and Myanmar (Oo, 2002). Menon (1964) remarked that the range of variation of characters between the populations of the Brahmaputra and Chindwin drainages is of no taxonomic significance. However, Zhang and Chen (2002) remarked that *G. kempi* is actually represented by the Brahmaputra river basin form and those specimens reported as *G. kempi* from the upper Irrawaddy river of Yunnan in China are described as *Garra tengchongensis*. But they have not included the specimens collected from the Chindwin-Irrawaddy River basin of Manipur and Myanmar in their study. A detail study on the identity of the specimens collected from the Chindwin-Irrawaddy river basin needs to be done to establish the correct identities and distributions of these species.

Ng (2005) redescribed *G. telchitta* and separated *G. botius* from it as a valid species based on the specimens collected from the Ganga river drainage in West Bengal, India. He designated neotypes of both the species to fix their taxonomy. Vishwanath *et al.* (2007) agree with him and reported both the species from the Brahmaputra basin of North East India. Many workers reported *G. telchitta* earlier from the Brahmaputra drainage in North East India (Kar and Sen, 2007; Vishwanath and Kosygin, 2002; Kar *et al.* 2002). Since the *G. botius* was earlier considered as junior synonym of *G. telchitta,* reexamination of the specimens reported as *G. telchitta* from the region needs to be done in order to establish the identity and distribution pattern of these fishes in the region.

Karmakar and Das (2007) reported *Conta conta* from Mizoram base on two specimens. However, Ng (2005) described a species of *Conta* from Brahmaputra drainage as *Conta pectinata*. Later, Vishwanath *et al.* (2007) recognized and included only *C. pectinata* under the genus, while reporting the fishes of the North East India and remarked that the fish is distributed only in Brahmaputra basin in Assam and Meghalaya. Therefore, more specimens of *Conta* from Mizoram need to be examined further to establish the correct identity.

Kar and Sen (2007) reported *Channa orientalis* Bloch and Schneider from the Tlawng river in Mizoram. However, *C. orientalis* is considered as an endemic species to Sri Lanka (Ekaratne, 2000; Vishwanath, and Geetakumari, 2009) and is often confused with *Channa gachua* (Hamilton). Some workers (Jayaram, 1999; Talwar and Jhingran, 1991; Sen, 1985) considered *Channa gachua* as a junior synomym of *Channa orientalis*. Vishwanath *et al.* (2007) recognized *C. gachua* as valid species and included in the list of fishes of North East India. Courtenay Jr. and Williams (2004) noted that *C. orientalis* differs from *C. gachua* in lacking pelvic fins and report of the former species from southern India and elsewhere are erroneous.

Identity of the specimens of Kar and Sen (2007) collected from the Tlawng river needs to be reexamined.

Figure 1.6. Diversity of fishes of Tlawng River, Mizoram showing percentage species composition in different families.

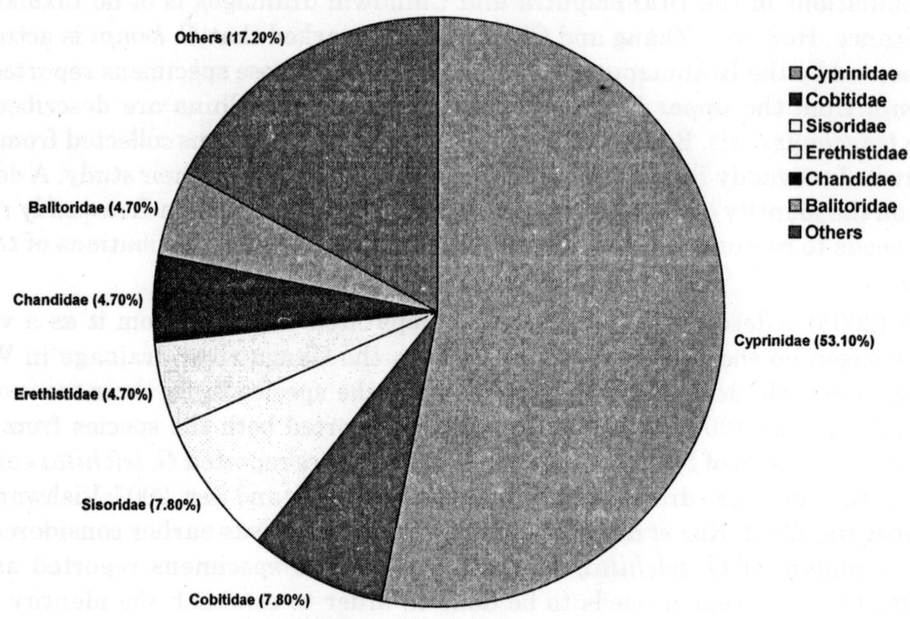

There are altogether sixty-four species which belongs to thirty-four genera and fifteen families known from the Tlawng river in Mizoram. Highest species composition was observed in the cyprinidae family (53.1%) followed by cobitidae (7.8%) and sisoridae (7.8%) (Figure 1.6). Based IUCN (2010), the fishes of the river consist of one endangered species (EN), three vulnerable (VU), two near threatened (NT). On the other hand, the river habours commercially important food fishes like *Tor tor, Labeo calbasu, L. rohita, Cirrhinus reba, Cirrhinus mrigala* etc., which shows the high fishery potentials of the river. Attempts may be made to introduce the *in situ* fish cultivation using scientific techniques. Fish fauna of the river is a mixture of endemic hill stream, Assamese and some widely distributed forms.

Conservation Strategies

The natural aquatic ecosystems are subject to considerable stresses due to changes in the environment and various human activities. These lead to decrease in fish population. Therefore, a need has arisen to take necessary steps so as to conserve the diverse fish genetic resources and their habitats on one hand and to rational and efficient utilization of fish stock on the other hand. Vishwanath *et al.* (2010) remarked that freshwater fishes of the Eastern Himalaya region are under anthropogenic pressures and a total of 70 species (13.9%) are considered threatened, while 46 (8.8%) are Near Threatened. Menon (2004) suggested that conservation of fish life is urgently needed in India both for economic and ecological reasons. He considered that 74 species of the Indian freshwater and marine fishes are either extinct or endangered, or vulnerable, or rear needing immediate protection. Kar and Sen (2007) pointed out that declaration of aquatic sanctuary of some portion of the Tlawng river near Venglai village could be ideal site. Since the river habours threatened and important fish species, necessary measures may be taken in order to save these genetic resources. Some measures for the conservation of the threatened fishes are suggested below:

1. **Ecological and biological investigations**: Detail knowledge on the ecology and biology of fishes is useful in fishery management and conservation. Therefore, investigation on the ecology, biology and population sizes of the threatened fishes may be conducted using scientific techniques.

2. **Maintenance of water quality ecological balance**: Proper maintenance of water quality and ecological processes is essential for the successful conservation of fishes and their habitats. Necessary measures may be taken up to control pollution of water with regular monitoring of water quality and habitat characteristics of the river.

3. **Afforestation and control of erosion**: Afforestation is a practical effective means of reducing soil erosion in the catchment area. The Tlawng river water is usually very turbid during the monsoon and early post-monsoon period due to heavy loading of soil particles from its catchment area. An extensive social forestry programme will effectively check the pressure on denudation of the available forest cover and practice of *Jhum* (shifting) cultivation may be controlled. This would ultimately help to retain vegetal cover and reduce the inflow of silt particles along with the runoff to the river which would be of great help for reducing the turbidity of the river water.

4. **Sustainable utilization of fish resources**: The fish resources should be wisely utilized. The challenge therefore is to conserve fish while providing sustained benefits to the local communities dependent upon these resources for sustenance.

5. **Implementation of domestic legislations**: The destructive fish harvesting techniques may be banned in the river. Laws on catching brooders may be formulated and implemented. The introduction of exotic species that impact upon indigenous fish and their habitat may be controlled.

6. **Establishment of fish sanctuaries**: Fish sanctuaries for the protection of the threatened species may be established at suitable portion of the river.

7. **Identification card preparation and awareness programmes**: Information and identification pamphlets of the threatened fishes may be prepared and distributed to all the stakeholders. Local communities should be encouraged to

participate in the conservation of the fishes. Proper awareness programmes on the status and importance of fishes and their habitat may be organized for the local communities, so that the concept of 'Social Fencing' would come into the existence. Once the understanding between the conservation authorities and local communities is established, the people themselves would start protecting the fish and their habitat.

[**Acknowledgements:** The author is grateful to the Director, Zoological Survey of India (ZSI), Kolkata for providing research facilities and encouragements; to Prof. W. Vishwanath, Department of Life Sciences, Manipur University for allowing me examine the holotype of *Garra manipurensis* in the Manipur University Museum of Fishes; to Dr. Ayub Ali of Veterinary College, Mizoram and Dr. L. Inaotombi Devi of RIPANS, Aizwal, Mizoram for their valuable helps during collection the fishes. The author is also thankful to Shri P. Verma, Photographer, Estuarine Biology Regional Centre, ZSI for taking the photographs of the fishes.]

REFERENCES

Arunkumar, L. and Basudha, Ch. 2003. A new species of mahseer fish (Cyprinidae: Cyprininae) from Manipur, India. *Aquacult.* 4(2): 271-276.

Courtenay, W.R., Jr. and Williams, J.D. 2004. *Snakeheads (Pisces, Channidae) – A Biological Synopsis and Risk Assessment.* U.S. Department of the Interior, U.S. Geological Survey circular: 1251, pp. 143.

Ekaratne, S. U. K. 2000. A review of the status and trends of exported ornamental fish resources and their habitats in Sri Lanka. Bay of Bengal Programme, Chennai,India.(hhtp://www.fao.org/waicent/faoinfo/fishery/bobp/website/homepage.htm., accessed on November 28, 2008).

Hora, S.L. 1921. Indian cyprinoid fishes belonging to the genus *Garra* with notes on related species from other countries. *Rec. Indian Mus.*, 22(5): 633-687.

Hora, S.L. and Mukerji, D.D. 1935. Fishes of the Naga Hills, Assam. *Rec. Indian Mus.*, 37(3): 381-404.

IUCN 2010. IUCN Red List of Threatened Species. Version 2010.4. <http://www.iucnredlist.org/>. Downloaded on 29 October 2010.

Jayaram, K.C. 1999. *The fresh water fishes of the Indian region.* Narendra Publishing House, Delhi, 551 pp.

Kar, D., Laskar, B.M., Mandal, M., Lalsiamliana and Nath, D. 2002. Fish genetic diversity and habitat parameters in Barak drainage Mizoram and Tripura. *Indian J. Environ. and Ecoplan.* 6(3): 473-480.

Kar, D., Dey, S.C., Mandal, M., Laskar, B. A. and Searnleana, L. 2000. Preliminary survey of the fish genetic resources of the rivers in Barak drainage, Mizoram and Tripura. In: *Proceedings of Lake 2000- International Symposium on Restoration of Lakes and Wetlands* (Ramachandra, T.V., C.R. Murthy and N. Ahalya eds.), Center for ecological Sciences, Indian Institute of Science, Bangalore. (ces.iisc.ernet.in/energy/water/proceed/section2/paper2.htm., accessed on January 28, 2009).

Kar, D. and Sen, N. 2007. Systematic list and distribution of fishes in Mizoram, Tripura and Barak drainage of north eastern India. *Zoo's Print Journal*, 22(3): 2599-2607.

Karmakar, A.K. and Das, A. 2007. Fishes. In: *Fauna of Mizoram*, State Fauna Series, 14 (DZSI ed.), Zoological Survey of India, Kolkata, pp. 507-535.

Kosygin, L. and Vishwanath, W. 1998. A report on fish diversity of Tizu river, Nagaland with some new records. *Ecol. Env. Cons.*, 4(4): 243-247.

Menon, A.G.K. 1964. Monograph of the cyprinid fishes of the genus *Garra* Hamilton. *Mem. Indian Mus.*, 14(4): 173-260.

Menon, A.G.K. 2004. *Threatened fishes of India and their conservation*. Zoological Survey of India, Kolkata, pp. 170.

Nebeshwar, K., Vishwanath, W. and Das, D.N. 2009. *Garra arupi*, a new cyprinid fish species (Cypriniformes: Cyprinidae) from upper Brahmaputra basin in Arunachal Pradish, India. *J. Threatened Taxa*, 1 (4): 197-202.

Nelson, J.S. 1984. *Fishes of the World*. Second Edition. John Wiley and Sons, New York, pp. 523.

Ng, H.H. 2005. *Conta pectinata*, a new erethistid catfish (Teleostei: Erethistidae) from northeast India. *Ichthyol. Explor. Freshwat.* 16(1): 23-28.

Ng, H.H. 2005. *Glyptothorax botius* (Hamilton, 1822), a valid species of catfish (Teleostei: Sisoridae) from northeast India, with notes on the identity of *G. telchitta* (Hamilton, 1822). *Zootaxa*, 930: 1–19.

Oo, W. 2002 Inland fisheries of the Union of Myanmar. In T. Petr and D.B. Swar (eds.) Cold Water Fisheries in the Trans-Himalayan Countries. FAO Fish. Tech. Pap. 431.

Sen, T.K. 1985. The fish fauna of Assam and the neighbouring North-Eastern States of India. *Rec. Zool. Surv. India*. Occasional Paper No. 64, pp. 216.

Talwar, P.K. and Jhingran, A.G. 1991. *Inland fishes of India and adjacent countries*. Oxford and IBH Publishing Co. Pvt. Ltd., New Delhi, 2 volumes, pp.1158.

Vishwanath, W., Lakra, W.S., and Sarkar, U.K. 2007. *Fishes of North East India*. National Bureau of Fish Genetic Resources, Lucknow, pp. 264.

Vishwanath, W. 1993. On a collection of fishes of the genus *Garra* Hamilton from Manipur, India with description of a new species. *J. Freshwater Biol.*, 5(1): 59-68.

Vishwanath, W. and Geetakumari, Kh. 2009. Diagnosis and interrelationships of fishes of the genus *Channa* Scopoli (Teleostei: Channidae) of northeastern India. *J. Threatened Taxa*, 1(2): 97-105.

Vishwanath, W. and Sarojnalini, Ch. 1988. A new cyprinid fish *Garra manipurensis* from Manipur, India. *Jpn. J. Ichthyol.* 35(2): 124-126.

Vishwanath, W. and Kosygin, L. 2002. Ecological survey of Barak and Jiri rivers of Manipur with special reference to Ichthyofauna. In *Ecology and Conservation of Lakes, Reservoirs and Rivers* (Kumar, A. ed.), 2: 205-209.

Vishwanath, W., Ng, H.H., Britz, R., Kosygin Singh, L. Chaudhry, S. and Conway, K.V. 2010. The status and distribution of freshwater fishes of the Eastern Himalaya region. Pp. 22-41. In *The Status and Distribution of Freshwater Biodiversity in the Eastern Himalaya* (Compilers Allen, D.J., Molur, S., Daniel, B.A.). IUCN, Cambridge, UK and Gland, Switzerland.

Vishwanath, W., Singh, H.T., Singh, O.S. and Sharma, M.G. 1987. First record of freshwater fishes *Garra gravelyi* and *G. kempi* in Manipur. *Indian J. Fish.*, 34(3): 362-364.

Yunfei, W. 1987 A survey of the fish fauna of the Mount Namjagbarwa region in Xizang (Tibet), China. p. 109-112 In S.O. Kullander and B. Fernholm (eds.) *Proc. V Congr. Europ. Ichthyol.*, Stockholm, Sweden.

Zhang, E. and Chen, Yi-Yu. 2002. *Garra tengchongensis*, a new cyprinid species from the upper Irrawaddy River basin in Yunnan, China (Pisces: Teleostei). *Raffles Bull. Zool.* 50(2): 459-464.

2

Future of River Dolphin (*Platanista gangetica gangetica*) in North East India

—*S.P. Biswas*[1], *Th. Pawlen Singha*[2], *L.P. Hazarika*[3] & *D. Baruah*[4]

[The dolphin is considered as a keystone species in a river system. The role of dolphin in aquatic system is similar to that of a tiger in a forest. They maintain the equilibrium of the riverine ecosystem by controlling the population of primary and secondary consumers. The dolphins are almost exclusively piscivorus; hence, abundant of fish in the habitat is a prerequisite for existence of dolphin. Females give birth to one calf at a time, and the mother dolphin exhibits high degree of parental care. Present count of river dolphin in the Brahmaputra, the Barak and their tributaries is estimated to be about 250 of which only about 10 individuals are found in the Barak system and the rest being in the Brahmaputra river system. The river dolphins prefer deep water and particularly favour counter-current pools, of eddies, that provide refuge from the swift current of the river. A total of three factors, either individually or in combination relates the availability of the dolphin in a particular river- adequate water cover, water quality and abundance of prey food. It is heartening that river dolphin has recently been declared as National Aquatic Animal, but mere declaration will not be of much help as far as conservation of this endangered mammal. The need of the hour is to adopt a holistic approach for the conservation of the habitat as well as the prey base of dolphin.

Key words: River dolphin, North East India, conservation, Barak and Brahmaputra river]

Introduction

The term dolphin is originated from Greek word *delphis* which means 'with a womb'. In other words dolphin was regarded as a fish with a womb. The dolphin is considered as a keystone species in a river system. The role of dolphin in aquatic system is similar to that of a tiger in a forest. They maintain the

[1] Department of Life Sciences, Dibrugarh University, Assam-786 004
[2] Department of Chemistry, Cachar College, Silchar-788 001
[3] Department of Zoology, Lakhimpur College, Assam
[4] Department of Botany, Lakhimpur Girls' College, Assam
Corresponding author: spbsdu@yahoo.com

equilibrium of the riverine ecosystem by controlling the population of primary and secondary consumers. Interesting folklores are associated with river dolphins. In fact, dolphins are a matter of curiosity to mankind from time immemorial.

Despites having a fish like body, dolphins are not fish, they are aquatic mammals related to whales and porpoises. Along with porpoises and whales, dolphins are included in the order Cetacea within the class Mammalia. Although sailors and fishermen use the word 'porpoises' to refer to any small dolphin, porpoises are, however, not dolphins. The order Cetacea has 86 living species, of which about 40 species are dolphins. Most of the dolphins found in world are found in marine water and only 4 of them are restricted to freshwater habitat. The dolphin found in the Ganga-Meghna-Brahmaputra systems is one such freshwater dolphin found in the world. The scientific name of the Gangetic dolphin is *Platanista gangetica gangetica.* It has various English names- Blind River Dolphin, Ganges Dolphin, Ganges River Dolphin, Gangetic Dolphin, Indian River-Dolphin, and South Asian River Dolphin. In Hindi it is known as *Susu* while in Assam it is known as *sihu, petukari, nasubi, susuk, foo-machh* in different parts of the state. The species was once abundant throughout its distribution including Assam, but wanton killing for its oil; habitat degradation and shrinkage have made the species extremely threatened so much so that it is now localized only in certain pockets of the Brahmaputra and Barak valley.

Platanista gangetica (Roxburgh, 1801), is the most primitive group of all river dolphins. It is closely related to five extinct families of dolphins: Prosqualondotidae, Squalondotidae, Squalodelphinidae, Waipatiidae, and Dalpiazinidae. These five families were widely distributed during the Oligocene and Miocene. It is hypothesized that they inhabited the estuarine regions created during the rise of the sea level in the middle Miocene and survived in rivers as the waters regressed in the late Miocene. Rice (1998) recognized two sub-species in the genus *Platanista,* and they are the Indus River dolphin, *Platanista gangetica minor* and the Ganges River dolphin, *Platanista gangetica gangetica.*

General Biology of Dolphin

The body of the Gangetic dolphin is brownish or slaty black in colour and stocky at the middle. They have a long, pointed beak, known as rostrum studded with more than 130 pointed and conical homodont teeth arranged in serial rows. No distinct neck visible, but a depression exists between head and body. The species has only a small triangular hump in place of the dorsal fin. The broad flippers have a crenellated margin, with visible hand and arm bones. The flukes are also broad. The front teeth are long and sharp and are visible even when the mouth is closed. The lower jaw is slightly longer than the upper jaw and in the adult females the snout is slightly curved upwards. Forehead is vaulted and compressed with a *blow-hole.* They are devoid of eye lens and therefore are virtually blind. Males are smaller than females and rostrum is also larger in females (Pilleri, 1970). Usually, length of an adult Ganges dolphin ranges between 1.5 and 2.6 m with a weight varies from 80 to 200 kg. The longest specimen of river dolphin in India was reported as 366 cm. The calves are about 70 cm long and dark chocolate brown at the time of birth but the colour fade as they grow older. As a part of the aquatic adaptation, dolphins lack body hair which is compensated by a thick layer of fat, known as *blubber* that helps in maintaining their body temperature even when water temperature goes down below 10^0C. The blubber is of about 5 cm thick just beneath the smooth skin which contributes to 40-45% of the total body weight. It is a rich source of oil for which dolphins are usually hunted.

The dolphin constantly swims in the river. As it is a mammal it breathes air with lungs and comes to the surface of water at regular intervals for breathing. The breathing takes place through a blowhole which is actually the 'nostril' of the animal lying on the top of its head. It is a highly complex organ equipped with air sacs, tongue and lips and complicated muscles. The blowhole is closed by a strong valve, which opens briefly when the animal surfaces or rolls up for a fast exhalation and inhalation. Below the blowhole there is a system of air sacs, which help in closing the blowhole and which also play a role in sound production. The dolphins come up the surface and breathe every 30 seconds rarely staying under for more than 2 minutes (Mohan, 1989).The surfacing mode of young dolphins is a little different from their parents. Perhaps, it requires time for them to learn the art of rolling as their parents. They are seen to surface and take breathe once in every 20 seconds or so. They generally break the silence of river water emerging almost vertical to little incline and going down again producing a sound like "*Chrup*". This mode of emerging often leads exposure of its full body in the air. They at times resort to acrobatic jump and somersault in air in its attempt to keep pace with its fast swimming mother. Very often large sized dolphins have been seen to remain under water for more than 5 minutes (Singha, 2009).

External ear in dolphin is absent and eyes are small and without crystalline lenses. That is why the Gangetic dolphins are virtually blind. They communicate among themselves and also detect prey through echolocation. They emit sounds of frequencies 40-50 kHz, sometimes going as high as 380 kHz. By listening to the echo and analyzing it, the dolphin estimates the distance and nature of the object. This process enables the dolphin to see their environment through sound. The echo carrying the acoustic image of surrounding environment is conducted to the middle ear of the dolphins mainly via the blubber which is an excellent sound conductor. The middle ear in dolphins serves two functions: one is to stiffen the sound transmission system and optimizing it for high frequencies. The other is to balance the pressure between the inner ear and the external environment. The pressure of a given sound in water is about 60 times as high as the same sound intensity in air.

The dolphins are almost exclusively piscivorus and it appears that they prefer elongated and scale-less fish. Invertebrates such as prawns and clams have also been found in their stomach contents. Hence, abundant of fish in the habitat is a prerequisite for existence of dolphin. Females give birth to one calf at a time, usually between April and September. They do not conceive as long as the baby takes milk of the mother. Gestation period is about 9 months. Calving of Ganges Dolphins apparently can occur at any time of the year, but there may be peaks between March and May. There might some births in September-October soon after the prospective mothers return to their residential spots from their monsoon migration spots. Dolphins give birth to their young ones in preferred spots where they congregate in dry months and these are their residential spots. Mother dolphin exhibits high degree of parental care, by exhibiting constant vigil and may become highly aggressive if someone disturbs her baby continuously.

Habitat, Feeding and Swimming Behaviour of River Dolphin

Hydrological factors such as the discharge, the velocity of flow, turbulence, depth etc. in association with geomorphologic features present, help in shaping habitat of river dolphins specially for leading their lives during the dry season. Smith (1993) identified primary and marginal habitats of Ganges dolphins, according to differences in physical characteristics and sighting frequencies. Primary habitats were characterized by an eddy counter-current system in the main river flow caused by a fine sand/silt point bar formed from sediment deposits of a convergent stream branch or tributary. Marginal habitats were characterized by a smaller eddy counter-current system caused by an upstream meander. The most preferred habitat for Gangetic dolphin is river confluence where the counter current of two streams makes the fish disorganized and they become easy prey for the awaiting dolphins (Reeves *et al*, 1993; Mohan *et al.*, 1998). The other preferred habitats are river meanders, deep pools and in and around sand bars. Ganges river dolphins live not only in the main channels, but also during the flood season, in seasonal tributaries, and the flooded lowlands. The distribution is said to be restricted only by the lack of water and by rocky barriers (Reyes, 1991). Therefore, riverine geomorphology is an important factor for distributional pattern of Gangetic dolphin. Dolphins may be found solitarily or in groups foraging in their feeding ground.

In the Brahmaputra system, the ideal dolphin habitat have the following characteristics: deep pools (at least 3 m) generally found at the confluence and meanderings of tributaries, sandy river bed, moderate water current (about 0.3-0.5 m/sec) and water temperature between 20-28°C, high dissolved oxygen (7.2-9.8 mg/L) and specific conductance (110-140 µS) and slightly alkaline water (pH 7.5-8.0) with high density of fish congregation (Biswas *et al.*,1997).

The dolphins are almost exclusively piscivorus (Sinha *et al.*, 1993) and it appears that they prefer elongated and scale-less fish (Biswas & Boruah, 2000). Invertebrates such as prawns and clams have also been found in their stomach contents. Hence, abundant of fish in the habitat is a prerequisite for existence of dolphin (Mohan *et al.*, 1997). Catfish and other carnivorous fishes like *Aorichthys, Clupisoma, Mastacembelus, Monopterus, Wallago, Xenentodon,* etc. Thus, they help in controlling the carnivorous and trash fish population. This is a unique example of how the natural population of primary predators in a riverine ecosystem is maintained.

Dolphins are champion swimmers; capable of swimming 15-20 km/hr even against a current velocity of 0.7-1.2 m/sec. The hydrological regime of river has a direct bearing on the movement and migration of the dolphin. It seems that mainly in search of food, dolphins move and undertakes local as well as seasonal migration. The dolphin strictly follows pre-spawning upstream migration of fish during June-July and return back to residential spots of the river with down river migration of post spawned migration of fish. These movements' synchronies with the seasonal rains (pre-monsoon and monsoon) and the seasonal recede of monsoons from the region. During the dry season from October to April, many dolphins leave the tributaries and congregate in the main channel. They may become isolated in pools and river branches during the dry season (Reeves and Brownell, 1989). The Ganges Dolphins often swim on their sides and follow fish in shallow waters less than half a meter deep. In 2000, one dolphin chased one carp of about 1kg wt. up to the shore of Gandhighat (Madhuramukh) in the Barak. The fish could not but jump on the shore to save itself from the jaws of the cetacean (Singha, 2009).

The Ganges Dolphin often spins 90° on their lateral axis and 180° on their longitudinal axis. During this side swimming, the body is oriented head down at a 10° from the bottom. The head sweeps up and down in a scanning motion. The flipper, which is thought to have an important tactile function, trails along or slightly above the bottom (Smith, 1993). The Barak as hill stream in Manipur flows in crystal clear water. In that clear water, the Manipuri should have noted the trailing of the flipper of dolphin since the animal is called *Nasubi* in Manipuri, meaning the one which trails or pokes the ground with ear or fin.

Distribution Pattern and Habitat Ecology of River Dolphin

The range of distribution of the dolphin spreads to mainstream and tributaries of Ganges-Brahmaputra-Meghna and Karnaphuli-Sangu river systems in Indian sub-continent (Jones, 1982; Shrestha, 1989). Anderson (1878) mapped the distribution range of the Ganges River Dolphin. Since then there has been a constant decline in the global distribution range of the dolphin. Historically, Ganges dolphins were found year-round in the Yamuna river approximately 400 km upstream to Delhi. However, dolphins were not reported in the Yamuna river above the Chambal river confluence during the dry season because upstream channels have become too shallow and polluted to support dolphins. In the Ganges River, roughly 100 km range has been lost in the last 200 years. Barrages and Dams erected in the Damodar river system have also reduced and fragmented dolphin habitat (Smith *et al.* 2000). More significantly, the Farakka barrage has isolated four sub-populations of Ganges dolphin. In the northern tributaries of the Ganges, of the six subpopulations that were isolated above or between barrages, three have been extirpated and one reduced to insignificant numbers. They have apparently been extirpated from the Son river, at least during the dry seasons (Sinha and Sharma 2003). Thus, throughout its range, the dolphin's habitat is severely fragmented, and the current population size has been reduced by an unknown amount compared to historical levels, but it is still thought to be large enough to be viable in the long-term if adequate conservation measures are taken soon (IWC 2000, Reeves *et al.* 2000).

Till the third quarter of 20th century Ganges river Dolphin was common and could be easily sighted almost in all Dahrs i.e. riverine pools, upstream and downstream of islands and at and near the confluences of the Barak as well as the Brahmaputra and their tributaries. The animal was also common in the major tributaries and even in wetlands (ox-bow lakes) in the rainy seasons.

Present Status of River Dolphin in the Brahmaputra and the Barak Valley

Present count of river dolphin in the Brahmaputra, the Barak and their tributaries is estimated to be about 250 of which only about 10 individuals are found in the Barak system and the rest being in the Brahmaputra river system. Dibru Saikhowa, Subansiri river and Kulsi River in the Brahmaputra system and Lalmati Dahr-Niyairgram Dahr in the Barak River harbours residential population of river dolphin besides migrating population of this cetacean group in the lower stretches of the Brahmaputra near Goalpara. Stray individuals are also seen in the confluences of some of the tributaries like Dehing, Dhansiri, and Manas rivers etc. Recent survey (2008-09) in lower Subansiri river and Dibru River inside Dibru Saikhowa National Park recorded 29 and 8 dolphins respectively whereas at Lalmati Dahr, the number being 8, a decrease from 14 individuals counted in 1999. Similarly, a recent count (2009) on the residential population of the Kulsi is found to be 9 (Islam, *Pers. Comm.*)

The dolphins of Subansiri prefer to stay in the deeper regions containing alluvial sediment in lean period within a limited territory. In contrast, during the rainy season, with increased flow, low level sand bars are submerged, and in heavy flood they may disappear entirely due to heavy discharge and velocity of the river and make the river highly fluviatile, enabling the dolphins to move freely in the entire downstream. Interestingly, they again return back to their original territory when the river sustained with low down pour. As a whole, dolphins are usually sighted in the confluences with the Brahmaputra (Jamuguri ghat), at the connective seasonal channels (near Jamuguri ghat), confluences with its tributaries like Ghagar river (near Haziramora), Ranganadi river (near Khabologhat), and Dikrong river (near Dhunaguri ghat). These junctions with comparatively deeper pools and rich fish diversity generate a suitable habitat for river dolphin. Given the choice, however, the river dolphins prefer deep water and particularly favour counter-current pools, of eddies, that provide refuge from the swift current of the river. A recent survey indicates that there are now 29 Ganges dolphins, up from 21 in 2006 in the downstream of Subansiri River.

Dibru-Saikhowa National Park is one of the few areas where residential population of river dolphin is still found. The population of course small, comprises about 15 including calves with a ratio 2:3 in favour of females. About half of the dolphin population in DSNP is localized in three river confluents viz. Memdubi, Balijan and Laikaghat. The water bodies in DSNP offer ideal habitats for river dolphin not only for foraging but also as a breeding ground of river dolphin. However, human interference like over fishing in dolphin habitats, *jeng* fisheries and overcoming in certain parts of the park compel.

The river dolphins prefer deep water and particularly favour counter-current pools, of eddies, that provide refuge from the swift current of the river. The deeper pools (between 4.0 m. and 15.3 m) of the river become the natural habitat of dolphins during lean months. A total of three factors, either individually or in combination relates the availability of the dolphin in a particular river- adequate water cover, water quality and abundance of prey food. Gradual increase in their population number in the last four years is an indicator of the healthy status of the Subansiri River for river dolphin.

In Barak river, dolphins are found to concentrate at two riverine spots- Lalmati Dahr ($24^0 49'28''$ N, $92^0 51'48''$ E) near Kashipur, 7 kilometer to the east of Silchar and at Niyairgram Dahr, about 9 kilometer to the south-east of Silchar in dry months. Lalmati Dahr-Niyairgram Dahr is a residential area of dolphins and they have been living here from remote past.

In rainy season, preferably in the first flood of a year, one or two dolphins have been observed visiting and foraging for a short period, from a few hours to one or two days, at the confluence of tributaries as well as meanders of the Barak River. The confluences where dolphins are often encountered in wet seasons are–Chirimukh, Sonaimukh, Badrimukh, Madhuramukh, Jatingamukh and Katakhalmukh. Similarly, the following meanders and Dahrs (deep pools) of the Barak are the sighting place of river dolphin in rainy seasons: Daibe Erel (Mutum Dahr) at Singerbond, Boiya Khaori Dahr near Sonai, Madhura Ghat, Annapurna Ghat, Gajambarer Dahr, near Silchar, Masimpur Dahr near Masimpur, Hatirhar Dahr near Surtara on Silchar-Badarpur Highway, Banaimullah Dahr near Sialtek, Srigauri Dahr near Srigauri, Malua Dahr near Haritikar. Beside these, some other spots especially mid-channel islands, bridge pillars at Samoonoong (Dilkhush), Topkhana near Gajambarer Dahr, Badarpur Ghat Railway bridge pillar, Padrighat near Badarpur Malua Dahr are temporary habitats for river dolphins in monsoon.

Ganges River dolphins live not only in the main channels, but also during the flood season, in seasonal tributaries, and the flooded lowlands (Jefferson *et al.*, 1993). The distribution is said to be restricted only by the lack of water and by rocky barriers (Reyes, 1991). In the Cachar Flood of September, 2007, three dolphins foraged for a whole day in the inundated paddy field of Roypur, a village within a radius of 10 km in the north-west from Silchar Town. There have been instances of the dolphin of the Barak rearing in pond at least for one year. In 1950, a dolphin calf, accidentally caught in a net from Chirimukh, was brought safely and kept reared for a year in a shallow pond (0.25 km. away from the catch site) of average depth of about 1.5m and size about 10m by 15m.

at a place called Deb Para, Chandrapur near Banskandi, 10 km to the east from Silchar. It was released in the Barak River after one year (Singha, 2009).

Threats of River Dolphin

Until 90s, freshwater dolphins throughout its distribution were intentionally caught by harpooning and other means mainly for its oil (Biswas and Boruah, 2000). The oil extracted from the blubber of killed dolphin by various crude methods. In one instance, *about* 35 tins (approx. 700 liters) of oils were collected from Narain *Dahr* in the upstream of the Barak River in 1985-86 (Biswas, 1995). In the Brahmaputra river an annual mortality of 10-15 was reported (Biswas and Michael, 1992). Although the intentional killing of dolphin has decreased considerably mainly due to alarming depletion in the number of dolphin in the Brahmaputra and the Barak system and also awareness programmes initiated by some individuals as well as a few NGOs. Nevertheless, occasional mortality of dolphins particularly calves have often been reported. The monofilament nylon gillnet is identified as the single most threat for dolphin. Siltation of river is also becoming a serious threat as it diminishes the adequate water cover required for the existence of dolphin. Over fishing is another factor that has an adverse impact on the prey base of this aquatic mammal (Boruah and Biswas, 2002). The construction of river dams in the Himalayan rivers complicates the situation further (Smith *et al.*, 2000; Boruah and Biswas, 2002). All these factors have made the Ganges river dolphin is in grave danger of extinction (Sinha, 1997; Smith *et al.*, 2001). Sources indicate that its population is declining rapidly at a rate of 10% annually from their different habitats and if present trends continue, these vulnerable Ganges dolphins may be wiped-out within a very short period. According to Behera (2005) the high mortality rate of Ganges dolphin in India is combined with low political will, absence of grass root support for river dolphin conservation, absence of proper co-ordination among members and lack of conservation awareness among people.

The recent decline of *Baiji,* the Chinese lake dolphin, has been attributed to excessive bycatch in fisheries, reduced abundance of prey due to over-fishing, and severe degradation of the Yangtze environment. It is believed that over-fishing and sub-aquatic sonar pollution (which interfered with the dolphin's sonar-based method of locating food), led to the extinction or near extinction of the animal.

Impact of Dams on River Dolphin

A lot of hydel projects are now under various stages of construction or already installed in N.E. India and there are public resentment and unrest still continued on the adverse impact in the downstream of the Barak (Tipaimukh dam) and lower Subansiri dam. The probable impact of Lower Subansiri dam on river dolphin needs special mention here. The Subansiri and its sub basins, act as important foraging ground of river dolphin. Secondary information reveals that the dolphins used to come to these habitats for feeding and breeding. The dam built in the adjacent Ranga river and the water coming out from the power house is released to the river Dikrong in a way through a tail race channel of 56 m length. Both Ranga and Dikrong are the tributaries of Subansiri. The habitat of Ranga river already been degraded by deposition of flushing sediments as well as reducing water discharge by the 405 MW Ranganadi Hydroelectric Project. This is a runoff the river project, where two rivers are used for production of hydropower, technically known as inter-basin channelization. The dam authority cleans the reservoir by flushing the deposited sediment and debris to the down stream of Ranga River. So, when the hydel project releases the extra waters and flushing the sediment and debris from its reservoir it produces tones of coarse and fine sediment and gets deposited in its down stream. This unnatural high sedimentation led to shrinkage of their foraging habitat by rising of riverbed as well as reducing the water table. Similarly, greatly reduced volume of water caused the dolphins' dry-season range to shrink considerably in the downstream. Due to water obstruction, the Ranga river becomes virtually dry by loosing its longitudinal connectivity in the low-water season, especially area between Pahumara ghat and Amtola, once best foraging habitat of river dolphin for adequate water cover and availability of preferred food. On the other hand, Dikrong river carries a substantial amount of sluicing sediment arriving from the same hydel project and the greatly reduced volume of water have the similar impact leading to shrinkage of dolphin foraging habitat. Environmental flows are not natural flows but aim to find a balance for meeting a variety of water needs, including those of ecosystems and downstream communities. Construction of dams for generation of electricity and irrigation in the Ganga river system has divided dolphin populations into small isolated subpopulations, preventing migrations and reducing food availability. The construction of the Farraka barrage and high degree of aquatic pollution has drastically depleted the pollution of river dolphin in the Ganga river (Sinha, 1997). Similar impact of Kaptai dam on river dolphin was also reported from Bangladesh (Ahmed, 1992).The population above the Kaptai dam in the Karnaphuli River disappeared over a period of 6 or 7 years after the completion of the dam (Smith *et al.*, 1998).

Strategy for Conservation of River Dolphin

The Yangtze river dolphin, *baiji's* imminent extinction is a national tragedy and an international disgrace. The story of the baiji's decline, tied as it is to the rampant degradation of a once-productive river system, needs to be told and re-told. Its message - that our freshwater resources are exhaustible and that a river's capacity for serving human needs is finite - can be applied in the context of the Ganges dolphins which are also in trouble and headed in the same direction as the *baiji.*

It is heartening that river dolphin has recently been declared as National Aquatic Animal, but mere declaration will not be of much help as far as conservation of this endangered mammal. The need of the hour is to adopt a holistic approach for the conservation of the habitat as well as the prey base of dolphin. Siltation of the river is causing the loss of adequate water cover for the animal particularly in the upstream of the Brahmaputra and the Barak and that is why dolphins are found in lesser number with every passing year.

In order to address the above-mentioned threats, both short and long term planning for restoration of river health, including maintenance of proper flow of the river in all seasons, implementation of Fisheries and other related Environment and Biodiversity Acts in letter and spirit, proper notification of dolphin inhabited areas, regular monitoring of dolphins in the Ganga-Brahmaputra-Meghna system, proper networking of all those involved in river dolphin conservation programme and creation of dolphin parks/sanctuaries. In short, formation of a nodal agency in the line of Tiger Project is necessary which will provide not only the logistic support for dolphin conservation programme at regional and national level, but also will be answerable for well-being of this endangered species.

REFERENCES

Ahmed, B. 1992. Impact of Kaptai dam and industries on Karnaphuli River in Bangladesh, with special reference to loss of habitat for the river dolphin, *Platanista gangetica*. Proceeding of the seminar on the conservation of river dolphins in the Indian sub-Continent, New Delhi, India, 18-19 Aug., 1992.

Anderson, J. 1878. Anatomical and zoological researches comprising an account of zoological results of two expeditions of western Yunnan in 1868 and 1875; and a monograph of two cetacean genera, *In Platanista and Orcella*. B. Quatrich, London, Vol.2.

Behera, S.K. 2005. Conservation of Ganges River Dolphin in upper Ganga River, Project Report, WWF India.

Biswas, S. P. 1995. Present status of river dolphins in the Barak (in Assamese), pp. 14-15 *Endangered River Dolphin*, Souvenir published on the occasion of the Workshop on River Dolphin, River Dolphin Protection Committee, Assam.

Biswas, S. P. and Michael, R.G. 1992. Fishery characteristics and the present status of fisheries of the River Brahmaputra. *Proc. Seminar on the Conservation of River Dolphins of the Indian Subcontinent*, New Delhi, 18–19 August 1986.

Biswas, S. P., Baruah, A. and Mohan, R.S.L. 1997, Current status of river dolphin (*Platanista gangetica*) in the river Brahmaputra. *Int. J. Ecol. Environ. Sci.*, 23, 357–361.

Biswas, S. P. and Boruah, S. 2000, Ecology of river dolphin (*Platanista gangetica*) in the Upper Brahmaputra. *Hydrobiologia*, 430, 97–111.

Boruah, S. and Biswas, S.P. 2002. Ecohydrology and Fisheries of the Upper Brahmaputra Basin. *The Environmentalist*, 22(2): 119-131.

Jefferson T.A., Leatherwood, S. and Webber, M.A. 1993. *FAO Species identification guide: Marine mammals of the world*. UNEP / FAO, Rome, 320 pp.

Jones, S. 1982. The present status of the Gangetic susu, *Platanista gangetica* (Roxburgh), with comments on the Indus susu, *P. minor* (Owen). FAO Advisory Committee on Marine Resources Research, Working Party on Marine Mammals. *FAO Fisheries Series*, 5, 97–115.

IUCN, 1996. *IUCN Red List of Threatened Animals*, IUCN, Gland, Switzerland and Cambridge, UK, p. 448.

IWC. 2000. Report of the standing sub-committee on small cetaceans. *Journal of Cetacean Research and Management* 2 (Supplement), 235-263.

Mohan, R. S. L. 1989. Conservation and management of Ganges River Dolphin *Platanista gangetica* in India. In *Biology and Conservation of the River Dolphins* (eds. W.F. Perrin *et al.*), IUCN Species Survival Commission Occasional Paper No. 3. IUCN, Gland, Switzerland, pp. 64–69.

Mohan, R. S. L., Dey, S.C., Bairagi, S.P. and Roy, S. 1997. On a survey of the Ganges River Dolphin, *Platanista gangetica* of the Brahmaputra river, Assam. *J. Bombay Nat. Hist. Soc.*, 94, 483–495.

Mohan, R. S. L., Dey, S.C. and Bairagi, S.P. 1998. On a residential dolphin population of the Ganges river dolphin, *Platanista gangetica* in the Kulsi river (Assam), a tributary of Brahmaputra. *J. Bombay Nat. Hist. Soc.*, 95, 1–7.

Pilleri, G. 1970. Observations on the behaviour of *Platanista gangetica* in the Indus and Brahmaputra rivers. *Invest. Cetacea*, 2, 27–60.

Reeves, R. R. and Brownell Jr., R.L. 1989. Susu *Platanista gangetica* (Roxburgh, 1801) and *Platanista minor* (Owen, 1853). In *Handbook of Marine Mammals* (eds. S.H. Ridgway and R. Harrison), Academic Press, London, vol. 4, pp. 69–99.

Reeves, R. R., Leatherwood, S. and Mohan, R.S.L. (eds.). 1993. *A Future for Asian River Dolphins: Report from the Seminar on the Conservation of River Dolphins of the Indian Subcontinent*, 18–19 August 1992, New Delhi, India, Whale and Dolphin Conservation Society, Bath, UK.

Reeves, R. R., Smith, B.D. and Kasuya, T. (eds.). 2000. *Biology and Conservation of Freshwater Cetaceans in Asia*. IUCN/SSC Occasional Paper No. 23, Gland, Switzerland and Cambridge, UK.

Reyes, J.C., Mead, J.G. and Van Waerebeek, K. 1991. A new species of beaked whale *Mesoplodon peruvianus*. *Marine Mammal Science* 7, 1-24.

Rice, D.W.1998. *Marine Mammals of the World*: Systematics and Distribution .Society for Marine Mammalogy, Special Publication No.4, Allen Press, Lawrence, Kansas.

Shrestha, T. K. 1989, Biolog status and conservation of the Ganges river dolphin *Platanista gangetica* in Nepal. In *Biology and Conservation of the River Dolphins* (eds Perrin, W. F. *et al.*), IUCN Species Survival Commission Occasional Paper No. 3. IUCN, Gland, Switzerland, pp. 70–76.

Sinha, R.K., Das, N.K., Singh, N.K., Sharma, G. and Ahsan, S.N. 1993. Gut content of the Gangetic dolphin, *Platanista gangetica*. *Invest. Cetacea*, 24, 317–321.

Sinha, R.K. 1997, Status and conservation of Ganges River dolphin in Bhagirathi–Hooghly River systems in India. *Int. J. Ecol. Environ. Sci.*, 23, 343–355.

Sinha, R.K. and Sharma, G. 2003. Current status of the Ganges river dolphin in the rivers Kosi and Son. *Journal of the Bombay Natural History Society* 100(1): 27-37.

Singha, Th. P. 2009. Habitat ecology and conservation strategy of the freshwater dolphin *Platisnista gangetica* Roxburgh in the River Barak in Southern Assam. Unpublished PhD thesis, Assam University, Silchar, India. 297p.

Smith, B. D., 1993. Status and conservation of the Ganges River dolphin (*Platanista gangetica*) in the Karnali River, Nepal. *Biol. Conserv.*, 66, 159–169.

Smith, B. D., Benazir, A., Edrise, A.M. and Braulik, G. 2001. Status of the Ganges river dolphin or shushuk *Platanista gangetica* in Kaptai Lake and the southern rivers of Bangladesh. *Oryx,* 35: 61–72.

Smith, B. D., Haque, A.K.M., Hussain, M.S. and Khan, A. 1998. River dolphins in Bangladesh: conservation and the effects of water development. *Environ. Manage.*, 22, 323–335.

Smith, B. D. and Reeves, R.R. 2000. Survey methods for population assessment of Asian river dolphins. In *Biology and Conservation of Freshwater Cetaceans in Asia* (eds. R.R. Reeves, B.D. Smith and T. Kasuya), Occasional Papers of the IUCN Species Survival Commission no. 23, IUCN, Gland, Switzerland, pp. 97–115.

Smith, B. D., Sinha, R.K., Zhou, K. Chaudhry, A.A., Renjun, L., Wang, D., Ahmed, B., Haque, A.K.M., Sapkoda, K. and Mohan, R.S.L. 2000. Register of water development projects affecting Asian river cetaceans. *In: Biology and Conservation of Freshwater Cetaceans in Asia* (eds. R. R. Reeves, B. D. Smith, and T. Kasuya). IUCN/SSC Occasional Paper No. 23, Gland, Switzerland and Cambridge, UK. pp. 22-39.

3

Ecology and Diversity of Fishes in the Lotic Habitats of Upper Brahmaputra Basin

—J.N. Das[1] & S.P. Biswas[2]

[The upper stretches of the Brahmaputra and its sub-basins along with numerous hill streams provide refuge to a diverse fish fauna and forms the major capture fisheries resource of the region, thereby sustaining the livelihood of thousands of indigenous fisher-folks. However, the natural stock of many of the commercially important fish species, particularly many cyprinids (*Chagunius, Semiplotus, Raiamas, Crossocheilus, Tor, etc.*) and giant catfishes (*Sperata, Pangasius, Silonia, Bagarius*, etc.) of the basin have been declined over the years, both natural causes and human activities found to contribute in one or the other way. The present communication deals with ecology and diversity of fishes in the lotic habitats of the upper Brahmaputra basin, besides highlights the causes responsible for depletion of natural fisheries of the region.

Key Words: Lotic habitats, upper Brahmaputra basin, habitat ecology, fish diversity]

Introduction

The Brahmaputra drainage system in North East India is one of the largest hydrographic basins in Southeast Asia and sustains a very rich and diverse aquatic gene pool, particularly fishes and as such the region is featured among the global hotspots of freshwater fish biodiversity (Kottelat and Whitten, 1996). The upper Brahmaputra basin encompassing parts of upper Assam and south-eastern Arunachal Pradesh, is a part of the North Eastern Himalayan global hotspot of freshwater fish biodiversity and represents a mosaic of varying lotic and lentic habitats with highly variable hydro-biological regime. The lotic habitats of the basin, mainly represented by the mighty Brahmaputra, its tributaries and distributaries in the alluvial valley of upper Assam along with innumerable fast flowing streams at relatively lower altitude in the south- eastern Arunachal Pradesh support a highly diversified ichthyofauna. Besides these, this stretch of the river is also an abode for other aquatic mega fauna including the giant freshwater dolphin and turtles.

[1]Department of Zoology, Sibsagar College, Joysagar, Assam 785 665 (India)
[2]Department of Life Sciences, Dibrugarh University, Assam 786 004 (India)
E-mail: dasjn123@yahoo.com/spbsdu@yahoo.com

Information on diversity and ecology of natural fish stock, particularly their distribution pattern, habitat preferences, population structure, biodiversity status, etc. has a great bearing on their sustainable utilization and is highly essential to adopt effective conservation measures. However, only a few reports are available on the diversity, distribution and abundance as well as ecology of fish species available in the lotic habitats of the region (Boruah, 1999; Biswas and Boruah, 2002, Das and Biswas, 2006; Das, 2007; Das and Biswas, 2008). Therefore, a detailed study on relevant aspects of diversity and habitat ecology of diversified fish fauna from the riverine and stream habits of upper Brahmaputra basin has been made.

Methodology

Study area

The present investigation was carried out in the upper Brahmaputra basin comprising of six districts of upper Assam, viz. Tinsukia, Dibrugarh, Sivasagar, Jorhat, Dhemaji and Lakhimpur and three districts of eastern Arunachal Pradesh viz. Lohit, Changlang and Tirap. This region is situated between 93°30'E to 96°40'E longitude and 26°15'N to 28°15'N latitude and covering an area of about 28,030 sq. km. The Brahmaputra river and its tributaries and distributaries form an extensive network throughout the region and drain the entire area. Besides these, innumerable snow fed stony bed torrential hill streams is also an important component of lotic habitats in this region.

Study period

The present study was carried out for three consecutive years from January 2006 to December 2008.

Collection

All the major lotic habitats of the study area were surveyed for species inventorization. Fishing nets of various mesh sizes and indigenous fishing gears were employed for collection of fishes from different parts of the study area.

Identification

Fish specimens were identified following Nelson (1994), Talwar and Jhingran (1991), Jayaram (1999) and Vishwanath (2000). Further, species/genus/family names were given following Kullander and Britz (2002) and Thomson and Page (2006).

Habitat ecology

Variation of the selected physico-chemical parameters of the lotic habitats (viz., River Burhidihing and Tengapani hill stream as the representative of riverine and hill stream habitats respectively) was studied seasonally (pre-monsoon: March- May, monsoon: June-August, post-monsoon: September-November and winter: December-February).

The physico-chemical parameters of the selected habitats considered for the present study were air and surface water temperature, transparency, specific conductivity, total dissolved solids (TDS), total suspended solids (TSS), water current, pH, dissolved oxygen (DO), free CO_2 and total alkalinity. Air as well as surface water temperature was measured by a maximum and minimum thermometer graduated up to 110 °C. Transparency was measured using Sacchi disc of 15 cm diameter. Similarly, specific conductivity was measured by a digital conductivity meter whereas water current was measured following Trivedy *et al.* (1987). TDS and TSS values were determined following APHA (1995). The pH was measured *in situ* with the help of a digital pH meter. On the other hand, dissolved oxygen, free CO_2 and total alkalinity values were estimated by titration method (APHA, 1995).

Statistical analysis

Data obtained from different experiments were statistically analysed using the SPSS software (Windows version 10.0).

Results and Discussion

Being a part of the fresh water fish biodiversity hotspot, upper Brahmaputra basin also harbours a unique variety of fishes. The fish fauna in the upper reaches of Brahmaputra is reported to be a mixture of Indo-Gangetic and Indo-Malyan origin (Biswas and Boruah, 2002). During the present investigation, altogether 97 species of fishes belonging to 66 genera and 21 families have been recorded (Table 3.1). Cyprinidae was found to be the most dominant among the families with 35 species followed by Bagridae (10 spp.), Erethistidae (7 spp.), Cobitidae 6 spp.), Sisoridae and Schilbeidae (5 spp. each), Siluridae (4 spp.), Amblycipitidae, Ambassidae, Nandidae and Mastacembelidae (3 spp. each). Rest of the families contributes one or two species each. Again, among the genera *Puntius* and *Mystus* are found to be the most dominant with 5 species each. *Barilius* (comprising 4 species), *Garra, Labeo, Botia, Ompok, Amblyceps* and *Erethistes* (3 species each) are among the other important genera.

Out of the 97 species recorded from the region so far, the species *Pangasius pangasius* has been enlisted as critically endangered (CR) as per CAMP (1998) conservation status whereas another 7 species, viz., *Chitala chitala, Tor putitora, Ompok bimaculatus, O. pabda, Eutropiichthys vacha, Pseudeutropius atherinoides* and *Sisor rhabdophorus* have been placed in endangered (EN) category indicating that these species facing a very high risk of being wiped out from their natural habitat.

Each and every aquatic ecosystem has its own hydro-biological features, which are largely influenced by the geology, topography and the climatic regime of the region where it is located. During the present investigation, the ecology of two different types of aquatic habitats, viz., River Burhidihing and Tengapani hill stream have been studied as the representative of riverine and hill stream habitats respectively. The available data showed marked variations in both physical and chemical parameters among different habitats as well as in different seasons of the same habitat (Table 3.2 and Table 3.3).

Temperature regulates various physico-chemical as well as biological activities and thus has enormous significance. Air and surface water temperature in different habitats showed wide seasonal fluctuations. Temporal variation of mean surface water temperature ranged between 17.1°C and 33.2°C for riverine habitat, and the same ranged between 13.2°C and 21.7°C for hill stream. During the present study, it was observed that the ambient temperature and surface water temperature followed the identical annual trends in the habitats, maximum and minimum being recorded in monsoon and winter, respectively. Again, mean surface water temperature was observed to be lower than the atmospheric temperature in all the seasons indicating a strong influence on the former by the later.

The transparency values in the two habitats under study showed significant fluctuation in different seasons, maximum being in winter and minimum during rainy months. In riverine habitat, the minimum transparency was recorded to be 13.2 (± 1.06) cm in monsoon whereas the maximum was of 43.1 (± 3.63) cm in winter. In case of hill streams, it was as low as 12.2 (± 0.12) cm in monsoon and as high as of 84.6 (± 3.42) cm in winter. The transparency has a great bearing on productivity of a water body through its effect on light penetration. The low value of transparency in monsoon was due to presence of higher amount of suspended solids in river and hill stream supplied by the sediment load from the catchment area. On the other hand, relatively higher value of transparency during winter could due to less input of solids by the surface run off and cessation of monsoon rain (Gupta *et al.*, 1996). Besides, settlement of suspended particles due to drop in current flow in river and hill stream could also be a contributory factor. In all the instances, the transparency exhibited strong and significant negative correlations with values of suspended solids.

The specific conductivity is an index of the amount of water-soluble salts present in water indicating the state of mineralization in an aquatic system (Das, 2000). In the present study, specific conductivity exhibited a moderate seasonal range of fluctuation (Table 2 and 3). In river water, conductivity found to fluctuate within a narrow annual range from 48.7 μS to 87.4 μS, with a maximum value of 81.1(± 3.34) μS in monsoon and minimum in winter (56.7 ± 2.54 μS). On the other hand, 61.1(± 2.72) μS and 79. 1 (± 8.31) μS were the recorded minimum (during winter) and maximum (during post-monsoon) values of conductivity in hill stream respectively. The conductivity values were positively correlated with that of total dissolved solids for lotic habitats. Similarly, a positive correlation was also observed between conductivity and dissolved oxygen in case of lotic habitats, corroborating the findings of Bhaumic *et al.* (2003).

In the present study the overall TSS level appeared to be high in case of riverine habitats whereas it was less in hill stream. Maximum amount of TSS (275.1 ppm for river and 115.3 ppm for hill stream) was recorded during the rainy months. The presence of higher amount of TSS during monsoon happens to be a natural phenomenon in this region as the water bodies receive huge input of suspended solids as surface run off from the catchment area following heavy monsoonal downpour. However, it was observed to decline gradually thereafter, minimum values being 91.1 ppm, and 28.2 ppm for river and hill stream respectively recorded in winter.

Measurement of dissolved solids indicates the total concentration of dissolved ions that has a wide bearing on the productivity (Das, 2000). Like the TSS, TDS concentration was also recorded to be in the higher side throughout the year, fluctuated widely in different seasons with a mean annual range of 64.8 – 204.6 ppm in river water and 28.2-83.5 ppm in hill stream. The maximum TDS concentration was recorded in rainy months in both the habitats thereby exhibits a similar trend of variation with that of the TSS. It shows a significant positive correlation with TSS ($r= 0.982$, $p<0.05$), air temperature ($r= 0.964$, $p<0.05$) and surface water temperature ($r=0.968$, $p<0.05$ levels) while exhibits negative correlation with transparency and conductivity. A similar range of both TSS and TDS was also recorded by Biswas *et al.* (2004).

Hill streams usually have a higher velocity of water due the nature of gradient. Similarly, the up-streams of rivers have faster current, which is gradually become slower as it proceeds towards the low gradient plains. The current flow was as high as in 1.10 ±0.08 m/sec during monsoon, whereas the minimum value was 0.24 ±0.05 m/sec during winter months in case of riverine habitat. For hill stream, the recorded maximum and minimum current flow was 1.32 ± 0.08 m/sec during monsoon and 0.45 ±0.04 m/sec during winter.

In river water, pH depicted alkaline nature for most part of the year, ranging between 6.8 and 7.8. During pre-monsoon it remained in the lower side and progressively increased thereafter with the maximum record in winter months. The alkaline nature of river water was also reported by earlier workers (Biswas and Boruah, 2000; Wakid, 2004) from the region. On the other hand, the pH of stream water remained alkaline throughout the year (mean annual range of 7.5 - 7.83) with a very minor seasonal fluctuation. In general, the water had a good buffering action and is also manifested by the findings of the present study.

Dissolved oxygen (DO) is a critical factor in natural waters and is significant both as regulator of metabolic process and indicator of water quality (Sultan *et al.*, 2003). The mean seasonal variation of DO content of river water was found between 9.3 ppm and 14.2 ppm and the same for stream water is between 13.7 ppm and 17.9. The minimum value of dissolved oxygen was recorded in winter while the highest concentration was recorded to be in monsoon for both the habitats. Higher concentration of DO in river and hill stream waters during monsoon may be due to turbulence and oxygenation resulting from high rainfall and mixing up of well aerated run off carrying by different feeder streams. These observations are in conformity with the findings of Gupta and Mehrotra (1991), Boruah (1999) and Jha and Barat (2003). A strong positive correlation was observed between DO with that of air and surface water temperature for both the habitats. Similarly, DO have showed a positive correlation with current flow for both riverine ($r= 0.922$) and stream habitats ($r= 0.967$; $p<0.05$).

Seasonal data of free CO_2, an indicator of biological respiration of aquatic systems, showed a narrow range of fluctuation in the habitats under study. It was recorded to be minimum in case of hill stream (with an annual range of 1.11-2.21 ppm) in comparison to riverine habitat (3.1- 8.1 ppm), the highest concentration being recorded in winter in both the habitats. The concentration of free CO_2 was greater in winter than other seasons possibly due to greater biological activities and abundant growth of algal bloom in shallow marginal areas of the river. Correlation studies showed highly significant negative relation between free CO_2 and that of DO and current flow. Again, free CO_2 recorded a positive correlation with that of total alkalinity and pH.

In the present study, the total alkalinity (TA) refers to bicarbonate alkalinity, as the carbonate alkalinity was recorded to be absent in both the habitats. For river water it was varied between 68.4 ± 0.12 ppm (in pre-monsoon) and 102.8 ± 0.11 ppm (in post-monsoon) whereas the same was observed to fluctuate between 72.8 ± 1.8 (monsoon) and 120.8 ± 4.45 ppm for hill stream water. Many workers (Roux *et al.*, 1998; Zeri *et al.*, 2000) also reported decrease in TA with increase in concentration of dissolved metals and are also evidenced by inverse relationship observed during present study. The values of TA were positively related to free CO_2 values for the two types of habitats under study. However, no significant relation was established between TA and pH during the present study.

Orth and White (1993) defined fish habitat as "place or a set of places in which a fish, or a fish population or fish assemblage can find the physical and chemical features needed for life, such as suitable water quality, migratory routes, spawning grounds, feeding sites and shelter from enemies and adverse weather". The total habitat available for fish is the combination of macro and microhabitats (Arunachalam, 2000), of which the former controls the distribution and abundance of the later, which in turn control the general pattern of species distribution. Again, different features of the habitat act as major determinants in distribution and abundance of fishes (Shelford, 1911). Habitat complexity in terms of substrates types, depth, flow and in-stream cover along with the existing temperature regime have been identified as the most important factors reported to influence the structure and function of fish assemblage in streams (Hora, 1922; Lagler, *et al.*, 1962; Gorman and Karr, 1978; Schlosser, 1982, Pusey *et al.*, 1995). Thus, the occurrence and abundance of a fish species is affected either by a single environmental factor or a multiplicity of all the habitat factors.

In the Brahmaputra system, the species composition altered seasonally depending on flood and the migratory behaviour of major groups of fishes (Biswas, 2003). Based on habitat preference and tolerance to various physico-chemical parameters of habitat, the fish species recorded from the lotic habitats of the upper Brahmaputra basin have been categorized in to six groups (Table 3.4). The first group comprises of *Garra, Balitora, Erethistes, Amblyceps, Olyra, Pseudecheneis* and *Glyptothorax* are the typical hill stream species that explore the stony bed torrential streams with higher concentration of dissolved oxygen and lower temperature regime found in the hills of Arunachal Pradesh. This group of fish is provided with or without special structural modifications to adapt in highly rheophilic environment existed in hill streams. This modification is more pronounced in case of *Garra* and *Balitora*. The presence of a series of transverse folds in the chest region helping them to attach to the stony substratum (Hora, 1922; Kosygin and Vishwanath, 2005). Compression of body in certain hill stream fishes is also a form of adaptation to minimize the affect of strong current of water.

The second group preferred to inhabit upland pool or upstream of flood plain rivers. These habitats are characterized by substrates made up of pebble and fine gravel with sand, moderate water velocity with relatively higher depth of water. Such upland pool dwellers include species such as *Barilius, Tor, Psilorhynchus* and *Sisor*.

On the other hand, open river forms comprises the third group, which are adapted to live in wider range of depth and velocity of water. They are found in the main rivers and their tributaries in the plains. These are characterized by moderate water current and temperature regime predominantly with sandy substratum. *Chitala, Gudusia, Aspidoparia, Ompok, Rasbora, Osteobrama, Hemibagrus, Rita, Ailia, Rhinomugil, Sicamugil, Mastacembelus,* etc. preferred to inhabit this sort of habitat. Again, the loaches (*Acanthocobitis, Botia, Lepidocephalichthys*) forms the fourth group which prefers to live in the shallow and slow moving rain-fed streams in the plains where sand and mud provides the ideal substratum for them.

Danio, Devario, Salmophasia, Puntius, Chanda, Pseudambassis, Glossogobius, Macrognathus, Tetraodon, etc. are widely distributed species which forms the fifth group and are adapted to live in both lotic (riverine and streams on lower altitude) as well lentic habitats including the seasonal water bodies such as low-lying paddy fields, ponds, etc. Species like *Danio, Devario, Puntius* and *Salmophasia* are also recorded from the torrential streams of the region as well. Arunachalam (2000) and Dobriyal and Singh (1988) also reported the occurrence of these species in higher altitudinal streams in Western Ghats and Garhwal Himalayas. The shallow pockets of lateral waters with pebbly bottom and lesser water velocity in these streams provided the required microhabitat or refuge for them.

Decline of species in the upper Brahmaputra basin within a decade is a matter of serious concern. Many a species commonly found earlier are no more easily available whereas many others are found to declined drastically in their abundance in the basin. Further, some abundant species distributed earlier over a larger range, are now found restricted or scattered in their distribution. Relative abundance of different species in the basin shows that species such as *Chitala, Rasbora, Balitora, Hemibagrus, Rita, Sisor, Rhinomugil* etc. found to be rare in occurrence. Some of these species are reported as commonly available in the basin a decade ago by previous workers (Biswas *et al.*, 1996; Pandey *et al.*, 1998; Biswas *et al.*, 2000). On the other hand, some of the species such as *Notopterus, Gudusia, Aspidoparia, Catla, Chagunius, Cirrhinus, Danio, Garra, Labeo calbasu, Puntius sarana sarana, P. gelius, Raiamas, Semiplotus, Tor, Parluciosoma, Botia, Sperata, Bagarius,*

Ompok, Ailia, Clupisoma, Eutropiichthys, Amblyceps, Erethistes, Gagata, Nangra, Olyra, etc. are either occasional or rare as far as their abundance is concerned.

Diverse type of aquatic habitats of this region has also been under constant stresses in many areas and the gradual acceleration in local species extirpation is seemingly associated with the steady increase in human interventions on aquatic resources, mainly due destruction and alteration of habitats, over exploitation and unsustainable fishing. The Brahmaputra and its major tributaries have been converted into highly braided channels in the plains with interspersing sandbars (locally called as *chars*), which could be attributed to severe aggravation and degradation of the rivers resulted from frequent shifting of river courses coupled with higher rate of siltation. Alluvial texture of soils coupled with heavy monsoon rain washed down enormous quantities of silts and debris to the plains and deposited in the riverbed as well as in the floodplain wetlands. Consequently, the riverbed rises with every passing year besides filling up of wetland areas and drainage congestion in the valley (Biswas and Boruah, 2002; Goswami and Das, 2003). This not only results in significant shrinkage of habitats but also induces alteration in their normal hydro-ecological quality of the Brahmaputra basin, making the riverine habitat as an unstable one for the fishes as well as other aquatic mega fauna. Similarly, floodplain wetlands (*beels*) are an integral part of riverine ecosystems and play an important role in sustaining the functions of a river and its biodiversity (Gopal, 2001) including fisheries. Loss of riverine connection to the adjacent *beels* due to heavy siltation, not only affect the "auto stocking' of *beels* by riverine species but also induces undesirable changes in the ecology of the *beels* through the enhancement of eutrophication process.

Excessive fishing pressure and unsustainable mode of exploitation of resources have been identified as the prime factors for fast depletion certain commercially important fish species of the region. Wanton killing of brood fishes during spawning season (violation of closed season) and juveniles during post monsoon season reflects the excessive fishing pressure on natural fish stock. Again, wide application of banned fishing gears like mosquito net or "kapda jal" round the year is a matter of great concern as such activities induce fish mortality in the smaller size groups that are to be recruited to the stock (Biswas *et al.*, 1995).

Table 3.1

A checklist of fish species recorded from upper Brahmaputra basin.

Order: Osteoglossiformes

Family: Notopteridae

- *Chitala chitala* (Hamilton)
- *Notopterus notopterus* (Pallas)

Order: Clupeiformes

Family: Clupeidae

- *Gudusia chapra* (Hamilton)
- *Hilsa (Tenualosa) ilisha* (Hamilton)

Order: Cypriniformes

Family: Cyprinidae

- *Amblypharyngodon mola* (Hamilton)
- *Aspidoparia jaya* (Hamilton)
- *A. morar* (Hamilton)
- *Barilius barila* (Hamilton)
- *B. barna* (Hamilton)
- *B. bendelisis* (Hamilton)
- *B. tileo* (Hamilton)
- *Catla catla* (Hamilton)
- *Chagunius chagunio* (Hamilton)
- *Chela laubuca* (Hamilton)
- *Chirrinus mrigala* (Hamilton)
- *C. reba* (Hamilton)
- *Crossocheilus latius latius* (Hamilton)
- *Danio rerio* (Hamilton)
- *Devario aequipinnatus* (McClelland)
- *D. devario* (Hamilton)
- *Esomus danricus* (Hamilton)
- *Garra gotyla gotyla* (Gray)
- *G. kempi* Hora
- *G. lissorhynchus* McClelland
- *Labeo calbasu* (Hamilton)
- *L. gonius* (Hamilton)
- *L. rohita* (Hamilton)
- *Osteobrama cotio cotio* (Hamilton)
- *Parluciosoma daniconius* (Hamilton)
- *Puntius conchonius* (Hamilton)
- *P. gelius* (Hamilton)
- *P. sarana sarana* (Hamilton)
- *P. sophore* (Hamilton)
- *P. ticto* (Hamilton)
- *Raiamas bola* (Hamilton)
- *Rasbora rasbora* (Hamilton)
- *Salmophasia bacaila* (Hamilton)
- *Semiplotus semiplotus* (McClelland)
- *Tor putitora* (Hamilton)

Family: Psilorhynchidae

- *Psilorhynchus balitora* (Hamilton)

Family: Balitoridae

- *Balitora brucei* Gray

Family: Cobitidae

- *Acanthocobitis botia* (Hamilton)
- *Botia dario* (Hamilton)
- *B. histrionica* (Blyth)
- *B. rostrata* Günther
- *Canthophrys gongota* (Hamilton)
- *Lepidocephalichthys guntea* (Hamilton)

Order: Siluriformes

Family: Bagridae

- *Batasio tengana* (Hamilton-Buchanan)
- *Hemibagrus (Mystus) menoda* (Hamilton)
- *Mystus bleekeri* (Day)
- *M. cavasius* (Hamilton)
- *M. dibrugarensis* Chaudhuri
- *M. tengara* (Hamilton)
- *M. vittatus* (Bloch)
- *Rita rita* (Hamilton)
- *Sperata (Aorichthys) aor* (Hamilton)
- *S. seenghala* (Sykes)

Family: Siluridae

- *Ompok bimaculatus* (Bloch)
- *O. pabda* (Hamilton)
- *O. pabo* (Hamilton)
- *Wallago attu* (Schneider)

Family: Schilbeidae

- *Ailia coila* (Hamilton)
- *Clupisoma garua* (Hamilton)
- *Eutropiichthys vacha* (Hamilton)
- *Pseudeutropius atherinoides* (Bloch)
- *Silonia silondia* (Hamilton)

Family: Pangassidae

- *Pangasius pangasius* (Hamilton)

Family: Amblycipitidae

- *Amblyceps apangi* Nath & Dey
- *A. arunachalensis* Nath & Dey
- *A. mangois* (Hamilton)

Family: Sisoridae

- *Bagarius bagarius* (Hamilton)
- *B. yarrellii* Sykes
- *Gagata cenia* (Hamilton)
- *Glyptothorax telchitta* (Hamilton)
- *Sisor rhabdophorus* Hamilton

Family: Erethistidae

- *Conta conta* (Hamilton)
- *Erethistes (Hara) hara* (Hamilton)
- *E. jerdoni* (Day)
- *E. pussilus* Muller & Troschel
- *Glyptothorax telchitta* (Hamilton)
- *Nangra nangra* (Hamilton)
- *Pseudecheneis sulcatus* (McClelland)

Family: Olyridae

- *Olyra longicaudata* McClelland

Order: Cyprinodontiformes

Family: Belonidae

- *Xenentedon cancila* (Hamilton)

Order: Perciformes

Family: Ambassidae/ Chandidae

- *Chanda nama* (Hamilton)
- *Pseudambassis baculis* (Hamilton)
- *P. lala* (Hamilton)

Family: Nandidae

- *Badis assamensis* Ahl
- *Badis badis* (Hamilton)
- *Nandus nandus* (Hamilton)

Family: Mugilidae

- *Rhinomugil corsula* (Hamilton)
- *Sicamugil cascasia* (Hamilton)

Family: Gobiidae

- *Glossogobius giuris* (Hamilton)

Family: Mastacembelidae

- *Macrognathus aral* (Bloch & Schneider)
- *M. pancalus* Hamilton
- *Mastacembelus armatus* (Lacepede)

Order: Tetraodontiformes

Family: Tetraodontidae

- *Tetraodon cutcutia* Hamilton

Table 3.2

Seasonal variations in certain physico-chemical parameters of riverine habitat in upper Brahmaputra basin

Parameters	*Season*			
	Pre-monsoon	*Monsoon*	*Post-monsoon*	*Winter*
1	*2*	*3*	*4*	*5*
1. Air temperature (°C)	23.1- 30.8 (25.3 ±2.32)	31.9 - 37.2 (34.5 ±2.34)	27.9 - 36.4 (32.8 ±2.12)	17.8 - 21.2 (19.6 ±0.84)
2. Water temperature (°C)	21.5 - 27.6 (25.1 ±1.27)	30.8 - 35.2 (33.2 ±2.52)	25.6 - 31.2 (28.6 +1.75)	16.6 - 19.8 (17.1 ±0.92)
3. Transparency (cm)	27.2 - 38.4 (34.0 ±2.21)	12.2 - 16.3 (13.2 ±1.06)	20.1- 25.8 (24.1 ±0.06)	32.6 - 54.2 (43.1 ±3.63)
4. Conductivity (mS)	61.2 - 7 7.4 (68.8 ±2.31)	71.5 - 87.4 (81.1 ±3.34)	69.8 - 86.1 (75.3 ±5.21)	48.7 - 62.3 (56.7 ±2.54)
5. Total suspended solids (ppm)	105.6 -145.3 (118.4 ±9.11)	231.3 - 314.8 (275.1 ±13.41)	198.2 - 256.8 (218.5 ±15.19)	87.2 - 112.3 (91.1 ±8.71)
6. Total dissolved solids (ppm)	98.5 - 123.6 (111.3 ±5.32)	175.3 - 243.8 (204.6 ±5.48)	107.9 -145.3 (125.3 ±6.85)	50.4 - 73.8 (64.8 ±4.05)

(Contd...)

1	*2*	*3*	*4*	*5*
7. Current flow (m/sec)	0.28 - 0.39 (0.34 ±0. 12)	1.04 - 1.25 (1.10 ±0.08)	0.42 - 0.71 (0.62 ±0.03)	0.14 - 0.32 (0.24 ±0.05)
8. pH	6.8 – 7.1 (6.9 ±0.24)	7.1 -7.6 (7.3 ±0.07)	7.3 -7.6 (7.4 ±0.81)	7.4 -7.8 (7.5 ±0.26)
9. Dissolved oxygen (ppm)	8.3 -13.4 (10.2 ±0.83)	9.5 -17.4 (14.2 ±2.22)	12.6 - 15.4 (13.4 ±1.20)	7.2 - 10.9 (9.3 ±0.23)
10. Free CO_2 (ppm)	4.2 - 7.6 (6.1 ±1.21)	2.4 - 4.7 (3.1 ±0.42)	5.3 - 8.6 (7.2 ±0.02)	7.6 - 10.2 (8.1 ±1.02)
11. Total alkalinity (ppm)	61.2 - 81.2 (68.4 ±4.12)	84.2 - 94.1 (88.4 ±2.98)	98.4 -118.2 (106.8 ±3.11)	86.1- 101.2 (91.1 ±2.58)

*Figures in parentheses indicate mean ±standard deviation.

Table 3.3
Seasonal variations in physico-chemical parameters of hill stream habitat in upper Brahmaputra basin

Parameters	*Season*			
	Pre-monsoon	*Monsoon*	*Post-monsoon*	*Winter*
1. Air temperature (°C)	17.8 - 21.2 (18.2 ±0.23)	22.1- 28.9 (24.3 ±1.21)	21.2 - 28.4 (23.3 ±1.34)	15.5 - 17.8 (15.9 ±0.21)
2. Water temperature (°C)	16.9 - 19.8 (17.3 ±0.21)	18.7 - 26.9 (21.7 ±2.88)	19.9 - 25.7 (21.6 ±0.49)	11.6 - 16.9 (13.2 ±1.13)
3. Transparency (cm)	56.1-71.3 (62.1±3.11)	10.5 - 15.3 (12.2 ±0.12)	27.8 - 41.0 (34.5 ±2.34)	75.9 - 93.5 (84.6 ±3.42)
4. Conductivity (mS)	65.8 - 77.8 (71.2 ±3.42)	62.2 - 86.3 (76.3 ±4.62)	66.6 - 91.2 (79.1±8.31)	57.2 - 7 0.3 (61.1±2.72)
5. Total suspended solids (ppm)	33.4 - 39.5 (37.8 ±.62)	102 - 122.7 (115.3 ±4.96)	74.2 - 88.9 (81.2 ±5.05)	21.2 - 30.7 (26.3 ±2.73)
6. Total dissolved solids (ppm)	37.7 - 45.2 (41.1±1.23)	76.3 - 97.4 (83.5 ±4.54)	57.3 - 69.2 (61.4 ±2.23)	26.1- 32.1 (28.2 ±1.28)
7. Current flow (m/sec) (0.56 ±0.04)	0.39 - 0.67 (1.32 ±0.08)	1.04 – 1.44 (0.87 ±0.13)	0.72 - 0.99 (0.45 ±0.04)	0.36 - 0.52
8. pH	7.5 - 8.3 (7.6 ±0.12)	7.3 - 7.7 (7.5 ±0.03)	7.8 - 8.1 (7.83 ±0.12)	7.6 - 7.9 (7.6 ±0.31)
9. Dissolved oxygen (ppm) (13.3 ±1.74)	10.4 – 15.3 (17.9 ±1.12)	16.2 - 20.6 (16.7 ±1.65)	15.8 – 19.2 (12.7 ±0.84)	11.6 - 14.6
10. Free CO_2 (ppm)	0.97 - 3.23 (1.89±0.82)	0.82 - 2.43 (1.11 ±0.26)	1.81 - 2.54 (1.99 ±0.13)	2.03 - 2.64 (2.21 ±0.18)
11. Total alkalinity (ppm)	81.6 - 92.4 (84.2 ±2.21)	68.6 - 79.8 (74.8 ±2.98)	85.6 - 102.2 (90.5 ±3.98)	108.1 - 142.8 (120.8 ±11.45)

* Figures in parentheses indicate mean ±standard deviation.

Table 3.4

Spatial distribution of fish assemblage (habitat preference) of recorded fish species from lotic habitats of Upper Brahmaputra basin

Fish group	*Habitat characteristics*				*Principal genera*
	Substrate type	*Current flow (m/sec)*	*Depth(m)*	*Temperature and DO*	
1. Hill stream dwellers and cobble	Stony bed; rock >1.0	Fairly high < 1m	Shallow	15.0 - 24.0°C High (> 12ppm)	*Balitora, Garra, Amblyceps, Erethistes, Olyra, Pseudecheneis, Glyptothorax*
2. Upland pool inhabitant gravel with sand	Pebble, fine 0.56 - 0.80	Moderate deeper Up to 2 m	Relatively	17.6 - 25.4°C 8-14 ppm	*Barilius, Psilorhynchus, Sisor*
3. Open river forms	Sand	Variable 0.24 – 1.10	Highly variable 1-5 m	19.6 – 34.5°C 6.0 – 10.0 ppm	*Aspidoparia, Chitala, Chela, Cirrhinus Gudusia, Megarasbora, Osteobrama, Rita, Hemibagrus, Mystus, Ompok, Ailia, Gagata, Rhinomugil, Sicamugil, Mastacembelus.*
4. Low land streamdweller	Sand and mud	Moderate 0.2-0.5	Shallow <2 m	18.0 – 30°C 4-8 ppm	*Acanthocobitis, Botia, Lepidocephalichthys*
5. Widespread species, no single habitat preference	Found more or less in all lotic and lentic water bodies, well acclaimed to highly variable regime of water current, depth, temperature, and DO levels, adapted to explore all types of substrates				*Danio, Devario, Salmophasia, Chanda, Puntius, Pseudambassis, Glossogobius, Macrognathus, Tetraodon*

Recent satellite surveys indicate a considerable decline in the forest cover in the catchment areas of the Brahmaputra basin due to unabated deforestation, land use conversion and land degradation. Shifting or "*jhum*" cultivation practiced traditionally by different ethnic tribes in the hilly areas of Arunachal Pradesh and Nagaland and felling of trees for timber are among the preeminent threats contributing significantly for the degradation of the basin. Similarly, removal of riparian cover in the plains for agricultural purposes is another example of harmful land use practices as it facilitates erosion of both topsoil as well as riverbank (Das *et al*, 2006).

Again, man-made obstructions to the free flow of rivers like dams, bridges, etc. tending towards elimination of the natural hydrologic and geomorphologic dynamism of the system. Isolation of the river from the alluvial plains also play pivotal role in habitat degradation and adversely affect the fisheries concerned. The proposed hydro-electric power projects (with an estimate to produce 25,000 MW power) in the rivers of Arunachal Pradesh including the major sub basins of the Brahmaputra, the Subansiri, Siang and Dibang is expected to bring drastic changes in the surface flow; nutrient composition as well as other associated hydro-biological parameters both upstream and downstream of the said river basins on completion.

Several impacts on the aquatic habitats and their fauna have already been witnessed in the region following the construction of embankments or dykes, spurs, and levees along the banks of the Brahmaputra and its major tributaries for flood control in the last 50 years. Such structural devices along the riverbanks prevents the backflow of water to the *beels* and seasonal inundation of low-lying areas,, thereby, results in shrinkage of feeding and spawning grounds of many endemic fish species including the Indian Major Carps (Biswas and Boruah, 2002).

Moreover, the fish habitats of this region is subjected to the adverse effects of agricultural run off containing pesticide and fertilizer residues and act as non-point source of aquatic pollution in the region(Biswas, 2003; Das *et al.*, 2006). Fishing by intentional (herbal and chemical) poisoning and liming becomes a major threat to in now a day in this region (Das and Biswas, 2006). Similarly, in the hilly areas, dynamiting and electro fishing, etc. are responsible for rapid decline in natural fish population (Biswas *et al.*, 1996; Biswas, 2003) and have pronounced adverse effects on the commercially desirable rheophilic ornamental fishes. Likewise, the wastewater from coalfields containing acids, accidental leakage during drilling and transportation of crude oils, disposal of untreated sewages, industrial effluents, etc. also intensify the water pollution problem.

Conclusion

The upper Brahmaputra basin is blessed with rich diversity of fishes. However, due to alteration of habitat and unsustainable mode of exploitation of resources leads to gradual depletion of aquatic biota of the region in general and fisheries in particular. The need of the hour is, therefore, to make an integrated approach for sustainable management and conservation of natural fish stock as well as other aquatic biota as these are important from both ecological and socioeconomic perspectives. Provision of occupational diversification for fisher communities would be helpful in lessening the pressure put by them on the already threatened bioresources to meet the additional requirement for their livelihood. Development of culture-based fisheries of commercially important fish species in the *beels* and other seasonal water bodies in a co-operative basis or forming self help groups could be an effective and viable alternative. Similarly, strict enforcement of relevant Fisheries Act as well as wildlife/biodiversity act(s) to check the violation of "closed season", intentional poisoning of water-bodies and operation of mosquito net would be helpful in augmentation of natural fish stock of the region.

REFERENCES

Arunachalam, M. 2000. Assemblage structure of stream fishes in the Western Ghats (India). *Hydrobiologia*, 430: 1-31.

APHA. 1995. *Standard Methods for the Examination of Water and Waste Water* (19th edn). American Public Health Association, New York, 522 pp.

Bhaumic, U., Mandloi, A.K., Paria, T. and Ojha, P. 2003. Ecology and production potential of Barnoo reservoir in Madhya Pradesh with suggestion for stocking as management tool. *J. Inland Fish. Soc.*, 35 (1): 58-67.

Biswas, S.P. 2003. Strategy for Conservation of Ornamental Fishes in the North Eastern Himalayan Region. In: *Participatory Approach for Fish Biodiversity Conservation in North East India* (eds. P.C. Mahanta and L.K. Tyagi), NBFGR, Lucknow, 181-192.

Biswas, S.P. and Boruah, S. 2000. Fisheries ecology of the North-Eastern Himalaya with special reference to the Brahmaputra River. *Ecol. Eng.*, 16: 39-50.

Biswas, S.P. and Boruah, S. 2002. Ecohydrology and Fisheries of the upper Brahmaputra basin. *The Environmentalist*, 22: 119-131.

Biswas, S.P., Baruah, A., Baruah, D. and Mohan, R.S.L. 1996. Present status of aquatic fauna in the upper stretches of the Brahmaputra. *Bull. Life Sci.*, 6: 25-32.

Biswas, S.P., Baruah, A. and Boruah, S. 2000. Conservation status of Fish resources in the Brahmaputra Basin. *Tropical Zoology*, 2 and 3: 125-131.

Biswas, S.P., Boruah, S. and Das, J.N. 2004. Prospect of aquaculture in the seasonal water bodies of Assam. *J. Assam Sci.* Soc., (in press).

Boruah, S. 1999. Investigation on certain aspects of hydrobiology in relation to fisheries in the upper stretches of the river Brahmaputra. *Ph. D. Thesis* (unpublished), Dibrugarh University, Assam, India.

CAMP. 1998. Report on the workshop on Conservation Assessment and Management Plans (CAMP) for Freshwater Fishes of India. Zoo Outreach and NBFGR, 156 pp.

Das, A.K. 2000. Limno-chemistry of some Andhra Pradesh Reservoirs. *J. Inland Fish. Soc. India*, 32 (2): 37–44.

Das, J.N. 2007. Current status, Distribution and Ecology of Ornamental Fish Diversity of Upper Brahmaputra Basin. Ph. D. Thesis (unpublished), Dibrugarh University, Assam, India, 238 pp.

Das, J.N. and Biswas, S.P. 2006. Diversity of Indigenous Ornamental Fishes in the Rivers of Upper Assam. In: *Proc. Nat. Sem. on Biodiversity Conservation and Future Concern*. NEBA, Gauhati University, Guwahati, 76-83.

Das, J.N. and Biswas, S.P. 2008. *A Handbook of Ornamental Fishes of Brahmaputra Basin.* EBH Publishers, Guwahati, India, 111pp.

Das, J.N. and Biswas, S.P. and Saikia, T. 2006. Restoration and Management of wetlands of Upper Assam for better fish production–A Review. *Bull. Life Sci.*, 12: 31-39.

Dobriyal, A.K. and Singh, H.R. 1988. Ecological Basis for Ichthyological Variation in two Hillstreams of Garhwal Himalaya. In: *The First Indian Fisheries Forum Proceedings* (ed. M. Mohan Joseph). Asian Fisheries Society, Indian Branch, Mangalore, 313-317.

Gopal, B. 2001. National Workshop on Conservation of Rivers and Floodplains in India–A Report. *Bulletin of the National Institute of Ecology*, 11: 69-74.

Gorman, O.T. and Karr, J.R. 1978. Habitat structure and stream fish communities. *Ecology*, 59: 507-515.

Goswami, D.C. and Das, P.J. 2003. The Brahmaputra River, India. *The Environmentalist Asia*, 11 (1): 9-14.

Gupta, A.K. and Mehrotra, R.S. 1991. Ecological studies on Watei Molds of Kurukshetra. *Current Trends in Limnology*, 1: 47-64.

Gupta, R. K., Sharma, M., Gorai, A.C. and Pandey, P.N. 1996. Impact of Coal Mining Effluents on the Physico-Chemical Characteristics of Raja Tank, Jharia (Dhanbad). *J. Freshwater Biol.*, 8 (2): 63-72.

Hora, S.L. 1922. Structural modifications in the fish of mountain torrents. *Rec. Indian Mus.*, 24: 31-61.

Jayaram, K.C. 1999. *The Freshwater Fishes of the Indian region.* Narendra Publishing House, Delhi, 551 pp.

Jha, P. and Barat, S. 2003. Hydrobiological study of lake Mirik in Darjeeling Himalayas. *J. Environ. Biol.*, 24 (3): 339-344.

Kosygin, L. and Vishwanath, W. 2005. Hill stream fishes of the northern part of Ukhrul district, Manipur. *J. Bombay Nat. Hist. Soc.*, 102 (1): 56-60.

Kottelat, M and Whitten, T. 1996. Freshwater biodiversity in Asia with special reference to fish. World Bank Tech. Paper No. 343, The World Bank, Washington, DC, 17-22.

Kullander, S.O. and Britz, R. 2002. Revision of the family Badidae (Teleostei: Perciformes), with description of a new genus and ten new species. *Ichthyological Exploration Freshwaters*, 3 (4): 295-372.

Lagler, K.F., Bardach, J.E. and Miller, R.R. 1962. *Ichthyology: The Study of Fishes.* John Wiley & Sons, Inc. New York, 545 pp.

Nelson, J. S. 1994. *Fishes of the World.* John Wiley & Sons Inc., New York, 600 pp.

Orth, D.J. and White, R. J. 1993. Stream habitat management. In: *Inland Fisheries Management in North America* (eds. C. Kohier and W. Hubert), *Am. Fish. Soc. Bethesda*, MD: 205-230.

Pandey, A.K., Boruah, A. and Biswas, S.P. 1998. On the ornamental fish fauna in the Brahmaputra drainage system. *India J. Fish.*, 45(1): 95-97.

Pusey, B.J., Arthington, A.H and Read, M.G. 1995. Species richness and spatial variation in fish assemblage structure in two rivers of the Wet Tropics of northern Queensland, Australia. *Environmental Biology of Fishes*, 42: 181-199.

Roux, L.L., Roux, S. and Appriou, P. 1998. Behaviour and speciation of metallic species Cu, Cd, Mn and Fe during estuarine mixing. *Mar. Poll. Bull.*, 36: 56-64.

Shelford, V.E. 1911. Ecological Succession: stream fishes and the method of physiographic analysis. *Biol. Bull.*, 21: 9-35.

Schlosser, I.J. 1982. Fish community structure and function along two habitat gradients in a headwater stream. *Ecol. Monogr.*, 52: 395-514.

Talwar, P.K. and Jhingran, A.G. 1991. *Inland Fishes of India and adjacent countries* (vol. I and II). Oxford-IBH Publ. Ltd., New Delhi, 1158 pp.

Thomson, A.W. and Page, L.M. 2006. Genera of the Asian Catfish families of Sisoridae and Erethistidae (Teleostei: Siluriformes). *Zootaxa,* 1345: 1-96.

Trivedi, R.K., Goel, P.K. and Trisal, C.L. 1987. *Practical Methods in Ecology and Environmental Science.* Enviro Media Publication, Karad, India, 197 pp.

Wakid, A. 2004. Studies on certain aspects of Ecology and Behaviour of the River Dolphin (*Platanista gangetica*) in Dibru-Saikhowa National Park, Assam. Ph. D. Thesis (unpublished), Dibrugarh University, Assam.

Vishwanath, W. 2000. *Fishes of North East India–A Field Guide to Species Identification.* NATP-NBFGR, Lucknow, 195 pp.

Zeri, C., Voutsinou-Taliadouri, F., Romanov, A.S., Ovsjany, E.I. and Moriki, A. 2000. A comparative approach of dissolved trace element exchange in two inter connected basins: Black Sea and Aegean Sea. *Mar. Poll. Bull.*, 40: 666-673.

4

Ornamental Fish Biodiversity of the Rivers of Tripura and Their Conservation

—*Sagar C. Mandal[1], Arun B. Patel[2], Pronob Das[3], Rupam Sharma[4] & A.U. Muzaddadi[5]*

[The North Eastern states of India are recognized as one of the hot spots of freshwater biodiversity of the world. A rich biodiversity of species, favourable climate, cheap labour and wide distribution make this region suitable for ornamental fish farming for small scale business among the low income fisher folk family. Tripura is very rich in floral and faunal biodiversity. The state is blessed with 134 species of fin fishes belonging to 83 genera under 37 families and 13 orders, which have tremendous export potential. However, the indigenous ornamental fishes which are available in the rivers of the state are being neglected due to lack of awareness. A well organized programme for collection, identification, selection, artificial feeding, captive breeding, rearing and marketing of indigenous ornamental fishes will be beneficial to the development of market as well as conservation of germplasm.

Key words: Ornamental fish, biodiversity, conservation, Tripura]

Introduction

The biodiversity encompasses all species of organisms, the ecosystems and ecological processes of which they are part. Biodiversity reflects the degree of native variety and is essential for overall development of environmental quality, the intrinsic worth of all species on earth (Ehrlich and Wilson, 1991; Wilson, 1988). The species diversity of an ecosystem is often related to the amount of living and non-living organic matter present in it. However, apparently, species diversity does not depend on the characteristics of an ecosystem in isolation but more on the interactions between the ecosystems (Kar, *et al.*, 2006). Fish constitute more than half of the total number of vertebrates in the world and they live in almost any conceivable aquatic habitats. About 21,723 living species of fish have been recorded out of 39,000 species of vertebrates in the world (Jayaram, 1999). Of these, 8,411 are freshwater species and 11,650 are marine.

[1,2,5] College of Fisheries, Central Agricultural University, Lembucherra, Tripura–799210, India
[3,4]Central Institute of Fisheries Education (ICAR), Seven Bungalows, Mumbai–400061, India
*Corrsponding author: scmandal02@gmail.com

Megabiodiversity hot spot areas have been considered to have the largest fraction of the world's species and with a high concentration of endemic species. About 2,500 species of fishes have been reported in India, of which, about 930 live in freshwater and about 1570 are marine (Kar, 2003). Two areas in India have been identified as megabiodiversity hot spot areas namely Western Ghat forests and Eastern Himalayan forests (Anon., 2003). India is one of the megabiodiversity countries in the world and occupies the ninth position in terms of freshwater megabiodiversity (Mittermeier and Mittermeier, 1997).

The North-East India comprising eight sister states of Arunachal Pradesh, Assam, Manipur, Meghalaya, Mizoram, Nagaland, Tripura and Sikkim is recognized by World Conservation Monitoring Centre (WCMC, 1998) as one of the hot spots of freshwater biodiversity of the world. These states with their unique topography, diverse physiographic features and varied watershed pattern are a lucrative field for ichthyological studies. With numerous hills, streams, rivers and lakes, the region is bestowed with fish species of both hill stream and plains. The region is mainly fed by two major drainage systems. The northern region is drained by Brahmaputra River and its tributaries and the southern region by Barak River and its tributaries. Eastern Region (Nagaland and Manipur) is also partly drained by Chindwin drainage. These river systems are blessed with many species of ornamental fishes such as barbs, danio, rasboras, loaches, catfish, gourami, snake heads, goby, etc. which are in high demand in the international market. This region is the home for many endemic loaches which have a very good international market demand and surely this region can be regards as 'Land of Loaches'. Most of the species of North-Eastern region possess high aesthetic values due to the attractive coloration pattern, graceful behaviour, peculiar body morphology and endemicity. The major shares of the country's ornamental fish export are captive collection from North-Eastern states comprising as much as 85% of total aquarium fish trade (Bhattacharjya and Choudhury, 2004).

Tripura represents a transition zone between the Indian, Indo-Malayan and Indo-Chinese sub-regions of Oriental Zoogeographical region and is very rich in floral and faunal biodiversity (Gupta, 2000). Aquarium fish has become a popular hobby and aquarium fish selling is a popular trade in Tripura state. The business is mostly confined to import of exotic ornamental fishes from other states of the country particularly from Kolkata, which are then sold out on the basis of temporary management. The indigenous ornamental fishes, which are available in the rivers of the state are being neglected due to lack of awareness. Most of these fishes are caught before their first maturity and sold in the market as food fishes. These indigenous ornamental fishes can easily be collected from the wild resources and can be cultured and reared for keeping in the aquarium. The skill of aquarium fish breeding, rearing and management could not be developed in the state sufficiently and systematically due to lack of specialized training for upgradation of skill for the said management.

Ornamental Fishes and Aquarium Trade

The traditional criteria of any fish to be categorized as ornamental fish are smaller in size, attractive look with pleasing coloration, adaptability to live in small confined areas like aquarium, compatibility to live with other fishes, ability to thrive on/accept artificial feed. However, Bhattacharjya and Choudhury (2004) opined that beautiful colouration is not a must for a fish in order to qualify to be called as an ornamental fish. For instance, a fish having a single dark streak or a blotch on its body can also be called as an ornamental fish. The natural dark colour with stripes, dots, blotches, brilliant silvery colour or with yellow or pinkish hues, all have made fishes to get classified as ornamental fishes. Even a slow moving or sedentary fish and somewhat having unique body, mouth, opercle or fin movements may be considered as an ornamental fish.

The bright external colour patterns of animals, plants and fish in particular has attracted the attention of biologists and aquaculturists since time immemorial (Goswami, 2005). It has been known that these magnificent colouring particles or fragments are due to the presence of carotenoids, which furnish the red, yellow, orange or green colouration of the fish. The fishes are perhaps the most cheerful living being next to birds and are by all account the most beautiful and fascinating aquatic creatures to watch (Biswas, *et al.*, 2007). They are graceful in their shapes, colouration, body forms and movements. As such keeping fishes as pets in an aquarium has emerged as an one of the most popular hobbies in recent years. However, its potential is yet to be fully realized in India as in India values of domestic trade of ornamental fishes is only worth Rs. 10 crores while the annual export is reported to be around Rs. 3 crores. Furthermore, the trade is mainly based on riverine collection of indigenous freshwater species (Gopakumar, 2005).

Indigenous Ornamental Fishes in North-Eastern States

Recent ichthyological studies showed 296 species of fishes in North-Eastern states belonging to 110 genera under 35 families (Vishwanath, 2002) accounting about 33.2 % of total Indian freshwater fishes. Based on trade potentiality, the fish species belonging to the genera *Notopterus, Amblypharyngodon, Danio, Devario, Esomus, Oreichthys, Parluciosoma, Rasbora, Botia, Pseudambassis, Badis, Polycanthus, Channa, Tetradon* are considered as excellent categories of ornamental fishes available in the region. Tripura is rich in variety of fish fauna with about 90 species, most of these are under different degrees of threat. Few attempts have been made so far to identify the endangered fish species of this state systematically (Ghosh and Lipton, 1982).

Topography of Tripura

Tripura with an area of 10,47,700 ha is a small land-locked hilly state. The state is blessed with diversified aqua-resources in the form of rivers, streams, lakes, reservoirs, swamps, ponds and mini-barrages totaling about 23,747.76 ha. The state is lying in the sub-Himalayan region within latitude of 24°32′ to 25°51′ N and longitude 91°10′ to 92°21′ E. The climate of the state is generally hot and humid. The average maximum temperature is 35° C during May-June and the average minimum temperature is 10.5° C during December-January. The average rainfall is 230 cm per annum. The monsoon generally starts during April and continues up to September. The favourable environment accompanied with varying geographical features has contributed for the creation of diverse flora and fauna including diversified native ornamental fishes.

The state is endowed with ten major rivers having a total length of 1200 km and water spread area of 5260 ha. The three major drainage systems of the state are (i) Gomati (ii) Manu and (iii) Khowai. The discharge of Gomati basin is the maximum. These rivers originate from hill streams and flow through plains into Bangladesh borders and are mostly annual in nature on both sides (Datta, 2000). There are other rivers like Feni, Muhuri, Dhalai, Juri, Howrah, Deo and Longai, which are rainfed and hence they carry enormous quantities of water only during rainy seasons but not adequate for agricultural activities for the rest of the year. The upper part of all these rivers passes through narrow gorges while their lower part became a broad sluggish river. River beds are shallow which causes heavy floods during rainy seasons. None of these rivers has undergone any great changes in its course in the past (Anon, 2001). Tripura lies within the earthquake zone and whenever an earthquake occurs, beds of rivers are raised, causing floods in the countryside. The pattern of drainage is merely a function of the topography and hence longitudinal streams occupy the northern half of the territory while dendratic in the west and south (Anon., 2001). Biswas (2009) highlighted the ten main rivers of Tripura with their catchments, lengths and combined catchments areas (Table 4.1).

Table 4.1

Rivers of Tripura, length of rivers and their catchments area

Serial No.	*River*	*Catchments*	*Length in Km (in Tripura)*	*Combined Catchments in sq. km*
1	Howrah	Baramura	53	570
2	Gomati	Baramura	165	2378
3	Khowai	Atharamura	70	1378
4	Dhalai	Atharamura-Longthorai	76	695
5	Manu	Longthorai	167	1979
6	Muhuri	Baramura-Devtamura	64	840
7	Juri	Jampui	79	586
8	Feni	Baramura-Devtamura	87	555
9	Longai	Jampui	98	342
10	Deo	SakhantlangJampui	100	328

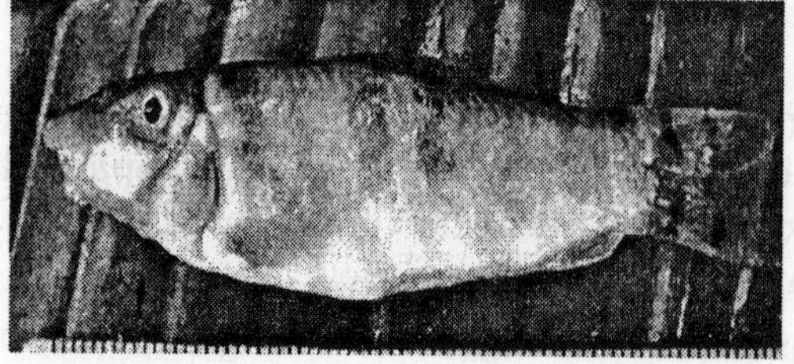

Botia dario

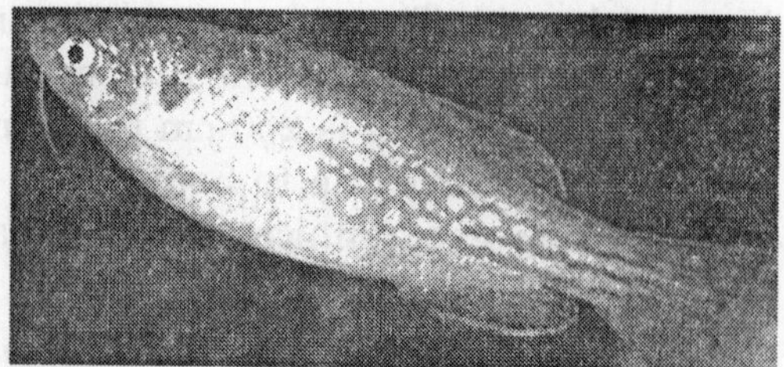

Danio dangila

Puntius sarana

Barilius barila

Mystus cavasius

Macrognathus aculeatus

Bagarius bagarius

Acanthocobitis botia

Figure 4.1. Important ornamental fishes found in the rivers of Tripura.

Prospects of Aquarium Fish Keeping in Tripura

As indicated above, a great variety of indigenous ornamental fishes are abundant in Tripura (Figure 4.1). A list important ornamental species of the state are enlisted in Table 4.2. The availability of indigenous ornamental fishes in natural water bodies is perceived to be declining rapidly and this issue needs to be addressed on high priority (Talukdar, 2004). A well organized program for collection, identification, selection, artificial feeding, captive breeding, rearing and marketing of ornamental fishes will be beneficial to the development of market as well as conservation of germplasm.

Table 4.2
Important ornamental fish available in the rivers and streams of Tripura

	Scientific name	*English name*	*Local name*	*Habitat*	*Trade name*
	1	*2*	*3*	*4*	*5*
1.	*Botia dario*	Necktie loach	Ranimach	Rivers, Lakes	Rani loach
2.	*Danio dangila*	Danio	Naptamach	Rivers, Wetlands	Moustached danio
3.	*Puntius sarana*	Olive barb	Sarputi	Rivers, Reserviour	Olive barb
4.	*Salmostoma bacaila*	Chela/ Long razor belly minnow	Katari	Rivers, Lakes	Razor belly
5.	*Psilorhynchus balitora*	Balitora minnow	Baile	Streams, Rivers	Banded torrent
6.	*Barilius barila*	Barred baril	Kiksa	Rivers	Bared trout
7.	*Acanthocobitis botia*	Striped Loach/ Zipper Loach	Gutum	Streams, Swamps, Wetlands	Leopard loach
8.	*Puntius gelius*	Golden dwarf barb	Tit puti	Wetlands, Rivers	Golden barb
9.	*Puntius phutunio*	Dwarf barb/ pigmy barb	Puti	Wetlands, Rivers	Dwarf barb
10.	*Parluciosoma daniconius*	Black line rasbora	Darkina	Streams, Swamps	NA
11.	*Lepidocephalicthys guntea*	Panther Loach or Platinum Line Loach	Gutum	Streams, Swamps, Wetlands	NA
12.	*Megarasbora elanga*	Bengal barb	Darkina	Rivers, Wetlands	Bengal barb
13.	*Gagata cenia*	Indian gagata/ yellow catfish.	Ghora tangra/ Jungla	Lakes , Rivers, Beels	Clown catfish

(Contd...)

1	2	3	4	5
14. *Mystus vittatus*	Striped dwarf catfish	Tengra	Rivers,Beels	Pyjama striped catfish
15. *Mystus bleekeri*	Dey's mystus	Tengra	Rivers, Beels	Guinea catfish
16. *Mystus cavasius*	Gangetic mystus	Sada gulaya	Rivers, Lakes	NA
17. *Aorichthys seenghala*	Long-whiskered catfish	Aair	Wetlands, Ponds, Rivers	Shovel mouth catfish
18. *Rita rita*	Rita	Ritha	Rivers	White catfish
19. *Pseudeutropius atherinoides*	Indian potasi	Aaoinnya	Rivers	Striped glass catfish
20. *Ailia coila*	Ailia	Kajuli	Beels, Lakes, Rivers	Gangetic ailia
21. *Sisor rhabdophorus*	Whiptail catfish	Bagair	Rivers	Whiptail catfish
22. *Erethistes pussilus*	Gangetic erethistes	NA	Rivers, Wetlands	Gangetic erethistes
23. *Nangra viridescens*	Huddah nangra	Nadir tengra	Rivers, Wetlands	NA
24. *Macrognathus aculeatus*	Lesser spiny eel	Baim	Rivers, Lakes	NA
25. *Mystus gulio*	Long whisker catfish	Guila tengra	Rivers, Wetlands	Whisker catfish
26. *Bagarius bagarius*	Dwarf goonch	Bagair	Rivers, Wetlands	Painted giant catfish
27. *Mystus tengra*	Tengra catfish	Tengra	Rivers, Lakes, Wetlands	Guinea catfish

Source: Nandeesha *et al.*, 2008 (Unpublished Intramural Research Project final document)

Major Constraints/Factors Hindering Sustainable Utilization of the Water Bodies of Tripura

1. Degradation of the water bodies due to processes of siltation, eutrophication and other human activities like paddy cultivation, diversion of water and pollution.
2. Lack of scientific expertise and gap in coordination between various organizations involved.
3. Poor attention to socio-economic issues associated with utilization of water bodies so as to not adversely affect the ornamental fishes.
4. Introduction of exotic, fast-growing fish species like *Clarias gariepinus*, *Pangasius sutchi*, *Hypophthalmicthys nobilis* and prolific breeders like *Oreochromis niloticus*, etc. may have caused natural ecological imbalance.
5. Natural fish genetic resources are neglected because of diversion of attention on the major carp aquaculture operations and their improvement.

6. During culture operation for eradication of weed fishes, attempts were made to remove the fish species of *Puntius* spp., *Mastacembalus* spp., *Channa* spp. *Chanda nama, Chanda ranga,* etc. may have attributed to the decline of the biodiversity.
7. Indiscriminate use of insecticides in agriculture which leaches to the fish ponds.

Strategies for Sustainable Utilization and Conservation of Water Resources of Tripura

The siltation problem can be solved by collaborative efforts of sister departments like Central Water Commission, Joint River Commission, State Flood Control and Management Department, State Fisheries Department, Soil Conservation Department, College of Fisheries and ICAR institutes. The vegetation cover on the slopes is much helpful in protecting siltation problem (Gautam and Singh, 1998). The problem of eutrophication may be tackled through combination of motivating people for disposing their domestic wastes in safe manner and effective framing and implementation of legislation to this effect. The motivational program may be managed by civic bodies in consultation with the relevant scientific departments and in collaboration with NGO's by involving people at the grass roots level and utilizing services of mass media and press. The industries, wherever they are located on the bank and shore of the water bodies, are to have pollution control wings for re-cycling and safe disposal of the waste products. The Central and State Pollution Control Boards are to strengthen their monitoring activities.

Insecticides and pesticides used in paddy cultivation have to be used with care since most of the residues find their way into both lentic and lotic systems either directly or indirectly. The part time fishermen involved in agriculture and other allied activities, may be given incentives through FFDA and other financial organizations so as to enable them to earn their livelihood only by fishing activities and small scale entrepreneurship/ ornamental fish culture. The various problems of rivers may also be taken up through extension activities of National River Action Plan of the rivers of North East India.

Many socio-economic issues like rehabilitation of fishermen, training of fishermen as well as revamping of coordination between all concerned organizations is required. The Fisheries Co-operative Societies have to be tackled through different mechanisms by the state governments, extension workers, universities, central government institutions and NGO's. The ICAR, College of Fisheries and NEC may take appropriate steps for channelizing the funds for such activities. Promotion of livelihood/employment generation option for the local communities through formation of SHGs, which shall undertake activities that will help them earn their livelihoods besides imparting protection to the fish biodiversity. The fishermen should be encouraged to undertake specific eco-tourism based activities. Detailed survey of flora and fauna and geographical/ecological features for different river system should be undertaken for the preparation of the Management Action Plan (MAP) for each important species for each river system. The awareness amongst the people and amongst the different stakeholders needs to be generated about the importance and necessity of water resources and the role they play in our every day life. People should be made aware of the threat of contamination of water resources by way of mixing of harmful chemicals, so that they remain safe source of ground water table. The people should be made to feel proud of the water bodies with a sense of oneness with them, which shall then be our accomplishment. As of now, conservation of biodiversity in Tripura is mainly the responsibility of the State Forest Department, which works through a network of 13 Forest Divisions and four Wildlife Protected Areas (sanctuaries) covering all habitat types (Gupta, 2000).

Conclusion

The world conservation strategy recognizes that the concept of biodiversity conservation is closely linked with sustainable development of both human and natural resources (Ramakrishnan, 1992). Conservation of biological diversity can not be achieved in isolation without addressing the needs of the large human populations who are totally or partially dependent on various river water resources for their survival. The problem of biodiversity conservation in Tripura is basically a problem of scarcity of resources. The problem is further compounded by the failure of development of appropriate technology and/or transfer of technological advances to the people in socially and culturally acceptable forms. It is also known that rivers and streams serve as mother for the water of lower reaches of the river streams. A slight degradation in the water of higher altitude rivers leads to a greater chance of pollution in lower reaches of the river system. Therefore, the control of deterioration in aquatic environment in the upper reaches is necessary for the sake of lower reaches.

The majority of indigenous ornamental fishes found in Tripura have tremendous demand in the international ornamental fish market (Bhattacharjya and Choudhury (2004). Therefore, the export potential for these indigenous ornamental fishes need immediate attention to explore the possibilities for creating alternative job avenue and income generation for unemployed youth of the state. The availability of indigenous ornamental fishes in natural resources is sharply decreasing, which needs immediate focused attention to conserve these valuable natural resources of the state. In the field of diversified aqua-resource management, the promotion of aquarium fish breeding and rearing with management of aquarium may provide sustainable and suitable trade opportunities in the state. The College of Fisheries, Central Agricultural University, Agartala, ICAR Centre for NER, Agartala and State Fisheries Department has also laid much stress on aquatic bio-diversity conservation and sustainable utilization of aqua-resources with people's participation.

REFERENCES

Anon. 2001. Soil, Natural Vegetation and Wildlife. *In*: Bareh, H.M. (Ed). *Encyclopaedia of North-East India*-8-Tripura. Mittal Publications, New Delhi, pp. 86-110.

Anon., 2003. Megadiversity, Hot Spots vis-à-vis Biodiversity (Indian Panorama). *In*: Chaudhuri, A. B. and Sarkar, D. D. (Eds.). Megadiversity Conservation-Flora, Fauna and Medicinal Plants of India's Hot Spots. Daya Publishing House, pp. 13-23.

Bhattacharjya, B. K. and Choudhury, N. 2004. Ornamental fish culture and trade in the North-East – An overview, *In*: Bhattacharjya, B. K. and Choudhury, M. (Eds.). Ornamental fish culture and trade in North-Eastern India, CICFRI, Barrackpore, Kolkata, India, pp.1-14.

Biswas, S. P., Das J. N., U.K. Sarkar and Lakra, W.S. 2007. *Ornamental Fishes of North East India: An Atlas.* NBFGR, Lucknow, 111p.

Biswas, P. 2009. Water and wetlands crises of Tripura. www.tripurainfo.com

Datta, M. K., 2000. Endangered fish species of Tripura need conservation, p. 83-87. In: *Fish Biodiversity of North East Indian* (eds. A. G. Punniah and U. K. Sarkar), NBFGR. NATP Publication, Lucknow.

Ehrlich, P. R. and Wilson, E.O. 1991. Biodiversity studies: Science and Policies. *Science*, 253: 758-762.

Gautam, A. and Singh, N. 1998. Conservation of Lotic Aquatic Resources of Himalaya: The Case of Garhwal Himalaya, pp. 188-199. In: *Conservation & Management of Aquatic Resources* (ed. A. Gautam), Daya Publishing House, New Delhi,.

Ghosh, S. K. and Lipton, A.P. 1982. Ichthyofauna of the N.E.H. Region with special refernce to their economic importance. *ICAR Spl. Bulletin* (ICAR Research Complex, Shillong), 1: 119-126.

Gopakumar, G. 2005. Ornamentals, pp. 150. In: *International Symposium on Improved Sustainability of Fish Production Systems and Appropriate Technologies for Utilization*, Cochin University of Science and Technology, Cochin; March 16-18.

Goswami, U. C. 2005. Influence of Certain Ecological Factors on The Pigmentation of Ornamental Fish, pp. 1-16. In: *Biodiversity, Ecophysiology and Conservation of Freshwater Animals* (ed. J. Ojha), Narendra Publishing House, Delhi.

Gupta, A. K. 2000. Shifting Cultivation and Conservation of Biological Diversity in Tripura, Northeast India. *Human Ecology*, 28(4): 606-6629.

Jayaram, K. C. 1999. *The Freshwater Fishes of the Indian Region*. Narendra Publishing House, New Delhi, 551pp.

Kar, D. 2003. Fishes of Barak drainage, Mizoram and Tripura, pp. 203-211, In: *Environment Pollution and Management* (eds. A. Kumar, C. Bohra and L.K. Singh), APH Publishing Corporation, New Delhi.

Kar, D., Nagarathna, A.V., Ramachandran T.V. and Dey S.C. 2006. Fish Diversity and Conservation Aspects in an Aquatic Ecosystem in Northeastern India. *Zoos' Print Journal*, 21(7): 2308-2315.

Mittermeier, R. A. and Mittermeier, C. G. 1997. Megabiodiversity: Earth's Biologically Wealthiest Nation. In: *Global Freshwater Biodiversity* (eds. D.E. McAllister, A.L. Hamilton and B. Harvery). Sea Wind, Cemex, Mexico City, 11: 1-140.

Nandeesha, M. C., Mandal S.C., Pradhan P.K. and Barman, D. 2008. Mapping of the indigenous ornamental fish resources of South Tripura district and identification of the potential species for commercial production. Unpublished Intramural Research Project final document, College of Fisheries, CAU, Agartala, India.

Ramakrishnan, P. S. 1992. *Shifting Agriculture and Sustainable Development: An Interdisciplinary Study from Northeastern India.* The Parthenon Publishing Group, USA.

Talukdar, P. 2004. Status Paper on Ornamental Fish in Tripura, pp.61-62. In: *Prospects of*

Ornamental Fish Breeding and Culture in Eastern and North-Eastern India (eds. R.C. Das, A. Sinha, S. Datta and S. Ghosh). National Seminar, 16-17 February.

Vishwanath, W. 2002. *Fishes of North-East India, a field guide to species identification.* Department of Life Science, Manipur University, 194 pp.

WCMC. 1998. Freshwater Biodiversity: A preliminary Global Assessment Document. A document prepared for the 4th Meeting of the Conference of the Practices to the Convention of Biological Diversity, World Conservation Monitoring Centre.

Wilson, E. O. 1988. *Biodiversity,* National Academy Press, Washington D.C.

5

Fishes of the Barak River System in Manipur

—*K. Nebeshwar*[1] & *W. Vishwanath*[2]

[A detail survey of the fish fauna of the Barak river and its tributaries in Manipur was carried out. The river system covers nearly 50% of the total state's catchment area and contains representatives of Assam, Mizoram, and endemic Manipur elements. The diversity is attributed to many reasons, viz., the geomorphology, consisting of hills, plateaus and valleys, resulting in the occurrence of a variety of torrential hill streams, rivers, and lakes in the altitudinal range from 250-2500 m asl. The river system is in the region of Indo-Chinese hotspot region and its fauna is also related with its counterparts in the Himalayan hotspot region. The study shows occurrence of 88 species belonging to 57 genera, 25 families and 11 orders. Order Cypriniformes represents maximum 45 number of species, followed by 23 species of Siluriformes, and 12species of Perciformes. Genera *Barilius*, *Puntius*, and *Garra* represent maximum 5 species each, followed by 4 species each of *Tor*, *Glyptothorax* and *Schistura*. The collection includes widely distributed forms, low altitude forms, shallow water form, and strong current forms. The study shows nine new records viz., *Ailia coila, Psilorhynchus balitora, Salmophasia phulo, Salmophasis bacaila, Tor mosal, Eutropichthys murius, Sperata seenghala*, and *Rita rita* and six new species viz., *Schistura tigrinum, S. minutes, Pterocryptis barakensis, Exostoma barakensis, Sisor barakensis,* and *Badis tuivaiei.*

Key words. Fish diversity, Barak drainage, Manipur.]

Introduction

The fish fauna of Manipur is of special interest in the fact that it is drained by two important drainages of the world, viz, Barak-Brahmaputra in the western half and Chindwin-Irrawaddy drainage in the central plain and eastern half of the state. The state is considered to be the meeting place of the Eastern Himalaya and Burmese region. The interest in the fish fauna of the Barak river system in Manipur is that it is a part of the Brahmaputra river system. The different ecological conditions in the river, consisting of highly gradient stream in the northern part at an altitude of about 2500 m asl where it originates at Lei Leiyai area in Senapati district and sandy and muddy bottom at the southern point at an altitude of about 250 m asl at Tipaimukh, suggest a variety of forms and diversity of species. From the origin, the

[1,2]Department of Life Sciences, Manipur University
Email: [1]knebeshwar@yahoo.com; [2]vnath54@yahoo.co.in

river follows a southern course, forms boundary between Manipur and Nagaland for some distance and then enters into the interior of the state of Manipur in Tamenglong district. It flows till it reaches Tipaimukh and then turns sharp towards north and follows along the western side of the Vangai range till it reaches Jirighat. The river then bends towards the west and enter Cachar district of Assam and Sylhet district of Bangladesh. It finally discharges itself into the old Brahmaputra bed to form the Meghna.

The river system covers nearly 50% of the total state's catchment area sharing fauna predominately with that of Indo-Gangatic fauna and endemic elements. The main tributaries of the Barak River are the Irang, Makru, Jiri and Tuivai. The Irang, the biggest tributary has its origin at Karong. It flows almost parallel to the Barak from the eastern side. The Irang has tributaries like, Tuipi, Iring, Iyei and Leimatak. The Leimatak rises in the Northern hills of Churachandpur district and receives the outflow water of the Loktak Electric Hydel Project. The Makru and Jiri rise from the southern part of the Barail range and flow almost parallel to each other. Both the tributaries meet the Barak from the west. The Tuivai rises in the hills of Churachandpur district, flows westward between Manipur and Myanmar and then between Manipur and Mizoram. It finally flows into the Barak in Tipaimukh. Above this, small streams like Akangki and Jaduki River originate from the Kadi hill range of Assam. The fish fauna of the Barak river system contains representatives of Assam, Mizoram, Myanmar and endemic Manipur elements. The rainy season is quite prominent in this region which starts from the month of June and continues upto October.

The first ever scientific publication on the fishes of Manipur was that of Chaudhuri (1912). He described two new species, viz., *Danio naganensis* (now *Devario naganensis*) and *Nemacheilus manipuriensis*, (now *Schistura manipurensis*) without giving detail information on the type locality of the fishes. Hora(1921) provided a list of 56 species which were collected from 4 localities viz. i) Loktak lake, ii) sluggish streams in the Manipur valley, iii) streams with rocky bed in the southern watershed and iv) streams with rocky bed in the northern watershed of the Naga hills. The northern watershed of the Naga Hills as referred in his report covered only small streams at various points on the road between the Manipur valley and Dimapur (Nagaland) which belongs to Barak-Brahmaputra drainage. In later years, Hora and Mukerji (1935) reported 44 species from the Brahmaputra drainage (Dayang valley) and Chindwin drainage (Tizu valley) of Nagaland of which 18 species were collected from Brahmaputra basin. Further report of Hora (1936) on the fishes of Naga Hills (Manipur) was all collected from the Barak River and its tributaries. Menon (1952) made a note on 19 species collected from different areas: i) Kanglatombi Field Typhus Station (briefly called Kanglatombi Camp), about 16 miles north of Imphal in the Dimapur road, ii) Kanglatombi Bazar, about 12 miles in the Imphal-Dimapur road, iii) Imphal market, iv) water-pools and streams on Imphal-Pallel road north of Imphal at 18-22. Later, Menon (1955) made extensive collections from the Loktak Lake, the Imphal River and the Barak River at different localities. The latest report on the ichthyofauna of Barak river system in Manipur is of Vishwanath and Kosygin (2002). They made a list of 55 species.

Despite all the above literature dealt on the fauna of the Brahmaputra and Chindwin drainage, the detailed general inventory of the region is not yet performed based on advanced fish taxonomy and there has been no specific reference of the fish diversity and list of fishes of the Barak River system in Manipur. Ichthyologists have been looking into the geographical history, river basin formations and early fossil appearances for correlation with the origin of fish and their evolution in time and space. The affinity and interrelationships of species have been related with longitudinal river valleys, inter basin connections, geological history etc. With the publication of the result of extensive surveys of fishes in the Indo-Chinese region, there are great changes in the nomenclature in the family and generic status after a phylogenetic study of many workers. The present concept is that fresh water fishes are distributed in a particular river basin except for some air breathing and hardy type. The congeners distributed in a separate different basin are proved to be a different species. Early workers had access to a very few and poorly preserved specimens. With the new concept, various revision studies of some highly variable and widely distributed forms of earlier days have now been shown to be aggregates of distinct species (Ng and Ng 2001; Ng and Rainboth, 2001; Kullander and Britz, 2002; Ferraris, 2004; Chakrabarty and Ng 2005; Ng 2006; Britz and Maclaine 2007; Ferraris and Vari, 2007; Ng and Kottelat, 2007, 2008 etc.).The fauna of this region is related with its counterparts in adjacent waters of the Himalayan and Indo-Chinese region. Kottelat (1989) indicated that the shared genera between the Salween and Chao Phraya and the Mekong are also shared with the Brahmaputra and the Ganges and are widely distributed in the south and south-east Asia. The most basic need for exchanging biological information, within and through all borders, is a consistent nomenclature which of course can only be reached by in-depth systematic research. In the last few years, extensive surveys had been done in *the Indo-Chinese* region and a little work on the *Himalaya* region. However, the surveys could not cover the regions of the North east India. A detail survey of the fish fauna of the Barak River and its tributaries in Manipur was carried out and studied in respect of taxonomic status, species diversity, and zoogeographical point of view.

Materials and Methods

Collection

Fishes were collected during the period from 2000 to 2004. Methods of collecting the fishes differ depending on their habit and size. The fishes were collected from Vanchengphai (Hmar village), Khunphung, Taizizang, Barak water fall, Barak Bridge, Leimatak River at Leimatak, Iyei River at Noney, Tuivai River, Jiri River at Jiri and more above from many areas below the water fall from a Barak rafting expedition. Fishes were caught by using gill nets of different mesh sizes, cast nets and indigenous gears of the area. Other means of fishing is the temporary damming of shallow portion of stream and diverting the water from the main stream to dry up. In the interior area, fishermen also use explosives, poisoning by bleaching powder to catch bigger sized fishes.

Preservation

Fishes were fixed and preserved in 10% formalin. Bigger sized fishes were given a narrow longitudinal incision on the abdomen and formalin was injected into the viscera with the help of hypodermic syringe.

Identification

Standard measurements and counts followed Kottelat (2001) Measurements were made with a dial caliper to the nearest 0.1 mm and body proportions were expressed as percentage of standard length (SL) and head length (HL). Transverse scales were counted as scales between lateral line and dorsal fin origin and between lateral line and pelvic fin origin. For small specimens counting of fins and scales were made under stereoscopic microscope, preferably after using a temporary surface stain (fountain pen ink).

Morphometric analysis for identifying the fishes were carried out generally with reference to Talwar and Jhingran (1991) and Jayaram (1999).For further confirmation the following literature were followed: viz., Jayaram (1991) for genus *Puntius* Hamilton-Buchanan; Menon (1964), Kullander and Fang (2004), Stiassny and Getahun (2007) for *Garra* Hamilton-Buchanan; Rainboth (1983) for *Psilorhyncus*; Rainboth (1986) for *Chagunius,* Menon (1987) and Kottelat (1990) for genus *Schistura* McClelland; and Misra (1976), Jayaram (1977a and b, 1979 and 1981) for Siluroid fishes. In addition to these, relevant literatures including original description of some fishes were followed accordingly. Morphometric data of the fishes were compared with that of the type and other specimen in the Zoological Survey of India, Calcutta (ZSI) or with specimens collected from other parts of Manipur deposited in MUMF.

The sequence of families is phylogenetic, following the classification of Nelson (1994) with a few deviation based on the recent studies.

Systematic Index of the Fishes:

Order: Osteoglossiformes
Family: Notopteridae
Genus: *Chitala* Fowler
C. chitala (Pallas)
Order: Anguilliformes
Family: Anguillidae
Genus: *Anguilla* Schrank
A. bengalensis (Gray)
Order: Clupeiformes
Family: Clupeidae
Genus: *Tenualosa* Fowler
T. ilisha (Hamilton)
Genus: *Gudusia* Fowler
G. chapra (Hamilton)
Order: Cypriniformes
Family: Cyprinidae
Genus: *Salmophasia* Swainson
S. bacaila (Hamilton)
S. phulo (Hamilton)
Genus: *Aspidoparia* Heckel
A. *maror* (Hamilton)
Genus: *Amblypharyngodon* Bleeker
A. mola (Hamilton)
Genus: *Barilius* Hamilton
B. barila (Hamilton)
B. bendelisis (Hamilton)
B. tileo (Hamilton)
B. sp.1
B. sp.2
Genus: *Chela* Hamilton
C. laubuca (Hamilton)
Genus: *Esomus* Swainson
E. danricus (Hamilton)

Genus: *Devario* Heckel
D. aequipinnatus (McClelland)
Genus: *Tor* Gray
T. mosal (Hamilton)
T. progenius (Hamilton)
T. putitora (Hamilton)
T. tor (Hamilton)
Genus: ***Neolissochilus*** Rainboth
N. hexagonolepis (McClelland)
N. hexasticus (McClelland)
Genus: ***Osteobrama*** Heckel
O. cotio (Hamilton)
Genus: ***Chagunius*** Smith
C. chagunio (Hamilton)
Genus: ***Puntius*** Hamilton
P. conchonius (Hamilton)
P. chola (Hamilton)
P. sarana (Hamilton)
P. sophore (Hamilton)
Genus: *Poropuntius* Smith
P. clavatus (McClelland)
Genus: *Bangana* Hamilton
B. dero (Hamilton)
Genus: *Crossocheilus* Kuhl & van Hasselt
C. latius (Hamilton)
Genus: *Garra* (Hamilton)
G. lissorhynchus (McClelland)
G. manipurensis Vishwanath & Sarojnalini
G. naganensis Hora
G. nasuta (McClelland)
G. sp.
Family: Psilorhynchidae
Genus: *Psilorhynchus* McClelland
P. balitora (Hamilton)
P. homaloptera Hora & Mukerji
Family: Balitoridae
Genus: *Balitora* Gray
B. brucei Gray
Genus: *Acanthocobitis* Peters
A. botia (Hamilton)
Genus:*Neonoemacheilus* Zhu& Guo
N. assamensis (Menon)
Genus: *Schistura* McClelland
S. chindwinica (Tilak & Hussain)
S. khugae Vishwanath & Shanta
*S.tigrinum*Vishwanath & Nebeshwar
S. sp.
Family: Cobitidae
Genus: *Botia* Gray
B. dario (Hamilton)
Genus:*Lepidocephalichthys* Bleeker
L. guntea (Hamilton)
Genus: *Cantophrys* Swainson
C. gongota (Hamilton)
Order: Siluriformes
Family: Bagridae
Genus: *Rita* Bleeker
R. rita
Genus: *Sperata* Holly
S. aor (Hamilton)
S. seenghala (Sykes)
Genus: *Mystus* Scopoli
M. bleekeri (Day)
M. cavasius (Hamilton)
M. tengra (Hamilton)
Family: Siluridae
Genus: *Pterocrytis* Pallas
P.barakensis (Vish. &Nebeshwar)
Family: Schilbeidae
Genus: *Ailia* Gray
A. coila (Hamilton)
Genus: *Eutropichthys* Bleeker
E. murius (Hamilton)
E. vacha (Hamilton)
Family: Amblycepitidae
Genus: *Amblyceps* Blyth
A. mangois (Hamilton)
Family: Sisoridae
Genus: *Sisor* (Hamilton)
S. barakensis (Vishwanth & Darshan)
Genus: *Gogangra* Roberts
G. viridescens (Hamilton)
Genus: *Gagata* Hamilton
G. cenia (Hamilton)
Genus – *Glyptothorax* Blyth
G. cavia (Hamilton)
G. manipurensis Menon
G. sp.1
G. sp.2
Genus *Exostoma* Blyth
E.barakensis Vishwanath & Joyshree
Genus: *Pseudocheneis* Blyth
P. sulcata (McClelland)
Family: Erethistidae
Genus: *Hara* Blyth
H. serratus (Hamilton)
Genus:*Erethistes* Muller & Troschel
E. pussilus Muller & Troshel.
Family: Chacidae
Genus: *Chaca* Gray
C. chaca (Hamilton-Buchanan)
Order: Mugiliformes
Family: Mugilidae
Genus: *Sicamugil* Fowler

S. cascasia (Hamilton-Buchanan)
Order: Beloniformes
Family: Belonidae
Genus: *Xenentodon* Regan
X. cancila (Hamilton)
Order: Cyprinidontiformes
Family: Aplocheilidae
Genus: *Aplocheilus* McClelland
A. panchax (Hamilton)
Order: Synbranchiformes
Family: Mastacembelidae
Genus: *Mastacembelus* Lacepede
M. armatus (Lacepede)
Order: Perciformes
Family: Chandidae
Genus: *Chanda* Hamilton
C. nama (Hamilton)
Genus: *Parambassis* Bleeker
*P. sp.*1.
Family: Sciaenidae
Genus: *Johnius* Bloch
J. coitor (Hamilton)
Family: Nandidae
Genus: *Badis* Bleeker
B. badis (Hamilton)
B. tuivaiei Vishwanth & shanta
Genus: *Nandus* Valenciennes
N. nandus (Hamilton)
Family: Gobiidae
Genus: *Glossogobius* Gill
G. giuris (Hamilton)
Family: Belontidae
Genus: *Trichogaster* Bloch & Schneider
T. fasciata Bloch & Schneider
T. labiosa Day
Family: Channidae
Genus: *Channa* Scopoli
C. marulius (Hamilton-Buchanan)
C. gachua
C. stewartii (Playfair)
Order: Tetraodontiformes
Family: Tetraodontidae
Genus: *Tetraodon* Linnaeus
T. cutcutia (Hamilton)

Results and Discussion

The main concerning factors of biodiversity of Manipur is that the state falls in North-east India, which is considered as one of the hotspots of the freshwater fish biodiversity in the world (Kottelat and Whitten, 1996). The diversity is attributed to many reasons, viz., the geomorphology, consisting of hills, plateaus and valleys, resulting in the occurrence of a variety of torrential hill streams, rivers, lakes and swamps; the drainage pattern which include the Ganga-Brahmaputra, Koladyne and Chindwin-Irrawaddy systems. Another important factor is the tectonic setting in the Indo-Chinese sub region caused by collision of Indian, Chinese and Burmese plates, resulting in the formation of the mighty Himalayas and Indo-Burma ranges. These factors suggest a distinctive nature in its piscine fauna. North East India forms part of two of the 34 biodiversity hotspots listed by Conservation International. They are the *Himalaya* and *Indo-Burma* (Roach, 2005).The Himalaya is the home of the world's highest mountains and deepest gorges. The mountains rise abruptly, resulting in a diversity of ecosystems. The Indo-Burma, also called Indochina bioregion includes portions from eastern India to Viet Nam. The whole of Arunachal Pradesh and Assam, north of the Brahmaputra and Sikkim belong to the Himalayas while Mizoram, Assam, south of the Brahmaputra, and Meghalaya, Nagaland and Manipur belong to the Indo-Burma.

The study report gives a comprehensive account on the fish fauna of the Barak River system in Manipur. The study shows occurrence of 88 species belonging to 57 genera, 25 families and 11 orders. Order Cypriniformes represents maximum number of species (45), followed by Siluriformes (23) and Perciformes (12). Out of the 57 genera, genera *Barilius*, *Puntius*, and *Garra* show maximum diversity represented by 5 species each, the next genus being *Tor, Glyptothorax* and *Schistura* with 4 species followed by *Mystus and Channa* with 3 species each.

The collection includes both groups of fishes with and without special adaptive devices. *Barilius, Tor, Neolissocheilus, Chagunius, Bangana, Devario, Salmophasia,* etc. are the genera with no special modification except for compression of body shape, rounding off and tapering towards the anterior and posterior extremities. On the other hand, species of the genera *Balitora, Glyptothorax, Pseudocheneis, Garra, Psilorhyncus, Neonoemacheilus, Schistura, Rita, Crossocheilus* etc. have special structural modifications. Similar observations in conformity with those of Hora (1922) were made in respect of the modifications of the fishes to adapt to torrential water habitats. The study site has an altitudinal variation from 200m asl to 2500m asl. Accordingly, the gradient of water in the Barak is 0.39°. The fishes may be grouped into the following groups in relation to the altitudinal and water current variations.

Widely distributed forms

The group includes *Barilius, Mastacembelus, Chitala, Anguilla, Neolissochilus, Tor, Schistura, Channa, Lepidocephalicthys, Glyptothorax, Xenentodon, Aplocheilus, Glossogobius, Johnius, Badis* and *Trigchogaster.*

Low altitude forms

The species which were recorded only at low altitude are *Chitala chitala, Tenualosa ilisha, Gudusia chapra, Aspidoparia morar, Barilius tileo, Salmophasia bacaila, Osteobrama cotio, Puntius sarana, Botia dario, Cantophrys gongota, Mystus cavasius, Ailia coila, Eutropichthys murius, E. vacha, Amblyceps mangois, Erethistes pussilus, Hara serratus, Chaca chaca, Sicamugil cascasia, Johnius coitor, Nandus nandus, Trichogaster fasciata, T. labeosa, Tetraodon cutcutia, Sperata seenghala.* These species were collected only from Jiri River at Jiri at an altitude of 250m asl. Of which *Nandus nandus, Eutropichthys vacha* and *Trichogaster fasciata* are also found in the Imphal valley and the Manipur River system, which is at an altitude ranging from 400-760m asl.

Shallow water forms

Species, which inhabit moderate to high currents but at comparatively shallow water, are *Anguilla bengalensis, Barilius bendelisis, Barilius* sp.1, *Puntius sp., Garra sp., Psilorhynchus balitora, Psilorhyncus homaloptera, Schistura* sp., *Glyptothorax* sp., *Pseudocheneis sulcata, Mastacembelus armatus, Badis badisi, B. tuivaiei.*

Strong current forms

Fishes, which live in strong currents, are *Pseudocheneis sulcata, Psilorhynchus balitora, P. homaloptera, Glyptothorax* sp., *Garra* sp. and *Balitora brucei.*

Most of the strong current forms have special modifications, which facilitates the fishes to live in strong current. *Pseudocheneis* sp., *Glyptothorax* sp. have a series of transverse folds on the chest and ventral side of first spinous pectoral and pelvic-fin rays, which are adapted to attachment to substratum to withstand strong water currents. Fishes of the *Garra* have a suctorial disc of semi-cartilaginous pad on the chin, which is also adapted to attachment to substratum. In some species of *Garra*, having proboscis on snout, the proboscis is also an adaptive apparatus.

In species of the genus *Psilorhynchus* and *Balitora brucei*, the paired fins are horizontally placed. They have with four to ten of pectoral and two of pelvic are simple and thick. These inflated paired fins can attach to the substratum to withstand strong water currents in torrential water habitat.

In *Garra* shifting of vent forward from the base of anal fin and development of proboscis on snout are interesting adaptations. The characters are also considered to be of great systematic and evolutionary significant (Menon, 1964).

Distribution pattern of tubercles and presence of uni (one pointed) or di (two pointed) or tri (three pointed) or tetra (four pointed) cuspid tubercles and sizes of the mental adhesive disc of the lower lip are not only having specific significance but also the adaptations of the species.

In the present study, the genus *Garra* is represented by 5 species, out of which *G. lissorhynchus, G. manipurensis, G .naganensis* are the forms without proboscis on snout and in the increasing order of adaptation to torrential water habitat as shown by the positions of vents. The species with proboscis on snout are *Garra nasuta* and *G. sp* in order of adaptation of torrential water habitat.

New species

The present collection includes 6 new species. They are *Schistura tigrinum, S. minutes, Pterocryptis barakensis, Exostoma barakensis, Sisor barakensis, Badis tuivaiei.* These species are described separately elsewhere.

Neolissocheilus hexasticus (McClelland) was already reported in Manipur as *Barbus hexasticus* from Senapati stream (Barak Basin), Southern watershed of Naga Hills (Chindwin basin and Itok stream near Chanderkhong in the Manipur valley (Hora, 1921). But Hora and Mukerji (1935) synonymised *Barbus hexasticus* Hora (nec McClelland), as synonym of *Barbus hexagonolepis* McClelland, presently known as *Neolissochilus hexagonolepis* (McClelland). As remarked by them, the species is very closely related to *B. tor* (Hamilton). So it was very difficult to define the precise systematic position of *B. hexasticus*. But Day (1878) remarked that the character of the interrupted groove behind the lip in *Barbus hexagonolepis* easily distinguishes from *B. hexasticus*. In *hexastichus*, the lower lip without or with a badly developed lobe, but having a shallow and continuous fold and sometimes pores on the cheeks. In the present study, the specimens collected from the Barak River fit to the characters mentioned by Day (1878) and the specimens have 9-10 gill rakers on the lower arm of first arm.

New record in Barak River

Ailia coila, Psilorhynchus balitora, Salmophasia phulo, Salmophasis bacaila, Eutropichthys murius, Sperata seenghala, Rita rita are recorded for the first time from Barak drainage in Manipur.

The present study shows that though there is great specific diversity, the genera of the Brahmaputa and the Chindwin drainage shows certain degree of similarity. Of the 88 species reported here, 18 are also distributed in the Chindwin drainage. The species are *Chitala chitala, Tor mosal, Tenualosa ilisha, Anguilla bengalensis, Puntius chola, P. sophore, Jhonius coitor, Rita rita, Xenentodon cancila, Glossogobius giuris, Trichogaster fasciata, T. labeosa, Channa marulius, C. gachua, Tetradon cutcutia, Eutropiichthys vacha, Amblypharyngodon mola, Esomus danricus.*

The restricted distribution of the fishes in the Barak River system shows that they evolved in the system comparatively late during the orogenic movements in the part of the world. As there is no water connection between Chindwin-Irrawaddy and Barak-Brahmaputra drainages as they are entirely separated by mountain ranges (Chaudhuri, 1919, Hora, 1921a and Hora and Mukerji, 1935), these fishes are not found in other parts of the Chindwin head waters of Manipur and Chindwin-Irrawady drainage. Other widely distributed forms probably distributed when there was water connection between the Tsangpo and the Chindwin-Irrawaddy as per the post Himalayan River system hypothesized by Gregory (1925). When connection between Tsangpo and Ganga-Brahmaputra established their connection, the fishes might have distributed to other parts of the country or vice-versa. Thus it is quite probable that Gondwana fresh water fishes might have already reached Myanmar through the Indus (Trans-Himalayan drainage) before it could reach Manipur. The present fishes of Manipur might have come via the Manipur River after the formation of the Manipur valley, while those of the western part might have reached via the Barak after the formation of Assam plains.

The diversity of species in this region is attributed to the recent geological history, especially the Himalayan orogeny (Kottelat, 1989). Peninsular India is essentially a part of the Gondwana where the freshwater fishes originated. During Miocene and later, India collided with Eurasia and the sea in the present northeastern part of India elevated and freshwater condition prevailed. Freshwater fishes from Indian mainland would have migrated towards Myanmar and other parts of South-East Asia only then. Thus the species of fishes in this region evolved only very recently from their old Gondwana ancestors. The diversity is attributed to the vigorous movements of Earth's crust and constant changes in the course of the water.

Local people inhabiting in the shore stretch of the Barak river catch fish mainly for their consumption and the fishes have been becoming the economic source of some ethnic groups inhabiting in the river shore of the Barak River. The people of Vanchengphai, a Hmar village catch fish and sell to Tamenglong head quarter and some people of Jiri, Awangkhul, and Noney live by fishing. Fishermen use cast-nets, gill-nets of different mesh size, rod and lines etc for catching fishes. Other techniques used are explosives, electric fishing, and damming, diverting and dewatering shallow portion of streams. Usually fishing period in some area is very long (i.e. whole year). The different ecological conditions prevailing in the Barak River is of great interest from taxonomic point of view since it would support different types of specialized as well as widely adaptive fishes.

REFERENCES

Britz, R. and Maclaine, J. 2007. A review of the eel-loaches, genus *Pangio*, from Myanmar (Teleostei: Cypriniformes: Cobitidae). *Ichthyological Exploration of Freshwaters*. 18 (1): 17-30.

Chaudhuri, B.L. 1912. XXXV. Descriptions of some new species of freshwater fishes from north-India. *Records of the Indian Museum* (Calcutta). 7 (5): 437-444, Pls. 38-41.

Chaudhuri, B.L. 1919. Report on a small collection of fish from Putao (Hkamti Long) on the northern frontier of Burma. *Records of the Indian Museum* (Calcutta). 16 (4): 271-287, Pl. 22.

Chakrabarty, P. and Ng, H.H. 2005. The identity of catfishes identified as *Mystus cavasius* (Hamilton, 1822) (Teleostei: Bagridae), with a description of a new species from Myanmar. *Zootaxa*. 1093: 1-24.

Day, F. 1878. The fishes of India; being a natural history of the fishes known to inhabit the seas and fresh waters of India, Burma, and Ceylon. The fishes of India; being a natural history of the fishes known to inhabit the seas and fresh waters of India, Burma, and Ceylon. Part 4: i-xx + 553-779, Pls. 139-195.

Ferraris, C.J., Jr. 2004. A new species of the Asian schilbid catfish genus *Clupisoma* from Myanmar, with a redescription of *Clupisoma prateri* Hora (Osteichthyes: Siluriformes: Schilbidae). *Zootaxa*. 437: 1-10.

Ferraris, C. J., Jr. and Vari, R. P. 2007. Revision of catfishes of the genus *Eutropiichthys*, with the description of two new species (Siluriformes: Schilbidae). Copeia 2007 (4): 866-885.

Gregory, J.W. 1925. The evolution of the river system of South-eastern Asia. *Scottish Geographical Journal*. 41: 129-141.

Hora, S. L. 1921. Fish and fisheries of Manipur with some observations on those of the Naga Hills. *Records of the Indian Museum* (Calcutta). 22 (3): 165-214, Pls. 9-12.

Hora, S. L. 1922. A structural modifications in the fish of mountain torrents. *Records of Indian Museum*. 24: 31-61.

Hora, S.L. 1936. On a further collection of fish from the Naga Hills. *Records of the Indian Museum* (Calcutta). 38 (3): 317-331.

Hora, S.L. 1941. Fishes collected by the Vernay-Hopwood upper Chindwin expedition, 1935. *Journal of Bombay natural History Society*. 42(3): 478-482, pl. 1.

Hora, S.L. and Mukerji, D.D. 1935. Fishes of the Naga Hills, Assam. *Records of the Indian Museum* (Calcutta). 37 (3): 381-404, pl. 7.

Jayaram, K.C. 1977a. Year of publication and type-locality of *Ompok Bimaculatus* (Bloch) (Pisces: Siluroidea). *Newsletter of the Zoological Survey of India*. 3 (1): 50-51.

Jayaram, K.C. 1977b .Aid to identification of siluroid fishes of India, Burma, Sri Lanka, Pakistan and Bangladesh. 1. Bagridae. *Records of the Zoological Survey of India. Miscellaneous Publication, Occasional Paper* No. 8: 1-41.

Jayaram, K.C.1979. Aid to the identification of sisorid fishes of India, Burma, Srilanka, Pakistan and Bangladesh, 3. Sisoridae. Occasional papers. *Zoological Survey of India*. 14:1-62.

Jayaram, K.C. 1981. The freshwater fishes of India, Pakistan, Bangladesh, Burma and Sri Lanka—a handbook. *Zoological Survey of India*. i-xxii + 1-475, pls. 1-13.

Jayaram, K.C. 1991. Revision of the genus Puntius Hamilton from the Indian region. *Records of Zoological survey of India, Occ. Paper No*. 135:178.

Jayaram, K.C. 1999. The freshwater fishes of the Indian region. *Narendra Publishing House, New Delhi*, 551pp.

Kottelat, M. 1989. Zoogeography of the fishes from Indochinese inland waters with an annotated check-list. *Bulletin Zoölogisch Museum, Universiteit van Amsterdam*. 12 (1): 1-55.

Kottelat, M. 1990. Indichinese Nemacheilines, a revision of nemacheiline loaches (pisces: Cypriniformes) of Thailand, Burma, Laos, Cambodia and southern Vietnam. Verlag. Dr. Friedrich Pfiel, Munchen, 262pp, Figs. 1-180.

Kottelat, M. 2001. *Fishes of Laos.*: 1-198., WHT Publications (Pte) Ltd., Pls. 1-48, figs. 1-65.

Kottelat, M. and Whitten, T. 1996. Freshwater Biodiversity in Asia with special reference to Fish: *World Bank tech. Paper No.* 343. The World Bank, Washington DC.

Kullander, S. O. and Fang, F. 2004. Seven new species of *Garra* (Cyprinidae: Cyprininae) from the Rakhine Yoma, southern Myanmar. *Ichthyological Exploration of Freshwaters* 15(3): 257-278.

Kullander, S. O. and Britz, R.2002.Revision of the family Badidae (Teleostei: Perciformes), with description of a new genus and ten new species. *Ichthyological Exploration of Freshwaters.* 13 (4): 295-372.

Ng, H. H. 2006. The identity of *Batasio tengana* (Hamilton, 1822), with the description of two new species of *Batasio* from north-eastern India (Teleosei: Bagridae). *Journal of Fish Biology.* 68: 101-118.

Ng, H. H. 2007a. *Akysis pulvinatus*, a new species of catfish (Siluriformes: Akysidae) from southern Thailand. *Zootaxa.* 1608: 51-58.

Ng, H. H. 2007b. *Botia udomritthiruji*, a new species of botiid loach from southern Myanmar (Teleostei: Botiidae). *Zootaxa.* 1608: 41-49.

Ng, H. H. and Ng, P. K. L. 2001.A revision of the akysid catfish genus *Acrochordonichthys* Bleeker. *Journal of Fish Biology.* 58 (no. 2): 386-418.

Ng, H. H. and Rainboth, W. J. 2001. A review of the sisorid catfish genus *Oreoglanis* (Siluriformes: Sisoridae) with descriptions of four new species. *Occasional Papers of the Museum of Zoology University of Michigan.* 732: 1-34.

Menon, A.G.K. 1955. Further observations on the fish fauna of the Manipur State. *Records of the Indian Museum* (Calcutta). 52 (1): 21-26.

Menon, A.G.K. 1964. Monograph of thecyprinid fishes of the genus Garra Hamilton, *Memoirs of Indian Museum.* 14(4): 173-260.

Menon, A.G.K. 1987. The fauna of India and the adjacent countries. Pisces. Vol. IV. Teleostei - Cobitoidea. Part 1. Homalopteridae. *Zoological Survey of India.* Vol. IV.: i-x + 1-259, Pls. 1-16.

Menon, M. A.S. 1952. On a small collection of fish from Manipur. *Records of Indian Museum.* 50: 265-270.

Nelson, J. S. 1994. *Fishes of the World.* 3rd edition. John Wiley & Sons.: i-xvii + 1-600.

Misra, K. S. 1976. Teleostomi: Cypriniformes; Siluri. Teleostomi: Cypriniformes; Siluri. 3: i-xxi + 1-367, Pls. 1-5.

Rainboth, W. J. 1983.*Psilorhynchus gracilis*, a new cyprinoid fish from the Gangetic lowlands. *Proceedings of the California Academy of Sciences* 43 (6): 67-76.

Rainboth, W. J. 1985.*Neolissochilus*, a new genus of South Asian cyprinid fishes. *Beaufortia.* 35 (no. 3): 25-35.

Rainboth, W. J. 1986. Fishes of the Asian cyprinid genus *Chagunius. Occasional Papers of the Museum of Zoology University of Michigan.* 712: 1-17.

Roach, J. 2005. Conservationists name nine new biodiversity hotspots. *National geographic News,* February 2, 2005.

Stiassny, M. L. J. and Getahun, A. 2007. An overview of Labeonin relationships and the phylogenetic placement of the Afro-Asian genus *Garra* Hamilton, 1922 (Teleostei: Cyprinidae), with the description of five new species of *Garra* from Ethiopia, and a key to all African species. *Zoological Journal of the Linnean Society* 150: 41-83.

Talwar, P. K. and Jhingran, A. G. 1991. *Inland fishes of India and adjacent countries.* In 2 vols. Oxford & IBH Publishing Co., New Delhi, Bombay, Calcutta. 1-2: i-xvii + 36 unnumbered + 1-1158, 1 map.

Vishwanath, W. and Kosygin. L. 2002. Ecological survey of Barak and Jiri rivers of Manipur with special references to ichthyofauna. Pp. 205-209. In *Ecology and Conservation of Lakes, Reservoirs and Rivers* (ed. A. Kumar), ABD Publishers, Jaipur.

6

Ecology and Diversity of Fish and Amphibian Fauna of Torrential Streams of Wokha District, Nagaland, India

—Sabitry Choudhury Bordoloi[1] & Nzano Humtsoe[2]

[Wokha district of Nagaland is richly endowed with numerous hill streams, rivers and lakes that serve as congenial habitats for amphibian and fish species. A survey of fish and amphibian fauna was taken up in twelve torrential streams and adjoining forest areas of Wokha district of Nagaland. It recorded 28 fish species and 21 amphibian species. Two species of fishes namely, *Amlyceps arunachalensis* (Nath and Dey, 1989) and *Tor progeneius* (McClelland, 1839) and 3 species of amphibia namely, *Rana leptoglossa, Rhacophorus rhodopus* and *Rhacophorus suffry* were recorded for the first time the Nagaland. The genus, *Leptolalax lateralis* is new record for India. The fauna of the Wokha district of Nagaland are complex and include species from Himalayan region, North East mountain region and a number of ubiquitous species of Indian sub-continent. Detailed ecological data of the habitat has been included.

Key words: Torrential stream, Wokha district, amphibian, fish, ecology.]

Introduction

Hill streams are unique ecosystems characterised by shallow, narrow channels with rocky beds and swift, turbulent water flow. The rocky beds form a useful hiding and anchoring substratum for aquatic biota including algal mats, periphyton, fishes and amphibians. Nature has provided the hill stream organisms with unique anchorage systems to combat swift, shooting and turbulent water flow. Hill stream ecosystem is less prone to severe pollution as it is far away from human habitation compared to lentic ecosystems. These important and unique ecosystems with abundant food and comparatively less competition for space and food are inhabited by organisms having special adaptation. These organisms have evolved mechanical devices to combat the force of current and have successfully adapted themselves to these unique ecosystems. Various structural modifications can be noted in these animals. There are various conditions that influence the life of organisms in hill streams such as rapid water current, shallow streams with high intensity of

---

[1,2]Resource Management and Environment Division, Institute of Advanced Study in Science and Technology, Paschim Boragaon, Guwahati–781035, Assam, India

sunlight, highly oxygenated water, abundant food, and places to breed and hide from predators (Hora, 1922).

Nagaland, a state of North East India is endowed with numerous hill streams that harbour rare and unique amphibian and fish fauna. Most of these hill streams remained unexplored due to remoteness and inaccessible condition of the hilly terrain. Presently the forest areas near the hill streams are being cleared for *jhum* cultivation. Human activities like cutting and felling of trees, clearing land for cultivation, use of fertilizers, extensive fishing using bleaching powder to kill fishes, etc., have put severe stress and strain on the natural environment of the area thereby leading to environmental degradation.

Various workers have reported on the fish and amphibian fauna of Nagaland. Earlier workers like Day (1889), Hora (1921b, 1936), Hora and Mukherji (1935) and Kosygin and Vishwanath (1998) have contributed to the fish fauna of Nagaland. Ao and Dey (2005) made for the first time a reassessment of the fish germplasm of Nagaland with an annotation on their sustainable management. Karmakar and Das (2006) under the State Fauna Series of Zoological Survey of India published a list of 108 fish species from Nagaland which also included Government farm and exotic fishes based on their own field collection (81 species) as well as from earlier Zoological Survey of India records (27 species). Their list included 40 new records for the state. Recently, Ao *et al.* (2008) published a book titled "Fish and fisheries of Nagaland" in which they listed 149 fish species collected from 12 different rivers of Nagaland.

Systematic study and publications on amphibian fauna of Nagaland are very few. Kiyasetuo and Khare (1986) reported a new genus of frog from Nagaland *Pterorana khare* which Dubois (1992) considered it to be a sub-genus of *Rana*. A year later they again reported the presence of *Rhacophorus bissaculus* Taylor from Nagaland (Kiyasetuo and Khare, 1987). Kiyasetuo in his thesis (1986, unpublished) recorded 21 species of frogs from Kohima district out of which sixteen (16) species have already been reported from other parts of North East India and the rest five (5) include two new records and three new species. Chanda (1994) recorded 10 species of frogs from Nagaland. Ao *et al.* (2003) recorded 32 species of frogs from three districts of Nagaland (Kohima, Mokokchung and Dimapur) that included 19 new records for the state as well as 5 new records for India. "Fauna of Nagaland" published by ZSI in 2006 recorded 10 species of frogs from Nagaland. Matthew and Sen (2007) reported a new species, *Xenophrys zunhebotoensis* from Zunheboto district, Nagaland. Humtsoe *et al.* (2008) reported a genus of frog, *Leptolalax lateralis* (Anderson, 1871).

Description of Study Area

Wokha district of Nagaland (25°55′-26°35′ N Latitude and 94°10′-94°25′E Longitude, has a geographical area of 1628 sq. km with a population of 1,61,098 (as per 2001 census) spread over 129 villages. The district altitude ranges from 160m MSL to 1970m MSL and exhibits sub-tropical to sub-temperate climate and vegetation with vast biodiversity. It is situated on the western part of the state and is bordered by Mokokchung on the North, Kohima on the South, Zunheboto on the East and Assam on the West. Wokha district falls under Brahmaputra basin. The landscape is a network of many small rivers, streams and rivulets. The Doyang river, one of the largest rivers of the state and an important tributary of the river Brahmaputra, runs across the entire length of the district. The Doyang Hydro Electric Project (under NEEPCO) is located in this district.

Wokha district is divided into two regions, demarcated by the Doyang river. Northern area called Lio and Southern hilly area called Nrung. According to the topography, the district is divided into three ranges, viz, the Upper Range, located in Nrung, the Middle and Lower Ranges, both located in Lio area. The Upper Range has high mountains and steep slopes. The highest altitude of Wokha district, Mt. Tiyi at 1969m MSL, falls within this range. The Upper Range comprises of high hills and is a catchment area for many streams and rivers. Many torrential streams, mostly perennial, originate from this range. The Middle Range has low hills with gentle slopes and flat lands. The Lower Range is located adjoining Assam and is less hilly with vast flat areas.

Vegetation is tropical semi-evergreen type. Climate is hot and humid at the lower foothills, moderate in middle range and cold in the temperate mountain ranges. The average temperature ranges from 6°C to 35°C. Rainfall is about 175-250 cm, most of it between May and October. Highest rainfall occurs during June-August (meteorological data for three years given in Table-6.1). Relative humidity ranges from 85% to 95% during rainy season. Soil is acidic in nature (pH 4.8-5.2) with high organic content and poor phosphate content. Soil texture varies from loam to silty coarse sand and sandy loam. Soil erosion is a common phenomenon.

All the streams in the study area originate from the mountains and flow through hilly terrain and cultivated lands; they carry crystal clear water. The stream beds are rocky, shallow and temporarily deep in some areas. The streams comprise submerged vegetation, rocks and huge boulders, steep waterfalls, rapids and pools. Human activities such as tree cutting, land clearing for cultivation, fishing, etc. prevails throughout the year.

The vegetation around the hill streams is dominated by trees such as *Alnus nepalensis, Rhus semiliata, Ficus hookeri,* etc., shrubs like *Rubus ellipticus, Pyrus pashia, Imperata cylindrica, Thysilolaena maxima, Zanthoxylum aromaticum, Bidens biternata, Solanum khasianum, Butlajia mycrostychia, Bambusa* sp. etc., herbs like *Eupatorium adinoforum, Artemisia nilagirica, Cresocephalum crepiloedes, Ageratum conyzoides, Fagopyrum cymosum, Polygonum molle, Polygonum chinensis, Polygonum orientalis, Urtica parviflora,* etc., climbers, parasitic plants, epiphytes and creepers.

Plants found attach to solid objects in the streams like rocks and boulders are mosses and liverworts, those rooted into the substratum like *Ranunculus* sp., *Potamogeton* sp., *Nitella* sp., *Chara* sp., etc., are found in the slow current areas. Phytoplanktons are represented by *Navicula, Fragilaria, Synedra, Pinnacularia, Cymbella, Nitzschia, Closterium, Scendesmus, Oedogonium Ulothrix, Cladophora, Spirogyra, Zygnema, Oscillatoria.*

Materials and Methods

Streams selected for study

Twelve streams, 7 along the National Highway 61 (Wokha-Mokokchung road) and 4 along Wokha-Merapani road of Nrung region and 1 stream from Lio Wokha were selected for the study (Figures 6.1-6.6).

Streams located along NH 61 in Nrung region:

1. Engorotchu
2. Kyotchu
3. Lungkitchu
4. Nhyatsutchu
5. Nitsutchu
6. Tchulumo
7. Tchupvu.

Streams located along Wokha-Merapani road in Nrung area:

1. Humtso Tsupvu
2. Sosirotchu
3. Vekhvurotchu
4. Yikhum Sanga.

Stream located at Lio area:

1. Tsurang river.

Seasonal analysis of abiotic parameters

The present investigation was carried out during August 2005 to August 2007. Water samples were collected from 12 hill streams during pre-monsoon, monsoon and post-monsoon and were analysed for temperature, pH, dissolved oxygen (DO), free carbon-dioxide (CO_2), total alkalinity and total hardness following standard methods (APHA, 2000). Heavy metal content of water was done with AAS. Atomic Absorption Spectrometric analysis of water was made at the Sophisticated Analytical Instrumentation Facility of the North Eastern Hill University, Shillong by using Perkin-Elmer-2380 Atomic Absorption Spectrometer with air-acetylene flame.

Collection and identification of fish and amphibian fauna

Fish and amphibian fauna of 12 streams were surveyed from August 2005 to August 2007. Fishes were collected by fisherman with some indigenous methods such as by bait and hook, hand picking, etc. The specimens collected were first narcotized with chloroform and then preserved in 6% formaldehyde solution with a small abdominal incision.

The frog specimens were collected by visiting various habitats. Frogs were caught by hand and also by using hand net. Frogs killed for food by locals were also recorded. The frogs were first narcotised with chloroform and stretched by keeping overnight and preserved in formaldehyde solution (adult in 8% and tadpoles in 10%).

Measurements were taken with standard dial vernier calliper. Fishes were identified following Talwar and Jhingran (1991); Jayaram (1999); Vishwanath (2002); www.fishbase.org (2010) and www.aquabase.org (2010). Conservation status was noted as per CAMP Report (1997). The threat categories re-evaluated by Jha *et al.* (2006) in Nepal as well as those re-assessed by Biswas and Sugunan (2008) were also referred. Identification of frog specimens was done with following Boulenger (1890 and 1929); Chanda (1994 and 2002); Dutta (1997), Dubois and Ohler (2000), Bordoloi *et al.* (2007) and Global amphibian assessment (GAA, 2009). Conservation status was evaluated as per GAA IUCN Red List Status (2009). The collected frog specimens were also compared with specimens previously collected from Nagaland (Ao *et al.*, 2003) and from other states like Arunachal Pradesh (Borah and Bordoloi, 2001 and 2003) and Manipur (Ningombam and Bordoloi, 2007). For each species the valid name (as per GAA 2009) with its authority is given. Number and sex of each specimen are listed with their collection number. The species are presented in order of families, genera and species group names. Wherever possible, local names have been mentioned.

Observations

Physico-chemical Characteristics

Physico-chemical parameters of the 12 streams were monitored for 2 years. Ecological parameters such as water temperature, pH, dissolved oxygen (DO), free carbon di-oxide (CO_2), total alkalinity and hardness of water are summarized in Table 6.2.

Temperature: During study period, the atmospheric temperature recorded around the 12 hill streams ranged from 14°C to 32°C and water temperature from 7.2°C-28°C. Maximum atmospheric temperature was recorded (25.2°C) during the month of August 2006 and minimum temperature (13.2°C -16°C) during the months of December and January (winter season). Similarly, the highest range of water temperature (26°C – 28°C) was recorded during summer seasons (April-August) and minimum (7.8°C) in the months of December and January.

pH: The pH of water was measured with narrow range pH paper as these areas are far from the city. It was observed that the pH range from 6.5-8.5 in all the streams. pH shows decreasing trend from the month of March to October (6.5-7.5) then it increases (7.5-8.5).

Dissolved Oxygen (DO): The dissolved oxygen ranged from 7.2-15.2 mg/l in all the streams. High level of dissolved oxygen (10-15.2 mg/l) was observed during winter seasons (October to march) and lower level of dissolved oxygen (7.2-9.6 mg/l) was observed during summer months (April to September) every year.

Free CO_2: Free CO_2 was observed in moderate quantity (4-8 mg/l). Maximum value of free CO_2 was recorded during the months of July, December and January (8 mg/l).

Total Alkalinity: Total alkalinity ranged from 38-68 mg/l and showed seasonal variations. Maximum value was observed during the month of December (56-68 mg/l) and minimum during the months of April, August and September (38-44 mg/l).

Figure 6.1. Vekhvurotchu stream.

Figure 6.2. Tchupvu stream.

Figure 6.3. Nhyatsutchu stream.

Figure 6.4. Humtso Tchupvu stream.

Figure 6.5. Engorotchu stream.

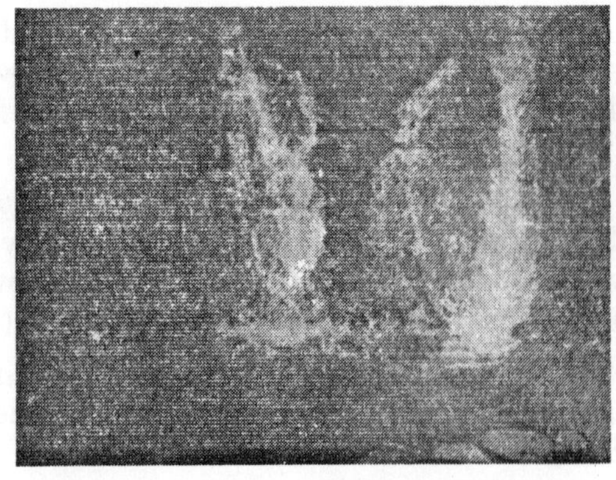

Figure 6.6. Nitsutchu stream.

Figure 6.7. *Psilorynchus sucatio*

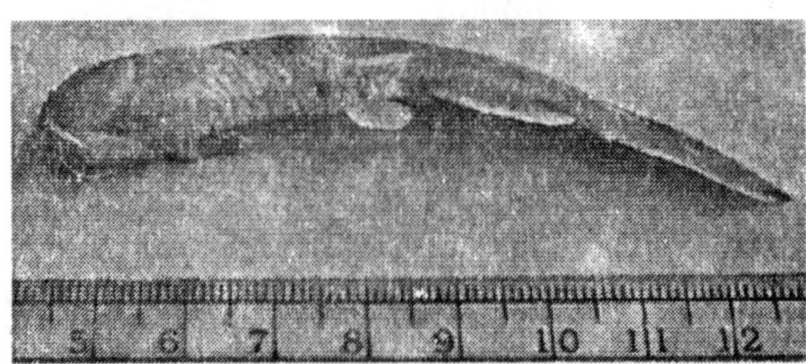

Figure 6.8. *Amblyceps arunachalensis*

Figure 6.9. *Rhacophorus suffry.*

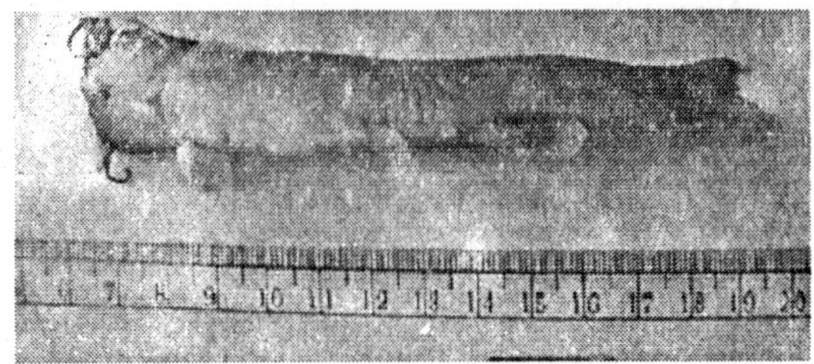

Figure 6.10. *Amblyceps apangi*

Figure 6.11. *Xenophrys major.*

Figure 6.12. *Rhacophorus bipunctatus.*

Table 6.1
Meteorological data of Wokha district, Nagaland (2005-2007)

Year	*Total rainfall in mm*	*Temp Max-Min°C*	*Average dew point*	*Average RH%*
2005	145.417	9.8-25.4	16.641	84.975
2006	131.83	9.4-28.9	16.245	74.16
2007	235.591	8.7-27.5	15.925	77.40

Table 6.2
Physico-chemical characteristics of 12 streams of Wokha district, Nagaland (Method followed after APHA, 2000)

Sl. No.	*Streams*	*Water temperature °C*	*pH*	*Dissolved oxygenMg/l*	*Free CO_2 Mg/l*	*Total alkali-nity Mg/l*	*Hard-nessMg/l*
1	Lungkitchu	7.8-18	6.5-8.5	7.2-13.6	4-8	38-60	58-120
2	Nhyatsutchu	8-19.2	6.5-8.5	7.6-15.2	4-8	38-60	56-112
3	Kyotchu	8-19.6	6.5-8.5	7.6-14	4-8	42-60	58-110
4	Engorotchu	7.8-19.6	6.5-8.5	7.2-13.6	4-8	46-60	56-106
5	Nitsutchu	7.4-19.2	6.5-8.5	7.2-14	4-8	38-60	56-110
6	Tchulumo	12.4-25.8	6.5-8.5	7.6-14.8	4-8	40-62	50-92
7	Tsupvu	8.2-25	6.5-8.5	8-15.2	4-8	46-68	54-102
8	Humtso Tsupvu	12-26	6.5-8.5	8-15.6	4-8	42-64	58-102
9	Sosurotchu	8.2-25	6.5-8.5	7.2-14.4	4-8	38-60	50-102
10	Vekhvurotchu	12-25	6.5-8.5	7-12.8	6-8	48-60	48-88
11	Yikhum Sanga	8.2-24	6.5-8.5	7.2-13.2	6-8	40-64	56-106
12	Tsurang	13.6-28	6.5-8.5	7.6-14.4	4-8	52-70	52-102

Table 6.3
Heavy metal Spectrum of the stream (in ppm) of Wokha district, Nagaland

Months	*Cu*	*Pb*	*Mn*	*Cd*	*Zn*	*Fe*	*Se*
August 2005	ND	ND	0.07	ND	0.05	0.35	1.40
September 2005	0.01	0.01	0.02	0.01	0.15	0.61	0.93
October 2005	0.01	0.04	0.04	0.02	0.15	0.39	2
November 2005	0.06	ND	0.03	0.03	0.03	0.41	0.025
December 2005	0.05	0.01	0.01	0.02	0.02	0.26	0.023
January 2006	0.06	0.05	0.02	0.03	0.07	0.19	0.016
February 2006	0.04	0.06	ND	0.03	0.08	0.25	0.023
March 2006	0.03	0.03	ND	0.02	0.04	0.18	0.018
April 2006	0.02	0.01	0.02	0.03	0.04	0.10	0.017
May 2006	0.04	0.01	0.02	ND	0.16	0.16	0.018

Abbreviations: ND- Not Detected; ppm- parts per million;

Table 6.4
List of fishes of Wokha district, Nagaland with their status (CAMP, 1997)

Scientific names	*Status*	*Remarks*
Order: Cypriniformes.		
Family: Cyprinidae		
Sub family: Cyprininae		
1. *Puntius sophore* (Hamilton, 1822)	Vu*	R
2. *Puntius ticto* (Hamilton, 1822)	LRnt	R
3. *Tor putitora* (Hamilton, 1822)	EN	R
4. *Tor progeneius* (McClelland, 1839)	DD	NR
Sub family: Rasborinae		
5. *Barilius barila* (Hamilton, 1822)	LRnt*	R
6. *Barilius barna* (Hamilton, 1822)	EN*	R
Sub family: Danioninae		
7. *Danio rerio* (Hamilton, 1822)	Vu*	R
8. *Danio dangila* (Hamilton, 1822)	EN*	R
9. *Devario aequipinnatus* (McClelland, 1839)	EN*	R
Sub family: Garrinae.		
10. *Garra kempi* (Hora, 1921)	Vu	R
11. *Garra lisorynchus* (McClelland, 1839)	Vu	R
Family: Psilorynchidae		
12. *Psilorynchus balitora* (Hamilton, 1822)	ne	R
13. *Psilorynchus homaloptera* (Hora & Mukherji, 1935)	Vu	R
14. *Psilorynchus sucatio* (Hamilton, 1822)	EN	R
Family: Balitoridae.Sub family: Nemacheilinae.		
15. *Schistura manipurensis* (Chaudhuri, 1921)	Vu	R
16. *Schistura vinciguerrae* (Hora, 1935)	EN	R
Family: CobitidaeSub family: Cobitinae.		
17. *Lepidocephalichthys berdmorei* (Blyth, 1861)	EN	R
18. *Lepidocephalichthys guntea* (Hamilton, 1822)	Vu*	R
19. *Somileptus gongota*, Hamilton, 1822 (Hamilton-Buchanan, 1822)	LRnt	R
Order: Siluriformes		
Family: Amblycipitidae		
20. *Amblyceps apangi* (Nath & Dey, 1989)	Vu	R
21. *Amblyceps arunachalensis* (Nath & Dey, 1989)	Vu	NR
Family: Sisoridae.		
22. *Exostoma labiatum* (McClelland, 1842)	ne	R
23. *Glyptothorax trilineatus* (Blyth, 1861)	Rare*	R
Family: Olyridae.		
24. *Olyra longicaudata* (McClelland, 1842)	EN#	R

(Contd...)

Scientific names	Status	Remarks
Order: Perciformes.		
Family: Nandidae.Sub family: Badinae		
25. *Badis badis* (Hamilton, 1822)	ne	R
Sub order: Channoidei.		
Family: Channidae.		
26. *Channa orientalis* (Bloch & Schneider, 1801)	EN*	R
27. *Channa punctatus* (Bloch, 1793)	Vu*	R
28. *Channa stewartii* (Playfair, 1867)	EN#	R

Abbreviations: DD: data deficient; EN: endangered; LC: least concern; ne: not evaluated; LRnt: lower risk near threatened; NT: near threatened; NR: new record: R: recorded earlier Vu: vulnerable; *as per Jha *et al.*, 2006; # as per Biswas & Sugunan, 2008.

Table 6.5
Categorization of fishes of Wokha district, Nagaland according to habitats

Sl. No	Name of the fishes	True hill stream forms	Semi-torrential/ intermediate forms	Migratory forms
1	*Puntius sophore*			+
2	*Puntius ticto*			+
3	*Tor tor*			+
4	*Tor progeneius*			+
5	*Barilius barila*			+
6	*Barilius barna*			+
7	*Danio rerio*			
8	*Danio dangila*		+	
9	*Devario aequipinnatus*		+	
10	*Garra kempi*	+		
11	*Garra lissorynchus*	+		
12	*Psilorynchus balitora*	+		
13	*Psilorynchus homaloptera*	+		
14	*Psilorynchus sucatio* (figure 6.7)	+		
15	*Schistura manipurensis*		+	
16	*Schistura vinciguerrae*		+	
17	*Lepidocephalichthys berdmorei*		+	
18	*Lepidocephalichthys guntea*		+	
19	*Somileptus gongota*		+	
20	*Amblyceps apangi*	+		
21	*Amblyceps arunachalensis*	+		
22	*Exostoma labiatum*	+		
23	*Glyptothorax trilineatus*	+		
24	*Olyra longicaudata*		+	
25	*Badis badis*			+
26	*Channa orientalis*			+
27	*Channa punctatus*			+
28	*Channa stewartii*		+	

Total Hardness: Total hardness of the streams ranged from 38-120 mg/l. High value of total hardness (80-120 mg/l) was recorded during winter months and lower during summer months. The highest value of total hardness was recorded (120 mg/l) in the month of January while minimum value (38 mg/l) recorded in the month of August.

Trace elements: Trace elements such as Cu, Cd, Fe, Mn, Pb, Se, Zn, etc were recorded in ppm (Table 6.3).

Faunal Biodiversity

The twelve hill streams surveyed were found to be rich in fish and amphibian fauna which are rare and unique to the ecosystem. Human interference like clearing of land for cultivation, cutting of trees, fishing exploitation during breeding season, use of chemicals for killing the fishes etc. were found to be the main causes of depletion of fauna.

Fish fauna: A list of fish fauna of the Wokha district of Nagaland is furnished in Table 6.4. In the present study 28 fish species belonging to 4 orders, 9 families and 14 genera were recorded from the hill streams.

The collected fish species were found to occupy diverse ecological regime in their distributional ranges. According to Sinha (1984), the fish genetic resources of NE India are a mixture of Indo-Gangetic plains, Burmese and Chinese forms. Out of the 28 fish species, 17 species are found in Burma (Myanmar) and 3 are found in Southern China (Yunnan). Species like *Amlyceps arunachalensis* (Nath and Dey, 1989) and *Tor progeneius* (McClelland, 1839) are found to be new record for the state of Nagaland.

The collected fish species can be categorised according to their habitats as true hill stream forms, semi-torrential forms and migratory forms (Table 6.5). Nine species of true hill stream fishes were recorded.

The true hill stream fishes are characterized by the following structural peculiarities. Body is dorso-ventrally flattened, ventral surface is flat and dorsal profile is slightly arched (eg. *Garra, Glyptothorax, Psilorynchus*). Eyes are generally small, pushed towards the upper surface and lie close to each other; subcutaneous in some (eg. *Amblyceps*). Paired fins are placed horizontally so that ventral surface is freely applied to rocks and they can also act as organs of adhesion. In some fishes the fins are covered with a thin film of membrane and the fin bases are swollen (eg. *Amblyceps, Exostoma*). Lips are usually thick and fleshy and modified in some to form suckers for adhesion (eg. *Exostoma*). A prominent adhesive disc on the throat region is present in *Garra,* while adhesive apparatus consisting of a fold of skin is present on the chest between the bases of pectoral in *Glyptothorax.* The mouth is usually ventral in position, crescentic in shape with strong jaws. The caudal peduncle is usually long and narrow. Fishes which are semi torrential such as *Schistura, Olyra,* etc. show lesser degree of morphological adaptation.

Systematic accounts of some rare fish species

Amblyceps apangi Nath and Dey, 1989
(Figure 6.10)

Materials examined: 140 males (29.4- 158.6 mm) and 105 females (Total length 42.3- 146.6 mm).

Distinguishing characters: D ii 5-6; A ii-iii 7; P I 6; V i 5. Body long, slender, sub-cylindrical, compressed posteriorly; abdomen rounded; head narrow and depressed, dorso-ventrally flattened, covered with thick skin; nostrils close together, eyes small, almost indistinguishable, sub-cutaneous; mouth terminal, upper jaw slightly longer than lower jaw; lips thick and fleshy, rictal folds well developed; barbels four pairs, one pair each of maxillary and nasal and two pairs mandibular; maxillary barbel longest and reaches beyond posterior base of pectoral fin; dorsal fin rayed, without spine; adipose fin continuous up to caudal fin but is not confluent with it; caudal fin truncate; all paired and unpaired fins covered by a thin membrane; lateral line present, complete, straight, with tubular openings, pores arranged in pairs, each pair separated by a short gap colour dull blackish to brownish as per habitat.

Remarks: This catfish is found in fast moving and torrential hill streams, and occasionally in slow moving ones and inhabit places with rocky beds, rocks and boulders.

Amblyceps arunachalensis Nath and Dey, 1989
(Figure 6.8)

Materials examined: Two males (Total length 88-85- 96.1 mm)

Distinguishing characters: D i 5; A ii 7; P i 5; V i 4. Body cylindrical with rounded abdomen; head small, snout rounded, eyes fairly small; mouth terminal, lower lip slightly longer than upper lip; jaws sub-equal, toothed; barbels four pairs, one pair each of maxillary and nasal and two pairs mandibular; rayed dorsal without spine, inserted above pectoral fins; adipose dorsal fins small, low originates from the opposite end of anal fin; caudal fin deeply forked; upper lobe longer than lower lobe; lateral line present, incomplete, ceases at the middle of pectorals; colour varies from blackish to brownish.

Remarks: New record for Nagaland; this catfish is found in fast moving and torrential streams, and occasionally in slow moving ones and inhabit places with rocky stream beds, rocks and boulders.

Amphibian fauna

A list of amphibian fauna of the Wokha district of Nagaland is furnished in Table 6.6. In the present study 21 species of anurans belonging to four (4) families and ten (10) genera were recorded. Out of which 4 are new records for the state. The genus *Leptolalax* is new record for India.

The habitats of amphibians are highly varied. Habitat requirement varies from species to species. Some have restricted distribution while others can tolerate wide variety of conditions. A high level of endemism occur both in specific and generic level in India (Inger, 1999). Two regions in India have rich concentration of amphibian fauna- Western Ghats and North Eastern India (Inger and Dutta, 1986).

Based on their habitats, the frogs can be divided into four types: (i) True forest species (ii) Lotic species (iii) Lentic species and (iv) Ubiquitous forms. Species of *Rhacophorus* (Figures 6.9 and 6.12) and *Polypedates* were found only in forested areas and their morphological and anatomical adaptations are meant for arboreal life. Species of *Amolops* and some species of *Rana* inhabit running water and they have comparatively flattened body, long legs and toes fully webbed with adhesive discs.

Out of the 21 species recorded, five species are Ubiquitous having distribution in peninsular India, South Western China and Myanmar, eleven species occur in the Himalayas, South Western China and Myanmar and six species occur only in South Western China and Myanmar. Three species are endemic to North East India. Thus the species recorded so far are closely related to the fauna of South West China and Myanmar. Thorough survey has become very essential in inaccessible areas so that a clear picture of amphibian distribution can emerge.

Table 6.6
List of amphibians collected along with IUCN status

Scientific name	*IUCN Status*	*Remarks*
Class: Amphibia Gray, 1825		
Order: Anura Fischer Von Waldheim, 1813		
Family: Bufonidae Gray, 1825		
1. *Bufo melanostictus* Schneider, 1799	LC	R
Family: Megophridae Rafinesque, 1815		
2. *Leptolalax (Lalax) lateralis* Anderson, 1871	NE	NR
3. *Xenophrys major* Boulenger, 1908	LC	R
4. *Xenophrys wuliangshanensis,* Ye and Fei, 1995	DD	R
Family: Ranidae Rafinesque, 1815		
5. *Amolops formosus* Gunther, 1876	LC	R
6. *Amolops gerbillus,* Annandale, 1912	LC	R
7. *Amolops marmoratus,* Blyth, 1855	LC	R
8. *Euphlyctis cyanophlyctis,* Schneider, 1799	LC	R
9. *Fejarvarya nepalensis,* Dubois, 1992	LC	R
10. *Fejarvarya terainsis,* Dubois 1984	LC	R
11. *Hoplobatrachus tigerinus* (Daudin, 1802)	LC	R
12. *Rana humeralis* (Boulenger, 1887)	LC	R
13. *Rana khare* (Kiyasetuo and Khare, 1986)	VU	R
14. *Rana livida* (Blyth, 1855)	DD	R
15. *Rana leptoglossa* (Cope, 1868)	LC	NR
Family: Rhacophoridae Hoffman, 1982		
16. *Polypedates megacephalus* Hallowell, 1861	LC	R
17. *Polypedates teraiensis,* Dubois, 1987	LC	R
18. *Rhacophorus maximus,* Gunther, 1858	LC	R
19. *Rhacophorus bipunctatus,* Ahl, 1927`	LC	R
20. *Rhacophorus rhodopus* Liu and Hu, 1960	LC	NR
21. *Rhacophorus suffry* (Bordoloi, Bortamuli and Ohler, 2007)	LC	NR

Abbreviations: DD: data deficient; LC: least concern; ne: Not evaluated; R: Recorded earlier; NR: New Record; Vu: Vulnerable.

Systematic accounts of some rare amphibian species

Leptolalax lateralis Anderson, 1871

Materials examined: Three males, one female, two young.

Morphological characters: Small sized frogs (largest specimen female with SVL-35.35); dorsal skin smooth and with dark spots, snout rough and granular; tips of fingers and toes rounded and with pads; webs rudimentary; dorsal colour brown or grey brown; latero-dorsal folds absent; ventral surface smooth and creamy white; two white spots, one each on both the chests; dorsum with a W-shaped band below the head and a dark triangular shaped band in between the eyes; hind limbs with dark cross bars; two white spots, one each on the thighs; string of white dots on the periphery ventrally and abdomen up to the flanks.

Remarks: New genus record for India; torrent living frogs; found in streams with rocky bottom; found under rocks and on tree branches.

Rana leptoglossa (Cope, 1868)

Materials examined: One male and two females.

Morphological characters: Large-sized frog (male SVL 51.1mm, female SVL 49.8-55.4mm); dorsally brown with small to large black spots or markings; sides sometimes black with white spots or marbling; tips of fingers and toes with small discs, webbing moderate; limbs with blackish cross-bands; posterior side of thigh marbled black and white ; ventral part white, spotted or marbled with brown; skin strongly granulated on the head and back; a strong and broad glandular dorso-lateral fold run above the tympanum to the hip, another glandular fold extents from posterior region of eyes upto the shoulder, followed by a glandule.

Remarks: New record for Nagaland.

Rhacophorus rhodopus Liu and Hu

Materials examined: One female.

Morphological characters: Rather small sized frog (female SVL 45.4mm) with pointed snout; dorsum smooth; tips of fingers and toes with discs, webbing on fingers reduced; a single round spot present behind the armpit; dorsum colour red brown or yellow with indistinct markings, usually with a dark brown X-shaped marking; posterior back with dark transverse stripes; flanks bright yellow; legs with cross-bars and webbing brick red.

Remarks: New record for Nagaland; arboreal, live in undisturbed forest area.

Rhacophorus suffry Bordoloi, Bortamuli and Ohler, 2007
(Figure 6.9)

Materials examined: One male.

Morphological characters: Moderate-sized frogs (SVL of male 43.25mm); dorsum smooth, between eyes granular and side of head smooth with fine granules, co-ossified skin absent; tips of fingers and toes with discs, webbing well developed, complete; dorsal and lateral part of head and body leafy green; upper part of flank leafy green lined with light yellow colour, lower part light green with few creamy white scattered spots; dorsal part of foot between fourth and fifth toes are green, others are orange red.

Remarks: New record for Nagaland; arboreal, found in undisturbed forest areas.

Discussion

Wokha district of Nagaland state has numerous hill streams all of which flow into the Doyang river, a tributary of the Brahmaputra river. Ichthyofauna and amphibian diversity of hill streams of Wokha has not been studied at all. Increasing stress on the aquatic systems due to natural as well as man-made causes such as deforestation, soil erosion, habitat degradation, etc. has necessitated the extensive field survey of fish and amphibian fauna along with habitat characteristics and ecological conditions of the streams.

Water quality of the streams is characterised by rich oxygen content (7.2-15.2 mg/l), comparatively lesser amount of free CO_2 (4-8 mg/l), moderate alkalinity (38-68 mg/l), soft to moderate hardness (38-120 mg/l) and pH ranges from 6.5-8.5. Detection of heavy metals like Cd, Cu, Fe, Mn, Pb, Se and Zn is believed to be due to anthropogenic activities in the upper reaches of the streams. The ecology of the study area was found to be conducive to harbour a good number of hill stream fishes and amphibians along with some migratory forms. Out of the 28 fish species recorded from the 12 streams, two species viz *Amblyceps arunachalensis* (Nath and Dey, 1989), and *Tor progeneius* (McClelland, 1839) were recorded for the first time from Nagaland.

Out of the 21 amphibian species recorded from the 12 streams and the adjoining forest areas, one species, *Leptolalax lateralis* (Anderson, 1871) is a new genus record for India and three species- *Rhacophorus suffry* (Bordoloi, et al. 2007), *Rhacophorus rhodopus* (Liu and Hu, 1960) and *Rana leptoglossa* (Cope, 1868) are recorded for the first time from Nagaland.

The fish fauna collected from the streams of Wokha district during our survey from 2005-2007 shows a mixture of Indo-Gangetic, Burmese and Chinese forms. Out of the 28 fish species reported here, 16 are also found in Myanmar and 2 in South China. *Schistura manipurensis* (Choudhuri, 1912) is endemic to Manipur and Nagaland (Vishwanath and Laishram, 2001), *Schistura vinciguerrae* (Hora, 1935) was reported from Brahmaputra drainage by Das *et al.* (2001), *Amblyceps apangi* (Nath and Dey, 1989) and *Amblyceps arunachalensis* (Nath and Dey, 1989) were first reported from Arunachal Pradesh by Nath and Dey (1989). In the streams a good number of *A. apangi* could be recorded.

The amphibian species also show a mixture of Indo-Chinese, Indo-Myanmarese and ubiquitous forms. The number of endemic species might be higher if thorough survey in inaccessible areas could be done. Most of the areas have remained unexplored due to inaccessibility. Further exploration might reveal the presence of rare species.

[**Acknowledgements:** The authors express their gratitude to Prof. Annemarie Ohler of Muséum National d'Histoire Naturelle, Département de Systématique et Evolution, USM 602 Taxinomie et Collections, 25 rue Cuvier, 75005 Paris, France for confirmation of identification of the frog species. One of the authors (N. Humtsoe) gratefully acknowledge the financial assistance from Rajiv Gandhi National fellowship scheme.]

REFERENCES

Ao, J. M, Bordoloi, S. and Ohler, A. 2003. Amphibian fauna of Nagaland with nineteen new records from the state including five new records for India. *Zoos Print Journal* 18(6): 1117-1125.

Ao, S, Dey, S.C. and Sarmah, S.K. 2008. *Fish and fisheries of Nagaland*. Department of Fisheries, Govt. of Nagaland, 234 pp.

APHA. 2000. *Standard method for examination of water and waste waters*. 20th edition, American Public Health Association, Washington D.C, 1193 pp.

Biswas, B.K and Sugunan, V.V. 2008. Fish diversity of Brahmaputra river system in Assam, India. *J. Inland. Fish. Soc. India*, 40 (1): 23-31.

Borah, M.M and Bordoloi, S. 2001. Nine new records of Amphibia (Anura) from Arunachal Pradesh, India. *J. Bombay Nat. Hist. Soc.*, 98 (1), 123-126.

Borah, M.M and Bordoloi, S. 2003. First record of *Hoplobatrachus crassus* (Jerdon, 1853) from North Eastern region in Assam and Arunachal Pradesh. *J. Bombay Nat. Hist. Soc.* 96 (1): 158-159.

Borah, M.M and Bordoloi, S.C. 2003. Altitudinal distribution pattern of amphibian fauna of Arunachal Pradesh with special reference to Dehang-Debang Biosphere Reserve, Arunachal Pradesh. *Himalayan Biosphere Reserve Bulletin*, 5 (1 and 2): 51-55.

Bordoloi, S., Tutul B. and Ohler, A. 2007. Systematics of the genus *Rhacophorus* (Amphibia: Anura): Identity of red-webbed forms and description of a new species from Assam. *Zootaxa*, 1653: 1-20.

Boulenger, G.A. 1890. The *fauna of British India including Ceylon and Burma. Reptilia and Batrachia*. Taylor and Francis, London.

Boulenger, G.A. 1920. A monograph of the South Asian, Papuan, Melanesian and Australian frogs of the genus *Rana*. *Records of Indian Museum*, Calcutta, 226 pp.

CAMP. 1997. 'Freshwater fishes of India'. Conservation assessment and management plan (CAMP) workshop, NBFGR, Lucknow, 22-26 September, 156pp.

Chanda, S.K. 1994. Anura of North East India. *Mem. Zool. Surv. India*, 18:1-143.

Chanda, S.K. 2002. Hand book of Indian amphibians. *Zool. Surv. Ind.* Kolkata, 335p.

Choudhury, N.K, Hussien, B., Baruah, M., Saikia, S. and Sengupta, S. 2002. Amphibian fauna of Kamrup district, Assam, with notes on their natural history. *Hamydryad*, 26 (2): 276-282.

Das, P, Bordoloi, S.C. and Dutta, A. 2001. Taxonomic variation in *Schistura vinciguerrae* (Hora, 1935) from Basistha river, a new record from Assam, India. *J. Bombay Nat. Hist. Soc.*, 98 (3): 459-461.

Dey, S.C and Ao, S. 2005. Reassessment of fish germplasm of Nagaland with an annotation on their sustainable management. *In Abstracts of Natl. Symp.* NBFGR (ICAR), Lucknow, 26-27 April 2005.

Dubois, A. 1992. Notes sur la classification des Ranidae (Amphibiens: Anoures). *Bull. Soc. Linn.Lyon*, 61 (10): 305-352.

Dubois, A. and Ohler, A. 2000. Systematics of *Fejarvarya limnocharis* (Gravenhorst, 1829) (Amphibia: Anura) and related species. Nomenclatural status and type-specimen of the nominal species *Rana limnocharis* Gravenhorst 1829. *Alytes,* 18 (1-2): 1-96.

Dutta, S.K. 1997. Amphibians of India and Sri Lanka (Checklist and Bibliography). *Odyssey Publishing House, Bhubaneshwar,* 342 pp +xxii.

Editor - Director, 2006. *Fauna of Nagaland, State fauna series* 12: 1-620 (Published by Director, Zoological Survey of India, Kolkata).

Hora, S.L. 1921b. Indian cyprinoid fishes belonging to the genus related species from other countries. *Rec. Inc. Mus.,* 22(5): 633-687.

Hora, S.L. 1922. Structural modification in the fishes of mountain torrents. *Rec. Inc. Mus.,* 24: 46-58.

Hora, S.L. 1936. On a further collection of fishes from the Naga hills. *Rec. Inc. Mus.,* 38(3): 317-331.

Hora, SL and Mukherji, D. D. 1935. Fishes of Naga Hills, Assam. *Rec. Inc. Mus.,* 38 (3): 317-331.

Humtsoe, L.N, Bordoloi, S., Ohler, A. and Dubois, A. 2008. Rediscovery of a long known species, *Ixalus lateralis* Anderson 1871. *Zootaxa,* 1921: 24-34.

Inger, R.F. 1999. Distribution of amphibians of Southern Asia and adjacent islands, pp 456-482. In: *Patterns of distribution of amphibians: A global prespective* (ed. Duellman, W.E.) John Hopkins University Press, Baltimore.

Inger, R.F and Dutta, S.K. 1986. An overview of the amphibian fauna of India. *J. Bombay. Nat. Hist. Soc.* 83: 135-146.

GAA. 2009. Global Amphibian assessment< www.globalamphibians.org> (accessed on November 16 2009).

Jayaram, K.C. 1999. *The freshwater fishes of the Indian region.* Narendra Publishing House, New Delhi, 551pp.

Jha, B.R, Herwig, W., Sharma, S. and Straif, M. 2006. Fish species, composition, number and abundance in different rivers and seasons in Nepal and re-evaluation of their threat category for effective conservation and management. *Journal of Ecology, Environment and Conservation,* 12: 25-36.

Karmakar, A.K and Das, A. 2006. *Fauna of Nagaland- Fishes. Zoological Survey of India, State fauna series No.* 12: 393-425.

Kiyasetuo and Khare, M.K. 1986. A new genus of frog (Anura: Ranidae) from Nagaland at the N-E hills of India. *Asian Expl.Sci,* 1: 12-17.

Kiyasetuo. 1986. Study on survey of toads and frogs of Kohima, Nagaland and certain aspects of ecology and development of *Rhacophorus leucomystax* (Kuhl). Ph.D. Thesis, North Eastern Hill University, Shillong, India.

Kiyasetuo and Khare, M.K. 1987. A new record of *Rhacophorus bissaculus* Taylor from north eastern hills of India. *J. Assam. Sci. Sco.,* 29(3):16-19.

Kosygin, L. and Vishwanath, W. 1998. A report on fish diversity of Tizu river, Nagaland, with some new records. *Eco. Env. & Cons.* 4 (4): 243-247.

Matthews, R. and Sen, N. 2007. Description of two new species of *Xenophrys* (Amphibia: Anura: Megophyridae) from North-East India. *Cobra,* 1(2).

Nath, P. and Dey, S.C. 1989. Two new species of the genus *Amblyceps* Blyth from Arunachal Pradesh, India. *Journal of Assam Science* 32(1):1-6.

Ningombam, B. and Bordoloi, S. 2007. Fauna of Loktak Lake Manipur, India with ten new records for the state. *Zoos' Print Journal* 22 (5): 2688-2690.

Sinha, M. 1994. Fish genetic resources of the Northeastern region of India. *Journal of Inland Fishery Societies of India*, 26: 1-19.

Talwar, P.K. and Jhingran, A.G. 1991. *Inland fishes of India and adjacent countries.* Oxford and IBH Publishing Co. Pvt. Ltd. N. Delhi, Vol. I and II: XIX+1158.

Vishwanath, W. 2002. Fishes of North East India: A field guide to species identification. Manipur University and National Bureau of Fish Genetic Resources, Lucknow, 198 pp.

Vishwanath, W. and Laishram, J. 2001. Fishes of the subfamily Nemacheilinae Regan (Cypriniformes: Balitoridae) from Manipur. *J. Bombay Nat. Hist. Soc.*, 98: 197-216.

7

Studies on Ecological Aspects of Manipur River System, India

—*Shantabala Devi Gurumayum*[1] & *U. C. Goswami*[2]

[Spatial and seasonal distributions of plankton, periphyton and macrobenthic fauna along with species diversity and productivity of the three tributaries of Manipur rivers viz., Khuga, Thoubal and Imphal were studied during the years 1999 and 2000. A comprehensive study of physico-chemical and sediment characteristics of the rivers was also incorporated. The details of the study are presented in the paper.]

Key words: Plankton, periphyton, diversity, macrobenthos, physico-chemical parameter, Manipur river]

Introduction

Manipur, situated in the northeastern corner of India, is included among the ten hot spot biodiversity regions of the world (Kottelat and Whitten, 1996). Despite its small size (22,327 sq. km), the state is home to a rich variety of fish genetic resources and has a distinct zoo-geographical identity. Hill area covers about 92% of the landscape, which encloses a central valley of about 1800 sq. km. The Loktak lake which is the biggest lake in the northeast India is situated in the lowest part of the valley. Known for its scenic beauty, the state has a total water resource of 52,959 ha, Garg *et al.*, (1998) and 2000 km of hill streams and plain rivers. The important river systems of the state are Manipur river system and Barak river system. The Barak drainage system is connected with the Barak-Brahmaputra river system of India, whereas the Manipur river is connected with the Chindwin-Irrawaddy drainage of Myanmar. The ubiquitous floodplain lakes of Manipur, associated with this drainage system form an important habitat for flora and fauna.

Scientific management of river water is very important now a day as it provides sustainable resources. Due importance have started giving for conservation and management of reverine ecosystem as most of the Indian rivers are either in under exploited or over exploited state. Water being the primary requisite to support aquatic life, its physico-chemical properties is responsible for maintaining the aquatic environment. Water and soil quality provide essential source of energy and nutrients, while morphometric features like depth, areas, volume, water level fluctuation etc.

[1]Zoological Survey of India, Gopalpur-on-Sea 761 002, Orissa
[2]Department of Zoology, Gauhati University, Guwahati, 781014, Assam

regulate the supply of energy and nutrients. So a thorough knowledge of the limnobiology of the rivers is very essential for its proper monitoring with particular references to the study of micro flora, which forms the basic in complex food chain. Plankton communities, on which the whole aquatic productivity depends directly or indirectly, play an important part in regulating the productivity of an ecosystem. In addition, use of plankton community structure analysis in water quality monitoring for evaluation of the pollution status of the water bodies may be used as a reliable tool in biomonitoring studies of aquatic systems. Similarly, the benthic organisms which inhabits the bottom regions of lakes and rivers, plays an important role in the trophic budget of water bodies, since they form an important link in the food chain.

River appearance changes from source to mouth. Slopes are steep in the headwater, coarser particles including gravel and boulders are typical of upland stream, while a finer and softer substrate and calmer water is found in large lowland rivers. This diverse geography of the rivers provides diverse variation in biota. River chemistry also changes temporally under the multiple influences of seasonal changes in discharge regime precipitation input and biological activity. Phenomenon of natural eutrophication occurs every year during flood. This natural eutrophication increases partly by man's activities resulting in what is regarded as cultural eutrophication and partly by morpho-edaphic features such as silt load, depth, water current, river bed, its slope and meandering behavior that is referred to as topographic eutrophication.

In spite of its importance very few studies have been done to unravel the dynamics of such a vital component of the aquatic ecology of open water in the rivers of northeast. And fewer scientific studies have so far been made on the ecology of the rivers of Manipur (Singh, 1987; Kosygin and Dharmendra, 2009 & 2010). In the present study, an attempt has been made to evaluate the ecological status and fishery potential of Manipur river system.

Study area

Manipur is situated at the eastern threshold of India with an expanse of 22,327 km.[2] The Manipur river is the main drainage in Manipur, which drains the central part of Imphal valley. The principal tributaries of the river are Iril, Thoubal, Nambol and Khuga. Imphal River originates from the northern part of Manipur (Kangpokpi) and flows southward to the eastern side of the Loktak lake, traverses along the Manipur valley towards south and flows out of the Indian boundary to Myanmar and joins Chindwin river. The Imphal river is the only outlet, which drains Manipur valley and is associated with number of lakes, oxbow lakes and swamps. The bottom bed of the Imphal river is lower than that of Loktak lake, so the river does not fall into the lake but receives the out flow of lake and forms a short course called Khodrak. Being the flood plain of Imphal-Manipur river and its tributaries, the central plain of Manipur has thick deposits of sand, silt and clay. Imphal river after being joined by Chakpi stream is known as Manipur river. It drains more hospitable terrain of the valley than Barak drainage, which drains mostly the hilly terrain. Total catchments area of Manipur river is 6,332 sq. km. Imphal river is been utilizing for irrigation, hydro power generations, water supply and for other domestic purposes. The river has a total discharge of 6,550 cusecs, which is partly utilized at the Imphal Barrage for irrigation (Trisal and Mnihar, 2004).

Materials and Methods

Three tributaries of Manipur river have been selected for the study viz., the Imphal, the Khuga and the Thoubal. Two sampling stations each for the Khuga and the Thoubal rivers at upstream and downstream were selected. For the Imphal River three stations namely upper, middle and downstream were selected. The upstream station for the Khuga River was in Churachandpur (55 km from Imphal) near the Khuga dam. The downstream station was at Kumbi (35 km from Imphal) when the rivers come to valley. The selected sampling site for upstream of Thoubal was at Litan (50 km from Imphal), while the downstream sampling site was at Thoubal bazar (20 km from Imphal). For the Imphal river sampling points were: upstream-Kaubruleikha (32 km from Imphal), midstream-Mayang Imphal (Imphal city) and downstream at Sugnu (45 km from Imphal) after joining all the tributaries and before flowing to Myanmar.

The present investigation was carried out from January 1999 to December 2000. Water samples of Manipur river were collected seasonally i.e. winter, pre-monsoon, monsoon and post-monsoon. Plankton samples were collected using plankton net (nylobolt no.25) by filtering 100 liters of water and preserved in 4% formalin. Periphyton samples from the rivers were collected by scratching one square centimeter from the stones/boulders, which were submerged in the river water for at least two times. Counting was done by drop method taking one drop in each slide at least for five times. Ekman dredge (15.2 x 15.2 cm) was used for collection of bottom biota. Samples were preserved in 10 % formalin for further study. Planktons, periphytons and benthic forms were identified up to genera with the help of standard books (Edmondson, 1959; Needham and Needham, 1966 and ICAR monograph series on algae).

The rate of primary production by phytoplankton in the water bodies were estimated by *in situ* incubation of water samples in 500 ml light and dark water bottles, as per the standard *"light and Dark bottle technique"* (Gaarder and Gran, 1927; Vollenweider, 1969) with an incubation period of three hour during day.

Physico-chemical analysis of water and sediments were carried out following APHA (1989), Welch (1948), Welcher (1989), Manivasakam (1985) and Jhingran *et al.* (1969). Water parameters like temperature, transparency, dissolved oxygen, free carbon dioxide, alkalinity, pH, total dissolved solids, and specific conductivity were analysed at the spot. Parameters like total hardness, calcium, chloride, magnesium, silicate, total iron, nitrate, phosphate, dissolved organic matter were analysed in the laboratory along with the sediments characteristics.

Results and Discussions

Plankton

Plankton communities, on which the whole aquatic productivity depends directly or indirectly, play an important part in regulating the productivity of an ecosystem. They are the integral part of aquatic food chains. Knowledge of the distribution of plankton in time and space and of the environmental condition which favours or impedes its development is fundamental to the scientific utilization of the natural waters.

Khuga river: In the Khuga river, upper stretch showed richer plankton population (43-363 ul^{-1}) in comparison to lower stretch (22-30 ul^{-1}). Seasonal and zonal fluctuation of different group of plankton has been depicted in Table 7.1. Thirty three plankton species belonging to myxophyceae, bacillariophyceae, chlorophyceae and zooplankton were recorded from the upper stretch while it was only 22 species in lower stretch. Plankton species observed in Khuga rivers are:

Myxophyceae	*Rivularia, Oscillatoria, Phormidium, oelospharium, Anabaena, Spirulina*
Bacillariophyceae	*Nitszchia, Gomphonema, Navicula, Synedra, Cocconeis, Surirella, Fragilaria, Denticula, Caloneis, Gyrosigma, Diatoma, Cymbella, Melosira, Amphora, Pinnularia*
Chlorophyceae	*Cladophora, Geminella, Rhapidonema Sederocelis, Chlorococcum, Chlorella, Chlorosarcinia, Schizothrix, Ulothrix,*
Dinophyceae	*Ceratium*
Cladocera	*Daphnia, Ceriodaphnia, Scapholeberis*
Ciliates	*Cyclochaeta*
Rotifera	*Trichocerca*
Rhizopoda	*Centropyxis, Difflugia, Vorticella*
Copepoda	*Cyclops, Acari*
Turbellaria	*Planeria*
Ephemeroptera	Mayflynymph.

In upper stretch seasonal fluctuation of plankton both in terms of density and composition was observed. Chlorophyceae dominated during winter (48.9%) and pre-monsoon (70%) but bacillariophyceae dominated during monsoon (58.3%) and post-monsoon (91.2%). Contribution of zooplankton (4.4-13.9%) and Myxophyceae (1.6-13.9%) were relatively low in all the seasons. The contribution of zooplankton, however, showed an increasing trend during monsoon (13.9%).

In the lower stretch, plankton population remained almost the same throughout the seasons (22-30 ul^{-1}) with greater dominance of zooplankton (77.3%) during winter and chlorophyceae during the remaining seasons; pre monsoon (56%), monsoon (58.6%) and post monsoon (76.6%). Dinophyceae was recorded during monsoon seasons contributing 13.8% of total count. Seasonal fluctuation in the density of plankton was not distinct as recorded in the case of upper stretch. Contribution of diatom sharply declined from upper stretch to lower stretch, 6.7% during post monsoon and 13.8% during monsoon. Moreover, no diatom could be observed during winter and pre-monsoon seasons. Zooplankton population showed a decreasing trend from winter to post-monsoon.

Thoubal river: In upper stretch of the Thoubal river, plankton density was relatively low (17-26 ul^{-1}) as compared to downstream, 66-144 ul^{-1} (Table 7.2). Main plankton groups encountered were myxophyceae, bacillariophyceae, chlorophyceae and rhizopods. Overall zooplankton population was very low (0-8.6%) No significant seasonal fluctuation could be observed in the upper stretch of the river. Chlorophyceae dominated during all the seasons, 73.9% (winter); 64% (pre monsoon); 70.6% (monsoon) and 100% (post monsoon). Bacillariophyceae population was relatively low (13.1-29.4%) with total absence during post monsoon. Myxophyceae was encountered during winter (4.3%) and pre monsoon (12%). Zooplankton represented by rhizopod was seen only during winter (8.6%) and pre monsoon (8%). Complete dominance of phytoplankton over zooplankton and complete dominance of chlorophyceae within phytoplankton groups were seen in the upper stretch of Thoubal river. Among chlorophyceae *Spirogyra* was the most dominant species, which showed luxurious growth. Sixty seven plankton species were found in the stretch.

In the lower stretch, zooplankton was present during all the seasons contributing to the tune of 6.0%, 4.3%, 11.1% and 3.1% in respective seasons. Among phytoplankton bacillariophyceae was predominant (79.1%, 53.6%, 62.5%, and 90.5% respectively). The contribution of myxophyceae was relatively low (1.5-11.8%). Percentage contribution of chlorophyceae in all the seasons, from upper to lower stretch, decreased significantly (10.4%, 33.4%, 11.8%, and 4.5%). Interestingly, the plankton peak was recorded during monsoon (144 ul^{-1}), which may be attributed to grater proliferation of bacillariophyceae (62.5%). A bacterium species was also encountered during this season. Thirty one (31) plankton species were encountered during different seasons. Plankton species observed in Thoubal river are:

Myxophyceae	*Rivularia, Oscillatoria, Phormidium, Coelospharium Spirulina, Anabaena*
Bacillariophyceae	*Navicula, Synedra, Surirella, Fragilaria, Tabellaria, Cymbella, Eucocconeis, Cyclotella Nitzchia, Gomphonema, Cocconeis, Gyrosigma, Diatoma, Melosira, Achanthis, Sauroneis*
Chlorophyceae	*Mougeotia, Cladophora, Spirogyra, Ulothrix, Desmidium, Protococcus Genicularia, Cladophora, Ankistrodesmus, Chlorococcum, Dactylococcus, Closteriopsis, Chaetophora, Sphaerozosma, Pediastrum, Chlorella, Schizothrix,*
Rhizopods	*Difflugia*
Zooflagilates	*Euglena*
Copepoda	*Nauplius, Diaptomus*
Cladocera	*Daphnia*
Bacteria	*Sphaerotilus.*

Imphal river: Imphal river showed distinct zonation in respect of plankton density. Upper stretch had the highest concentration of plankton (66-480 ul^{-1}), followed by lower stretch (40-48 ul^{-1}) and the middle stretch (19-24 ul^{-1}) (Table 7.3). Marked seasonal fluctuations in plankton population were clearly seen at upper stretch as compared to middle and lower stretch where density of plankton remained almost the same throughout the seasons. Main phytoplankton groups observed at the upper stretch of the Imphal River belong to myxophyceae, bacillariophyceae and chlorophyceae. A sharp decline in plankton density was observed during monsoon season (66 ul^{-1}). Bacillariophyceae dominated irrespective of seasons in the upper stretch. The contribution of chlorophyceae was relatively lower (5–21%). Myxophyceae was found to fluctuate within a range of 2--25% with greater abundance during monsoon seasons. Zooplankton was represented by rhizopods and ciliates and its contribution to the total plankton count was low (1-4.2%). Forty one (41) plankton species encountered during different seasons.

In the middle stretch plankton density declined to 19-24 ul^{-1} along with the species richness. Only 18 plankton species was encountered. During winter chlorophyceae (56.5%) was dominant followed by bacillariophyceae (30.4%) and myxophyceae (8.6%). Abundance of zooplankton was low (4.4%) represented by rotifers only. During pre monsoon, chlorophyceae (60%) was dominant followed by myxophyceae (30%) and bacillariophyceae was absent. Zooplankton population was 10% of the total plankton count, which however increased to 63.2% during monsoon. Chlorophyceae showed its total absence during monsoon. Myxophyceae and bacillariophyceae contributed to the tune of 26.3% and 10.5%, respectively. During post monsoon bacillariophyceae (41.7%) and chlorophyceae (58.3%) were the main plankters as no myxophyceae and zooplankton could be recorded during the season. Overall phytoplankton dominated over zooplankton in this stretch.

At the lower stretch of Imphal river, the main phytoplankton groups observed were myxophyceae, bacillariophyceae, chlorophyceae, dinophyceae and chrypsophyceae. Number of plankton species observed increases to twenty two (22). However, zooplankton taken over the phytoplankton population in this stretch of the river with a percentage contribution ranging between 36.5-91.6%. During winter both phytoplankton and zooplankton contributed almost equally, the former being marginally higher (54.7% and 45.3%). Among phytoplankton bacillariophyceae (30.9%) was most dominant followed by chlorophyceae (16.6%) and myxophyceae (7.0%). Zooplankton was represented by rotifers (14.2%), cladocerans (16.6%) and copepods (12.0%). During pre monsoon zooplankton (52.2%) dominated over phytoplankton (47.8%). Among phytoplankton chlorophyceae (21.8%) was the major component while among zooplankton rotifers (34.8%) dominated over copepods (7.4%). One species each of dinophyceae and chrypsophyceae were observed contributing to 4.3% each in the total plankton. During monsoon seasons zooplankton (91.6%) completely dominated over phytoplankton (8.4%). The contribution of phytoplankton was a mere 4.2% consisting of myxophyceae and bacillariophyceae. Among zooplankton copepods (72.8%) dominated over rotifers (14.6%) and cladoceras (4.2%). Greater proliferation of bacillariophyceae was recorded during post monsoon (57 %). Myxophyceae and chlorophyceae were completely absent in this season. Zooplankton population (36.5%) has the lowest contribution in this season of this stretch of the river. A bacterium species was also recorded during post monsoon (7.5%). Plankton species observed in Imphal rivers are:

Myxophyceae	*Anabaena, Coelospherium, Oscillatoria, Phormidium, Spirulina*
Bacillariophyceae	*Navicula, Cymbella, Gyrosigm Eucocconeis, Synedra, Eunotia, Fragilaria, Diatoma, Melosira, Gomphonema, Cocconeis, Surirrella, Melosira, Diatomella, Gomphonies, Eunotia, Tabellaria, Pinnularia, Neidium, Mastogloiea, Frustulia, Anomoneis*

Chlorophyceae	*Schizogonium, Closterium, Chaetophora, Pleurotanium, Ulothrix, Mougeotia, Protococcus, Crucigenia, Spondylosium, Spirogyra, Zygnema, Gonatozygon, Microspora, Nephrocytris, Chlorococcum, Chlorella, Raphidonema, Schizothrix, Goniochloris, Coelospharium, Chlorochytrium Ankistrodesmus, Chlorococcum,*
Dinophyceae	*Ceratium*
Chrypsophyceae	*Dinobryon, Tribonema,*
Zooflagilates	*Polytoma, Chilomonas*
Rhizopoda	*Centropyxi, Arcella*
Ciliates	*Femtonia.*
Rotifera	*Horaella, Keratella, Polyartha, rotifer sps.,*
Copepoda	*Cyclops, Mesocyclop*
Cladocera	*Daphnia, Ceriodaphnia*
Bacteria	*Sphaerotilus.*

In the present study, the plankton population was relatively low in all the rivers (17-480 ul^{-1}). Similar observations have also been reported from other lotic systems like Ganga (110-445 ul^{-1}) by Asif *et al.* (1998), Banerjee *et al.* (1998) in Damodor (16-604 ul^{-1}). Much high value has however, been reported by Khan *et al.* (1998) from river Ganga (400-3900 ul^{-1}) and Jhingran and Pathak (1988) reported up to 8,8840 ul^{-1}. Lower plankton population (22-51 ul^{-1}) has also been observed from headwaters of river Brahamaputra by Pathak *et al.* (2001). Generally in rivers plankton population was found low as rivers plankton were always in a continuous flow. In small, fast flowing streams, sloughing of attached autotrophs is the primary source of plankton, (Swanson and Bachman, 1976). However, in sluggish, lowland streams, the residence time of mass can be sufficient for true plankton to colonize and reproduce. Under such condition phytoplankton and zooplankton almost always are present (Hynes, 1970). Generally, in the upper portion of the rivers phytoplankton indicates the dominance over zooplankton. In lower stretch, however, the relative abundance of zooplankton showed much improvement.

In the present study *Oscillatoria, Spirulina, Chlorella, Melosira, Synedra* and *Sphaerotilus* were identified as indicator of organic pollution. *Rotifers, ciliates* were observed to be pollution tolerant among zooplankton. Presence of such species during the study is an indication of mild pollution of river water. A sewage bacterium (*Sphaerotilus*) observed in the lower stretch of the Thoubal and the Imphal reflected influx of domestic waste from the surrounding areas. But the low plankton density (22-480 ul^{-1}) and absence or almost negligible contribution of pollution tolerant species and low myxophyceae count (0-30%) for all the rivers showed more inclination towards olligotrophic rather than eutrophic condition of the water. Presence of *Ceratium* and dinobryon which are the olligotrophic indicating species in the Khuga and the Imphal rivers further support the findings. Poor plankton population in all the rivers and presence of minimum seasonal variation may probably be attributed to flowing condition of water of the rivers.

Periphyton

Rivers provide a system of continuous movement where any suspended particle will, often within a few days, be carried to the sea. Such a situation is far from ideal for plankton and makes phytoplankton a relatively minor element. So periphytons have gained much importance in rivers compared to lakes under such conditions. The distribution of periphyton in different stretches of Manipur river system is depicted in Table 7.4.

Khuga river: In upper stretch of the Khuga river, periphyton population ranged between 3,055 - 37,160 u cm^{-2} and showed lowest density during monsoon. Maximum periphyton density was observed during pre monsoon. Bacillariophyceae played the dominant role (57.4-76.4%) in all the seasons (Table 7.4). Chlorophyceae showed its maximum value during pre monsoon (37.1%) while myxophyceae during monsoon (40.4%). Some lower animal group like *Difflugia, Acanthocystis, Ploisoma, Ooikomonas, Arcella* and zooflagilates were also seen during winter, pre and post monsoon contributing 2.2%, 0.3% and 4.1% of total population. A total of 57 species belonging to different plant and animal group during periphyton analysis encountered in the upper stretch of the Khuga river. At lower stretch, periphyton populations were 9,471 u cm^{-2}, 8,860 u cm^{-2} and 9900 u cm^{-2} for winter, pre and post monsoon. There was no monsoon sample. Bacillariophyceae was the dominant group (57.2-58.7%). The contribution of chlorophyceae increased, and ranged from 38.4-40.9% and contribution of myxophyceae was low (1.5-2.9%). No animal group was observed during the study.

Number of periphyton species (57)) observed in upstream was higher than that of downstream (20). To sum up, in the Khuga river, bacillariophyceae showed overall dominance (57.2-76.5%) in the periphyton community irrespective of seasons and zonal variations. Contribution of chlorophyceae increased from upper to lower stretch. Myxophyceae was all the time low except during monsoon in upper stretch, where it elevated to 40.4% of total count.

Thoubal river: In the Thoubal river, periphyton population ranged between 4,947 -53,450 u cm^{-2} in upper stretch. Bacillariophyceae dominated during winter (57.1%), monsoon (69%) and post monsoon (96.1%). During pre monsoon Chlorophyceae (50.8%) dominated over bacillariophyceae (48.6%) and myxophyceae (0.6%). No animal group could be traced during the study. In total 33 algal species encountered both at upper and lower stretches of the Thoubal River. In lower stretch periphyton population was low ranged between 8,395 u cm^{-2}-16,320 u cm^{-2}. During winter bacillariophyceae (73.5%) dominated. There was no monsoon sample. Chlorophyceae dominated over (66.2%) bacillariophyceae (32.1%) and myxophyceae (1.6%) during pre monsoon but during post monsoon bacillariophyceae (95.6%) dominated over chlorophyceae (19%) and myxophyceae (1.8%). Some rhizopods were encountered during post monsoon and its contribution was 0.7% of the total count.

Imphal river: Periphyton population ranged between 8,974 -26,510 u cm^{-2} in the upper stretch of the Imphal River. Number of species encountered during the study was 45. Density wise, pre monsoon had the maximum and monsoon had the lowest population of periphyton. Bacillariophyceae (43.1-92%) dominated over chlorophyceae (2-54%) and myxophyceae (1-10.8%) irrespective of seasons. Some animal groups were also encountered during pre monsoon contributing about 1.5% of total population. In the middle stretch, periphyton density was 4,984 u cm^{-2}, 5,670 u cm^{-2} and 4,320 u cm^{-2} for winter, pre monsoon and post monsoon respectively. There was no sample for the monsoon season. Number of species encountered at this stretch drastically reduced to eighteen only. Like upper stretch, bacillariophyceae (80.6-87.5%) dominated over chlorophyceae (12.5-19.4%), but no myxophyceae and animal group could be traced in all the seasons. At the lower stretch periphyton population ranged between 3,276-8050 u cm^{-2} and showed maximum during winter and pre monsoon, minimum value was recorded during monsoon. Here bacillariophyceae 90-93.0% dominated over chlorophyceae (1.0-10%) and myxophyceae (1.9-5%). Only a single species of protozoa (*Polytoma*) was observed during post monsoon.

Over all in the Imphal river bacillariophyceae played the dominant role in the periphyton community irrespective of topographical variation. Periphyton showed maximum density during pre monsoon and minimum during monsoon. Myxophyceae population was all the time and all the location low (0-10.8%). Chlorophyceae population was also low (1-19.4%) except during monsoon season (54%) at upper stretch. Animal group was meager 0-1.5% and found only in pre and post monsoon seasons of upper and lower stretch. Number of species available decreased from upper stretch (45 species) to middle stretches (18 species), which again improved at lower stretch (30 species).

It may be referred from the present study that, periphyton population was high (3,055-53,450 u cm^{-2}) in all the rivers and showed seasonal fluctuation. A much low value (1441 u cm^{-2}) was observed by Laal et al. (1982) from different stretches of river Ganga. But Gurumayum *et al.* (2000) observed much higher value from the hill streams of Arunachal Pradesh (71,107-85,642 u cm^{-2}). Bacillariophyceae was the dominant group in most of the seasons of all the rivers despite zonal variation (43.1-96.1%). Dominance of bacillariophyceae in periphyton count was also noticed by some workers (Kawosa, 2001; Laal *op. cit.*, Sukumaran *et al.* 1996). Periphyton density increased during winter and pre monsoon in all the rivers irrespective of topographical variation. Rise in periphyton population during winter season was also noticed by Shrivastav, *et al.*, (1985); Sukumaran *et al.*, (1996) in lontic habitats. Upper stretches (3053-53, 450 u cm^{-2}) of the rivers have higher population than the middle and lower stretches (3,276-16,320 u cm^{-2}). Increase periphyton density in upstream of rivers is an expected phenomenon where the environment is more suitable for such organisms. Similar observation was made by Gurumayum *et al.* (2000) from the upstream of the tributaries of the Brahamaputra river. Animal groups were almost untraceable and appeared only in some seasons. Presence of animal groups in periphyton count was also reported by Sugunan (1991) from Nagarjunasagar Reservoirs and Sukumaran *et al.* (1996) from a lentic habitat of Karnataka.

In the present case also periphyton population were much more than plankton population and upstream showed the maximum density. Increase in density of periphyton in upstream may be attributed to favourable environmental condition like high dissolve oxygen, low carbon dioxide content and higher transparency. In hill streams periphyton play the major role in primary productivity, as the environment is favorable for periphyton community to propagate. Periphyton needs substratum for their growth which is easily available in the forms of boulders and rocks and the transparent water makes the sunlight to penetrate up to bottom and it makes the periphyton to take over the function of plankton. Whereas plankton are always in the mercy of water current in lotic ecosystem but when it get sufficient residence time in slow downstream rivers, it also contributes significantly to the productivity of the water body. Dominance of diatom in most of the count both in plankton and periphyton may be attributed to their non sessile free floating nature. Higher contribution of diatoms in both the plankton and periphyton count in the rivers corroborated with findings of Fritsch (1935), who suggested that the attachment of diatom is facilitated with the help of mucilage often secreted by them and this group can also withstand variable changes of light intensity, temperature and other adversities. Dominance of diatom in the river water are also seen by the worker like Gupta (1985); Pathak (2001); Nath (2001) in other lotic water bodies. Presence of almost same genera in both plankton and periphyton might be due to slaughter of diatom from the substrate. The importance of silicate for the proliferation of diatom is well documented by many workers, in the present observation silicate range of 2.2-7.0 mgl^{-1} was favorable for efficient growth of the diatom.

Species Diversity

Plankton

For many years, algae have been used as bio-indicators for water quality monitoring. It enables quick, easy and authentic information even at the micro level of impurity. Besides, using plankton as an index of water quality, in respect of industrial and municipal waste, have also been carried-out by many workers. Species diversity is the ratio of the number of species and importance value of individual (Odum, 1971). It tends to be low in physically control ecosystem and high in biologically controlled ecosystem. Such bio mathematical tools now a day become complementary to the convenient physico-chemical techniques for characterising the aquatic ecosystem. Species diversity of plankton and periphyton was calculated by applying the Shannon Weaver diversity index, Shannon and Weaver (1963). The species diversity index and evenness index at the different location of the rivers was depicted in Table 7.5.

In the present study, species diversity index was high (3.7-4.3) in all the rivers and evenness indices were also near unity (0.74-0.96). Much lower values were obtained from many aquatic ecosystems. Acharjee *et al.* (1995) reported value of 1.3-1.9 from a *beel* in Assam. Khan *et al.* (1998), observed $\hat{H}$ value of 0.2-2.4 in the river Ganga. Similar value was reported from Krishna river by Sugunan (1991). The findings of the value of species diversity index reflected that the water of the three selected rivers of Manipur is more or less free from pollution and still maintaining stable aquatic ecosystem. Presence of organic pollution tolerant species like *Oscillatoria, Spirulina, Chlorella, Melosira, Synedra* and *Sphaerotilus* were observed. Rotifers contribution was highest in the downstream. These species were observed to be pollution tolerant among plankton. Arora (1966), described rotifers as an indicator of organic pollution and their presence in downstream showed enrich water quality. Presence of eutrophy indicator species like *Oscillatoria* and *Anabaena* in negligible quantity besides very low myxophyceae population in all the rivers (0-30%) reflects less load and less enriched nature of the waters. A sewage bacterium (*Sphaerotilus*) observed in the middle and lower stretch of the Thoubal and the Imphal reflected influx of domestic waste from the surrounding areas. But the low plankton density (22-480 ul^{-1}) and absence or almost negligible contribution of pollution tolerant species and low myxophyceae count (0-30%) for all the rivers showed uncontaminated condition of the water. Presence of *Ceratium* and dinobryon which are the olligotrophic indicating species in the Khuga and the Imphal rivers further support the findings.

From the results of the study of the index value and other qualitative and quantitative abundance of plankton, it can be summarised that the water bodies are more or less free from heavy pollutions.

Periphyton

The species diversity index, evenness index, number of species and total population of periphyton observed at the different location of three rivers was depicted in Table 7.6.

Average periphyton species diversity index ($\overline{H}$) and evenness index (J) recorded from the upstream of Khuga river were 3.6 and 0.77 respectively. For downstream average and J values were 3.0 and 0.74 respectively. ($\overline{H}$) and (J) values were minimum during monsoon seasons and periphyton density and number of species encountered also decreased along with them. Species richness and average diversity and evenness indices changed from upstream to down stream. In up stream a total of 57 periphyton species was observed while in down stream it was only 20. In Thoubal river, the average species diversity index $\hat{H}$ and J value for upper stretch were 3.9 and 0.95 respectively. Species diversity index and evenness index also varied according to the species richness, $\hat{H}$ value was maximum during winter and pre monsoon and minimum in monsoon. 33 algal species encountered at upper stretch of the Thoubal river. In lower stretch, average $\overline{H}$ and J value were 3.1 and 0.72 respectively. Their values were maximum during post monsoon. A total of 45 periphyton species were encountered during the study. $\hat{H}$ and J value decreased from upstream to downstream. Average species diversity index, value at upper stretch of Imphal river was 4.2. Middle and lower stretch have the similar $\hat{H}$ value (3.4). Evenness index, J values for the three stretches were 0.77, 0.94 and 0.83 respectively. Number of species available decreased from upper stretch (45 species) to middle stretches (18 species), which again improved at lower stretch (30 species). $\hat{H}$ and J value of periphyton lowered during monsoon irrespective of topographical variation.

As we go from upstream to down stream, species diversity indices values decreases but not exceeding the undesirable level (3.0-4.2) given by Shannon and Weaver (1964). In the present study, 'J' values were almost near unity (0.70-0.99) and seasonal change has little effect. Its value was high in upstream of the Thoubal and the Khuga rivers where as in the Imphal river, middle stretch (0.94) have the maximum J value. $\overline{H}$ value decreased during monsoon for all the stretches (Table 6) which were corroborated with the findings of Sugunan (1991) on the river Krishna. Higher periphyton density, high $\overline{H}$ value and dominance of bacillariophyceae in the upstream of the rivers suggested more towards olligotrophic condition of the rivers stretch. Downstream decreased in periphyton densityin all the rivers is on the line of anticipation. But decreased in $\overline{H}$ value along the downstream reflected a more enriched water quality in comparison to upstream. To sum up $\overline{H}$ and J study of periphyton suggested overall unpolluted state of all the three rivers of Manipur.

Benthos

In aquatic system, the benthic macro-invertebrate community is most often investigated due to the absence of mobility and sensitivity towards physico-chemical stresses. Benthic organisms provide a valuable indicator of past and present water quality conditions because of their long life histories and central position in the food chain (Mackenthum, 1966; Cairns and Dickson, 1971).

Khuga river: Distribution of different species of macro invertebrate benthos in both the stretches of the Khuga river of is depicted in Table 7.7. The total macro invertebrate benthos population fluctuated between 129-1,255 um^{-2}. Clear seasonality in species composition and population density could be seen; pre monsoon has the maximum concentration and post monsoon season having the lowest concentration. A total of 8 genera were available in upstream, where as in downstream only 5 genera were encountered. In the Khuga river, at upper stretch, macro benthos population ranged between 129-1,255 um^{-2} and was maximum at pre monsoon and minimum during post monsoon. Such wide variation of macro benthos in healthy stream was also encountered by Robinson *et al.,* (2001) in a glacial stream of Swiss Alps. At lower stretch, benthos population ranged between 130-850 um^{-2} and was maximum at pre monsoon. Monsoon sampling could not be done. Abundance oligocheate in downstream may be due to occurrence of soft clayey soil of the river which serves as the most suitable niche for the growth and propagation of oligocheate which are efficient deposit feeders. Overall in the Khuga river upper stretch showed richer benthos population than the lower stretch. Presence of oligocheate in lower stretch indicated enrich water quality in the downstream.

Thoubal river: In the Thoubal river, benthos population ranged between 173-1,100 um^{-2} and was maximum in pre monsoon in both the station. Distribution of different species with groups of macro invertebrate benthos in the Thoubal river is depicted in Table 7.8. In upstream, 12 genera of macrobenthic fauna were encountered and in downstream only 8 genera was recorded. At upstream, macro zoobenthic population was 800 um^{-2} during winter, 1,100 um^{-2} during pre monsoon. The benthic population decreased to 216 um^{-2} during monsoon season. During post monsoon, the population was 433 um^{-2} At downstream, macro invertebrate zoobenthic population was 750 um^{-2} during winter and 1,082 um^{-2} during pre monsoon. During post monsoon only a single species of mollusca (*Bythemia*) represented the whole benthic fauna and its population was 173 um^{-2}. Monsoon sampling could not be collected due to over flowing of the river. Thoubal river as a whole was rich in macro invertebrate benthos, both population and species diversity. Over all benthos population was richer in upper stretch than downstream and pre monsoon showed the maximum density. Molluscan fauna predominant in both the stretches. Presence of representative of almost all the benthic fauna including the predator groups like megaloptera, diptera and odonata reflects the full utilization of different niches of the habitat. Thoubal river was found to be rich in predator population in both the stretches, a large number of predator like *Cingymula, Hydropsychae,* megalopteran nymph, dragon fly nymph, dipteran and coleopteans were common in the river.

Imphal river: In the Imphal river, total macro invertebrate zoo benthos population was low and ranged from 85-390 um^{-2}, pre monsoon showing the maximum concentration in upper and lower stretch. Distribution of benthic fauna in different stretches of the Imphal river are represented in the Table 7.9. Benthos population was represented by 3 genera in upper and middle stretch where as in lower stretch 6 genera were observed. At upper stretch, macro invertebrate zoobethos benthos population was maximum during pre monsoon, 390 um^{-2}, represented by trichoptera only where as during monsoon, having lowest count (85 um^{-2}) was represented by only mollucs. Post monsoon was represented only by may fly nymph (130 um^{-2}). At middle stretch, benthos population was maximum, 216 um^{-2} at post monsoon. In winter the total population was 170 um^{-2} whereas during pre monsoon the lowest count was observed 130 um^{-2}. There was no monsoon sample.

At lower stretch, benthos population was 300 um^{-2} during winter, 350 um^{-2} during pre monsoon, 173 um^{-2} during monsoon and 346 um^{-2} during post monsoon. An adult insect, *Aphelocherium* was also encountered (12.4 %) during the study. In the Imphal river species richness and density of macro invertebrate benthos population tends to increase from upstream to downstream of rivers. Overall macro benthos population has lowered in all the stretch of the Imphal river in comparison to the Khuga and the Thoubal river.

In the present study, the macro invertebrate benthos population was low in all the rivers, which ranged between 85-1255 um^{-2}. Similar observation of low benthos density (162-290 um^{-2}) was also made in tributaries of river Brahmaputra by Pathak *et. al.* (2001). According to Theinemann (1925), a lake bed producing 1000 individuals or less per meter square is usually regarded as poor in production, while a density of 2000 um $^{-2}$ indicates a productive water body, so it can be inferred that the three selected rivers were poor in macro invertebrate benthos population. But in general lakes harbors good benthos population than rivers. In comparison to other water bodies of India like Damodar (Banerjee *et. al.* 1998), tributaries of Brahamaputra river (Pathak *et al.* 2001) etc., macro invertebrate benthos population of the present finding was much richer. The present findings support the contention of Patrick (1970) that the rigorous and unpredictable stream environment is largely responsible for low equitability and low diversity in benthic community. The occurrence of trichoptera in plenty further support the contention of Marlier (1962), who explain the great abundance of benthic fauna due to availability of large quantity of food in suspension form which is swept by the current into the nest of these omnivorous creatures. To sum up, macro invertebrate population in the Khuga and the Thoubal rivers were poor to medium where as in the Imphal river all the three stretches showed much lower macro invertebrate population.

Productivity

Aquatic habitat being complex system with interaction among different biotic and abiotic components, showed significant fluctuation with respect to surroundings and have a considerable influence on the productivity of the water body. The primary productivity of a water body is the manifestation of its biological production. The study of primary productivity of a water body is being used by many workers as one of the important factor to measure the productivity of water body. The values of GPP and respiration of the three rivers are depicted in Table 7.10.

In the Khuga river productivity was low and fluctuated between 29.17 - 500.00 mg C m^{-3} day^{-1}. Respiration rate ranges between 17.5 - 450 mg C m^{-3} day $^{-1}$ Productivity was maximum during pre monsoon at upstream, whereas in downstream, post monsoon season showed the maximum value. Upstream has the higher productivity value than downstream. Productivity value was low (30.2-875 mg C m^{-3} day^{-1}) in Thoubal River and respiration fluctuates from 17.5-460 mg C m^{-3} day^{-1}. Maximum productivity was encountered during post monsoon at upstream. During this season, chlorophyceae was the dominant component (100%) in the plankton count. Productivity of the river decreased drastically when the river comes to plain, this may be attributed to variation in the ecosystem. In the Imphal River, productivity was low and ranged from 21.88 - 800.00 mg C m^{-3} day^{-1} and respiration fluctuates between 8.75-600%. Productivity was higher in upper (66.67-156.26) and middle stretch (100-800 mg C m^{-3} day^{-1}). Monsoon season showed minimum productivity value; however in other seasons, ranges of fluctuation of productivity value were minimum. At downstream productivity (21.88-30.2 mg C m^{-3} day^{-1}) decreased drastically and showed almost constant value throughout the seasons.

There was appreciable variation in phytoplankton primary productivity of the selected rivers of Manipur, which may be due to their different limnological parameters. Overall, the productivity was low for all the rivers (21.88–875 mg C m^{-3} day^{-1}) and considerable zonal and seasonal fluctuations were observed during the study, which can be correlated with the low plankton count (19-480 u l^{-1}). The rate of gross energy fixation in the rivers like Ganga, Yamuna, Ghaghra, Son, Brahmaputra and its main tributaries (Pathak 1999) were within the range of 1752 to 5060 cal m^{-2} day $^{-1}$ which is much higher than the present findings. Lowered primary productivity by phytoplankton was observed during monsoon season in all the rivers irrespective of zonal variation which is corroborated with the findings of earlier workers like Acharjee (1997); Mishra *et al.* (1992), which may be correlated to the disturbance by floods. In some case, respiration has reached nearly cent percent and even more as seen during post monsoon of middle stretch of the Khuga River (90%) and during monsoon seasons in Thoubal and Imphal river (120%). These values indicate a shift in photosynthesis oxygen production where the respiration exceeds production. Such an event takes place when dark clouds gather or due to rainfall. Going through the above observations, the three selected rivers of Manipur viz., Khuga, Thoubal and Imphal rivers can be classified as poorly productive with observed average production values of 87 mg C m^{-3} day^{-1}, 175.7 mg C m^{-3} day^{-1} and 75 mg C m^{-3} day^{-1} respectively.

Physico-chemical Parameters of Water

Water being the primary requisite to support aquatic life, its physico-chemical properties are responsible for maintaining the aquatic environment. Water maintains its equilibrium and act like a strong buffer. The quality of water is evaluated by its limno-chemical features and is dependent upon the concentration of different organic and inorganic compounds, minerals and living and dead organisms which again influence the productivity of the water bodies. A mere fluctuation in the concentration of different parameters may create problem for the organism in first trophic chain, which again may influence the productivity of the whole water bodies. Many researchers have correlated various biotic and abiotic parameters of the water bodies. The results of average physic- chemical characteristics of the three rivers are depicted in Table 7.11. The common feature in respect of water quality of the three rivers were alkaline pH, high DO, silicate and DOM and low CO_2, nutrients, calcium, TDS and specific conductivity. Poor average value of PO_4 (0.012-0.055 mgl^{-1}), NO_3 (0.135-0.526 mgl^{-1}), specific conductivity (62-182.8 μmho mhos cm^{-1}), TDS (27 31.6-90.5 mgl^{-1}), chloride (17.24-25.03 mgl^{-1}) and calcium (5.08-30.6 mgl^{-1}) reflected that there is very low amount of organic waste and practically there is no discharge of industrial waste. But considerable fluctuations were observed in respect of CO_2, DO, transparency, Ca, Mn and hardness along the Imphal river which, can be attributed to domestic activity carried on in the river. Imphal river, when flows through densely inhabited areas, activities like washing of cloth, utensil and bathing in the river could be observed. The increase in chemical parameters may also be attributed to impact of other tributaries of the Imphal River including Khuga and Thoubal. To sum up, rich dissolved oxygen (6.4-9.3 mgl^{-1}), neutral to alkaline pH (7.2-7.5), low to medium alkalinity (36-75.3 mgl^{-1}), rich DOM (1.01-1.9 mgl^{-1}), high silicate (3.6-4.4 mgl^{-1}), low free CO_2 (1.2-4.2 mgl^{-1}) all reflects unpolluted condition of the three rivers.

Table 7.1
Composition of plankton (ul^{-1}) at different stretches of the Khuga river during 1999–2000

Location/	*Upper stretch (Khuga Dam)*				*Lower stretch (Kumbi)*			
Seasons	*1*	*2*	*3*	*4*	*1*	*2*	*3*	*4*
Phytoplankton								
Myxophyceae	7 (2.5)	6 (1.6)	6 (13.9)	6 (1.9)	3 (13.6)	2 (8)	0	2 (6.7)
Bacillariophyceae	152(44)	86(23.7)	25(58.3)	287(91.2)	0	0	4(13.8)	2(6.7)
Dinophyceae	0	0	0	0	0	0	4 (13.8)	0
Chlorophyceae	167(48.9)	254(70)	6(13.9)	8(2.5)	2(9.1)	14(56)	17(58.6)	23(76.6)
Sub total	326(95.6)	346(95.3)	37(86.1)	301(95.6)	5(22.7)	16(64)	25(86.2)	27(90)
Zooplankton								
Rhizopods		0	0	0	3(13.6)	2(8)	0	3(10)
Zooflagilata	11 (3.2)	10(2.8)	0	14(4.4)	0	0	0	0
Ciliata	0	0	0	0	3 (13.6)	1 (4)	0	0
Rotifera	1 (0.3)	2 (0.5)	0	0	0	0	0	0
Cladocera	0	1(0.3)	0	0	8(36.4)	5(20)	4(13.8)	0
Copepoda	3 (0.9)	4 (1.1)	0	0	3 (13.6)	1 (4)	0	0
Ephemeroptera	0	0	3 (6.95)	0	0	0	0	0
Turbellaria	0	0	3 (6.95)	0	0	0	0	0
Sub total	15(4.4)	17(4.7)	6(13.9)	14(4.4)	17(77.3)	9(36)	4(13.8)	3(10)
Total	341	363	43	315	22	25	29	30

1: winter; 2 pre monsoon; 3: monsoon; 4: post monsoon.
Number within parentheses represent the percentage composition.

Table 7.2
Composition of plankton (ul^{-1}) at different stretches of the Thoubal river during 1999–2000

Location	*Litan (Ukhrul)*				*Thoubal bazar*			
Seasons	*1*	*2*	*3*	*4*	*1*	*2*	*3*	*4*
Phytoplankton								
Myxophyceae	1 (4.3)	3(12)	-	-	3 (4.4)	6 (8.9)	17(11.8)	1 (1.5)
Bacillariophyceae	3 (13.1)	4 (16)	5 (29.4)	-	53 (79.1)	37(53.6)	90 (62.5)	60 (90.9)
Chlorophyceae	17(73.9)	16(64)	12(70.6)	26(100)	7(10.4)	23(33.4)	17 (11.8)	3(4.5)
Sub total	21(91.4)	23(92)	17(100)	26 (100)	63(94)	66(95.7)	124(86.1)	64 (96.9)
Zooplankton								
Rhizopoda	2 (8.6)	2 (8)	-	-	1 (1.5)	-	-	1 (1.5)
Cladocera		-	-	-			12 (8.3)	-
Copepoda		-	-	-	3 (4.5)	3 (4.3)	2 (1.3)	1 (1.5)
Sub total	2 (8.6)	2(8)	-	-	4 (6)	3(4.3)	16 (11.1)	2(3.1)
Bacteria		-	-	-			4 (2.8)	-
Total	23	25	17	26	67	69	144	66

1: winter; 2 pre-monsoon; 3: monsoon; 4: post-monsoon.
Number within parentheses represent the percentage composition.

Table 7.3
Composition of plankton (ul^{-1}) at different stretches of the Imphal-Manipur river during 1999–2000

Location	*Kaubruleikha (Upper Stretch)*				*Mayang Imphal (Middle Stretch)*				*Sugnu (Lower Stretch)*			
Seasons	1	2	3	4	1	2	3	4	1	2	3	4
Phytoplankton												
Myxophyceae	23 (5.3)	34 (10)	16 (25)	10 (2)	2 (8.6)	6 (30)	5 (26.3)	-	3 (7.1)	4 (8.7)	2 (4.2)	-
Bacillariophy-ceae	370 (85.8)	286 (84)	34 (52)	41 (85.5)	7 (30.4)	-	2 (10.5)	10 (41.7)	13 (30.9)	4 (8.7)	2 (4.2)	23 (57)
Chrypsophyceae	-	-		-		-	-	-	-	2 (4.3)	-	-
Dinophyceae	-	-	-	-		-	-	-	-	2 (4.3)	-	-
Chlorophyceae	26 (6)	17 (5)	14 (21)	40 (8.3)	13 (56.5)	12 (60)	-	14 (58.3)	7 (16.6)	10 (21.8)	-	-
Sub total	419 (97.2)	337 (99)	54 (98)	460 (95.8)	22 (95.6)	18 (90)	7 (36.8)	24 (100)	23 (54.7)	22 (47.8)	4 (8.4)	23 (57)
Zooplankton												
Rhizopoda	4 (1)	4 (1)	2 (2)	(1.05)	5	-	-	-	1 (2.3)	-	-	3 (7.5)
Zooflagilata	8 (1.8)	-	-	10 (2.1)		-	-	-		-	-	-
Ciliata	-	-	-	5 (1.05)		-	-	-		-	-	-
Rotifera	-	-	-	-	1 (4.4)	2 (10)	-	-	6 (14.2)	16 (34.8)	7 (14.6)	-
Cladocera	-	-	-	-		-	5 (26.4)	-	7 (16.6)	-	2 (4.2)	11 (28)
Copepoda	-	-	-	-		-	7 (36.8)	-	5 (12)	8 (7.4)	35 (72.8)	-
Sub total	12 (2.8)	4 (1)	2 (2)	20 (4.2)	1 (4.4)	2 (10)	12 (63.2)	-	19 (45.3)	24 (52.2)	44 (91.6)	14 (36.5)
Bacteria	-		-	-		-	-	-		-	-3	(7.5)
Total	431	341	66	480	23	20	19	24	42	46	48	40

1: winter; 2: pre monsoon; 3: monsoon; 4: post monsoon.
Number within parentheses represent the percentage composition.

Table 7.4
Seasonal distribution of periphyton (u cm^{-2}) in different stretches of the Khuga, the Thoubal and the Imphal rivers

Rivers	*Location*	*Seasons*	*Myxo-phyceae*	*Bacillario-phyceae*	*Chloro-phyceae*	*Animal group*	*Total Peri-phyton*
1	*2*	*3*	*4*	*5*	*6*	*7*	*8*
		Winter	1,347 (3.9)	23,254 (68.3)	8,671 (25.4)	763 (2.2)	34,035
Khuga River	Upper stretch	Pre M	300 (0.8)	22,950 (61.8)	13,810 (37.1)	100 (0.3)	37,160

(Contd...)

1	2	3	4	5	6	7	8
		Monsoon	1,235 (40.4)	1,755 (57.4)	65 (2.2)	-	3,055
		Post M	2,000 (6.1)	25,050 (76.5)	4,350 (13.3)	1,350 (4.1)	32,750
		Winter	217 (2.2)	5,425 (57.2)	3,829 (40.4)		9,471
	Lower stretch	Pre M	260 (2.9)	5200 (58.7)	3400 (38.4)	-	8,860
		Monsoon				-	-
		Post M	150 (1.5)	5,700 (57.6)	4,050 (40.9)	-	9,900
		Winter	275 (0.6)	27,829 (57.1)	20,634 (42.3)		48,738
Thoubal River	Upper stretch	Pre M	300 (0.6)	26,000 (48.6)	27,150 (50.8)	-	53,450
		Monsoon	-	3,414 (69.0)	1,535 (31.0)	-	4,947
		Post M	150 (0.6)	40,250 (96.1)	1,500 (3.3)	-	41,900
		Winter	232 (2.06)	8,249 (73.5)	2,731 (24.3)		11,212
	Lower stretch	Pre M	140 (1.6)	2,700 (32.1)	5,555 (66.2)	-	8,395
		Monsoon	-		-	-	-
		Post M	300 (1.8)	15,600 (95.6)	315 (19)	105 (0.7)	16,320
		Winter	1,367 (7.4)	16,725 (90.5)	372 (2.0)		18,464
Imphal River	Upper stretch	Pre M	2,884 (10.8)	21,088 (79.5)	2,160 (8.2)	378 (1.5)	26,510
		Monsoon	260 (2.9)	3,870 (43.1)	4,844 (54.0)	-	8,974
		Post M	120 (1)	11,200 (92)	830 (7)	-	12,150
		Winter	-	4,185 (83.9)	799 (16.0)		4,984
	Middle stretch	Pre M	-	4,570 (80.6)	1,100 (19.4)	-	5,670
		Monsoon	-		-	-	-
		Post M	-	3,780 (87.5)	540 (12.5)	-	4,320
		Winter	157 (1.9)	7,168 (89)	725 (9.0)		8,050
	Lower stretch	Pre M	-	7,236 (90)	804 (10)	-	8,040
		Monsoon	-	3,024 (92.3)	252 (7.7)	-	3,276
		Post M	380 (5.0)	7,160 (93.0)	80 (1.0)	40 (1.0)	7,660

Figures within parentheses indicate % composition.

Table 7.5
Species diversity and evenness indices of plankton in three rivers of Manipur

Name of the rivers	*Location*	*Number of sps.*	*Species diversity Index*	*Evenness index*
Khuga	Upper stretch	33	3.7	0.74
	Lower stretch	22	4.0	0.91
Thoubal	Upper stretch	68	4.3	0.93
	Lower stretch	32	4.1	0.83
Imphal	Upper stretch	41	4.1	0.92
	Middle stretch	18	4.0	0.96
	Lower stretch	22	4.2	0.78

Table 7.6
Periphyton species diversity index ($\overline{H}$) and evenness index (J) of three rivers of Manipur

Rivers	*Location*	*Seasons*	*Number of species*	*Total Periphyton*	*Species diversity index*	*Evenness index*
1	*2*	*3*	*4*	*5*	*6*	*7*
Khuga river	Upper stretch	Winter	33	34,035	4.0	0.78
		Pre M	35	37,160	4.0	0.79
		Monsoon	12	3,055	2.5	0.71
		Post M	41	32,750	4.2	0.8
			57	**26,750 (avg.)**	**3.6 (avg.)**	**0.77 (avg.)**
	Lower stretch	Winter	32	48,734	4.6	0.95
		Pre M	20	8,860	3.4	0.78
		Monsoon	-	-	-	-
		Post M	13	9,900	2.6	0.71
			20	**9,410 (avg.)**	**3.0**	**0.74**
Thoubal River	Upper stretch	Winter	32	48,738	4.6	0.95
		Pre M	30	53,450	4.6	0.94
		Monsoon	9	4,947	3.1	0.99
		Post M	21	41,900	4.0	0.92
			33	**37258**	**3.9**	**0.95**
	Lower stretch	Winter	20	11,212	3.0	0.72
		Pre M	19	8,395	2.9	0.70
		Monsoon	-	-	-	-
		Post M	23	16,320	3.4	0.76
			33	**11,975**	**3.2**	**0.73**
Imphal River	Upper stretch	Winter	30	18,464	3.6	0.77
		Pre M	34	26,510	3.8	0.74
		Monsoon	5	8,974	1.7	0.73
		Post M	23	12,150	3.8	0.85
			45	**16,524**	**4.2**	**0.77**

(Contd...)

1	*2*	*3*	*4*	*5*	*6*	*7*
		Winter	15	4989	3.4	0.94
	Middle	Pre M	18	5,670	4.1	0.98
	stretch	Monsoon	-	-	-	-
		Post M	8	4320	2.7	0.90
			18	**4993**	**3.4**	**0.94**
		Winter	18	8050	3.4	0.80
	Lower	Pre M	22	8040	4.2	0.76
	stretch	Monsoon	5	3,276	1.9	0.85
		Post M	27	7,660	4.1	0.87
			30	**6,756**	**3.4**	**0.83**

Table 7.7
Distribution of different species with groups of macro invertebrate benthos in the Khuga river

Seasons	*Macrobenthos*	*Upper stretch* um^{-2}	*Lower stretch* um^{-2}
Winter	Mollusca	60 (10)	41 (9)
	Trichoptera	390 (65)	126(28)
	Odonata	150 (25)	135 (30)
	Oligocheata	-	148 (33)
	Total	600	450
Pre monsoon	Mollusca	43 (3.4)	150 (17.6)
	Trichoptera	1169 (93.2)	300 (35.3)
	Odonata	43 (3.4)	150 (17.6)
	Oligocheata	-	250 (29.5)
	Total	1255	850
Monsoon	Mollusca	260 (100)	-
	Total	260	-
Post monsoon	*Coleoptera*	43 (33.3)	-
	Odonata	43 (33.3)	-
	Hemiptera	43 (33.3)	-
	Oligocheata	-	130 (100)
	Total	129	130

Figure within parentheses is the percentage composition

Table 7.8
Distribution of macro invertebrate benthos in the Thoubal river

Seasons	*Group*	*Upper stretch* um^{-2}	*Lower stretch* um^{-2}
	Mollusca	360 (45)	200 (27)
Winter	Odonata	120 (15)	-
	Trichoptera	320 (40)	90 (12)
	Diptera	-	217 (29)
	Coleoptera	-	243 (32)
	Total	800	750

(Contd...)

Seasons	*Group*	*Upper stretch um^{-2}*	*Lower stretch um^{-2}*
Pre monsoon	Mollusca	350 (31.8)	203 (28)
	Odonata	200 (18.2)	43 (4)
	Trichoptera	550 (50)	130 (12)
	Diptera	-	260 (24)
	Coleoptera	-	346 (32)
	Total	1100	1082
Monsoon	Mollusca	130 (60.2)	-
	Ephemeroptera	86 (39.8)	-
	Total	216	-
Post monsoon	*Mollusca*	130 (30.1)	*173 (100)*
	Diptera	43 (9.9)	-
	Trichoptera	43 (9.9)	-
	Megaloptera	217 (50.1)	-
	Total	433	173

Figures within parentheses indicate the percentage composition.

Table 7.9
Distribution of different species with groups of macro invertebrate benthos in the Imphal river

Seasons	*Group*	*Upper stretch um^{-2}*	*Middle stretch um^{-2}*	*Lower stretch um^{-2}*
Winter	Trichoptera	187 (75)	-	-
	Mollusca	63 (25)	42 (25)	90 (30)
	Diptera	-	102 (60)	150 (50)
	Coleoptera	-	-	60 (20)
	Odonata	-	26 (15)	-
	Total	250	170	300
Pre monsoon	Trichoptera	390 (100)	-	-
	Diptera	-	86 (67)	200 (60)
	Mollusca	-	43 (33)	150 (40)
	Total	390	130	350
Monsoon	Mollusc	85 (100)	-	86 (50)
	Coleoptera	-	-	87 (50)
	Total	85	No sample	173
Post monsoon	Ephemeroptera	130 (100)	-	-
	Odonata	-	86 (40)	43 (12.4)
	Coleoptera	-	-	217 (62.8)
	Insects	-	-	43 (12.4)
	Hemiptera	-	-	43 (12.4)
	Diptera	-	130 (60)	-
	Total	130	216	346

Figure within parentheses is the percentage composition.

Table 7.10

Phytoplankton Primary Productivity (in mg C m^{-3} day^{-1}) at upper, middle and lower stretches of Khuga, Thoubal and Imphal river

Location	*Seasons*	*Khuga*		*Thoubal*		*Imphal*	
		GPP	*Respiration*	*GPP*	*Respiration*	*GPP*	Respiration
Upper stretch	Winter	101.39	47.00	578	293	113.89	63.75
	Pre M	208.34	87.50	760	460	156.26	75.00
	Monsoon	66.67	36.25	100	120	66.67	62.50
	Post M	29.17	17.5	875	300	118.75	53.75
Middle/Lower stretch	Winter	383.3	311	90.20	56.6	550	440
	Pre M	250.0	214	166.67	125.0	800	600
	Monsoon	400.0	270	73.96	27.5	100	120
	Post M	500.0	450	30.20	17.5	750	600
Lower stretch	Winter					27.00	13.95
	Pre M					30.20	15.62
	Monsoon					21.88	8.75
	Post M					29.17	17.50
Average NPP		87 mg Cm^{-3} day^{-1},		175.7 mg Cm^{-3} day^{-1}		75 mg C m^{-3} day^{-1}	

Table 7.11

Average values of Physico-chemical parameters in upper and lower stretches of the Khuga, the Thoubal and the Imphal rivers

	Khuga		*Thoubal*		*Imphal*		
Location	*Upper*	*Lower*	*Upper*	*Lower*	*Upper*	*Middle*	*Lower*
Water temperature (^{0}C)	20.1	22.3	20.25	19.6	21.0	22.2	18.87
pH	7.4	7.3	7.4	7.4	7.5	7.4	7.2
DO (mgl^{-1})	9.3	7.3	8.1	7.0	6.4	6.7	7.1
Free CO_2 (mgl^{-1})	3.4	3.4	1.2	3.4	2.0	3.7	4.2
Total alkalinity (mgl^{-1})	64.96	65.3	75.3	74.5	36.0	62.7	47.9
Transparency (cm)	16.8	12.5	34.1	7.5	18.8	16	9.7
Specific conductivity (mhos cm^{-1})	144.1	133.3	182.8	140.9	62.0	138.16	110.1
TDS (mgl^{-1})	73.2	90.5	82.4	71.9	31.6	78.36	52
Hardness (mgl^{-1})	72.6	75.1	97.5	65.4	36.3	76.69	59.6
Calcium (mgl^{-1})	8.72	25.12	30.6	8.72	5.08	21.81	12.61
Magnesium (mgl^{-1})	12.4	24.1	28.6	10.6	5.7	26.05	19.85
Total Iron (mgl^{-1})	0.436	0.34	0.213	0.253	0.123	0.526	0.256
Chloride (mgl^{-1})	20.33	20.63	17.24	25.03	18.8	22.3	19.1
Silicate (mgl^{-1})	4	4.4	4.4	4.6	3.6	3.7	4.066
Phosphate (mgl^{-1})	0.013	0.016	0.055	0.012	0.029	0.046	0.038
Nitrate (mgl^{-1})	0.423	0.248	0.135	0.526	0.28	0.435	0.227
DOM (mgl^{-1})	1.35	1.07	1.014	1.906	1.46	1.229	1.8

Table 7.12
Seasonal variation in sediment quality of the Khuga, Thoubal and Imphal rivers

Parameters	*Khuga*				*Thoubal*				*Imphal*			
	Texture			*pH*	*Texture*			*pH*	*Texture*			*pH*
	Sand	*Silt*	*Clay*		*Sand*	*Silt*	*Clay*		*Sand*	*Silt*	*Clay*	
Upper stretch												
Winter	87.9	5.1	6.0	6.6	99.3	0.3	0.4	6.6	97	1	2	6.5
Pre-M	89.0	5.0	6.0	6.4	100	0	0	6.5	98	1	1	6.5
Monsoon	86.0	5.5	8.5	7.0	98	1	1	7.1	97	1	2	6.5
Post-M	88.7	5.0	6.2	6.5	100	0	0	6.4	98	0	2	6.5
Middle/Lower stretch												
Winter	93.4	5.5	1.1	6.3	91.5	3	5.5	6.5	91.1	8.0	0.9	6.6
Pre-M	94.5	5.0	0.5	6.2	91.5	3.0	5.5	6.5	91	8.0	1	6.5
Monsoon	91.0	6.5	2.5	6.6	92	2.5	5.5	6.52	90	8.5	1.5	7.0
Post-M	94.9	5.1	0	6.1	91	3.5	5.5	6.5	92	7.7	0.3	6.3
Lower stretch												
Winter									91.5	3.3	5.2	6.6
Pre-M									91.7	3.2	5.0	6.4
Monsoon									90.5	3.5	6.0	6.7
Post-M									92.5	3.2	4.25	6.5

Sediments

Bottom sediment of a lotic ecosystem is highly heterogeneous system, consisting of both inorganic and organic material; it has numerous direct and indirect effects on the biota of running waters. It provides a surface to cling to or burrow in, shelter from current, material for construction of cases and tubes, and refuge from predators. Though the waters of the river are always moving still the water flowing above is governed to some extent by the existing bottom soil quality, as soil characteristic will be reflected more or less on the quality of the water remaining in contact with the soil. Coarse particle including gravel and boulders are typical of upland stream, while a finer and softer substrate often is found in larger lowland rivers. Their diverse geography provides for almost unlimited variation in the biota.

Seasonal variation in sediment quality in different stretches of the Kuga, Thoubal and Imphal rivers are depicted in Table 7.12. Overall pH in the three rivers of Manipur ranged between acidic to neutral (6.21-7.1). Pathak *et al.* (2001) also reported similar value from the different stretches of river Mahanadi (6.7-7.2). But the same author found a range value of 7.0-7.5 from the tributaries of river Brahmaputra. Soil texture in all the rivers studied was dominated by sand irrespective of seasons and topographical variation. Generally, river sediments are dominated by sand as it is mainly derived from the sediments brought from the headwaters by the weathering of rocks. In the present finding sand percentage fluctuated between 86-100% but Singh *op. cit.* reported much lower value of sand percentage (70-91%) from the different stretches of river Ghagrha. However, Pathak *op. cit.* reported 96-99% sand from the different stretches of Mahanadi. Silt and clay percentage fluctuated between 0-8.5% in these rivers. Maximum silt was observed in middle stretch of Imphal river and maximum clay was seen in upstream of Khuga river. Increase in silt and clay composition and decrease in sand percentage in downstream of Imphal and Thoubal rivers were clearly seen. Whereas, in Khuga river there was increased in sand percentage in downstream. In the rivers, percentage of clay showed the maximum value and silt percentage was also high. Increase silt and clay values at downstream were seen by Singh *op.cit.* in Ghagrha river. The same author reported a much higher value of clay and silt (5-19% and 4-13% respectively) from the same river. Whereas Pathak *op.cit.* reported much lower range of silt (0.5-2.5%) and clay (0.5-2.0%) from the different stretches of Mahanadi river.

Bottom sediments of the three rivers showed more inclination towards acidic character (pH: 6.21-7.1). pH values were slightly increased during monsoon in all the rivers. Neutral pH values were recorded only during monsoon seasons at upstream of Khuga and Thoubal rivers. Whereas for Imphal river neutral value was recorded at middle stretch during monsoon. Soil textures of the three rivers were sandy in character and effect of seasonal change was low. However, decrease in percentage of sand could be seen during monsoon in almost all the sites. Contribution of silt and clay were minimum, which reflected true riverine characteristic.

[**Acknowledgement:** We are grateful to the Indian Council of Agricultural Research, New Delhi for financial assistance under A. P. Cess Fund project on Ecology and Fisheries of the rivers of NE states and to the Director, CIFRI (ICAR), Barrackpore for providing necessary facilities.]

REFERNCES

Acharjee, B. 1997. Ecological studies and productive potential of some beels in lower Brahmaputra basin, Assam, Unpublished Ph.D. Thesis, Gauhati University, Guwahati

Acharjee, B. Choudhury, M. and Dutta, A. 1995. Phytoplankton species diversity indices in Dighali Beel, Assam, India. *Env. and Eco.*, **13** (3): 660-662

Ansari, S. A. 1973. Economics Geography of Manipur, *Trio Book Store, Manipur.*

APHA. 1989. Standard methods for the examination of water and wastewater. *American Public health Association.* Washington D. C.

Arora, H. C. 1966. Response of rotifera to variation in some ecological factors. *Proc. Indian Acad. Sci.,* (B) 63 (2):57-66.

Asif A. Khan, Aftab Alam and Rajeev K. Gaur. 1998. Seasonal variation in the abundance and composition of phytoplankton in the river Ganges at Narora, U. P. *J.Inland Fish. Soc. India,* 30 (1): 79-86.

Banerjee, R. K., Mukhopadhyay, M. K., Pandit, P. K., Karmakar, H. C., Mitra, K, Chatterjee, S. K., Das, B. B. and Sengupta, A. 1998. River Damodar and its environment. Bull. No. 79. Central Inland Capture Fisheries Research Institute, Barrackpore, India.

Cairns, J. Jr. and Dickson, K. L. 1971. A simple method for the biological assessment of the effects of waste discharge on the aquatic bottom dwelling organisms, *J. Wat. Pollut. Control Fed.* 43: 755-772.

Desikachary, T. V. 1959. ICAR Monograph on Algae, *ICAR*, New Delhi. 532 p.

Edmonson, W. T. 1959. *Freshwater Biology*, 2nd Edition. John Wiley and Sons. Inc New York, 1248 p.

Fritsch F. E. 1935. *The Structure and Reproduction of the Algae*, Vol. I. Cambridge University Press.

Garder, T and Gran, H. H. 1927. Investigation on the production of plankton in Oslo F. Jord. *Rapp. P, V. Reun. Cons. Perm. Int. Explor. Mer.*, 42: 1-48 (1 and 2) 1985.

Garg, J. K., Singh, T.S. and Murthy, T. V. R. 1998. *Wetlands of India. Project Report of Nation-wide Wetland Mapping Project*, sponsored by the Ministry of Environment and Forest, Govt. of India. ISRO, Ahmedabad. India. 239 pp.

Gupta, B. P. and Singh, S. N. 1985. An interesting observation on a monospecific bloom of a diatom *Attheya zachariasi* brun from a man made lake. *J. Inland Fish. Soc. India*. 17(1and 2):84-86.

Gurumayum, S. D., P. Daimari and M. Choudhury. 2000. Abundance and quality composition of periphyton communities in hill stream of Arunachal Pradesh, *J. Environ. Res.*, 10 (2): 113-116.

Hutchinson, G. E. 1957. *A treatise on limnology*. John Wiley and Sons Inc. New York. Vol. I.

Hynes, H. B . N. 1970. *The Ecology of Running Water*, University of Toronto Press.

Iyengar, M. O.P. and Desikachary, T. V.1981.Volvocales, *ICAR Monograph on Algae*, ICAR, New Delhi: 532 p.

Jhingran, A. G. and Pathak, V. 1988. Impact of man-induced environmental modification on productivity potential and energy dynamics of river Ganga. *J. Inland Fish. Soc. India*, **20 (2)**: 43-53.

Jhingran, V. G, Natarajan, A. V., Banerjee and David, A. 1969. Methodology on reservoir Fisheries: Investigation in India. *Bull. Cent. Inland Fish. Res. Inst.*, Barrackpore, India, **1**: 109 p.

Kawosa, M. S. 2001. Occurrence of periphyton on artificial substrate in a rural lake, Kashmir. *Geobios*, 28 (2 and 3): 137 –139.

Khan, M. A., Panwar, R. S., Mathur, A. and Rekha Jetly, 1998. Studies on plankton communities and primary production in middle stretch of river Ganga in context of cultural eutrophication. *J. Inland Fish. Soc.* India, 30 (10): 47-68.

Kosygin, L. and Dhamendra, H. 2009. Physico-Chemical and Biological Characteristics of Awang Khujairok River, Manipur, India. *Env. & Ecol.*, 27 (1):82-85, 2009.

Kosygin, L. and Dhamendra, H. 2010. Fish Diversity and Physico-Chemical Characteristics of the Upstream of Imphal River, Manipur, India. *Env., & Eco.*, 28 (2B):1328-1332, 2010

Kottelat, M. and Whitten, T., 1996. *Freshwater Biodiversity in Asia with Special Reference to Fish*. World Bank, Washington, DC, pp. 17-22.

Laal, A. K., Sarkar, S. K. and Sarkar. A. 1982. Side off – shoot sewage fed channel of river Ganga as tropical natural waste stabilization system. *J. Inland Fish. Soc. India*, 14 (2): 73-79.

Mackenthum, K. M. 1966. Biological evaluation of polluted streams, *J. Wat. Pollut. Control Fed.* 38: 241-47.

Manivasakam, N. 1985. *Physico-chemical examination of water, sewage and industrial effluents*. Pragati Prakashan, Meerut (U. P.). 275 pp.

Marlier, G. 1962. General des Trichopteres de l'Afrique. Muses Royal de l' Afrique centrate Tervuren, Belgique. *Ann. Sci. Zool.* 109:1-263.

Mishra , S. R. and Saksena, D. N. 1992. The primary productivity of phytoplankton in a sewage collecting morar (Kalpi) river at Jaderųa Bundha, Gwalior, Madhya Pradesh. *J. Inland Fish. Soc. India*, 24 (1): 61-68.

Nath, D. 2001. Water and soil characteristics of the Narmada estuary before commissioning of Sardar Sarobar dam. *J. Inland Fish. Soc. India*, 33 (2): 37-41.

Needham, G. J. and Needham, P. R. 1966: *A guide to freshwater Biology*, 5th Edn. Holden Day Inc. San Fransisco, 108 p.

Odum, R. P. 1971. *Fundamental of ecology*. Third edition. B. N. B. Sauders Co, Philadelphia. 178p.

Pathak, V. 1999. Energy dynamics in the management of riverine system. Summer School on Ecology, Fisheries and Fish stock assessment in Indian river. CIFRI, Barrackpore, 14 July to 12 August 1999. Pp 29-34.

Pathak, V., Mahavar, L. R. and Sarkar, A. 2001. Ecological status and production dynamics of a stretch of river Mahanadi. *J. Inland Fish. Soc. India*, 33(1):25-31.

Pathak, V., Sarkar, A, Mahavar, L. R and Bhattacharijya, B. K. 2001. Ecological status and fish production potential of Siang, Dibang and Lohit-the three forerunner of river Brahmaputra. *J. Inland fish. Soc. India*. 33 (2):23-28.

Patrick Ruth. 1970 Benthic stream communities. *Am. Scient.* 58:546-49.

Philipose, M. T. 1967. Chlorococcales *ICAR Monograph on Algae*, ICAR, New Delhi. 365 p.

Ramanathan, K. R. 1964. Ulotricals. *ICAR Monograph on Algae*, ICAR, and New Delhi: 188 p.

Robinson, C. T., Uchlienger, U and Hieber, M. 2001: Spatiotemporal variation in macroinvertebrate assemblages of glacial stream in Swiss Alps. *Freshwater boil.*, 46: 1663-1672.

Shannon, C. E. and Weaver, W. 1963. *The mathematical theory of communication*. Urban Univ. Illionis. U S.A.

Shrivastava, N. P., Desai, V. R. 1985. Studies on periphyton in a hydal impoundment, Rihand (UP). *J. Inland Fish. Soc. India*, 17 (1 and 2): 35-38.

Singh, H.T. (1987). Impact of Loktak Hydro-Electric project and current land use pattern on the riverine fishery of Manipur. In: *Symposium on the impact of current land use pattern and water resources Development on riverine fisheries*, Barrackpore, 27-27.

Singh, H. P. and Mahaver, L. R. 1997. Preliminary observation on soil sediments of river Ghaghra-a tributaries of river Ganga. *J. Inland Soc. India* , 30 (2), 74-78.

Sugunan, V. V. 1991. Changes in the phytoplankton species diversity indices due to artificial impoundment in river Krishna at Nagarjuna sagar. *J. Inland Fish. Soc. India*, 23 (1): 64-74.

Sugunan, V.V. and Pathak, V. 1986. Temporal and spatial distribution of periphyton in Nagarjunasagar reservoir, Andhra pradesh, India and new method for periphyton collection from Indian reservoirs. *J. Inland Fish Soc., India*, 18(2): 20-23.

Sukumaran, P. K., Shankuntala, K and Parameswaran, S. 1996. Production of periphyton in a lentic habitat in Banglore. *J. Inland Fish. Soc. India*, 28 (1): 28-32.

Swanson, C. O. and Bachman, R. W. 1976. A model of algal export in some Iowa streams. *Ecology*, 57: 1076-80.

Thienemann, A. 1925. DER-See-als Levenainheri. *Naturwissanschauftem*, 27: 89-660. (Not consult in original).

Trisal C. L. and Th. Manihar. 2004. *Loktak-The atlas of Loktak*. Wetlands Int. and Loktak development Auth., New Delhi, India.

Vollenweider, R. A. 1969. A manual of methods for measuring primary production in Aquatic Environments. *IBP Handbook*. No. 12. Blackwell Scientific Publ. Oxford, 213p

Welch, P. S. 1948, *Limnological methods*, Mc Graw Hills book Co. Inc., New York. 471 p.

Welcher, F. J. 1989, *Standard methods of chemical analysis* D. Van Nostrand Co. Inc., New York, 1283-2613.

8

Preliminary Study on Parasitic Protozoan Diversity of Fishes of Koladyne River, Mizoram

—*Th. Hemananda*[1] & *N. Mohilal*[2]

[Freshwater biodiversity conservation is considered to be one of the priority agenda during the International Decade for Action on "Water for Life". Ichthyoparasites are also considered one of the important part of aquatic biodiversity. Investigation on Ichthyoparasites during 2008 – 09 at Koladyne River near Kawlchaw Bridge, Kawlchaw, Lawntlai Dist. Mizoram revealed the presence of protozoan parasites, which belong to three genera viz, *Myxobolus, Trichodina* and *Tripartiella.* The first one belongs to Phylum Myxozoa and the later two belongs to Phylum Ciliophora. This is the first record of protozoan parasites from Koladyne River, Mizoram. Detailed morphological descriptions are provided.

Key words: Ichthyoparasites, Koladyne River, *Myxobolus, Trichodina, Tripartiella.*]

Introduction

Northeast India comprises of the eight sister states viz, Arunachal Pardesh, Assam, Manipur, Meghalaya, Mizoram, Nagaland and Tripura and Sikkim. The region belongs to two of the 34 biodiversity hotspots listed by conservation International viz, the Himalaya and Indo-Burma (Roach, 2005). The rich fish diversity of the region is attributed to many reasons, viz., the geomorphology, consisting of hills, plateaus and valleys, resulting in the occurrence of a variety of torrential hill streams, rivers, lakes and swamps; the

Figure 1. A view of Koladyne River, Mizoram

[1,2]Department of Life Science, Manipur University, Canchipur 795 003, Manipur
Email:mohilaln@yahoo.co.in; hemanandat53@gamil.com

drainage pattern which include the Ganga-Brahmaputra, Koladyne and Chindwin-Irrawaddy systems. Mizoram belongs to the Indo-Burma hotspot. Koladyne River is one of the bigest rivers flowing through Mizoram. It originates from Western face of Chin Hills in Myanmar bordering Mizoram and Manipur (India), flows northward along the IndoMyanmar border in eastern side of Mizoram state, then turns south to flow back into Myanmar and finally drains into Bay of Bengal. Koladyne drainage is located between Ganga-Brahmaputra and Chindwin-Irrawaddy drainage. This drainage is separated from the Barak-Brahmaputra basin in India by the Chittagong hill tract and from the Chindwin-Irrawaddy basin in Myanmar by the north-south extension of Arakan Yoma hill range. As the river is isolated from both drainages, the faunal diversity of Koladyne river is of special interest and many new findings are expected.

Freshwater biodiversity is the over-riding conservation priority during the International Decade for Action-'Water for Life'- 2005 to 2015. Inland waters and freshwater biodiversity constitute a valuable natural resource, in economic, cultural, aesthetic, scientific and educational terms. Their conservation and management are critical to the interests of all humans, nations and governments. Yet this precious heritage is in crisis. Fresh waters are experiencing declines in biodiversity far greater than those in the most affected terrestrial ecosystems, and if trends in human demands for water remain unaltered and species losses continue at current rates, the opportunity to conserve much of the remaining biodiversity in the freshwater will vanish before the 'Water for Life' decade ends in 2015 (Dudgeon et al., 2006). Fishes, one of the most important part of freshwater biodiversity, are the most diverse vertebrate taxa and are distributed in a range of aquatic environment. Of the total recorded species of fish, freshwater fishes constitute over 40%. But more than 20% of the freshwater fish species are at risk of extinction, making them the world's most threatened vertebrates after amphibians (Moyle and Leidy, 1992; Bruton, 1995). As a consequence, the global freshwater biodiversity is declining more rapidly than the terrestrial biodiversity (Ricciardi and Rasmussen, 1999). The main threats to freshwater fishes are due to habitat alteration, pollution, overexploitation and introduction of non-native species causing their global decline.

Besides parasites also cause immense losses and their role in the decline of healthy fishes has increased with the global warming processes. In this light ichthyoparasitological Research work particularly on protozoan parasites has become important. A study was carried out for fish protozoan parasites during 2008 – 09 in Koladyne River in order to explore the available protozoan parasite species. The available species are described in accordance with Lom and Arthur (1989); Lom and Dyková (1992); Lom (1963); Wellborn (1967) and Arthur and Lom (1984).

Materials and Methods

Fishes of 10 – 15 cm size were collected from local fishermen inhibiting near the bank of Koladyne River near Kawlchaw Bridge. Since immediate transfer of host fishes to the laboratory was impossible, gills and body smear were made on clean, grease free slides and air dried at the spot of collection. Such slides were carefully packed in slide boxes and brought to the laboratory for examination. Airdried myxozoan positive slides were stained with Giemsa after fixation in acetone free absolute methanol. Slides with trichodinid ciliates were impregnated with Klein's dry sliver impregnation technique (Klein, 1958). Measurements were done with the aid of a calibrated ocular micrometer. Photomicrographs were made with the help of CCD camera fitted to Magnus Phase-contrast microscope.

Results

The result reveals the occurrence of three protozoan species belonging to the genera *Myxobolus, Trichodina* and *Tripertiella.*

***Myxobolus* sp.**

The genus *Myxobolus* was first established by Bütschli (1882). The diagnostic characters of this genus are -. spores ellipsoidal, ovoid or rounded in calvular view, and biconvex in sutural view, shell valves smooth. Presence of two mostly pyriform polar capsules. Posterioly, the sutural ridge may extend into crescentic ridge. Binucleate sporoplasm, often with iodinophilous vacuole presente. Trophozoites as a rule large, polysporic with pansporoblast formation; histozoic in fresh water fishes, some in marine fishes rarely in amphibians. Detailed description of the *Myxobolus* sp obtained during the present survey is given below.

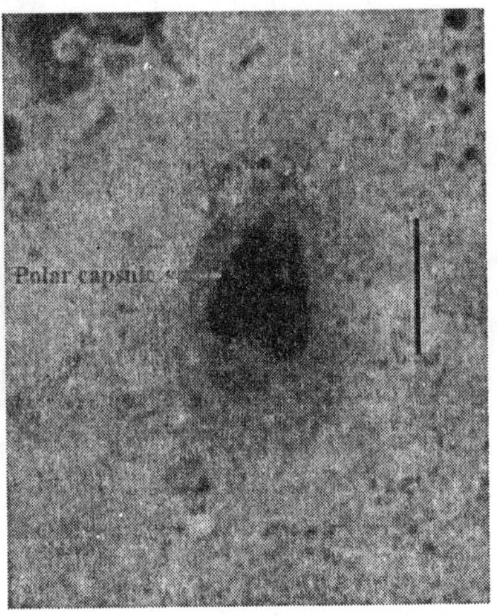

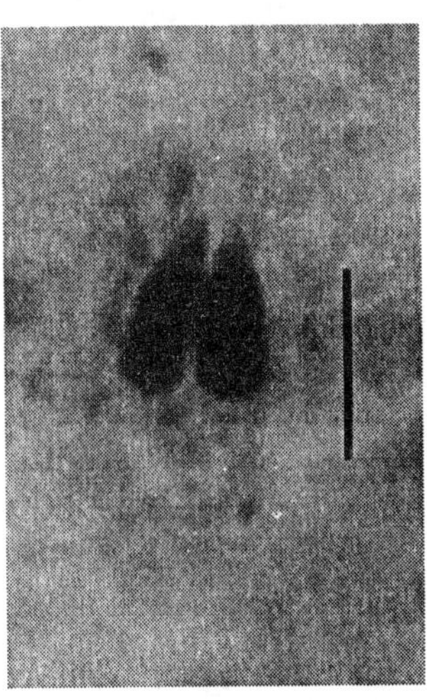

Photograph of *Myxobolus* sp. Bar-8.36μm

Cyst: Not found. Vegetative stages are not observed.

Spore: Mature spores are pyriform shaped in valvular view with narrow, bluntly pointed anterior end and rounded posterior end. The widest part is at the middle. Spore measure 12.26 – 14.9 μm (13.3 ± 0.68) in length and 7 – 8.4 μm (7.7 ± 0.49) in breadth. The two shell valves are thick, smooth and uniform with a thick straight sutural line with no ridge. No parietal fold and mucus envelope is seen.

Two anteriorly placed polar capsules are equal, large and pyriform with broadly rounded posterior and pointed anterior end. Each measure 7.7 – 9.1 μm (8.36 ± 0.41) in length and 2.8 – 3.5 μm (2.97 ± 0.30) in breadth. The two polar capsule are slightly convergent at the anterior part but never cross. Inside the polar capsule, the polar filament makes 7 – 12 spiral coils. A tubular like structure is present at the anterior tip as a common opening. The two polar capsules are so thick that they occupy nearly $^2/_3$ of the spore cavity.

The extra-capsular cavity is occupied by scanty sporoplasm filled with homogenous mass of sporoplasmic granule. Inside the sporoplasm two big sporoplasmic nuclei are observed. Each measure 0.7 – 1.4 μm (1.15 ± 0.37) in diameter. No iodinophilous vacuole is observed.

This *Myxobolus* species is obtained from the gill of *Puntius* sp; *Glyptothorax* sp and *Labeo* sp.

***Trichodina* sp.**

The genus *Trichodina* was first established by Ehrenberg (1838). The diagnostic characters of this genus are - the denticulate ring composed of denticles with straight or curved blades, distinct rays of various shapes and lengths, central parts lacking. The adoral spiral describes an arc of 330-450° (Raabe, 1963; Bychovskaya – Pavlovskaya *et al*, 1964). Detailed description of the *Trichodina* sp obtained during the present survey is given below.

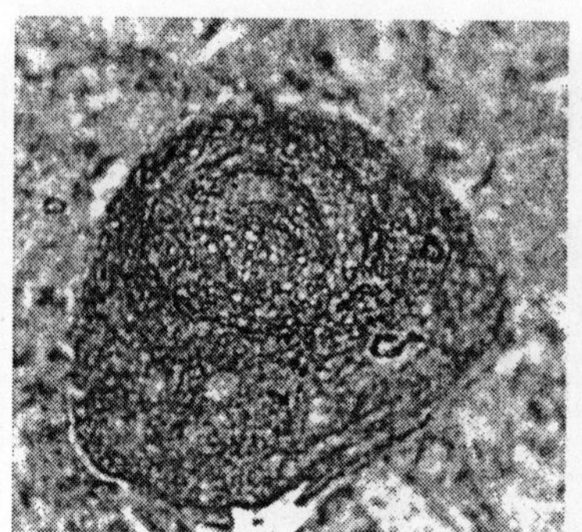

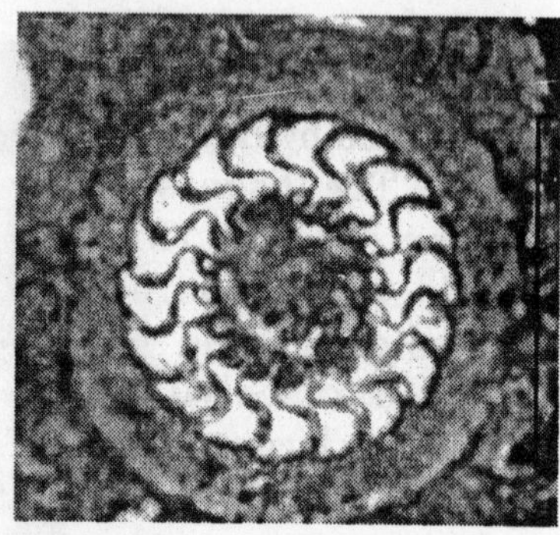

Photomicrograph of *Trichodina* sp. Bar-38.4μm

Descriptions: Medium sized trichodinid with flattened, disc-shaped body. Adhesive disc concave and surrounded by a finely striated border membrane. Central area of adhesive disc bears a clear, sometimes diffused, in which almost devoid of any argentophilic particles and surrounded by a circular perimeter. Denticulate ring consists of 18 – 22 denticles with 6 – 7 radial pins per denticle.

Table 8.1
Morphometry of 10 specimens of *Trichodina* sp. obtained in the present study

Characters	*Range*	*Mean*	*SD*	*SE*	*CV (%)*
1	*2*	*3*	*4*	*5*	*6*
Diameter of body	46.5 – 50	48.25	1.123	0.355	2.327
Diameter of adhesive disc	35.5 – 40	38.4	1.445	0.457	3.763
Dimension of body .....denticulate ring	23.5 – 28	25.85	1.643	0.519	6.355
Central area	9.5 – 11.5	10.6	0.624	0.197	5.886

(Contd...)

1	*2*	*3*	*4*	*5*	*6*
Width of the border membrane	4.5 – 5.5	5.05	0.415	0.092	8.217
Number of denticles	18 – 20	18.7	0.9	0.284	4.812
Number of radial pins/denticle	6 – 7	6.3	0.458	0.144	7.269
Dimension of denticle ——span	11.5 – 13.5	12.55	0.610	0.193	4.860
................length	4.5 – 6	5.35	0.55	0.122	10.280
Dimension of denticle components— length of the ray	4.6 – 6	5.3	0.509	0.161	9.603
length of the blade	3.5 – 4.5	3.95	0.35	0.110	8.860
width of the central part	2.5 – 3.5	3	0.447	0.141	14.9
Adoral ciliary spial	390^0 – 400^0				

Blade of denticle broad, filling most of space between y and y + 1 axes. Distal margin of blade flat, running parallel to border membrane. Tangent point slightly lower than distal margin. Anterior margin forms angle with y axes. Posterior margin indentation forms small semilunar arch with y axes. Deepest point of semilunar arch lies at same level as apex. Apex conical, rarely extends beyond y axes. Apical depression well developed, never impregnates with silver. Anterior apophysis of blade prominent, notch appears in between apex and anterior apophysis. Blade connection broad, not clearly distinguishable from blade. Posterior blade apophysis not clearly visible. Central part robust, broadly triangular with rounded tip. Tip extends more than halfway past, sometimes touches y-1 axis. Section of central part above x-axis sloping at steep angle. Indentation in lower central part absent. Connection between ray and central part thin. Ray slightly curved in posterior direction. Ray apophysis not prominent, but central groove clearly visible. The adoral ciliary spiral makes 390^0 - 400^0.

This *Trichodina* species is obtained from *Labeo* sp.

***Tripartiella* sp.** *Tripartiella*

The genus was first established by Lom (1959). The diagnostic chatacters of this genus are - the denticles consist of blades, rays and central parts; blades are stunted more or less obliquely backwards; their narrow bases by which they join central parts of denticles extend anteriorly in projections which may be short and thin or wide and knee like. These projections fit well into the corresponding notch in preceding denticle. Thus the denticles are wedged together

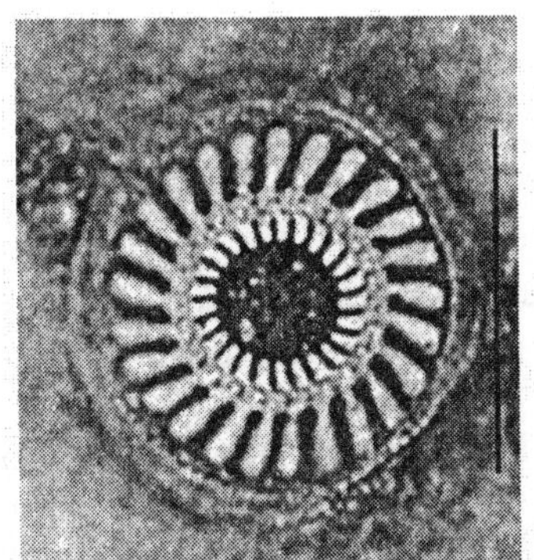

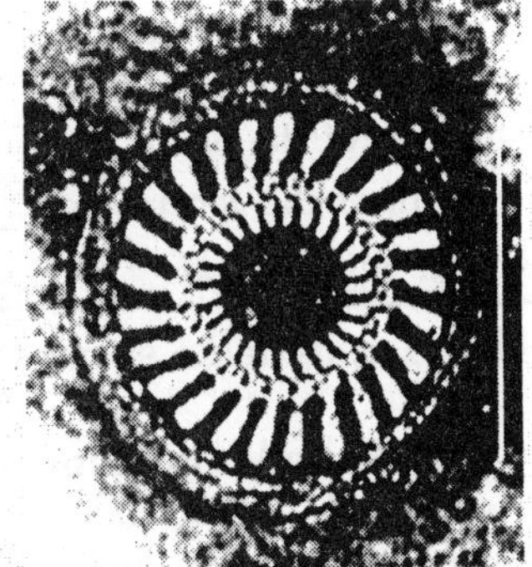

Photomicrograph of *Tripartiella* sp. Bar-19.4μm

both by central conical parts and anterior projections. The adoral spiral of cilia makes turn of 180-270 degrees (Lom, 1959; Lom and Haldar, 1977). Detailed description of the species obtained during the present survey is given below.

Descriptions

Body disc shaped. Adhesive discs dark. Uniquely shaped denticles. The blade is elongated, slanted obliquely backwards with parallel lateral margins, which are constricted at either end, but a sort of constriction is prominent at the mid length of blade. The distal margins of the blades are almost rounded. Tangent point is flat and is situated lower than the distal margin. Anterior projection of the blade is reduced to an insignificant tubercle. Well developed posterior notch is present. The section connecting anterior projection and central part is well – formed and curved towards the central area. The central part of the denticle is short, knee like and bent. The ray is short, stumpy and directed towards the centre of adhesive disc. Horse shoe shaped macronucleus observed. The adoral ciliary spiral makes a turn of 180^0-200^0.

Table 8.2

Morphometry of 10 specimens of *Tripartiella* sp. obtained in the present study

Characters	*Range*	*Mean*	*SD*	*SE*	*CV (%)*
Diameter of body	16.5 – 24	22.09	2.170	0.686	9.823
Diameter of adhesive disc	17 – 21.5	19.4	1.428	0.451	7.360
Dimension of body .....denticulate ring	8.5 – 11.5	10.4	0.943	0.298	9.067
Central area	3.5 – 7	4.35	1.001	0.316	23.011
Width of the border membrane	1.5 – 2	1.85	0.229	0.072	12.378
Number of denticles	20 – 25	22.4	1.854	0.586	8.276
Number of radial pins/denticle	5 – 6	5.6	0.489	0.154	8.7
Dimension of denticle ——span	4.5 – 5.5	4.64	0.535	0.169	11.530
................length	2 – 3	2.5	0.387	0.122	15.48
Dimension of denticle components— length of the ray	0.5 – 1.5	0.95	0.35	0.110	36.842
length of the blade	3.5 – 4	3.7	0.244	0.077	6.594
width of the central part	.5 – 1.5	0.95	0.35	0.110	36.842
Adoral ciliary spial	190^0 - 200^0				

This *Tripartiella* species was obtained from *Labeo* sp.

[**Acknowledgement:** The authors express heartiest greatfullness to Prof. W. Vishwanath Singh, Life Sciences Department, Manipur University for his kind helps during the survey work. The authors also acknowledge Dr. K. Nebeshwar Sharma, Dr. A. Darshan Singh and Geetakumari, Kh for their kind helps in identifying the host fishes.]

REFERENCES

Arthur, J. R. and Lom, J. 1984. Trichodinid Protozoa (Ciliophora: Peritrichida) from freshwater fishes of Rybinsk Reservoir, USSR. *J. Protozool.* 31: 82 – 91.

Bruton, M. N. 1995. Have fishes had their chips? The dilemma of threatened fishes. *Environmental Biology of Fishes*, 43: 1-27.

Bütschli, O. 1882. Myxosporidia. In Bronn's *Klass Ordn., des Tierreiches, Protozoa.* 1: 590 – 603.

Bykhovskaya-Pavlovskaya, I. E., A. V., Dubinina, M. N., Izyumova, N. A., Smirnova, T. S., Sokolovskaya, I. L., stein, G. A. Shulman, S. S. and Epstein, V. M. 1964. Key to parasites of freshwater fish of the U.S.S.R. Jerusalem: Isreael Program for Scientific Trnslation, IPST Cat. No. 1136, 919 pp.

Dudgeon, D., Arthington, A.H., Mark, O. G., Zen-Ichiro, K., Duncan, J. K., Christian, L., Robert, J. N., Anne-Helene, P. R., Doris, S., Melanie, L. J. S. and Caroline, A. S.. 2006. Freshwater biodiversity: importance, threats, status and conservation challenges. *Biol. Rev.* 81: 163-182.

Ehrenberg, C. G. 1838. Die Infusionsthierchen als vollkom nene Organismen. Ein Blick in das tiefere organische Leben der Natur, Leipzig: Leopold Voss. 547 pp.

Klein, B. M. 1958. The dry silver method and its proper use. *J. Protozool.* 5: 99 – 103.

Lom, J. 1959. On the systematic of the genus *Trichodinella* Srámek-Hušek (=Brachyspira Raabe). *Acta Parasitol. Polon.* 7: 573 – 590.

Lom, J. 1963. On the system of the family Urceolariidae Stein. *Proc. 1st. Int. Cong. Protozool., Prague, Aug. 1961. Progress in Protozoology.* pp. 516 – 518.

Lom, J. and Arthur, J. R. 1989. A guideline for the preparation of species descriptions in Myxosporea. *J.Fish Dis.* 12: 151 – 156.

Lom, J. and Dyková, I. 1992. Protozoan Parasites of Fishes. *Development in aquaculture and fisheries science* 26, Elsevier, Amsterdam.

Lom, J. and Haldar, D. P. 1977. Ciliates of the genera *Trichodinella, Tripartiella* and *Paratrichodina* (Peritricha, Mobilina) invading fish gills. *Folia Parasitol.* 24: 193 – 210.

Moyle, P. B. and Leidy, R.A. 1992. Loss of biodiversity in ecosystems: evidence from fish faunas: In P. L. Fiedler and S. K. Jain, (editors), *Conservation biology: the theory and practice of nature conservation, preservation and management.* Chapman and Hall, New york. 1245-1262.

Raabe, Z. 1963. Systematics of the family Urceolariidae Dujardin, 1841. *Acta Protozool.* 1: 121 – 138.

Ricciardi. A. and Rasmussen, J. B. 1999: Extinction rates of North American freshwater fauna. *Conservation Biology.* 13: 1220-1222.

Roach, J. 2005. Conservationists name nine new biodiversity hotspots. *National Georaphic News*, February 2, 2005.

Wellborn, T. L. Jr. 1967. *Trichodina* (Ciliata: Urceolariidae) of freshwater fishes of the southern United State. *J. Protozool.* 14: 399–412.

9

Study on Certain Ecological Aspects on Namsang Stream, Arunachal Pradesh

—Rajdeep Dutta[1] & Amalesh Dutta[2]

[The river Namsang, also known as "Chathju" originates from Jankhrum (1211 metres m.s.l) in Eastern part of Tirap district, Arunachal Pradesh. The river traverses a meander course of C.50 km in Tirap district of Arunachal Pradesh before it debouches into the plains near Upper Dihing Reserve Forest (120 metres m.s.l) to join the river Buri Dihing in Assam. In the present study, physico-chemical regime of Namsang stream was investigated for two years in four different seasons viz., pre-monsoon (March- May), monsoon (June- September), retreating monsoon (October and November) and winter (December- February) during 2007-09. River water with fast water current showed temperature (16.5-24.5^{0}C), TDS (57-105 mg/liter), pH (6.25-7.0), Alkalinity (32-55 mg/liter), Hardness (40-64 mg/liter), DO (8.80-16.64 mg/liter), Calcium(as $CaCO_3$) (22-29 mg/liter), Magnesium (as $CaCO_3$)(14-40 mg/liter), Chloride (as Cl)(5-18.5 mg/liter), Silicate (as SiO_2) (6.2-9 mg/liter),Iron (as Fe) (0.24-7.52mg/liter), Nitrate (as NO_3) (0.35-0.84 mg/liter), Phosphate (as P) (0.05-0.21mg/liter), Free carbon dioxide (4.27-7.21 mg/liter, Dissolved Organic Matter (17 to 23.5 mg/liter)) and Specific conductivity (21.09-90 micro mhos/cm). The physico-chemical characteristics of the Namsang stream were in general conducive for aquatic habitat. The investigation also revealed unpolluted status of the stream.

Key words: Physico-chemical parameters, Namsang stream, Arunachal Pradesh.]

Introduction

Arunachal Pradesh is very rich in its riverine resources. The state has 2000km of rivers and streams suitable for fish culture of which only 5% (about 100km) is utilized for fish culture. The present average productivity of these rivers is 50-400kg/km. (Nath & Dey, 2000).

The river Namsang, also known as "Chathju" originates from Jankhrum (1211 metres msl) in Eastern part of Tirap district, Arunachal Pradesh at a longitude of $95^0 39'$ E and latitude of 27°6′N. The river traverses a meander course of C. 50km in Tirap district of Arunachal Pradesh before it debouches into the plains near Upper Dihing Reserve Forest (120 metres msl) to join the river Buri Dihing in Assam.

1,2Department of Zoology, Gauhati University, Guwahati–781 014, Assam
Email:rajdeep18@rediffmail.com

From the point of origin Namsang river starts flowing towards east west direction. After covering a distance of about 18km it takes a northwardly turn at a longitude of $95^0 30/$ E to finally meet the river Buridihing after traversing another 32km distance.

Materials and Methods

The present study was conducted for a period of 2 years from 2007-2009. Sediment sample was collected, dried, powdered and sieved through a 20mm sieve and used for physical and chemical analysis following Jhingran *et al.* (1969). Water samples were collected from one station of the river (Longitude 95° 38′ E, Latitude 27° 14′ N, Altitude 151 metres m.s.l) during 4 different seasons viz. pre-monsoon (March to May), monsoon (June to September), retreating monsoon (October - November) and winter (December - February). Physico chemical parameters such as water temperature, pH, free carbon dioxide and dissolved oxygen were measured and recorded at the spot. For laboratory analysis the water samples were collected in plastic bottles with specific precaution and carried to the laboratory where other water quality parameters viz. alkalinity, total dissolved solids, total hardness, calcium, chloride, magnesium, silicate, total iron, nitrate, phosphate and dissolved organic matter were analysed as per the standard methods (APHA, 1989).

Results and Discussion

Sediment and water quality parameters of Namsang stream have been presented in Table 9.1 and 9.2.

Sediment quality

Physical composition of sediment reflected dominance of sand contributing 87%. Higher sand percentage of the present study corroborated with that of Pathak *et al.* (2001a) from the three headwaters of the river Brahmaputra (Siang, Dibang and Lohit – 97-99.5 %). Choudhury (2002) also reported sandy bottom in the rivers of Arunachal Pradesh. Sediment pH found to be neutral in nature (pH 7.07). Choudhury (2002) reported a pH range of (6.0-7.5) in the river soil of Arunachal Pradesh. Organic carbon content, available N, P_2O_5 and K_2O were found to be low in the sediment of Namsang stream. Low nutrient values were also reported by Pathak *et al.* (2001a) in Siang, Dibang and Lohit rivers of Arunachal Pradesh. Skeletal part found to be gravely (54.16 %). These are rounded and surrounded fragments (all kinds of rocks) less than 3 inches diameter called gravel. The gravels are formed due to bruising of stone fragments while transported by water from upper ridges.

Water quality

Water Temperature: Water temperature ranged between 16.5 and 24.5 ^{0}C, minimum during winter and maximum in the monsoon indicating clear-cut seasonal variations. Daimari *et al.* (2005) has recorded a temperature range of 7.0 to 24.0 ^{0}C of the Subansiri river of Arunachal Pradesh.

pH: pH of the river water fluctuated in the range of 6.25 (pre-monsoon)- 7.0 (retreating-monsoon). pH was found to be slightly acidic in premonsoon and monsoon seasons whereas it was in the slightly neutral side during retreating monsoon and winter. Variations in pH may be attributed to increase and decrease in biogenic activities of the system. Daimari *et al.* (2005) reported a pH range of 7.4-8.1 in Subansiri river of Arunachal Pradesh.

Dissolved oxygen: The dissolved oxygen content ranged between 8.80 mgl^{-1} (Pre-Monsoon) and 16.64mgl^{-1} (Winter). The dissolved oxygen content may be considered as conducive for aquatic habitat. Gurumaym *et al.* (2000a) also observed the same trend in Subansiri river of Arunachal Pradesh with maximum in winter (15.2 mgl^{-1}) and minimum during monsoon (7.64 mgl^{-1}). High DO value was also noticed from unpolluted rivers of India like Siang (10 mgl^{-1}), Dibang (9.02 mgl^{-1}), Lohit (9.73 mgl^{-1}) and Mahanadi (8.0 mgl^{-1}) by Pathak *et al.* (2001a, 2001b).

Total Dissolved Solids (TDS): TDS value was 57-105 mgl^{-1}. The minima was seen during monsoon while the maxima was recorded in winter. TDS value was found to be higher than the values already reported from this region. Daimari *et al.* (2005) found a TDS range of 51.5- 82.6 mgl^{-1} in Subansiri stream while Pathak *et al.* (2001a) reported the value of the rivers Siang (71.3 mgl^{-1}) and Dibang (50.6 mgl^{-1}).

Alkalinity: Alkalinity values were found to be low, the maxima was recorded during winter (55 mgl^{-1}) while the minima was observed in monsoon (32 mgl^{-1}). Lower alkalinity value of hilly streams was also reported by earlier workers. Gurumayum *et al.* (2000b) observed a range of 15- 82.60 mgl^{-1} in some selected rivers of Meghalaya. Pathak *et al.* (2001a) recorded an average alkalinity of 62.67 mgl^{-1} and 35.72 mgl^{-1} in rivers Siang and Dibang of Arunchal Pradesh respectively. An alkalinity range of 33.98- 66.7 mgl^{-1} was recorded in the river Subansiri by Daimari *et al.* (2005). According to Unni (2003), Indian rivers with an alkalinity range of 50-100 mgl^{-1} can be considered as low or unpolluted and the present finding revealed unpolluted status of the Namsang river.

Hardness: In Namsang stream hardness was found to be maximum during retreating monsoon (64 mgl^{-1}) and minimum in premonsoon (40 mgl^{-1}). Daimari *et al.* (2005) reported similar values (30.3- 81.75 mgl^{-1}) from Subansiri river of Arunachal Pradesh. Average hardness value was found to be 60.8 mgl^{-1} in Siang river and 38.4 mgl^{-1} in Dibang river of Arunachal Pradesh by Pathak *et al.* (2001a).

Calcium: Calcium concentration was recorded in the range of 22-29 mgl^{-1}. Minimum value was seen during monsoon while the maximum was observed during winter. According to Ohle (1934), calcium concentration of less than 10 mgl^{-1} is poor and more than 25 mgl^{-1} is rich. So the overall calcium concentration of the studied river can be categorised as medium. Similar medium range of calcium (4.1- 49.50 mgl^{-1}) was observed by Choudhury (2002) in the rivers of Arunachal Pradesh. Same author has reported medium range of calcium content in the rivers of Manipur (8.4- 64.4 mgl^{-1}) and Meghalaya (6.0-54.5 mgl^{-1}).

Magnesium: Magnesium content was observed to vary between 14 and 40 mgl^{-1}. Peak value was recorded during the period of retreating monsoon while the nadir was observed during pre monsoon. Earlier magnesium content in the rivers of Arunachal Pradesh observed by Choudhury (2002) ranged (0.13-49.50 mgl^{-1}). Same author has reported low range of magnesium content in the rivers of Tripura (4.8- 7.3 mgl^{-1}) and (2.4-21.8 mgl^{-1}) from the rivers of Mizoram.

Chloride: Lower range of chloride concentration was observed in the river. Highest (18.5 mgl^{-1}) chloride concentration was seen during winter while the lowest value (5 mgl^{-1}) observed during premonsoon and monsoon. Low chloride value reflects that there was minimal amount of organic waste of animal origin and practically no discharge of municipal and industrial waste. Choudhury (2002) observed higher values of Chloride concentration (14.7-32.2 mgl^{-1}) in Tirap river of this region. Nath (2001) observed comparatively lower (2-14 mgl^{-1}) chloride content in the Narmada river.

Silicate: The minimum concentration of silicate was found during retreating monsoon (6.2 mgl^{-1}) while the maximum was during premonsoon (9 mgl^{-1}). Similar values (4-6 mgl^{-1}) were also recorded by Gurumayum *et al.* (2000a) in Subansiri river of Arunachal Pradesh. According to Lund (1954) Silicate content should be above 0.5 mgl^{-1} for rich growth of diatoms. In this river, the value was 6.2-8.05 mgl^{-1} which reflects the productive nature of the river.

Iron: Iron content ranged between 0.24 and 7.52 mgl^{-1}, being the minimum during retreating monsoon while the maximum was found during premonsoon. Ellis (1942) opined that presence of more than 2mgl^{-1} of iron was unfavourable for fish. But its concentration more than 0.1 mgl^{-1} is essential as it plays an important role in metabolic process. Iron content observed by Choudhury (2002) in the rivers of northeast were Arunachal Pradesh (0.09 -4.7 mgl^{-1}), Manipur (0.02- 0.93 mgl^{-1}), Meghalaya (0.02- 1.0 mgl^{-1}) and in Mizoram (0.2-0.93 mgl^{-1}).

Table 9.1
Sediment quality of Namsang river

	Parameters	
Physical features	pH	7.07
	Sand (%)	87
	Silt (%)	3.57
	Clay (%)	7.14
	Texture class	Sandy
Chemical features	EC (dSm^{-1})	0.226
	Organic Carbon %	0.312
	Available N (Kg/ha)	103.81
	Available P_2O_5 (Kg/ha)	17.95
	Available K_2O (Kg/ha)	77.28
	Free $CaCO_3$ %	1.5
	Skeletal parts (Gravelly %)	54.16

Table 9.2
Seasonal variations in physico-chemical characteristics of water in the Namsang river. Except where mentioned, other values are given in mgl^{-1}

Seasons	*Premonsoon*	*Monsoon*	*Retreating Monsoon*	*Winter*
Water Temperature (^{0}C)	23	24.5	22.25	16.5
pH	6.25	6.8	7	6.9
Dissolved oxygen	8.80	9.55	12.88	16.64
TDS	75	57	87.5	105
Alkalinity (as $CaCO_3$)	36	32	47.5	55
Hardness (as $CaCO_3$)	40	47	64	63
Calcium (as $CaCO_3$)	26	22	24	29
Magnesium (as $CaCO_3$)	14	25	40	34
Chloride (as Cl)	5	5	8	18.5
Silicate (as SiO_2)	9	7.2	6.2	6.6
Iron (as Fe)	7.52	1.29	0.24	0.27
Nitrate (as NO_3)	0.55	0.84	0.58	0.35
Phosphate (as P)	0.05	0.07	0.21	0.05
Dissolve organic matter	19.5	23.5	18.5	17
Carbondioxide as CO_2	7.21	5.36	6.02	4.27
Specific conductivity (micro mhos/ cm)	76	90	66	21.09

Nitrate: Nitrate values were found to be varying between 0.35 and 0.84 mgl^{-1} being the minima and maxima during winter and monsoon respectively. Ranges of nitrate content in the different rivers of northeast observed by Choudhury (2002) were Arunachal Pradesh (trace – 0.85 mgl^{-1}), Manipur (trace- 1.02 mgl^{-1}), Meghalaya (trace – 0.55 mgl^{-1}), Mizoram (0.02 – 0.06 mgl^{-1}) and in Tripura (trace – 0.21 mgl^{-1}). Similar lower nitrate value was observed by Pathak *et al.* (2000 & 2001b) in the tributaries of Brahmaputra and Mahanadi.

Free CO_2: Free CO_2 content ranged between 4.27 and 7.21 mgl^{-1}, being the minimum in winter while the maximum was found during premonsoon. Similar observation was made by Nath (2001), in the river Narmada. Nath (1998), observed higher CO_2 values in the rivers affected by the industrial effluents like Ganga (1-12 mgl^{-1}), Jhningran (1988) also observed higher values (1.5-8.8 mgl^{-1}) from the same river. Singh (1999) also detected higher CO_2 values (3.5 mgl^{-1}) from the river Yamuna. Choudhury (2002) observed CO_2 values within the range of (0.9-3.8 mgl^{-1}) in the rivers of Arunachal Pradesh during premonsoon, retreating monsoon and winter season.

Phosphate: Phosphate content showed a very narrow range (0.05-0.21 mgl^{-1}). Highest values seen during retreating monsoon whereas the lowest was during premonsoon and winter periods. Lower values of phosphate (0.00- 0.03 mgl^{-1}) was also reported from healthy tributaries of Brahmaputra by Pathak *et al.* (2001a). According to Choudhury (2002) phosphate content in the rivers of northeast ranged from (trace – 0.15 mgl^{-1}). Lower PO_4 value further confirms the unpolluted status of Namsang stream.

Dissolve Organic Matter: Highest dissolved organic matter was observed during monsoon (23.5 mgl^{-1}) while its value was lowest during winter (17 mgl^{-1}). Similar observation was made in the river Mahanadi and tributaries of Brahmputra by Pathak *et al.* (2000 & 2001b). He opined that as the river flows through the deep forest, falling of tree leaves in the river water increases its carbon content. Dissolved organic matter observed by Choudhury (2002) in the rivers of northeast were ranged (0.52-1.73 mgl^{-1}) in Arunachal Pradesh, (0.44-1.73 mgl^{-1}) in Manipur, (0.24-2.52 mgl^{-1}) in Meghalaya, (1.1-4.6 mgl^{-1}) in Mizoram, (1.2-1.9 mgl^{-1}) in Nagaland and (1.1-5.1 mgl^{-1}) in Tripura.

Specific conductivity: Specific conductivity values were highest (90 micro mhos/ cm) during monsoon and lowest in winter (21.09 micro mhos/ cm). Specific conductivity values observed by Pathak *et al.* (2001a) in the rivers of Arunachal Pradesh were ranged Siang (148 micro mhos/ cm), Dibang (127.0 micro mhos/ cm) and Lohit (226.0 micro mhos/ cm). Similar observations made by Choudhury (2002) in the rivers of Northeast ranged (45.0-198.3 micro mhos/ cm) in Arunachal Pradesh, (54.0-265.4 micro mhos/ cm) in Manipur, (49.3-182.3 micro mhos/ cm) in Meghalaya, (78.5-160.8 micro mhos/ cm) in Mizoram, (93.6-322 micro mhos/ cm) in Nagaland and (85.1-188.1 micro mhos/ cm) in Tripura. Similar lower values of specific conductivity were observed from the unpolluted rivers of India by Pathak *et al.* (2001b) and Singh *et al.* (1999).

The sediment quality of Namsang stream reflected neutral pH, sandy bottom soil having less % of silt and clay and low nutrient content, which is the sign of less productive nature of water bodies.

Poor concentration of carbon dioxide, alkalinity, specific conductivity, TDS, chloride, hardness, calcium, magnesium, iron, DOM, nitrate, phosphate and slightly acidic to neutral pH, rich DO and silicate indicated oligotrophic condition of the river. Going through the above observations, it can be concluded that Namsang river can be classified as poorly productive, free from pollution and other anthropogenic activities.

REFERENCES

APHA, 1989. *Standard methods for the examination of water and wastewater.* American Public Health Association Washigngton D.C. 1134pp.

Choudhury, M. 2002. *Ecological status and fishery potential of six northeastern states Arunachal Pradesh, Manipur, Meghalaya, Mizoram, Nagaland and Tripura.* Final report NERC, CIFRI, Guwahati.

Daimari, P., Choudhury, M. and Dutta, A. 2005. Ecology and Fishery of River Subansiri (Arunachal Pradesh). *Environment & Ecology,* 23 (1):49-54.

Ellis, M.M. 1942. Fresh water impoundments. *Trans. Am. Fish. Soc.,* 71 (1968) 80-98.

Gurumayum, S.D., Daimari, P., Goswami, B.S. Sarkar, A. and Choudhury, M. 2000a. Ecology of the river Subansiri in Arunachal Pradesh. *J. Inland Fish. Soc. India,* 33(2): 50-54.

Gurumayum, S.D., Daimari, P., Goswami, B.S., Sarkar, A. and Choudhury, M. 2000b. Physico-chemical qualities of water and plankton of selected rivers in Meghalaya. *J. Inland Fish. Soc. India,* 34(2): 50-54.

Jhingran, A.G. and Pathak, V. 1988. Impact of man-induced environmental modifications on productivity potential and energy dynamics of river Ganga. *J. Inland Fish. Soc. India,* 20(2): 43-53.

Jhingran, V.G., Natarajan, A.V., Banerjee, S.M. and David, A. 1969. *Methodology on reservoir fisheries: Investigation in India.* Bull. Cent. Inland. Fish. Res. Inst., Barrackpore, India, 1:109pp.

Lund, J.W.G. 1954. The seasonal cycle of the plankton diatom *Melosira* (HER) Kutz. Subarctica, *O. Mul. J. Ecol.,* 42: 151-179.

Nath, D. 1998. Zonal distribution of nutrients and their bearing on primary production Hooghly Estuary. *J. Inland Fish. Soc. India,* 2: 64-74.

Nath, D. 2001. Water and soil characteristics of the Narmada estuary before commissioning of Sardar Sarobar dam. *J. Inland Fish Soc. India,* 2: 64-74.

Nath, P. and Dey, S.C. 2000. *Fish and fisheries of North Eastern India (Arunachal Pradesh).* Narendra Publishing House, Delhi, 217pp.

Ohle, W. 1934. Bioactivity production and energy utilization of Lakes. *Limnol. Oceangr.,* 2: 139-149.

Pathak, V., Sarkar, A., Mahavar, L.R. and Bhattacharjya, B.K. 2001a. Ecological status and Fish production potential of Siang, Dibang and Lohit- the three forerunners of river Brahmaputra. *J. Inland Fish. Soc. India,* 33(2): 23-28.

Pathak, V., Mahavar, L.R. and Sarkar, A. 2001b. Ecological Status and production dynamics of sretch of river Mahanadi. *J. Inland Fish. Soc. India,* 33(1): 25-31.

Pathak, V., Choudhury, M., Laal, A.K., Bhattacharjee, B.K., Sarkar, A. and Mahavar, L.R. 2000. *Ecology and production dynamics of river Brahmaputra with special emphasis on its tributaries.* Bull. No. 97, Central Inland Fisheries Research Institute, Barrackpore, India.

Singh, H.P., Mahavar, L.R. and Mishra, J.P. 1999. Limnochemical characteristic of river Ghaghara in U.P. *J. Inland Fish. Soc. India,* 31(10): 43-53.

Unni, K.S. 2003. *Ecology of Indian Rivers.* International Book Distributors, Dehra Dun, 312pp.

Pre-damming Investigation on River Morphology and River Bed Sediment Composition of the Downstream of Subansiri River, Assam

—*Debojit Baruah*[1*], *Ranjit Dutta*[2], & *Sarada Kanta Sarma*[3]

[Subansiri river has a considerable width and depth with meandering in nature throughout its journey in Assam. In the dry spell the downstream of the river created a numbers of channels, pools and meandering cut-off (static water body). The river has greater volume of water and high speed of water current, carrying huge amount of sand, gravel and woody debris to its downstream. In its initial 10-17 km downstream from the proposed 116 m high dam site, the river bed is composed of sand and gravel, beyond which the bed is exclusively composed of alluvial silt. National Hydroelectric Power Corporation (NHPC) has already been started the construction of the 2000 MW Lower Subansiri Hydroelectric Project close to the Assam-Arunachal Pradesh border near Gerukamukh in Dhemaji district of Assam. After operation of this hydel project, it will be catastrophic for the water and sediment regimes and the downstream morphology of the Subansiri river.

Key words: Downstream, Sediment, Subansiri river, Dam.]

Introduction

River ecosystems are adapted to the natural hydrological regime and many components of these systems rely on floods for the exchange, not just of water, but also energy, nutrients, sediments and living organisms (Acreman, 2000). *River morphology* is the shape or form of a river along with its length and width. Rivers are flow in *channels,* the area within the banks of the river. A river when it carries more water than the channels can accommodate (the bank full discharge), and the river spills over its banks on to the floodplain areas. The *wetted edge* of a stream is the surface of the bed and banks covered by water and are the zone where erosion and deposition can occur. The *floodplain area* is relatively flat-lying part of the

[1,2] Department of Botany, Lakhimpur Girl's College, P.O. Khelmati, Assam 787031, India
[3] Department of Botany, Gauhati University, Guwahati, Assam 781 014, India
*Corresponding author: debojitbaruah@rediffmail.com, ran.jit2@yahoo.com.

valley next to the channel. The greater the flood discharge, the larger the area of the floodplain that is inundated.

Sediment is important to determining the morphology of a river system. One of the most important functions of river is the transport of sediment and nutrient. Rivers has naturally evolved and changes their shapes by eroding, transporting and depositing sediment. The movement of sediment in rivers and their valleys determines the course of the river, the shape of the channel bottom, the locations of pools and riffles (light rapids) along the river and the materials that make up the bed of the river (WRM, 2000). These factors influence the types of wildlife habitat available in the river ecosystem.

The main aim of the present study was to investigate the river morphology and the composition of sediment transported by the Subansiri river before being regulated by the 2000 MW Lower Subansiri Hydroelectric Project (Figure 10.4, 10.13.

Methodology

Study Area

The study was conducted on both sides of the downstream of river Subansiri, located in Assam from Gerukamukh (dam site) to Jamuguri (where the river meet with river Brahmaputra). The survey area was arbitrarily divided into five sites namely (A) Gerukamukh, (B) Chowaldhowa (C) Khabolo (D) Dhunaguri and (E) Jamuguri (Figure 10.1a and 10.1b). The lower basin has a length of approximate 130 km during the dry season (recorded in December, 2008).

River Morphology

Morphological structure of the downstream like meandering nature, water depth and width of the river channels at different sites were recorded and calculated the maximum and minimum values.

Physical status of the river

Physical status like longitude, latitude, mean sea level, river bed slope of each sampling sites were measured with the help of GPS.

Riverbed sediment composition

Riverbed sediment composition of the entire downstream was measured by collection of sediment and classified according to Fischenich (1999) and Konkel (2001). Water flow and sediment discharge data were collected from secondary sources (WAPCOS, 1993; Goswami *et al.*, 1999; EIA Report, 2002).

Results

The Physical status of the river Subansiri like total length in Tibet and India (Arunachal Pradesh and Assam), Mean Sea Level (MSL), maximum & minimum depth and width of the downstream, along with the longitude and latitude of the sampling sites are presented in the Table 10.1. Surveys indicated that at the Gerukamukh, where the river enters in to Assam,the elevation was measured 99 m MSL. Throughout the downstream the total slope of the river is 40 meter with maximum inclination was found in between Gerukamukh and Chowaldhowa (23 m).

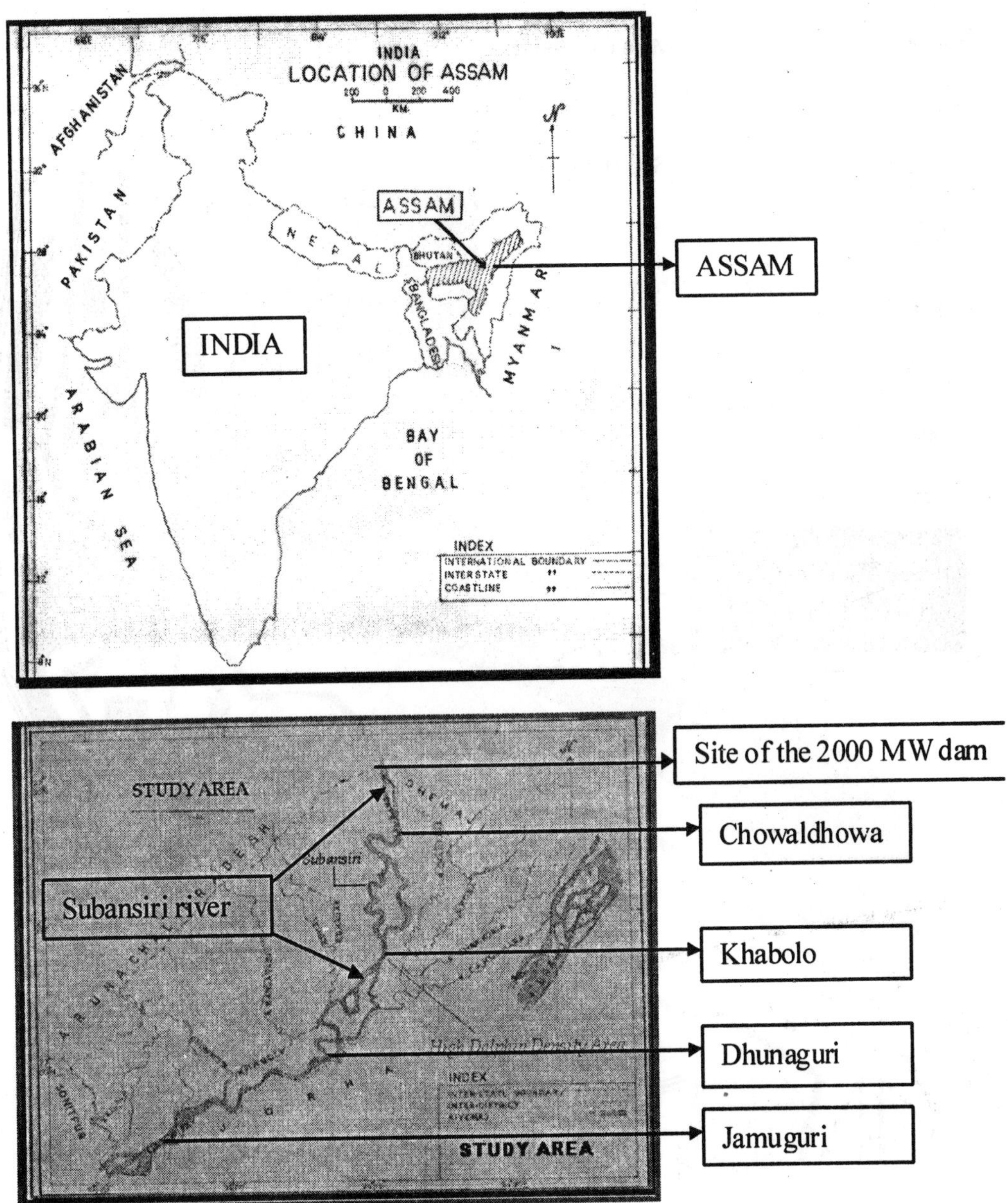

Figure 10.1a. Location map of Assam.
Figure 10.1b. Map of Subansiri river showing the study sites.

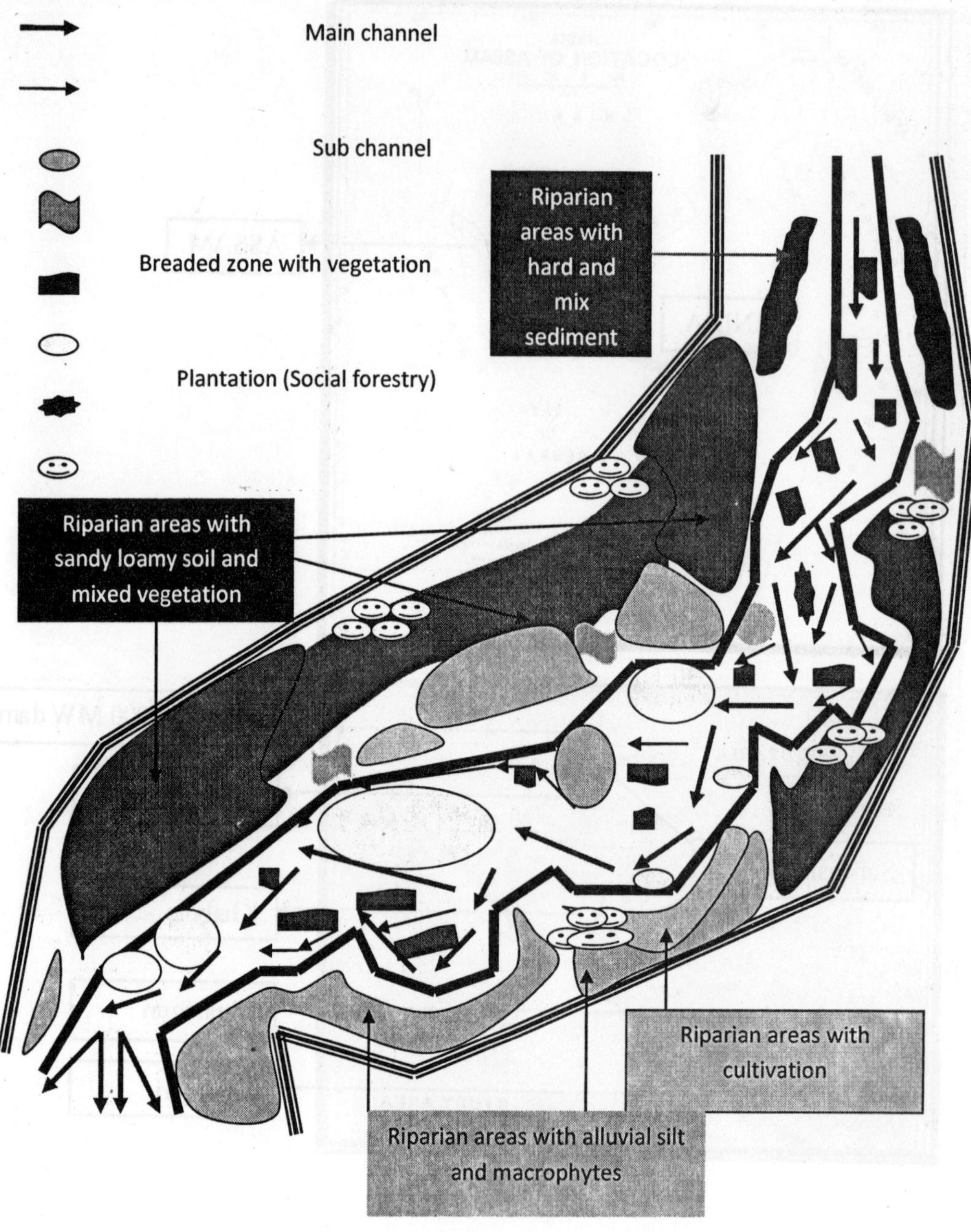

Figure 10.2. Schematic diagram of the downstream and riparian morphology of the Subansiri river.

Figure 10.3. Upstream of Subansiri river.

Figure 10.4. Dam site before construction of dam.

Figure 10.5. Sandbar.

Figure 10.6. River bank

Figure 10.7. River bank Stabilization.

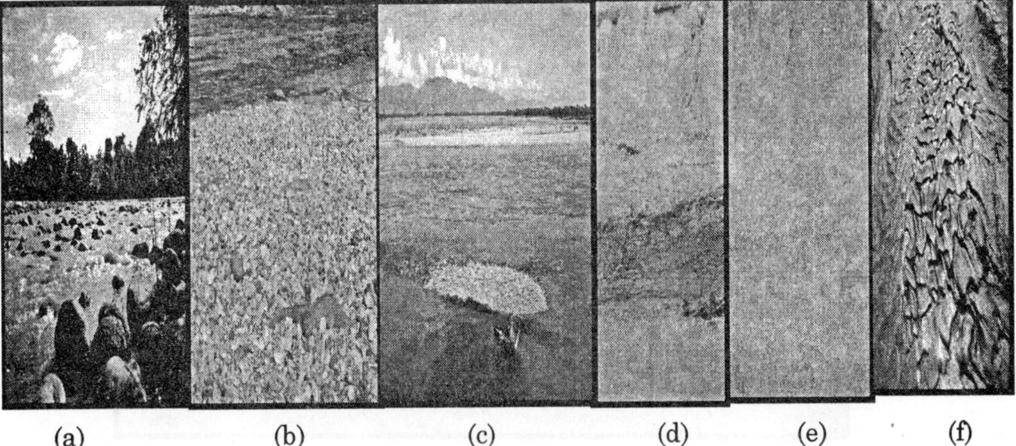

(a) (b) (c) (d) (e) (f)

Figure 10.8. Transportation of (a) boulders, (b) cobble, (c) pebble, (d) silt, (e) sand and (f) clay.

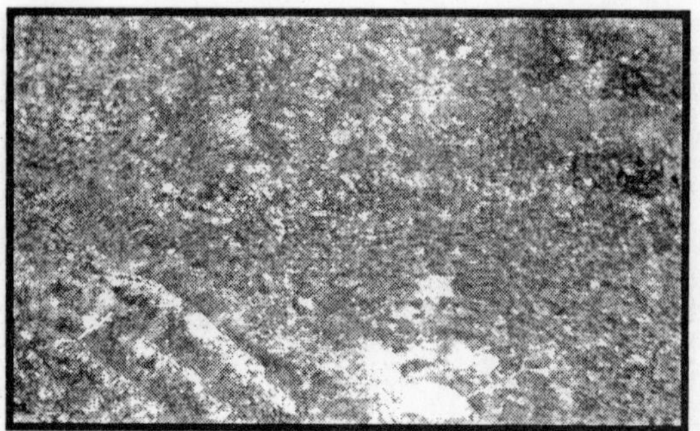

Figure 10.9. River bed with gravel and sand.

Figure 10.10. Sandy river bed.

Figure 10.11. Large woody debris.

Figure 10.12. Small woody debris.

Figure 10.13. Dam site during temporary flow diversion in 2008.

Downstream of river Subansiri is very wide with a considerable depth during rainy season but in the dry spell the river is shrinking and the depth is reduced. Maximum depth 4.9 meter and minimum width 200 meter were recorded near Jamuguri, while minimum depth 0.2 meter and maximum width 850 meter were recorded near Khabolo. The maximum and minimum daily discharge in the dam site (Gerukamukh) was 188 cumec/sec and 12,024 cumec/sec respectively (EIA Report, 2002). The lower reaches of the river is too much meandering in nature and create a numbers of channels in the dry period. The river bank of Subansiri river in the downstream consists of fine sand and silt. Therefore the river banks are vulnerable to erosion but in some areas riparian flora stabilizes these (Figure 10.6-10.7).

Subansiri river carries a mixture of large and small objects (Table 10.2), including boulders (larger than 256 mm in diameter), cobbles (64-256 mm in diameter), pebbles (2-64 mm in diameter), sand (0.06 to 2 mm), silt (0.002-0.6 mm) and clay (finer than 0.002 mm) (Figure 10.8). The river bed is mainly consists of fine and medium size sand with an exception in the first 10-17 km (site 1), where it is consist of hard sediment (mixture of cobbles, pebbles and boulders) (Figure 10.2, 10.9, 10.10). Bed-rock and bed-sand were the predominant sediments at the Chowaldhowa and Khabolo respectively. Bed-fine sediment and clay were pre dominant at Dhunaguri and Jamuguri. Beside sand and gravel, the river also carries natural debris such as Large Woody Debris and Small Woody Debris (Figure 10.11-10.12).

Discussion

All Himalayan rivers are glacier fed and perennial. The Subansiri river, the largest tributary of the mighty river Brahmaputra is also a Himalayan river (Figure 10.3). It is sustained by snowmelt run-off, the ablation of glaciers and monsoon rainfall. The change in geomorphic characters of Subansiri river has been recorded from 1919 onwards. Goswami *et al.* (1999) has reported four primary causes for the morphological changes of Subansiri river. Meandering bends of the river is more pronounced in dry season due to its low water carrying capacity as well as sand bar development through siltation in the original river course. Under low water-flow condition, the downstream shows 2-4 channels separated by sand bars, a type of platform generally known as "braided" as suggested by Wiebe (2006). During the rainy season, these low level sand bars are submersed and in heavy floods they may disappear entirely due to heavy discharge and high level of the river water. The pattern of channels is quite unstable and changes frequently, usually under flood condition. The changes are accompanied by erosion of flood plain river banks and deposition of sand bars (Figure 10.5).

Subansiri river in its downstream naturally evolves and changes their shapes by eroding, transporting and depositing sediments of its own, or alluvial fan deposits created by the numbers of feather streams emerging from the foot hills of Eastern Himalayas. It carries a mixture of large and small objects. During floods, when the river flow is dominant, it get the capacity for transportation of large and small objects, including boulders, cobbles and pebbles up to 10-17 km downstream. A sharp reduction of the river bed from Gerukamukh (99 m MSL) to Chowaldhowa

(76m MSL) also helps in easy movement of these hard sediments with sand, silt and clay. The mean daily water discharge of the Subansiri river at Gerukamukh is 13, 884, 2 m^3/sec (Goswami *et al.*, 1999). The sediment carrying capacity of the river is maximum during rainy wet period because the velocity and volume of the water flow of the river is greater in this time. When a river carries more sediment then its erosive force is more pronounced. High water flows is a dominant factor to control the river channels and floodplain morphology, influencing processes of sediment transport and erosion (Acreman, 2000).

Table 10.1

Physical status of the downstream of Subansiri river

Total length (km)	*Total length Length in Tibet (km)*	*Length in India (km)*	*Length in Arunachal Pradesh (km)*	*Length in Assam (km)*	*Total number of feeder stream*
530	200	330	200	130	10
Location	Gerukamukh (A)	Chowaldhowa (B)	Khabolo (C)	Dhunaguri (D)	Jamuguri (E)
Longitude	N-27°33′	N-27° 26′	N-27° 03′	N-27° 00′	N-26° 50′
Latitude	E-94°15′	E-94° 15′	E-94° 07′	E-94° 01′	E-93°48′
Mean sea level (m)	99	76	65	64	59
Slope of the river bed (m/km) between	(Initial spot)	2.3 (A&B)	0.2 (B&C)	0.03 (C&D)	0.12 (D&E)
Approximate distance (km) between	0	10 (A&B)	45 (B&C)	35 (C&D)	40 (D&E)
Approximate Maximum width of the river in dry spell (m)	74	118	107	113	89
Approximate Maximum river depth in dry spell(m)	6.0	9.3	11.5	13.6	15.3

Table 10.2

Sediment type in the downstream of Subansiri river

SedimentTypes	*Gerukamukh to Chowaldhowa*	*Chowaldhowa to Khabolo*	*Khabolo to Dhunaguri*	*Dhunaguri to Jamuguri*
Hard sediment	Bed Rock	NIL	NIL	NIL
Hard sediment	Sand	NIL	NIL	NIL
Hard sediment	Sand/Gravel	NIL	NIL	NIL
Hard sediment	Sand/Rock	NIL	NIL	NIL
Hard sediment	Gravel	NIL	NIL	NIL
Mixed sediment	Sand/Silt/Clay	Sand/Silt/Clay	Sand/Silt/Clay	Sand/Silt/Clay

The average annual sediment yield at Chowaldhowa is 94.83×10^3 m tones (WAPCOS, 1993). This was comparatively higher than other rivers in this region and it could be due to the *'Jhum'* cultivation and illegal felling of trees in its upstream (Arunachal Pradesh). These two anthropogenic effects have destabilized the upper riparian zone and the top soil. During rainy seasons the fertile top soil are eroded. In the recent years, yield of sediments in the downstream showed increasing trend which results in gradual up-liftmen of riverbed by the process of sedimentation. Subansiri river alter their channels within the floodplain by scouring sediment from the bed and banks and transporting it to the downstream. The sediment removed from any one stretch of the river is replenished by sediment from upstream.

River flow and sediment regimes are the physical elements and driving factors of river system (State Water Report of Victoria, 2004-2005). The development and utilization of rivers especially when damming off, alter river flow and sediment regimes as a result the functions of the river system and environment has been changed. Flow and sediment regimes in rivers have obvious dual character, which can not only promote environment and society development but also destroy them. Damming of river in its upstream will affect these two factors in the downstream because free-flowing river meets an impoundment behind a dam. The river uses the energy of its flow to carry sediment in the water, so as the flow nearly stops in the impoundment, coarse silt, sand and gravel settle on the bottom of the impoundment. Finer silt and clay suspended in the water are carried out into the impoundment and deposited in the reservoir. This process results in large amounts of sediment accumulating in the impoundment and consequently, the downstream of the river becomes "starved" for sediments and woody debris.

[**Acknowledgement:** Financial assistance provided by University Grant Commission, New Delhi, India in the form of Major Research Project [Project No: F No33-137/207(SR)-2008] to the first author is gratefully acknowledged.]

REFERENCES

Acreman, M. 2000. *Managed flood releases from reservoirs: Issues and guidance.* WCD Thematic Review, 2(1): 1-88.

EIA Report. 2002. *EIA report for Lower Subansiri project.* WAPCOS, Water and Power Consultancy Services, India.

Fischenich, C. 1999. "Glossary of stream restoration terms," *EMRRP Technical Notes Collection* (ERDC TN-EMRRP-SR-01), U.S. Army Engineer Research and Development Center, Vicksburg, MS.

Goswami, U., J. N. Sarma and A. D. Patgiri. 1999. River channel changes of the Subansiri in Assam, India. *Geomorphology*, 30 (3): 227-244.

Konkel, D. 2001. *The Aquatic Plant Community in the Chippewa River.* http://dnr.wi.gov/fish/reports/final/buffalochippewariver2001.pdf.

State Water Report of Victoria. 2004-05. *Securing the Environmental Health of Water Resources.* pp. 271-279.

WAPCOS. 1993. *Morphological Studies of River Brahmaputra.* N.E. Council, Water and Power Consultancy Services (India), New Delhi.

Wiebe, H. 2006. River Flooding and Erosion in NE India, Exploratory Consideration of key Issues, Draft for discussion-March 2006, Background Paper No. 04. *Study on Natural Resources, Water and the Environment Nexus for Development and Growth in Northeast India.*

WRM. 2000. Dam repair or removal: A decision making guide: Module 2, Dam and river Ecosystem basics: Morphology and Sediment. URL http://www.ies.wisc.edu/research/wrm.

11

Status and Functional Roles of Some Giant Riparian Amphiphytes in the Subansiri River Ecosystem of North East India

—*Debojit Baruah[1*], Ranjit Dutta[2], & Sarada Kanta Sarma[3]*

[Abundance, distribution and functional roles of some giant amphiphytes along with certain water quality parameters of the downstream of river Subansiri were investigated during 2008 prior to the operation of 2000MW Lower Subansiri Hydroelectric Project in the Assam-Arunachal Pradesh border of North East India. Descriptor scale has been applied on giant amphiphytes (height >1.5 meter) except *Impereta cylindrica* (L) Beauv. to assess their abundance and distribution in different habitats. Riparian amphiphytes provide many ecological benefits to their habitats. The proposed dam may alter the downstream ecological integrity of the river Subansiri by changing its flow regime, flow patterns, water chemistry and limiting downstream movement of sediments and biological materials.

Key words: Dam, riparian amphiphytes, river ecosystem, Subansiri river.]

Introduction

Amphiphytes are the primary autotrophs of riparian corridors, which usually grow in shallow and partly inundated river banks with aerial emergent plant parts. These are a natural part of most river systems and provide many benefits to fishes, wildlife and people (Westlake, 1975, Wetzel, 2000). In a river ecosystem, life depends directly or indirectly on the primary producers in the aquatic food chain, converting the basic chemical nutrients of water and soil into plant matter, which become foods for all other lives. Primary producers also provide many economical benefits along with the trapping of sediment or nutrients (Ambasht, 1970).

Stream diversion for hydroelectric power can adversely affect the riparian vegetation in several ways (Kondolf *et al.*, 1985). Reduced flows in a diverted reach may (a) lower the stream water surface below levels that mosses and other submerged

[1,2] Department of Botany, Lakhimpur Girl's College, P.O. Khelmati, Assam, 787 031, India
[3] Department of Botany, Gauhati University, Guwahati, Assam 781 014, India
*Corresponding author: debojitbaruah@rediffmail.com

and emergent plant species depend, (b) lower the alluvial water table below the rooting depths of some species, and (c) reduced natural periodic flood scour and annual high flows. As a result the area, density, composition and species diversity of riparian vegetation can be affected by altering stream-flow. While the dependence of riparian vegetation on stream-flow is generally accepted, efforts to quantify this dependence have been plagued with difficulties (Kondolf *et al.*, 1985).

Desiccation of downstream floodplains after impoundment has been documented by many workers, which includes reduced the downstream flood plain areas because it no longer receives sufficient overspill from dam (Begg, 1973; Chude, 1979; Ita and Mohammed, 1980), no longer floods (Moxon, 1969; Pople and Rogoyska, 1969). Kariba dam reduced flood magnitude of the Zambezi River by an average of 24 percent in eight out of ten years (1970–80 period) (SWECO/SWED, 1981). Cahora Bassa dam prevents flooding of the lower Zambezi floodplain, resulting in drying of flood-plain lakes (Anon, 1975).

The proposed 2000 MW Lower Subansiri hydroelectric Project is doubtlessly of substantial importance towards the riparian zones of the downstream. Therefore, an attempt has been made to investigate some water quality parameters related to river health and riparian amphiphytes and to know the status of giant amphiphytes as well as their functional roles in various aspects before regulate the river by the proposed large dam.

Methodology

Study area

The river Subansiri continue its journey through the Himalayan range inside Indian territory in the North Eastern part for 200 km and enters into the plains of Assam (Figure 11.1) through a gorge near Gerukamukh of Dhemaji district. Its total length is 520 km and drains a basin of 37,000 square km. The total length of the river within Assam is about 130 km which flows through the Lakhimpur district in between 26°50' - 27°35'N and 93°41' - 94°23' E. For a systematic monitoring and study of water quality parameters of the downstream from the proposed dam site (Figure 11.2, 11.3) up to the place of confluence of the river Subansiri with the river Brahmaputra of Subansiri river have been divided into four site namely Chowaldhowa (27°26' N and 94°15' E, Altitude 75 m), Khabolo (27°03'N and 94°07'E, Altitude 64 m), Dhunaguri (27° 00' N and 94° 01' E, Altitude 64 m) and Jamuguri (26°50' N and 93°48' E, Altitude 59 m) to cover as much possible area of the river (Figure 11.2). For the study of giant amphiphytes, survey was done in both the sides of the riverbanks in the riparian zones in the entire downstream.

Methods

Certain physico-chemical properties of the river like water temperature, current flow, transparency, pH, total dissolved solids, total suspended solids, dissolved oxygen (DO) and free carbon dioxide (CO_2) were measured folloeing APAH (1992). Identification of giant amphiphytes were done with following standard methods (Dutta, 1985; Khan and Halim, 1987; Cook, 1996; Press *et al.*, 2000) and by consulting different herbaria. In order to unify data on giant amphiphytes, a 5-grade scale or descriptor scale (Melzer, 1999 and Stelzer *et al.*, 2005) has applied which are as follows.

1. Very rare (covers up to 5% of a sampling area)
2. Rare (covers >5–25% of a sampling area)
3. Common (covers >25–50% of a sampling area)
4. Frequent (covers >50–75% of a sampling area)
5. Abundant and dominant (covers >75% of a sampling area).

Information regarding various ecological functions of these floras has been collected through direct observation and discussion with experienced villagers in the selected areas. Data were exposed to several statistical treatments and a conventional descriptive statistic based on normal distribution has been given.

Results and Discussion

The average values of certain physico-chemical parameters of the river at different sampling sites are given in Table 11.1.

The pH is a key indicator of water quality and controls the biological functions of aquatic ecosystem on one hand and on the other hand it is responsible for nutrient mobilization from their source (soil). The river water was slightly alkaline in nature which ranged from 7.44 mg/l to 7.58 mg/l in all the stations. We have recorded very low transparency in summer seasons (May-August) but with a substantial amount of dissolved oxygen. This may be due to either contribution of dissolved oxygen by submerged photosynthetic parts of the giant macrophytes as well as other aquatic plants like algae and true hydrophytes to some extent or may be due to high turbulence of water in the upstream as the river is still unregulated. On the contrary, the organic and inorganic substances in the form of suspended (TSS) and dissolved solids (TDS) providing a steady and continuous supply of nutrients to the giant amphiphytes which help in luxuriant growth during their active growth period. Various plant parts, those breaks off from aquatic vegetation form decomposed organic matter in the heterotrophic food webs in the ecosystem. The free carbon dioxide represents the positive balance between producer and consumer in the river system. All the studied physico-chemical properties of the water exhibit the healthy status of the downstream of the river Subansiri and influenced positively upon the riparian amphiphytes.

During the study, 14 amphiphytes species were recorded from the riparian zone of the river, out of which, 13 were considered as giant riparian amphiphytes. These species are inundated by water for at least 2-4 months in a year with fluctuating water levels ranges from 0.15 m to 1.20 m.

Descriptor scale measurement identified that *Oxystelma esculentum* (L. *f.*) Sm and *Salix tetrasperma* Toxb. species are very rare; *Calamus tenius* Roxb. and *Typha angustifolia* Linn. as rare; whereas *Vetiveria zizanoides* Stepf, *Themda villosa* Dor et Jack, *Arundo donax* Linn., *Impereta cylindrica* (L) Beauv, *Phragmites karka* (Retz.) Trin., *Microstegium ciliatum* (Trin.) A. Camus. as common species; *Tamarix dioica* Roxb. and *Ipomoea carnea* Jacq. are frequent species while, *Saccharum spontaneum* Linn. and *Saccharum arundinaceum* Retz. are abundant and dominant giant amphiphytes in the study area (Table 11.2).

Luxuriant growth of few species like *T. dioica, S. spontaneum and S. arundinaceum* indicate its high degree of tolerance to heavy water current and fluctuating environment as they are capable of enduring in long-term water saturation with high water current as well as in the extreme drought condition particularly in the sandy riverbanks. These species exhibits its high adaptive characters to the changing environment and established themselves as more frequent and abundant species. Most of these giant amphiphytes except *T. dioica, T. angustifolia* and *I. carnea,* having dual methods of propagation (seed and vegetative part) which help these species to grow easily in dry and sandy habitat. Furthermore, recovery rates of these species were found to be more from their underground vegetative propagules like rhizomes, old root-stalks, runners and suckers. Low disturbance from grazing animals also helps in flourishing growth of these species in the downstream riparian zone.

Species like *I. carnea, S. spontaneum, V. zizanoides, T. dioica* stabilize the riverbank with the help of their root system (Figure 11.6). They reduce river bank erosion by providing bottom stability and diminishing wave action. Baruah *et al.* (2000) reported that these three species having tremendous soil binding capacity by their root system when planted in an eroded riverbank of the Brahmaputra river system in Majuli of North East India. Ambasht (1970) has studied the functional role of three species of riparian macrophytes and concluded that *Saccharum* species exhibited the highest conservation value (CV) for water, soil and phosphorous which are very effective in checking erosion in the river Ganga. *T. dioica* and *I. carnea,* the two non-grass amphiphytes in the riverbank mitigate river bank erosion and attenuate peak flow. These two species also provides fuel wood to the riverbank inhabitants.

The species of *Saccharum* are good fodder for cattle. Importance of this species as fodder is very significant particularly during the floods, when most of the areas remain inundated and scarcity of food occurs among the domestic animals. *M. ciliatum, I. cylindrica,* and species of *Saccharum, Vetiveria* are widely used by the floodplain inhabitants to making roof and walls of their thatched houses. *Oxystelma esculentum* and *Salix tetrasperma,* which remain confined strictly in a meandering cut-off (static water body), signifies their low tolerance to heavy water current. Out of these, the former one is a climber and has a high adaptive ability in high as well as low water saturated conditions.

The natural flow pattern of the river helps to increase the effectiveness of hydrochory (transportation of plant parts through water) of the seeds and other vegetative propagules of some of these giant amphiphytes. So, their distribution is found to be present throughout the downstream river bank and riparian zones, which are usually formed by the deposition of silt. *Saccharum, Tamarix* and *Arundo* are some perfect examples of this category. These riparian amphiytes has other

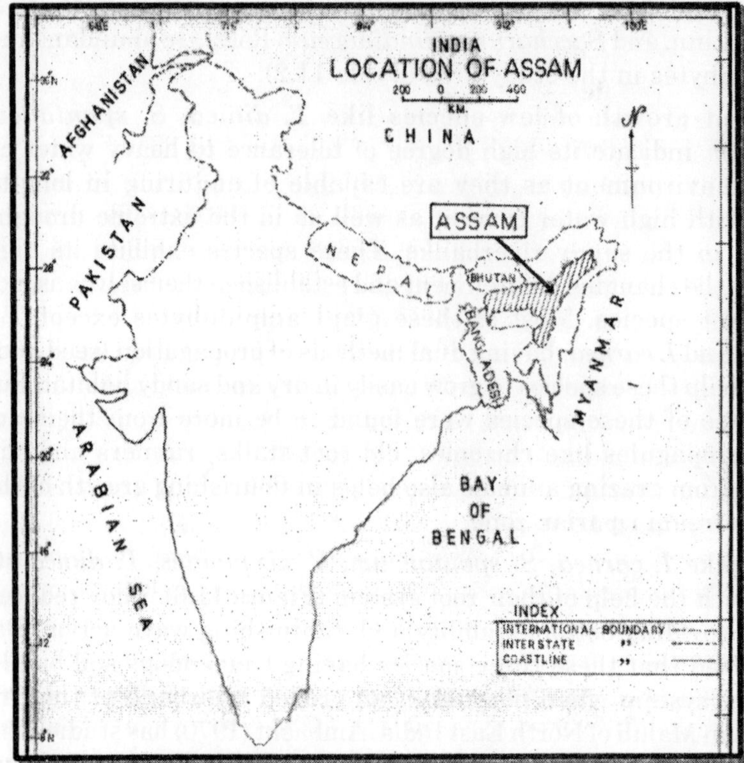

Figure 11.1. Location map of Assam.

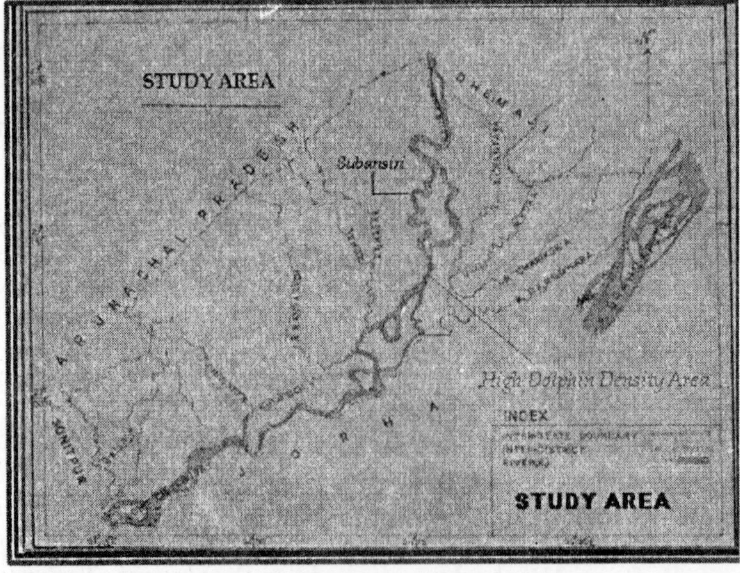

Figure 11.2. Map of Subansiri river showing the study sites.

Table 11.1
Statistical treatment and a conventional descriptive statistics of certain physico-chemical parameters of Subansiri river

Statistics	*Sites*	*Mean*	*Standard Error*	*Median*	*Standard Deviation*	*Skewness*	*Kurtosis*	P_{25}	*95% CL*
Water temperature (^0C)	A	18.52	1.23	18.70	4.28	0.101	-1.151	14.45	2.72
	B	18.70	1.23	19.15	4.26	-0.180	-1.322	15.15	2.71
	C	18.50	1.41	18.60	4.90	0.140	-1.628	13.15	3.11
	D	18.55	1.26	18.20	4.39	-0.057	-1.293	15.15	2.79
Transparency (cm)	A	38.61	5.47	35.05	18.97	0.703	-0.881	-22.60	12.05
	B	42.29	5.69	34.50	19.72	0.908	-0.747	26.85	12.53
	C	43.90	5.52	38.35	19.12	0.566	1.240	28.00	12.15
	D	45.32	6.08	35.65	21.08	0.749	1.044	28.65	13.39
Current flow (m/s)	A	1.91	0.15	2.05	0.52	-0.307	-1.350	1.35	0.33
	B	1.73	0.13	1.85	0.45	-0.467	-1.109	1.30	0.28
	C	1.79	0.11	1.85	0.38	0.075	-0.548	1.45	0.24
	D	1.70	0.12	1.70	0.43	-0.325	-0.393	1.45	0.27
pH	A	7.44	0.09	7.40	0.33	0.334	-1.219	7.15	0.21
	B	7.55	0.10	7.65	0.38	-0.341	-1.182	7.20	0.24
	C	7.53	0.10	7.65	0.35	-0.298	-1.583	7.15	0.22
	D	7.58	0.10	7.65	0.37	-0.159	-1.966	7.20	0.24
Total Dissolved Solids (ppm)	A	162.41	3.26	163.00	11.31	-0.125	-0.098	153.50	7.19
	B	166.25	4.22	164.50	14.64	-0.031	0.084	160.00	9.30
	C	165.08	4.61	169.50	16.00	-0.573	-0.719	154.00	10.16
	D	166.83	4.60	170.00	15.93	-0.400	-1.068	151.50	10.12
Total Suspended Solids (ppm)	A	29.00	1.84	29.50	6.38	0.431	1.234	25.00	4.15
	B	28.91	2.28	30.00	7.91	0.086	-0.971	21.51	5.02
	C	29.75	2.22	30.50	7.70	-0.234	-0.915	24.00	4.89
	D	29.41	1.04	28.50	3.62	1.855	4.091	27.00	2.30
Dissolved oxygen (mg/l)	A	8.61	0.23	8.70	0.82	-0.113	-0.982	7.95	0.52
	B	9.70	0.30	9.40	1.05	0.521	-0.333	9.05	0.67
	C	10.33	0.32	10.20	1.14	0.339	-0.303	9.60	0.72
	D	9.86	0.47	9.80	1.64	0.125	-1.448	8.35	1.04
Free carbon dioxide (mg/l)	A	4.71	0.30	4.80	1.04	-0.152	-1.063	4.05	0.66
	B	4.98	0.39	5.00	1.31	0.363	-0.930	4.05	0.88
	C	5.20	0.35	5.25	1.22	-0.066	-0.385	4.25	0.77
	D	4.25	0.22	4.25	0.78	-0.054	0.095	3.90	0.49

Table 11.2

Abundance and distribution of giant riparian amphiphytes in the Subansiri river ecosystem

Sl.No.	*Scientific Names*	*Abundance*	*Family*
1	*Arundo donax* Linn.	Common	Poaceae
2	*Calamus tenius* Roxb.	Very rare	Aracaceae
3	*Impereta cylindrica* (L) Beauv.	Common	Poaceae
4	*Ipomoea carnea* Jacq.	Frequent	Convolvulaceae
5	*Phragmites karka* (Retz.) Trin.	Common	Poaceae
6	*Saccharum arundinaceum* Retz.	Abundant and Dominant	Poaceae
7	*Saccharum spontaneum* Linn	Abundant and Dominant	Poaceae
8	*Tamarix dioica* Roxb.	Frequent	Poaceae
9	*Themda villosa* Dor et. Jack	Common	Poaceae
10	*Typha angustifolia* Linn.	Rare	Poaceae
11	*Vetiveria zizanoides* Stepf.	Common	Poaceae
12	*Microstegium ciliatum* (Trin.) A. Camus	Common	Poaceae
13	*Salix tetrasperma* Toxb.	Very rare	Salicaceae
14.	*Oxystelma esculentum* (L. *f.*) Sm.	Very rare	Asclepiadaceae

Figure 11.2. Dam site before construction of the dam (2003).

Figure 11.3. Dam site during the construction of the dam (2008).

Figure 11.4. Sediment traped by the macrophytes

Figure 11.5. Sediment deposition in the riparian zone.

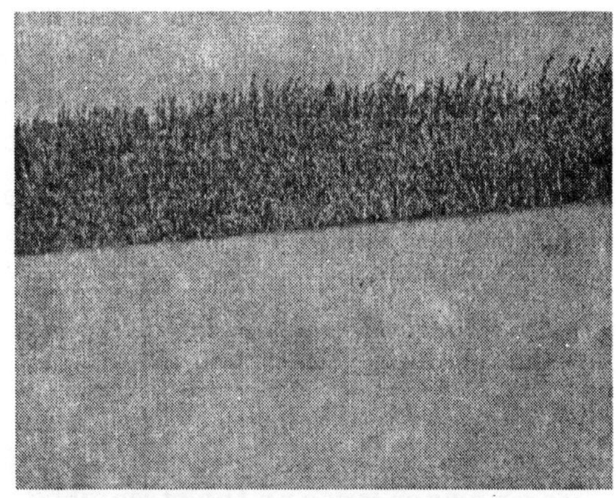

Figure 11.6. Riverbank stabilization by the riparian amphiphytes in the study areas.

functional aspect as well, as they act as habitat for juvenile prawn and periphytons which prefer bushy and shady places for colonization. During summer, the submerged parts of such amphiphytes providing an extra habitat for periphytons and maintaining the food chain (Srivastava, 2000).

In the lotic environment of Subansiri river, rooted submersed vegetation is temporary, but at least some emergent parts of all the giant riparian amphiphytes becomes submersed during seasonal flooding. The submersed parts act as sediment trapper and enhancing the deposition of silt and debris (Figure 11.4, 11.5). This rich organic content layer is an ideal habitat for benthic community particularly for rotifer, worms, larvae, nematodes etc. Bottom feeder fish species and some juvenile cat fish like *Mystes* are feed upon such benthic community. Because of temporary inundation, the submersed vegetative parts rotten very quickly and become rich source of bacteria and then infusorians and create a suitable environment for some small fishes like *Danio, Puntius* etc.

During flooding, aggregation of different micro and macro invertebrates notably grasshopper, locust, gyrotelpa, spider, ants etc. were observed in the emergent parts of the giant amphiphytes. Flooding is also responsible for vertical movement of soil micro invertebrates from soil as well as from the under storied herbs to the emerging parts. Though, most of them are able to climb to a safer places, some species are unable to move to the safe places and become predation of insectivores fishes like *Mystus, Clupisoma, Eutropicthys* etc. So, these amphiphytes though acts as a protector for many terrestrial micro and macro invertebrates from seasonal flooding by providing safe habitat, simultaneously they also maintain the aquatic food chain of the river by providing insects to the carnivores fishes.

Dam constrain a river's flow regime by storing water in reservoir, limiting riverbank floods and regulating and minimizing the magnitude, frequency, duration and timing of natural flows (Baxter, 1997). Reduction of water table height can be lethal to seedlings of these riparian amphibian plants and thus may limit or prevent recruitment of new species. Similar findings are also reported earlier by Rood and Mahoney (1990); Segelquist *et al.* (1993). Dams also affect sediment transport through a river system, as sediment entering a reservoir is stored behind dams, resulting in sediment-starved conditions in the downstream (Graff, 2002).

In our investigation, deposition and trapping of sediments in the soil, by the amphiphytes were noticed in its entire downstream. This is due to the unregulated condition of the Subansiri river, which transports enormous amount of nutrients from the upstream in the form of sediments to the downstream. However, the proposed 2000 MW dam will create a physical barrier and the entire downstream will be deprived from natural nutrients. This will also creates the potential for channel incision, bank erosion, change in channel platform and development of a coarse armor layer on the streambed as fine sediments wash the downstream. The damming of river Subansiri may encourage the loss of present habitats of the riparian amphiphytes. The destruction of a particular type of habitat will result in the disappearance of certain species. As the availability and diversity of habitat are major determinants of whether a given system is acceptable to a specific group of biota or not, knowledge of the availability and quality of habitats is very important in an overall assessment of ecosystem health (RHP, 2006).

Conclusion

Aquatic plants and riparian vegetation are an important element in tropical streams, but are frequently overlooked in limnological studies. Though, truly aquatic species are few, Himalayan river plants nonetheless play an important role in stream functions. Literatures of previous works advocated that river impoundment is the loss of pulse-stimulated responses at the water-land interface of the riverine system. Therefore, when the 2000MW Lower Subansiri Hydroelectric Project will be operated, elimination of these riparian amphiphytes could be possible from the Subansiri river ecosystem by the loss of connection of downstream flood plain from the river as well as by altering flow patterns, water chemistry, limiting water table, downstream movement of sediment and biological materials.

[**Acknowledgement:** Financial assistance provided by University Grant Commission, New Delhi, India in the form of Major Research Project (Project No: F No33-137/207(SR)-2008) to the first author is gratefully acknowledged.]

REFERENCES

Ambasht, R.S., 1970. Conservation of soil through plant covers of certain alluvial slopes in India. *Proc. Nat. Acad. Sc. X1 Tech. Meeting*, Switzerland, 44 - 48.

Anon, A., 1975. What's happening to the Zambezi? *Afr. Wildl.*, 29(2): 18–21.

APHA, 1992. *Standard Methods for the Examination of Water and Waste Water*. 18th edition. Am. Public Health Ass., Washington, D.C., 1193 pp.

Baruah, D., Biswas, S. P. and Hazarika, A. 2000. An Experimental studies of soil conservation using herbaceous plants in Majuli Island, Assam, India. *The Environmentalist*, 20: 19-27.

Baxter, R. M. 1997. Environmental effects of dams and impoundments. *Annual Review of Ecology and Systematics*, 8: 255-283.

Begg, G.W. 1973. The biological consequences of discharge above and below Kariba Dam. *Proc.Int.Congr.Large Dams*, 11: 421–30.

Chude, L.A. 1979. Fish and fisheries of Lake Ndakolowu - A floodplain downstream of Kainji Dam. *Annu.Rep.Kainji Lake Res. Inst*, 79: 41-42.

Cook, C.D.K. 1996. *Aquatic and Wetlands plants of India*, Oxford University Press.

Dutta, A.C. 1985. *Dictionary of Economic and Medicinal Plants*. Assam Printing Works: Jorhat, Assam, India, 23 pp.

Graff, W.L. 2002. Dam removal, research, status and prospects. In: *Proceedings of the Heinz Center's Dam Removal Research Workshop*. The H. John Heinz III Center for Science, Economics and the Environment. Washington, D.C.

Ita, E.O. and Mohammed, A. 1980. A preliminary report of expeditions to the inlet and outlet channels of Lake Ndakolowu (Tatabu) and a proposal for refilling the lake. *Annu.Rep.Kainji Lake Res. Ins.*, 80: 29–31.

Khan, M.S. and Halim, M. 1987. *Aquatic Angiosperms of Bangladesh*. Bangladesh National Herbarium.

Kondolf, G.M., Webb, J. W. and Felando, T. 1985. Basic Hydrologic Studies for Assessing Impacts of Flow Diversions on Riparian Vegetation, Examples from Streams of the Eastern Sierra Nevada, California, USA. *Environmental Management*, 11(6): 757-769.

Melzer, A. 1999. Aquatic macrophytes as tool for lake management. *Hydrobiologia*, 396: 181–190.

Moxon, J. 1969. *Volta, man's greatest lake*. London, Andre Deutsch Ltd., 256 pp.

Pople, W. and Rogoyska, M. 1969. The effect of the Volta River hydroelectric project on the salinity of the Lower Volta River. *Ghana J. Sci.*, 9(1): 9–20.

Press, J.I., Shrestha, K. K. and Suttan, D. A. 2000. *An annotated check list of Flowering plants of Nepal*. London British Museum of Nature History and Kathmandu: CDB, Tribhuban University, 430 xiii.

RHP; 2006. River *Health Programm, South Africa*. The River Health Programme, Indicators and Indices. htm. mailto:iwre@icon.co.za.

Rood, S.B. and Mahoney, J. M. 1990. Collapse of riparian popular forests downstream from dams in western prairies: probable causes and prospects for mitigation. *Environmental Management*, 14: 451– 464.

Segelquist, C.A., Scott, M. L. and Auble, G. T. 1993. Establishment of *Populus deltoids* under simulated alluvial groundwater declines. *American Midland Naturalist*, 130: 274- 285.

Srivastava, V. K. 2000. Effect of riparian vegetation on the river ecology and aquatic biodiversity with special reference to Arunachal Pradesh (India) - A review, pp. 135-146. In *Sustainable management of Forests-India* (eds. M.L. Khan and A. Arunachalam), International Book Distributors, Dehradun.

Stelzer, D., Schneider, S. and Melzer, A. 2005. Macrophyte based assessment of lakes – a contribution to the implementation of the Water Framework Directive in Germany. *Internat. Rev. Hydrobiol.*, 90(2): 223–237.

SWECO/SWED. 1981. *Cahora Bassa hydroelectric power scheme*. Stage II. Preliminary report on ecology. Stockholm, SWECO.

Westlake, D.F. 1975. Macrophytes in river ecology. In: *Studies in Ecology* (ed. B.A. Whitton), Vol. II, Blackwell Scientific Publications, Oxford, UK.

Wetzel, R.G. 2000. *Limnology, Lake and River Ecosystems*, 3rd edition. Academic Press: San Diego, CA, USA.

12

Ornamental Fishes of Hill Stream Channels of Subansiri Drainage System of Assam, North East India: Potential for the Aquarium Fish Trade and Conservation Issues

—Bikramaditya Bakalial[1], Sabita Borah[2], & Lakhi Prasad Hazarika[3]

[North East India is rich in fish biodiversity. Many freshwater fish species have been exploited from the region for the ornamental fish trade. To conserve and sustainably exploit the wild fish population, measures should be adopted to protect this natural resource. Already, several ornamental species have been severely overexploited, e.g. *Danio* sp. *Colisa* sp, *Channa* sp. etc but the extirpation of local populations occurs for many reasons, including deforestation, and not just because of fishing for the trade. Still there are many fish species having great ornamental potential. A total of 24 potential ornamental fishes from different hill stream channels of Subansiri drainage system have been identified. Their natural habitats and adaptability in the aquarium habitat are discussed. During one year in the changing environment of aquarium, fishes are thrive well with 100% survivality except 25% mortality of *Danio* sp. Captive breeding of these species not only help in conservation but also beneficial for the aquarium fish industry, environment, people's livelihoods.

Keywords: Subansiri drainage system, hill stream, ornamental fish and conservation.]

Introduction

A long narrow depression shaped by the concentrated flow of a stream and covered continuously or periodically by water is known as hill stream channel, while riffle is a reach of stream that is characterized by shallow, fast-moving water broken by the presence of rocks and boulders. In the Assam - Arunachal Pradesh border at the foot hills of Eastern Himalayas, the natural water flow in the form of feeder stream exhibits all these characteristics in the Subansiri drainage system. The unregulated Subansiri River during its course in Assam of North East India with

[1,2] Department of Zoology, Lakhimpur Girl's College, Lakhimpur, Assam, India
[3] Department of Zoology, North Lakhimpur College, Lakhimpur, Assam, India

an approximate length of 130 km provides diverse aquatic habitat (Hazarika *et al*, 2008). Still, there is vast unexplored potential for indigenous ornamental fishes in the Subansiri drainage system. Baruah *et al.* (2008) reported some potential aquarium plants from Subansiri flood plain while Hazarika *et al.* (2008) reported ornamental fish diversity of unregulated Subansiri River and discussed the probable impact of proposed dam on them.

In this paper, we document the fish fauna having ornamental value of different hill stream channel of Subansiri drainage system, assess their patterns of distribution with regard to different habitat types, bed materials and suggest suitable conservation strategies through aquarium culture.

Methodology

Sampling sites: Four different hill streams namely *Kakoi, Kona, Ranga* and *Dirgha* of Subansiri drainage system were chosen for the study.

Sampling of fishes: Samplings were carried out at all the streams on a monthly basis during 2008-09. Experimental fishing was carried out both by members of the project team as well as by using the expertise of local fisher folk. Different types of gear including scoop nets, drag nets (all with varying mesh sizes), traps and other local contrivances were used for collecting the fishes allowing us to sample a range of fish sizes and minimize the bias due to specific gears.

Study of stream habitats: We classified hill stream habitat types according to Konkel (2001) and Fischenich (1999) Armantrout (1999); Arunachalam (1999). Identification of fishes was made by following the standard literatures of Talwar and Jhingran (1991), Jayaram (1999); Data regarding threats faced by the fish fauna were obtained from both primary (direct observations and interactions with local community) and secondary (journal articles, reports, books, and internet search tools sources.

Aquatic environment: Four parameters viz. Water Temperature by using mercury thermometer, pH by pen type digital pH meter, DO by Winkler's modified method (APHA, 1989), FCO_2 according to Welch (1952) of the hill streams were recorded for twelve months.

Study of fishes in the changing environment: The glass aquariums (size 0.90 x 0.46 x 0.30 m) were used. Before plantation of aquatics, the bottom of the aquariums is filled with thoroughly washed coarse sand (3 cm) and pebbles (2 cm) and water level of 35 cm depth. Saplings and rhizomes of the plants and selected fishes were treated with methylene blue (deep for 1-5 minutes, 5 drops in 4 liters of water), oxy tetracycline capsule (500 mg in 4 liters of water) as antifungal and antibacterial agent respectively. Besides these, the plants were also immersed into non-iodized common salt solution for 5 minutes (1 teaspoon non-iodized NaCl dissolved in 4 liters of water) to avoid contamination from both fungi and bacteria prior to the introduction of plants in the aquariums. After introduction of submersed plants, the aquariums were placed in a room with only diffused sunlight. 1/3 of the water was changed at every fifteen days interval by siphoning and the levels maintained when necessary. Representative specimens of all fish species were trap carefully and within six hours transferred to the laboratory and kept in glass aquarium. To know their tolerance in the changing environment observation was made for twelve months.

Results

Rivers can have *riffles,* which are characterized by relatively steep gradients and shallow depths, and *pools,* which are characterized by relatively gentle gradients and greater depths. Riffles are found where coarse sediment has been deposited in the channel. Pools can form behind natural obstructions in the channel, such as logs deposited on the bed. As one might expect, a river's flow is faster in riffles than pools. Free-flowing rivers usually have pools and riffles, which provide diversity in habitat types in a river.

In the hill stream channels, six different meso-habitats such as –a) Coarse woody debris (CWD), b) Wind-throw (a tree or trees uprooted or felled by the wind), c) Riffle (a reach of stream that is characterized by shallow, fast-moving water broken by the presence of rocks and boulders) , d) Rootward (the mass of roots associated with a tree adjacent to or in a stream that provides refuge for fish and other aquatic life) e) pools, and f)Space between boulders, assembles of drift wood have been identified from where thirteen small and one big hill stream fishes having aquarium values were collected, identified and selected for kept in the aquarium. Each species has been presented along with their scientific names and respective families (Table 12.1). Fish species are confined to six different habitats indicates their preferential habitats. Most of the fishes prefer to stay in the pools and rootward. The estimated slope of the hill stream in the foothill regions are in the range of 1:25 to 1:33 where the average water current is estimated the range of 2.5 m/s to 3.75 m/s.

In the hill stream channel the ranges of water temperature, pH, DO and FCO_2 in between 11^0-34^0C, 7.6-8.9, 7.8-13.8 mg/l, 2.3-4.6 mg/l respectively (Table 12.2).). Fishes like *Danio devorio, D. aequipinnatus, D. dangila, Schistura arunachalensis, Barilius bendelisis, Barilius tileo, Barilius sachra, Barilius barna, Aspidoparia morar, Amblyceps mengois, Salmophasia phulo, Oreichthys* sp. *Chaudhuria indica, Olyra longicaudata, Badis assamensis, Badis badis, Channa aurentimaculata, Channa barca, Garra gotyla gotyla, Garra annandalei, Semiloptes gongota, Psilorhynchus gracilis, Puntius ticto, Glyptothorax cavia,* of hill stream channel prefers pools, rootward coarse woody debris, riffle. Fishes like *Danio devorio, D. aequipinnatus, D. dangila, Aspidoparia, Barilius bendelisis, B. barna, B. tileo, Salmophasia, Oreichthys* sp. *Puntius* sp. exclusively found in pools while *Schistura arunachalensis, Garra, Psilorhynchus, Semiloptes* found in riffles only. *Chaudhuria indica, Olyra longicaudata, Badis assamensis, Badis badis, Channa* sp. *Amblyceps* were found rootward habitat exclusively.

In the changing environment (aquarium), the ranges of water temperature, pH, DO, FCO_2 were in between 11^0-32^0C, 7.2-7.9, 5.2- 8.1 mg/l, 2.3-4.6mg/l respectively (Table 12.2). In the aquarium condition the value of dissolved oxygen fluctuates diurnally i.e. maximum in afternoon to evening while minimum in late night and early morning but no such fluctuation recorded in stream environment.

The artificial oxygen supply was not provided to fishes those kept in the aquarium except in the months from November to January. In these months the fishes exhibited irritating behaviour in the early morning (4-6 a.m.) and become normal by 9-10 a.m. During the experimental period of twelve months, no mortality was recorded in all the 24 species except death of one individual each of *Danio devario, Amblyceps, Olyra* and *Badis* out of 96 experimental individuals in four aquariums.

Feeding: Feeding was done by artificial prepared feed (Tokyu baby pellet), dry tubifex worm and live earthworms only. Approximately by 5% of their body weight, twice a day (morning and evening) alternatively pellet and worms were given. Acceptability and compatibility to artificial food was immediate and high in case of *Danio, Salmophasia, Puntius* and *Barilius* fishes moderate in *Schistura, Garra, Channa* and low in *Badis, Olyra, Amblyceps* and *Chaudhuria* species.

Among all these experimental fishes *Chaudhuria, Olyra* and *Amblyceps*, prefer rocky substratum of the aquarium environment and sluggish in behaviour. Because of predatory nature *Channa* species are not compatible with other species particularly barbs.

The present threats for this potential are illegal wanton exploration (fishing of juveniles and brooders, use of very small mesh sized net, blasting, liming, and herbal poisoning, electro-fishing etc), habitat degradation (aquatic pollution, sand mining, gravel mining etc). The most probable impact will be created by the 2000 MW Lower Subansiri Hydroelectric Project.

Discussion

Fishes of hill stream channels having ornamental values are small in size and fast mover against water current. Other morphological characteristics like cylindrical or flattend body, presence of burble, adhesive organs enable these fishes to adapt well in the hard riverbed structure with fast moving water. Natural flow and physical objects like boulders, uprooted trees, debris, shallow pool etc. are provided different types of micro & meso habitats for these native fishes. These various natural micro and meso habitats may be responsible for attractive colouration of the fish fauna in comparison to riverine system of plains. Rich riparian zone provide ample natural nourishment and thus maintaining the aquatic food chain besides creating shelter. So, most of the fishes used uprooted tree, shallow polls, boulders etc. for their shelter and feeding. The movement of fishes to the riffle, main stream was also observed. They used to move almost all the habitats for feeding. They generally preferred to used pools rather than riffles and main channel for breeding to avoid the damage of eggs and spawns.

Observed aquatic environment of hill stream channel revels that slightly basic pH value, with low temperature in dry spell and moderate temperature in wet spell along with high dissolve oxygen content and low FCO_2 contents throughout the year with negligible variations may be due to river channel continuum process.

In hill stream dissolved oxygen comes from a variety of sources. The action may be due to high turbulence and water tumbling over rocks helps mix oxygen in the atmosphere with moving water. Aquatic plants, partly submerged portions of

Table 12.1
List of the hill stream fishes of Subansiri River

Sl. No.	*Scientific name*	*Family*	*Habitat*
1.	*Danio devorio*	Cyprinidae	pools
2.	*Danio aequipinnatus*	Cyprinidae	pools
3.	*Danio dangila*	Cyprinidae	pools
4.	*Puntius ticto*	Cyprinidae	pools
5.	*Barilius bendelisis*	Cyprinidae	pools
6.	*Barilius tileo*	Cyprinidae	pools
7.	*Barilius sachra*	Cyprinidae	pools
8.	*Barilius barna*	Cyprinidae	pools
9	*Aspidoparia morar*	Cyprinidae	pools
10	*Oreichthys sp.*	Cyprinidae	pools
11	*Salmophasia phulo*	Cyprinidae	pools
12	*Garra annandalei*	Cyprinidae	riffles
13	*Garra gotyla gotyla*	Cyprinidae	riffles
14	*Olyra longicaudata*	Olyridae	rootward
15	*Badis assamensis*	Nandidae	rootward
16	*Badis badis*	Nandidae	rootward
17	*Channa aurentimaculata*	Chanidae	rootward
18	*Channa barca*	Chanidae	rootward
19	*Chaudhuria indica*	Choudhuriidae	rootward
20	*Amblyceps mengois*	Amblyciptidae	rootward
21	*Semiloptes gongota*	Cobitidae	riffles
22	*Psilorhynchus gracilis*	Psilorhynchidae	riffles
23	*Schistura arunachalensis*	Balitoridae	riffles
24	*Glyptothorax cavia*	Sisoridae	pools

Table 12.2
Physico-chemical parameter of hill streams

Habitat	*In hill streams*		*In aquarium*	
Parameters	*Min*	*Max*	*Min*	*Max*
Temperature (^{0}C)	11	34	11	32
pH	7.6	8.9	7.2	7.9
DO_2 mg/l	7.8	13.8	5.2	8
FCO_2 mg/l	2.3	4.6	2.3	4.6

Pool

Riffle

Gravel bars

Windfall

Wind through

Wildlife tree

Off channel pool

Rootward

Debris

Figure 12.1 Hill stream channel and riparian flora in different altitudes.

riparian plants, filamentous and non filamentous algal blanket of the rocky beds (Figure 12.1) also release oxygen into the water as a byproduct of photosynthesis during daylight hours. The values observed in the present investigation can be taken as positive indicators of the aquatic environment for survival of these fishes. Small amounts of carbon dioxide are a sign of low organic material and high dissolved oxygen. The low FCO_2 data shows the positive balance between producer and consumer in the hill stream channels.

In the aquarium, the fishes are able to cope with the changing environment. Their well adapted nature in the aquarium environment shows their ability to withstand in the changing aquatic environment of aquarium. Poor growth of aquatic plants in the aquarium unable to maintained dissolve oxygen for survival of the fishes in three winter months due to the reduction in rate of photosynthesis activities and thus reducing the dissolved oxygen content particularly in the early morning.

100% survivality most of the experimental species and except 25% mortality of *Danio devorio, Amblyceps, Badis* and *Olyra* species indicates their high adaptability in aquarium environment. Therefore these fishes may be regarded as aquarium fishes and one can use for sustainable development through in-vitro culture. Captive breeding of these species will help in conservation of these in one hand and on the otherhand low-cost enterprise can be developed for earning livelihoods.

Presently, illegal wild collection of fishes and harvesting, felling of trees, sand and gravel mining are the main threats for these potentials and degradation of their natural habitats. Deforestation and resultant habitat degradation can lead to a decline in exogenous food sources including insects and their larvae as well as reduction in leaf debris due to shade loss. Canopy cover, which has been shown to play an important part in fish community distributions, was found to be lacking in many sampling locations, probably due to deforestation (Marsh-Matthews and Matthews, 2000). Such habitat changes could therefore significantly alter the community structure as well as lead to a decline in species diversity (Raghaban et al., 2008). It is also important to note here that the sustainable utilization of aquarium fish resources is an important factor to include when forest conservation in general is considered (NgPeter and Tan, 1997). The economic value of conserving forest systems usually involves calculations of forest products and by-products, almost always leaving out the value of aquarium fish (Peane and Moran, 1994). This sustainable byproduct of an intact forest can add substantially to the overall calculations.

In future, the 2000 MW Lower Subansiri Hydroelectric Project, which has begun its construction near Assam-Arunachal Border will more catastrophic for these fishes through various alteration in the downstream drainage systems. Alterations to invertebrate and macrophyte communities, caused by hydropower schemes, can lead to reduced shading and food availability for fish (Petts, 1998).

Conclusion

Ninety five per cent of our ornamental fish export is based on wild collection. Majority of the indigenous ornamental fish trade in India is from the North Eastern states and the rest is from Southern states which are the hot spots of ichthyodiversity in India. This capture based export is not sustainable and it is a matter of concern for the industry and sustainability of native species. In order to sustain the growth, it is absolutely necessary to shift the focus from capture to culture based development. Most of the fish species grown for their ornamental importance can be cultured successfully. Organized trade in ornamental fish depends on assured and adequate supply of demand, which is possible only by mass culture. Simultaneously, it will help in conservation and sustainable use of potential aquarium fishes.

[**Acknowledgement:** Financial assistance provided by UGC, New Delhi, India in the form of Major Research project [Project No: F No35-124/2008(SR)] to the second author is gratefully acknowledged.]

REFERENCES

APHA 1989. *Standard method for the experimentation of water and waste water*, American Public Health Association, New York.

Armantrout, N. B. 1999. *Glossary of aquatic habitat inventory terminology*. American Fisheries Society, Bethesda.

Arunachalam, M. 1999. Methods for fish habitat inventory in streams/rivers. *Proc workshop Germplasm inventory and gene banking of freshwater fishes*. National Bureau of Fish Genetic Resources, Lucknow, India

Arunachalam, M. 2000. Assemblage structure of stream fishes in the Western Ghats. *Hydrobiology*, 430: 1–31.

Fischenich, C. 1999. *"Glossary of stream restoration terms," EMRRP Technical Notes Collection* (ERDC TN-EMRRP-SR-01), U.S. Army Engineer Research and Development Center, Vicksburg, MS.

Jayaram, K. C. 1999. *The freshwater Fishes of the Indian region*. Narendra Publishing house, New Delhi. 551 pp.

Konkel, D. 2001. *The Aquatic Plant Community in the Chippewa River*. http://dnr.wi.gov/fish/reports/final/buffalochippewariver2001.pdf

Marsh E. and Matthews, W.J. 2000. Geographic, terrestrial and aquatic factors: which most influence the structure of stream fish assemblages in the Midwestern United States? *Ecol. Fresh Fish*, 9: 9–21.

Ng Peter K. L. and H. H. Tan 1997. Freshwater fishes of Southeast Asia: potential for the aquarium fish trade and conservation issues. *Aquarium Sciences and Conservation*, 1: 79-90.

Peane, D. and Moran, D. 1994. *The Economic Value of Biodiversity*. IUC publications, Geneva, 172 pp.

Petts, G.E. 1988. *Impounded Rivers: Perspectives for Ecological Management*. John Wiley and Sons Inc.

Raghavan, R., Prasad, G., Anvar Ali, P.H. and Pereira, B. 2008. Fish fauna of Chalakudy River, part of Western Ghats biodiversity hotspot, Kerala, India: patterns of distribution, threats and conservation needs. *Biodivers Conserv.*, 17: 3119–3131.

Talwar, P. K. and Jhingran, A. G. 1991. *Inland Fishes of India and adjacent countries*, vol I & II. Oxford & IBH Publishing Co. Pvt. Ltd., New Delhi.

Trivedy, R.K., Goel, P.K. and Trisal, C.L. 1987. *Practical Methods in Ecology and Environmental Sciences*. Enviro Media Publication, Karad (India). pp.123-126.

Welch, P.S. 1952. *Limnology*, McGraw Hill Book Company, Inc., New York.

13

Fish Diversity and their Preferential Habitats of Subansiri Flood Plain in North East India: Hydroelectric Project and Conservation Needs

—*Lakhi Prasad Hazarika[1], Sabita Borah[2], Bikramaditya Bakalial[3] & S.P. Biswas[4]*

[The investigation has been carried out in order to explore the existing fish fauna of the feeder streams and floodplain wetlands of Subansiri Drainage System of Assam according to their habitat preference. Riverine fishes exhibit their preference in response to riverbed structure, availability of food and to some extent riparian vegetation. Hill stream fishes prefer river bed with hard sediments for breeding, spawning and feeding in the downstream whereas some fishes exclusively prefer vegetation rich riparian wetlands. The existence of River dolphin *Platanista gangetica* maintains the balance between herbivore and carnivores fishes in the entire downstream acting as top consumer. The presence of diverse ichthyofauna indicates the healthy river ecosystem of unregulated Subansiri river. The regulation of the river by the ongoing 2000 MW Hydroelectric Project would be a threat for downstream habitats and riparian or floodplain wetlands as well as the native fishes exist therein. Necessary steps to maintain the downstream flow regime is utmost essential to save the habitats and native fishes of different conservation status.

Keywords: Subansiri river, downstream, habitat, fish diversity, dam.]

Introduction

Large rivers significantly support the earth's aquatic biodiversity. Migratory fish species are vulnerable during their life cycle due to river damming, and about 20% of the world's fresh water fishes are estimated near extinction or in urgent need of conservation (Rai, 1989). Therefore, to sustain biodiversity and fisheries in rivers requires sustainable management of both the habitats and systems of exploitation. For the management of large river fisheries, the concerned authorities

[1] Department of Zoology, North Lakhimpur College, Lakhimpur, Assam 787031, India
[2,3] Department of Zoology, Lakhimpur Girl's College, Lakhimpur, Assam 787031, India
[4] Department of Life Sciences, Dibrugarh University, Dibrugarh, Assam, 786004, India

should be involved in the planning before the damming of rivers for electricity generation, in order to maintain and conserve river fish, fisheries and their environment accordingly (Rai, 1989).

The micro and meso climate provided by the different zones of a river attract different type of fish species depending on their preferential habitat, which in turn depends largely on factors like variability of nourishment, dissolved oxygen, light, flow etc. in different zones of a river (Vyas *et al.*, 2009). The organization of fish assemblage reflects both biotic and abiotic sets of features occurring in a particular movement which influence the distribution, abundance and species interactions, most of them direct or indirect partitioned (Wootton, 1998). Again, wetlands connected with river systems are called riparian wetlands, *which* are found in lowlands adjacent to river systems (WRM, 2000). These areas usually occupy the floodplain and are periodically flooded by the river. Riparian wetland ecosystems can have high biological productivity, species diversity, and density.

Subansiri river, the largest tributary of the mighty Brahmaputra is a Himalayan river originating from the western part of Mount Pararu (5059m height) in Tibetan Himalayas. It is sustained by snowmelt run off, the ablation of glaciers and monsoon rainfall. After flowing about 190 kms in Tibet, it enters into India and continues its journey through the Himalaya region of India for 200 kms and enters the plains of Assam near Gerukamukh of Dhemaji district. Its total length is 520 kms and drains a basin of 37000 sq km (Baruah *et al.*, 2008). Himalayan Rivers provide different gradient of habitat heterogeneity from its headwater to mouth for colonization of aquatic flora & fauna (Srivastava, 2007).

The Northeast India has been identified as India's 'future powerhouse'. According to a study of the Central Electricity Authority (CEA) of India (2001), probability of construction of a number of 168 large hydro electric projects having potential of 63,328 MW have been identified only in the Brahmaputra river basin. This includes 22 projects having potential of 15,191 MW in the Subansiri river basin.

The 2000 MW Lower Subansiri Hydel Project, proposed to come up at Gerukamukh on the Assam-Arunachal border is the first large project to be taken up in the Subansiri basin. Except for a few accounts of diversity, not much information is available on the fisheries of this river, especially those on the patterns of fish species distribution. So, the present study is aimed at assessing the preferential habitat of fish species with regard to the probable influence of river on such habitats and the fish fauna existing therein. This in-depth study will help in preparing a database of freshwater fishes of lower Subansiri drainage system and can be taken as a guard data after regulation of the river by the said hydel project. The goal of this paper is to assess probable disruptions in aquatic and riparian habitats caused by dams.

Study Area

For a systematic monitoring and study of fish diversity, entire downstream (from dam site to its confluence with mighty Brahmaputra) have been divided into four sectors namely Sector I (Gerukamukh or Dam site to Chowal dhowa Ghat,), Sector

II (Chowal dhowa Ghat to Khabolo Ghat, Sector III (Khabolo Ghat to Dhunaguri Ghat) and Sector IV (Dhunaguri Ghat to Jamuguri Ghat or confluence) (Figure 13.1). Along with the river channel, other fish habitat i.e. riparian wetlands which are influenced by the Subansiri river in the downstream drainage system have alsol been studied to know the preferential habitats of fish fauna.

Materials and Methods

Prior to the study, a thorough survey was made using proper surveying techniques to cover as much as possible areas as well as represent the maximum fish composition of the downstream floodplain of the Subansiri river. The collections were made at different locations during Dec. 2007 to June 2009.

For substrate composition, both visual observations and sampling were carried out. Structural composition of riverbed was determined according to Konkel (2001) and Fischenich (1999).

The fishes were collected using hired fishermen with cast and monofilament gill nets of different mesh sizes. The fish samples were also collected from different fish landing centers, local fish markets etc and preserved at 4% formaldehyde. The detailed taxonomic examination either in fresh or preserved specimens were made following the key of Talwar and Jhingran (1991) and Viswantath (2002).

Results and Discussion

Fish diversity in the downstream: The diversity of the ichthyofauna with their preferential habitats are depicted in Table 13.1. A total of 155 fish species belonging to 30 Families and 10 Orders have been identified in the total 130 km downstream floodplain of Subansiri river.

In the first sector of the river (Figure 13.2), most of the collected fish species are hill stream fishes. The riverbed is composed of boulders, gravels and pebbles and thus the river provided almost same habitat as in the upstream with exception only in width and current flow. Here the river approximately 3-5 times width in comparison to the up stream of the dam site. Riverbed structure, current flow, availability of algal and planktonic matter together make a healthy feeding, breeding and spawning ground for the hill stream fishes and exhibit their abundance in this sector. Baruah (2007) reported high dissolved oxygen content, slightly basic pH value, high transparency with substantial numbers of phyto and zooplankton. Fishes like *Tor* spp., *Labeo* spp., *Barilius* spp., *Neolissochielus* spp; *Garra* spp., *Glyptothorax* spp. *etc* use this habitat as their feeding and breeding ground. Giant catfish like *Bagarius bagarius* also use such habitat for their breeding. Other riverine fishes like *Catla, Wallago, Mystus, Ompok, Clupisoma* etc do not prefere this segment as their habitats and territory because of hard river bed composition and comparatively fast flowing and cold water. This type of riverbed structure in this segment is due to the sharp inclination coupled with high water discharge with more current velocity which encouraged in the transportation of boulder, cobble, pebble and sands.

The other three segments (II, III, & IV) with an approximate length of 110-113 km have alluvial silt type of river bed structure (Figure 13.3-13.5). Rivers is comparatively wide and deep in these sectors and has many seasonal and perennial feeder streams along with several floodplain lakes connected to it. The riverine fishes are distributed evenly in these sectors. The fourth sector is comparatively rich in fish species composition possibly for the movement of fishes from the greater Brahmaputra River to the Subansiri. In these stretches of downstream, the dominated fishes are – Indian Major Carps (IMCs), *Notopterus* spp., *Aspidoparia* sp., *Puntius* sp., *Barilius* spp., *Salmostoma* sp., *Chela* sp., *Esomus* sp., *Rasbora* sp., Catfishes like *Mystus* sp., *Aorichthys* sp., *Wallago* sp., *Ompok* sp., *Clupisoma* sp., *Eutropiichthys* sp., *Pseudotropius atherinoides* etc. (Figure 13.8-13.14).

Riparian wetlands and fish diversity: The entire downstream is rich in floodplains lakes or riparian wetlands (locally -*beels*) (Figure 13.6-13.7). The ecology of these wetlands are related with the river Subansiri as they made linked with the river at least for 4-6 months during the flood seasons (from last of April to the mid of September). They are rich in fish diversity as they provide varieties of microhabitat to the fishes. The most dominant fishes of riparian wetlands are *Notopterus notopterus, Labeo* spp., *Cirrhinus* spp., *Wallago attu, Mystus* spp. *Salmophasia bacaila, Puntius* spp., *Rasbora daniconius, Chela* spp., *Lepidocephalus guntea. Clarias batrachus, Heteropneustes fossilis, Nandus nandus, Anabas testudineus, Colisa* spp., *Channa* spp., *Monopterus cuchia* etc. These wetlands are the main breeding ground of most of the Indian Major Carps and many other fishes.

Hill stream fishes in the downstream

Many hill stream fishes have been collected in the downstream of river Subansiri which may perhaps came in the fast running water of its tributary Dikrong and Ranganadi originating from the hills of Arunachal Pradesh.

Composition of herbivorous and carnivorous fishes: Though the more diversity of piscivorous giant cat fish species are in the run, ecological pyramids exhibit upright directions in terms of numbers, biomass and energy by maintaining the balance between herbivores and carnivores fishes in each trophic levels. This ecological balance is due to the presence of substantial numbers of River dolphin in the Subansiri, which act as a top consumer in the river ecosystem and playing a significant role in maintaining the balance between the respective trophic levels. River dolphin consumed various fish species like *Clupisoma, Eutropiichthys vacha, Aorichthys aor, Aorichthys seenghala* and also *Notopterus chitala* and controls the population of carnivorous fish species. Hazarika *et al.* (2008) reported substantial numbers of this IUCN red listed animal in the downstream of Subansiri river.

Possible Impact of the 2000 MW Lower Subansiri Hydroelectric Project on fishes and their habitats. Some hill stream fishes used initial downstream stretch of Subansiri for feeding and breeding, where river bed is composed of sand and gravel along with high dissolve oxygen content and food sources. The dam can present itself as a physical barrier eliminating "a fish way" and thus, passage to spawning grounds is lost. Lower Subansiri dam will create obstacle in downstream migration of such hill stream fishes and thus these fishes may lost their existence in the river system. Downstream river bed degradation can also lead to the loss of important in-

Figure 13.1. Map showing the geographic location of Subansiri river.

Figure 13.2. Habitat at sector I.

Figure 13.3. Fishing operation in sector II.

Figure 13.4. Experimental trip at sector III.

Figure 13.5. Fishing at sector IV.

Figure 13.6. Community fishing at adjoining wetlands.

Figure 13.7. Fishing at riparian wetland.

Figure 13.8. *Aorichthys spp.*

Figure 13.9. *Pangasius sp.*

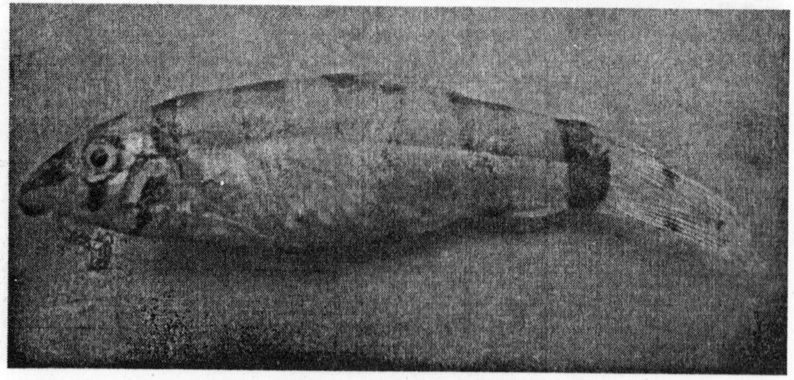

Figure 13.10. *Botia sp.*

Figure 13.11. *Mystus* sp.

Figure 13.12. *Tor sp.*

Figure 13.13. *Puntius sp.*

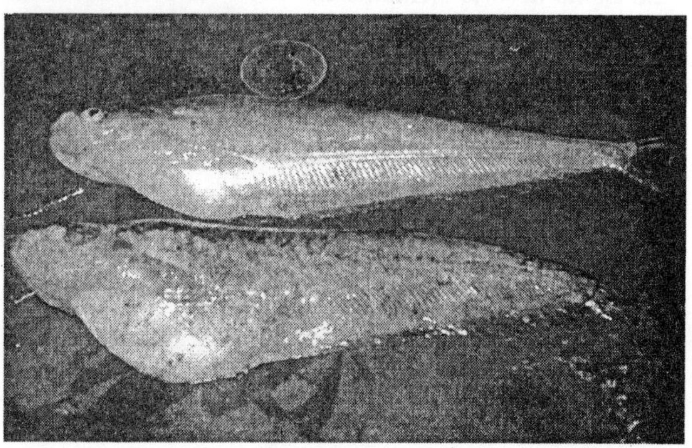

Figure 13.14. *Ompok spp.*

Table 13.1
List of fish and their preferential habitat

Scientific names	*Preferential habitat*	*Abundance*
Family: Notopteridae		
1. *Notopterus chitala* (Hamilton)	River	+
2. *Notopterus notopterus* (Pallas)	River, wetland	++
Family: Anguillidae		
3. *Anguilla bengalensis* (Gray)	River	*
Family: Clupeidae		
4. *Gudusia chapra* (Hamilton)	Large River	+
5. *Hilsa ilisha* (Hamilton) +	Large River	
6. *Goniolosa manmina* (Hamilton)	Rivers floodplain lakes	+
Family: Cyprinidae		
7. *Amblypharyngodon mola* (Hamilton)	Wetlands	+++
8. *Aspidoparia jaya* (Hamilton)	River	+++
9. *Aspidoparia morar* (Hamilton)	River	+++
10. *Barilius barila* (Hamilton)	Large river, hill streams	++
11. *Barilius barna* (Hamilton)	-do-	+
12. *Barilius bendelisis* (Hamilton)	-do-	++
13. *Barilius tileo* (Hamilton)	-do-	++
14. *Barilius shacra* (Hamilton)	-do-	+
15. *Barilius vagra* (Hamilton)	-do-	++
16. *Chagunius chagunio* (Hamilton)	Upstream river	++
17. *Chela cachius* (Hamilton)	River, wetlands	+++
18. *Chela laubuca* (Hamilton)	-do-	+++
19. *Semiplotus semiplotus* (McClelland)	River	++
20. *Danio aequipinnatus* (McClelland)	River, hill streams	+++
21. *Danio dangila*(Hamilton)	-do-	+++
22. *Danio devario* (Hamilton)	-do-	+++
23. *Danio rerio* (Hamilton)	-do-	++
24. *Esomus danricus* (Hamilton)	Wetlands, rivers	+++
25. *Bengala elanga* (Hamilton)	River	++
26. *Cirrhinus mrigala* (Hamilton)	River, streams, wetlands	+++
27. *Cirrhinus reba* (Hamilton)	River	++
28. *Catla catla* (Hamilton)	River, wetlands	+++
29. *Labeo angra* (Hamilton)	-do-	++
30. *Labeo bata* (Hamilton)	-do-	++
31. *Labeo boga* (Hamilton)	River	++

(Contd...)

Scientific names	*Preferential habitat*	*Abundance*
33. *Labeo calbasu* (Hamilton)	River, wetlands	+++
34. *Labeo dero* (Hamilton)	Hill streams	+++
35. *Labeo dyocheilus* (McClelland)	Large river, hill streams	++
36. *Labeo gonius* (Hamilton)	River, wetlands	+++
37. *Labeo nandina* (Hamilton)	-do-	++
38. *Labeo pangusia* (Hamilton)	River	++
39. *Labeo rohita* (Hamilton)	River, wetlands	+++
40. *Crossocheilus latius latius*(Hamilton)	Stream & river with rocky bed	++
41. *Crossocheilus burmanicus* (Hora)	-do-	*
42. *Garra annandalei* (Hora)	-do-	++
43. *Garra kempi* (Hora)	-do-	++
44. *Garra lamta* (Ham Buch)	-do-	+
45. *Garra gotyla gotyla* (Gray)	-do-	+++
46. *Garra nasuta* (McClelland)	-do-	+
47. *Garra rupicola* (McClelland)	-do-	++
48. *Neolissochilus hexagonolepis* (McClelland)	Upstream of River	++
49. *Oreichthys cosuatis* (Hamilton)	River	++
50. *Osteobrama cotio cotio* (Hamilton)	River, wetlands	+++
51. *Puntius chola* (Hamilton)	-do-	+++
52. *Puntius conchonius* (Hamilton)	Wetland, streams	+++
53. *Puntius gelius* (Hamilton)	Wetlands	+++
54. *Puntius sarana sarana* (Hamilton)	River	+++
55. *Puntius sophore* (Hamilton)	River, wetlands	+++
56. *Puntius ticto* (Hamilton)	River, wetlands	+++
57. *Puntius terio* (Hamilton)	River, wetlands	++
58. *Rasbora rasbora* (Hamilton)	Wetland, river	+++
59. *Parluciosoma daniconius* (Hamilton)	Wetlands	+++
60. *Securicola gora* (Hamilton)	Stream, river, wetlands	++
61. *Salmostoma bacaila* (Hamilton)	-do-	+++
62. *Salmostoma phulo* (Hamilton)	-do-	++
63. *Schizothorax progestus* (Day)	Upstream river	++
64. *Schizothoraichthys niger* (Heckel)	-do-	++
65. *Tor putitora* (Hamilton)	-do-	++
66. *Tor mosal* (Hamilton)	-do-	++
67. *Tor tor* (Hamilton)	-do-	+
Family: Psilorhynchidae		
68. *Psilorhynchus balitora* (Hamilton)	Hill streams	++

(Contd...)

Scientific names	*Preferential habitat*	*Abundance*
69. *Psilorhynchus gacilis* (Rainboth)	-do-	++
Family: Balitoridae		
70. *Balitora brucei* (Gray)	-do-	++
71. *Neomacheilus botia* (Hamilton)	Streams, rivers	+++
72. *Nemacheilus devdevi* (Hora)	-do-	+
73. *Nemacheilus corica* (Day)	-do-	++
74. *Nemacheilus reticulofasciatus* (Singh)	-do-	+
75. *Schistura beaveni* (Gunther)	-do-	+
76. *Schistura scaturigina* (McClelland)	-do-	+
Family: Cobitidae		
77. *Botia dario* (Hamilton)	River, wetlands	+++
78. *Botia histrionica* (Blyth)	-do-	++
79. *Botia rostrata* (Gunther)	-do-	++
80. *Lepidocephalus goalparensis* (Pillai & Yazdani)	-do-	++
81. *Lepidocephalus guntea* (Hamilton)	-do-	+++
82. *Lepidocephalus berdmirei* (Blyth)	-do-	++
83. *Cantophrys gongota* (Hamilton)	-do-	++
Family: Bagridae		
84. *Batasio batasio* (Hamilton)	River, streams	++
85. *Batasio tengana* (Hamilton)	Torrential streams	++
86. *Aorichthys aor* (Hamilton)	River, wetlands	+++
87 *Aorichthys seenghala* (Sykes)	-do-	+++
88. *Mystus bleekeri* (Day)	-do-	+++
89. *Mystus cavasius* (Hamilton)	-do-	+++
90. *Mystus menoda* (Hamilton)	-do-	+
91. *Mystus pulcher* (Choudhuri)	-do-	+
92. *Mystus tengara* (Hamilton)	-do-	++
93. *Mystus vittatus* (Bloch)	-do-	+++
94. *Rita rita* (Hamilton)	-do-	++
95. *Rama chandramara* (Hamilton)	River	++
Family: Siluridae		
96. *Ompok bimaculatus* (Bloch)	River, wetlands	++
97. *Ompok pabda* (Hamilton)	-do-	+
98. *Ompok pabo* (Hamilton)	-do-	++
99. *Wallago attu* (Bloch & Schneider)	-do-	+++
Family: Schilbeidae		
100. *Ailia coila*(Hamilton)	-do-	+++
101. *Clupisoma garua* (Hamilton)	-do-	++

(Contd...)

Scientific names	*Preferential habitat*	*Abundance*
102. *Eutropiichthys vacha* (Hamilton)	-do-	+++
103. *Pseudeutropius atherinoides* (Bloch)	River	+++
104. *Silonia silondia* (Hamilton)	-do-	+
Family: Pangasiidae		
105. *Pangasius pangasius* (Hamilton)	River, wetlands	*
Family: Amblycipitidae		
106. *Amblyceps mangois* (Hamilton)	Torrential streams	+
Family: Sisoridae		
107. *Bagarius bagarius* (Hamilton)	Large rivers,, hill streams	+
108. *Bagarius yarrelli* (Sykes)	-do-	++
109. *Conta conta* (Hamilton)	Hill streams	++
110. *Erethistes pussilus* (Muller & Troschel)	-do-	++
111. *Erethistoides montana* (Hora)	-do-	+
112. *Gagata cenia* (Hamilton)	River, hill streams	++
113. *Gagata gagata* (Hamilton)	-do-	++
114. *Glyptothorax cavia* (Hamilton)	-do-	+
115. *Glyptothorax gracilis* (Gunther)	-do-	++
116. *Glyptothorax trilineatus* (Blyth)	-do-	+
117. *Hara hara* (Hamilton)	-do-	++
118. *Hara jerdoni* (Day)	-do-	+
119. *Nangra nangra* (Hamilton)	-do-	++
120 *Pseudecheneis sulcata* (McClelland)	Torrent streams	+
121. *Sissor rabdophorus* (Hamilton)	Large rivers	+
122. *Laguvia sp.*	Hill streams	+
Family: Chacidae		
123. *Chaca chaca* (Hamilton)	Wetlands	++
Family: Olyridae		
124. *Olyra longicaudata* (McClelland)	Small hill streams	++
Family: Heteropneustidae		
125. *Heteropneustes fossilis* (Bloch)	Wetlands	+++
Family: Clariidae		
126. *Clarius batrachus* (Linnaeus)	-do-	+++
Family: Belonidae		
127. *Xenentodon cancila* (Hamilton)	River, wetlands	+++
Family: Synbranchidae		
128. *Monopterus cuchia* (Hamilton)	Wetlands	+++
Family: Channidae		
129. *Channa barca* (Hamilton)	-do-	++

(Contd...)

Scientific names	*Preferential habitat*	*Abundance*
131. *Channa gachua*(Hamilton)	River, streams	+++
132. *Channa marulius* (Hamilton)	wetlands	+++
134. *Channa punctatus* (Bloch)	wetlands, small rivers	+++
135. *Channa stewarti* (Playfair)	-do-	++
136.*Channa striatus* (Bloch)	-do-	++
Family: Ambassidae		
137. *Chanda nama* (Hamilton)	River, wetlands	+++
138. *Parambassis ranga* (Hamilton)	-do-	+++
139. *Pseudambassis baculis* (Hamilton)	-do-	++
Family: Badidae		
140. *Badis badis* (Hamilton)	Wetlands, hill streams	+++
Family: Nandidae		
141. *Nandus nandus* (Hamilton)	Wetlands	+++
Family: Anabantidae		
142. *Anabas testudineus* (Bloch)	Wetlands, streams	+++
Family: Gobiidae		
143. *Glossogobius giuris* (Hamilton)	River, wetlands	+++
Family: Osphronemidae		
144. *Colisa lalia* (Hamilton)	-do-	+++
145. *Colisa sota* (Hamilton)	-do-	+++
146. *Colisa fasciatus* (Schneider)	-do-	+++
Family: Mugilidae		
149. *Rhinomugil corsulaa* (Hamilton)	Large river	++
150. *Sicamugil cascasia* (Hamilton)	-do-	+
Family: Chouhduriidae		
151. *Choudhuria indica* (Yazdani)	Small hill streams	+
Family: Mastacembelidae		
152. *Macrognathus aral* (Bloch & Schneider)	River, wetlands	+++
153. *Macrognathus pancalus* (Hamilton)	-do-	+++
154. *Mastacembelus armatus* (Lecepede)	-do-	+++
Family: Tetraodontidae		
155.*Tetraodon cutcutia*	-do-	+++

Note: * = rare += occasional; ++ = seasonal; +++ = common

stream spawning grounds for hill stream migratory fish because sands and gravels will be trapped behind the dam and unable to establishing breeding and spawning habitat for hill stream migratory fishes, thus important habitat conditions can be affected. Furthermore, downstream habitat conditions can decline because these sediments no longer provide important organic and inorganic nutrients in its downstream leading to which may resulting in poor growth of phytoplankton which in tern causes depletion in dissolved oxygen contents as well as food sources of fishes. Likewise, slow moving waters and temperature changes caused by reservoirs can provide deteriorate environments for the fishes in the downstream. Release of reservoir water with low dissolve oxygen content and temperature along with reduced water discharge particularly in the dry spell to the downstream could be the other principal causes of downstream fish habitat degradation. When ecosystem changes occur, a new pattern of biological activity and equilibrium is likely to emerge. With this new equilibrium come changes to the plants, fish that populate these areas.

Adams (2000) suggested that dams impact fish in three ways. Firstly, dams can affect the physical stability of river channels. River bed degradation downstream of dams can also lead to the loss of important in-stream spawning grounds for fish. Secondly, dams can affect fisheries through impacts on water quality. Water released from low outlets in a dam (e.g. turbines) tends to be cold and may be deoxygenated, or rich in hydrogen sulphide. Thirdly, and most importantly, dams affect fishing communities by changing natural flooding patterns. During dam operation an artificial high flood with turbulent turbid water along with high velocity of current may destroy the aquatic biota. Flood will also bring the exotic cultured fishes to the natural habitat of native fishes creating an unexpected interspecies competition and led to the lowering of native species. Simultaneously, damming of water will create dry like situation in the downstream and there is a chance of degradation or elimination riparian wetlands and thus may affects the natural fish stock.

In this tropical floodplain river, the impacts of dams on natural flood regimes may drastically reduce fish populations in both river channel and floodplain. Many floodplain fish are stimulated by rising seasonal flood flows to move into the floodplain wetlands to breed in the warm organically rich water. As the flood subsides, fish move back to the river channel. Floodplain and wetland ecosystems develop in response to flow patterns and the nature of the landscape between the river and wetlands or floodplains (WRM, 2000). Damming of Subansiri may reduce flood peaks, ultimately fish fail to move or breed; change the flooding patterns, which will lead to changes in habitat and vegetation of riparian wetlands and the diversity of native fishes. The alteration of flow regimes is often claimed to be the most serious and continuing threat to ecological sustainability of rivers and their associated floodplain wetlands (Naiman *et al.,* 1995; Sparks, 1995; Lindquist, 1998;). While the obvious and often irreversible impacts of large impoundments are now well recognized, there is also growing awareness of the pivotal role of the flow regime as a key driver of the ecology of rivers and their associated floodplain wetlands (Junk *et al..* 1989; Poff *et al.,* 1997; Richter *et al.,* 1997; Puckridge *et al.* 1998; Hart and

Finelli 1999). They also reported that dams constrain a river's flow regime by storing water in reservoirs, limiting over bank floods, and regulating and minimizing the magnitude, frequency, duration, and timing of natural flows. Consequently, the food web and productivity of species adapted to dynamic flow conditions is considerably modified by dams.

Sediment transport through a river system is also greatly affected by dams. Most sediment entering a reservoir is stored behind dams, resulting in sediment-starved conditions downstream (Graf, 2002) and a principal cause of habitat fragmentation. A river system with a confined sediment flow often has reduced habitat diversity due to the decreased nutrient flux and loss of habitat (American Rivers, 2002; Graf, 2002).

Dams also cause populations of fish species to be isolated from one another. This change in community structure can lead to local extirpations of some riverine fish species because the isolated areas can no longer be repopulated from individuals traveling up or downstream. Flow variability and habitat structure at smaller spatial scales within the drainage network can also influence fish assemblages. In addition, the dam has altered the thermal regime such that the growth, reproduction and survival of native fishes have almost certainly been affected (Bulkley *et al.*, 1981).

Conclusion

Unregulated Subansiri provides a healthy river ecosystem and riparian wetlands of downstream floodplains for fishes and other aquatic biota. But 2000 MW Lower Subansiri Hydroelectric project in the river will create a physical barrier and thus a significant alteration to the downstream river ecology can be expected. Instream barrier will changes the water quality, quantity, sediment load and encouraged habitat fragmentation, results in loss of many native fish species in its downstream.

This in stream barrier may cause changes in the rate of water level fluctuation in the downstream because when commissioned, hydropower dam will operate "hydro peaking" mode to provide energy when it is most needed. Water will trapped behind the dam during hours when energy demand is low and released only when energy demand is high. While this may sound like an intelligent way to provide energy on demand, it means that the river downstream of the dam is exposed to near-dry conditions while water is filling up behind a dam, followed by large surges of water released to produce energy. These fluctuations can occur several times a day below some hydropower dams.

Because of the importance of a dynamic flow regime to the health of a river, primary goal should be to establish long-term water management agreements for in stream flows that provide ecosystem protection year-round. This includes flows ranging from minimum flows in the summer months to winter and spring pulse flows that are beneficial and necessary for the long-term maintenance of the "house" in which fish live.

Beyond the need for adequate minimum in stream flows, there is also the need for avoiding sudden and un-natural high flows. For example, when a hydropower project goes "on line" or "off line" for power generation, there are sudden increases or decreases of river flows below the dams. Instantaneous flow declines may strand fish and their food source. So, to conserve the river habitat as well as the native fish fauna following measures should be adopted for other proposed hydro-electric projects in the entire North East India - i) EIA study: In-depth environmental impact assessment on downstream ecology and biodiversity. ii) Compensation flow: The compensation water released from the dam must set so that the river flow downstream does not drop below its natural (unregulated) minimum. However, this does not necessarily mean that it is set to benefit or maintain ecological interest. Compensation releases do not always reflect natural fluctuations in river flow, and can result in higher flows during dry periods than would naturally occur. iii) Freshets: Flood flows are considered to be important, particularly for migratory fish, so artificial flood flows known as 'freshets' may be released. Freshets help to encourage movement of fish upstream and mitigate ecological damage caused by artificially prolonged low flows, such as drying out of spawning grounds. iv) Fish passes: Dams clearly present a formidable physical barrier to the movement of migratory fish, fish passes in dams are therefore essential for fish to reach their spawning grounds and nursery areas. v) Minimize downstream water quality impacts of storage release: The development of responses in dam design (e.g. fish ladders), dam operation (the release of artificial floods) and downstream river management (e.g. in-stream, flow diversion structures) to minimize adverse impacts of control on fish stocks.

[**Acknowledgement:** Financial assistance provided by UGC, New Delhi, India in the form of a Major Research project to the second author is gratefully acknowledged.]

REFERENCES

Adams, W. 2000. *The Downstream Impact of Large Dams: Thematic Review* I.1: World Commission on Dams.

American Rivers, 2002. The ecology of dam removal: a summary of benefits and impacts. Available:http://www.americanrivers.org/site/DocServer/damremoval.pdf?docID=494. (March 2006).

Baruah, D. 2007. *An ecological investigation of the downstream of Subansiri River Basin: Pre-impact study of the 2000 MW lower Subansiri Dam.* Final Technical Report submitted to UGC, New Delhi.

Baxter, R. M. 1977. Environmental effects of dams and impoundments. *Ann. Rev. of Ecol. and Syst.* 8:255-283

Bulkley, R. V., C. R. Berry, R. Pimentel and T. Black 1981. *Tolerances and preferences of Colorado River endangered fishes to selected habitat parameters.* Colorado River Fishery Project Final Report Part 3, US Fish and Wildlife Service, Bureau of Reclamation, Salt Lake City, Utah (cited in Converse and others 1998).

Fischenich, C. 1999. Glossary of stream restoration terms, *EMRRP Technical Notes Collection* (ERDC TN-EMRRP-SR-01), U.S. Army Engineer Research and Development Center, Vicksburg, MS.

Hart, D. D. and C. M. Finelli. 1999. Physical-biological coupling in streams: The pervasive effects of flow on benthic organisms. *Annual Rev. of Ecol. and Syst.* 30:363–395.

Konkel, D. 2001. *The Aquatic Plant Community in the Chippewa River,* http://dnr.wi.gov/fish/reports/final/buffalochippewariver2001.pdf.

Naiman, R. J., J. J. Magnuson, D. M. McKnight and J. A. Stanford 1995. *The freshwater imperative: A research agenda*. Island Press, Washington, DC. 165 pp.

Poff, N. L., and D. D. Hart. 2002. How dams vary and why it matters for the emerging science of dam removal. *BioSci.* 52:659-668.

Poff, N. L., J. D. Allan, M. B. Bain, J. R. Karr, K. L. Prestegaard, B. D. Richter, R. E. Sparks and J. C. Stromberg 1997. The natural flow regime. *BioSci.* 47:769–784.

Puckridge, J. T., F. Sheldon, K. F. Walker, and A. J. Boulton. 1998. Flow variability and the ecology of large rivers. *Marine and Freshwater Research.* 49:55–72.

Rai, A.K. 1989. Final Progress Report (16 July 1984–31 December 1989), Inland Fisheries Project, Kathmandu: Fisheries Development Division, National Agricultural Research and Service Centre, Ministry of Agriculture, Government of Nepal.

Richter, B. D., J. V. Baumgartner, R. Wigington and D. P. Braun 1997. How much water does a river need ? I 37:231–249.

Sparks, R. E. 1995. Need for ecosystem management of large rivers and floodplains. *BioScie.* 45:168–182.

Srivastava, V. K. 2000. Effect of riparian vegetation on the river ecology and aquatic biodiversity with special reference to Arunachal Pradesh (India)-A review: *In Sustainable management of Forests-* India: Khan M. L. and Arunachalam A. (Eds), International Book Distributors, Dehradun, pp: 135-146.

Talwar, P. K. and A. G. Jhingran, 1991. *Inland Fisheries of India and Adjacent Countries.* Oxford & IBH Publ. Com., New Delhi: vol. 1 & 2.

Vyas,V. V. Parashar and D. Damde, 2009. Fish Biodiversity and Preferential Habitats for Fishes in Selected Stretch of Narmada River. *Nat. Env. And Pollut Tech.*, vol. 8 (1): 81-89

Viswantath, W. 2002. *Fishes of North East India, A field guide to species Identification.* NATP, NBFGR, and Lucknow. 158 pp.

Welcomme, R. L. 1979. *Fisheries ecology of Floodplain Rivers.* Longman, London, 317 pp. (cited in Petts 1984).

Wootton, R.J. 1998. *Ecology of teleost fishes.* Kluwer Academic Publication, London. 385 pp.

WRM, 2000. *Dam repair or removal: A decision –making guide: Module 2: Dam and river Ecosystem basics: Morphology and Sediment.* URL http://www.ies.wisc.edu/research/wrm00/educmorph.htm

14

Biodiversity of Mosquitoes in the Brahmaputra Valley of Assam

—*Prafulla Dutta*[1], *Siraj Ahmed Khan*[2] & *Beauty Mahanta*[3]

[Mosquito fauna along with their biodiversity in the state of Assam, India has been studied and findings have been discussed with the previously published records. A checklist of 160 mosquito species under 16 genera as recorded from Assam has been prepared and presented on the basis of the recent taxonomical nomenclature.

Key words: Biodiversity, Mosquito, Ecosystem, Brahmaputra valley, Assam, Disease vector.]

Introduction

Mosquitoes are the most important among other arthropod vectors of human disease in the tropic and are notoriously responsible for causing much greater misery to mankind than all other combined together (Tandon, 1998). Mosquitoes are dipterous flies belonging to the family culicidae and are currently the most extensively studied group of haematophagus insects. The role of mosquitoes in the health and well being of man is both direct and indirect. Mosquitoes act as vector of protozoan, helminth and viral agents. Besides they cause discomfort, annoyance and blood loss to the host.

Most of the mosquito faunistic studies in India (Dhanpal and Naik, 1986; Nagpal and Sharma, 1987; Khame and Khaliwal, 1988) have been done in relation to the geographic location. These studies provide information on the distribution of mosquito species in different regions or states (Rajavel *et al.*2001). Bentley, C.A. (1902) and Challam (1923) appear to have been the first to record mosquitoes from northeast region of India. Bentley carried out Anopheline survey in Tezpur area of Assam and Challam carried out the survey in Kamrup district of Assam during a malaria investigation in a sugarcane estate. Study of mosquitoes in Assam received more impetus due to the development of industrial projects and tea gardens. Christophers (1925) conducted mosquito surveys in highly malarious industrial zones in this region. Strickland (1929) and Ramsay (1930) studied relative malaria infectivity of some Anophelines in Cachar district. Macdonald and Chowdhury (1931) conducted mosquito surveys during malaria investigation in some tea gardens. In

[1,2] Regional Medical Research Centre (ICMR), Post Box-105, Dibrugarh, Assam
[3] Department of Zoology, Duliajan College, Dibrugarh, Assam

1932, Gupta and Mazumdar carried out mosquito survey in Kachugaon, Goalpara district and Rice and Savage (1932) conducted Anopheline survey in tea gardens of Upper Assam. Christophers in 1933 and Barraud in 1934 published monographs of Indian Anophelines and Culicines, respectively. Rice and Mohan (1936) studied bionomics of *Anopheles minimus* in cold weather. Vishwanathan *et al.*, (1941) studied Anophelines longitudinally for 12 months to find important malaria vectors. Das *et al*, (1979) conducted a mosquito survey in some places of Darrang district. Sarkar *et al*, (1981) conducted a survey of mosquito fauna at 7 places of Dibrugarh district, viz. Chabua, Dinjan, Duliajan, Ledo, Lekhapani, Margherita and Tinsukia. These authors recorded a total of 27 species belonging to 8 genera *viz. Anopheles* (6 sps.), *Aedes* (4 sps.), *Armigeres* (3 sps.), *Culex* (8 sps.), *Coquillettidia* (1 sps.), *Mansonia* (3 sps.), *Malaya* (1 sps.) and *Toxorhynchites* (1 sps.). Sarkar *et al*, in 1984 conducted survey of mosquito breeding in plantain, arum and pineapple leaf axils at Tezpur in Sonitpur district. The highest number of mosquito species were recorded in arum leaf axils (7 sps.) followed by plantain leaf axils (5 sps.) and pineapple leaf axils (4 sps.). In 1984, Das *et al*, studied mosquito species of animal sheds in Darrang and Cachar districts of Assam. These authors reported a total of 37 species representing 8 genera. Majority of mosquito species (36 sps.) was collected from cattlesheds followed by sties (29 sps.) and stables (16 sps.). Malhotra in 1985 conducted faunistic survey of mosquitoes from different localities of Assam and reported 50 mosquito species representing 10 genera. The genus *Anopheles* was most dominant represented by 16 sps. followed by *Culex* (13 sps.), *Aedes* (8 sps.), *Armigeres* (3 sps.), *Coquillettidia* (2 sps.), *Mansonia* (4 sps.) and *Ficalbia, Malaya, Toxorhynchites* and *Tripteroides* (1 sps. each). Nagpal and Sharma (1987) conducted a mosquito survey and reported 58 species of 8 genera as *Aedes* (9sps), *Anopheles* (27sps), *Armigeres* (3sps), *Coquillettidia* (1Sps), *Culex* (13sps), *Malaya* (1sps), *Mansonia* (3 sps) and *Toxorhynchites* (1 sps). Dutta *et al*, (1989) reported 12 anopheline species from Dibrugarh district of Assam and *Anopheles dirus* was incriminated as vector for malaria transmission in a malaria endemic focus under Tengakhat PHC. Subsequently, in 1991 Dutta *et al*, conducted epidemiological observations on malaria in 44 villages of Tengakhat PHC. During this study, 16 anophelines were recorded including *An minimus, An philippinensis, An annularis,* and *An dirus* which are known malaria vectors. However, *An dirus* was found to be sporozoite positive during the study. Nandi *et al*, (1992) reported 19 species of anophelines in three villages in Boko PHC in Lower Assam. *An minimus* was found to be the principal vector in malaria transmission. However, *An dirus* and *An fluviatilis* were also incriminated. Wajihullah *et al.* 1992 also reported incrimination of *An minimus* as malaria vector from Sonapur area in lower Assam. In 1994, Dutta *et al.* collected six species of *Anopheles* from indoor resting sites in some localities of Boko PHC out of which *An minimus* was found to be sporozoite positive. In another survey *An minimus* was also incriminated from the KarbiAnglong district of Assam by Dutta and Mahanta,1995. Prakash *et al.*1996 also reported the incrimination of *An minimus* in lower Assam from Tamulpur area. Dutta *et al.*, (1997) reported mosquito survey of Goalpara district and 21 5pecies were detected and *An minimus* was incriminated as malaria vector. Khan *et al.*, (1998) reported mosquito survey of Assam (Upper Brahmaputra valley) and recorded 91 species under 15 genera. Prakash *et al,* (1998) made a report of mosquito survey in forest fringed area of Assam along with an updated list of mosquitoes of Assam. Fifty species of mosquitoes under 11 genera were recorded including *Culex peytoni, Culex bailyi* and *Heizmania reidi* for the first time, the updated list had 101 species of mosquitoes under 14 genera. Dutta *et al*(1999) made a survey in the world's largest river island, Majuli, Assam and recorded 28 mosquito species under 6 genera. Subsequently, Bhattacharyya *et al* (2000, 2003, 2004a, 2004b, 2005, 2008) described nine new species records viz, *Armigeres zoloensis, Culex quadripalpis, Culex mammilifer, Uranotaenia novobscura, Aedes ostentatio , Ur (Psc.) dibrugarhensis, Ve. (Neo.) assamensis, Oc.(Muc.) laniger and Ae. (Stg.) krombini* from the state.

The monograph on Indian *Anophiline* and *Culicine* published by Christopher (1933) and Barraud (1934) are the only documents with coverage of the mosquitoes at some parts of Assam. The Comprehensive recent information on distribution of Mosquito species in N.E. Region of India is very meagre. The documents published by captain P. J. Barraud and Sir Richard Christopher way back in 1934 and 1933 respectively have now become out dated. It was noted earlier that of the 239 described Indian *Culicine* mosquitoes, larvae of 96 species were unknown and many additional species awaited discovery (Barraud, 1934).

The study on mosquitoes is becoming necessity because of the emerging mosquito borne diseases. Among the mosquito borne diseases, malaria is the major health problem followed by Japanese encephalitis and some other viral disease in Assam. The border areas with different hill states are highly malaria prone contributing about 64.7% malaria positive cases (74.8% *P.falciparum* cases) as this state has 71% of the total population of NE region (Mohapatra et al, 1998). Japanese encephalitis is becoming a disease burden in upper Assam areas in recent years causing high morbidity and mortality during summer season (Baruah *et al,* 2002). Some dengue seropositive cases were also reported from upper Assam area (Mahanta *et al,* 1996, Baruah and Mahanta, 1996). Filariasis is also a problem in teagarden labourers, which constitute a major population of Assam (Dutta *et al,* 1995). The present study has depicted mosquito species diversity of the state of Assam in the light of recent taxonomical nomenclature.

Materials and Methods

Study area: The study has been conducted in Assam, which is located in the northeastern part of India and has an area of 30,285 square miles (78,438 square kilometres). It is bounded to the north by the kingdom of Bhutan and the state of Arunachal Pradesh; to the east by the states of Nagaland and Manipur; to the south by the states of Mizoram and Tripura; and to the west by the state of Meghalaya and West Bengal and Bangladesh.

Assam is mostly plain consisting of the large river basins besides the river Brahmaputra and the river Barak and their tributaries. The world's largest river island, Majuli is situated in the river Brahmaputra. The state is bounded by hills on all sides, which also include the Karbi Anglong and North Cachar hills. Agriculture is the mainstay of the people of the state, paddy cultivation occupying the prime position. Extensive tea plantations are also common adding to the scenic beauty of the state. Cultivation of the tea plant is done in large commercial operations. The plant, a species of evergeen (*Camellia sinensis*), is valued for its young leaves and leaf buds, from which the tea beverage is produced. Evergreen rainforests as well as deciduous forests are abundant.

The areas surveyed during the present study are – Khotkhati (Karbi Anglong), Majuli, Dhemaji, Lakhimpur, Sodiya, Goalpara, Tezpur, Jamuguri, Boko, Agia, Chariduar, Rangia, Golaghat, Bongaingaon, Mongoldoi, Saraipung, Dikom, Dibrugarh, Pengeri, Duliajan, Tengakhat, Digboi, Bokakhat, Telem, Dhubri and Karimganj (Figure 14.1).

Entomological survey: Mosquito immature collections were made from different breeding habitats using WHO standard dippers. The spoon was also used for collection of mosquito larvae from small water habitats or any other container habitats. These were reared separately in plastic photo-film vials till the emergence of adults and were processed for proper identification. CDC miniature light-traps were also operated in bovine-sheds for sampling of adult mosquitoes. Mosquitoes collected were identified following the standard keys (Christophers, 1933; Barraud, 1934; Puri, 1955; Wattal and Kalra, 1961; Harrison and Scanlon, 1975; Rao, 1984) and mosquito nomenclature (Knight and Stone, 1977; Reinert, 1975, 2001).

Results and Discussion

A total of 68 mosquito species under 11genera have been recorded in the present study from the state of Assam (Table 14.1-14.2). In Monsoon season, a total of 44 species and in post-monsoon season 56 species of mosquitoes were detected. However, this centre has reported an additional 26 species in recent studies (Khan *et al.*, 1998; Bhattarcharyya *et al.*, 1995, 1998, 2000, 2003, 2004a, 2004b, 2005, 2008). Thus, the cumulative number of mosquito species identified by this centre in Assam adds up to 94. This includes 26 species reported for the first time viz., *An. hodgkini, An. lasteri paraliae, Ae. harveyi, Ae. ostentatio, Ae. prominens, Ae. pseudoalbopictus, Ar. digitatus, Ar. zoloensis, Cx. alienus, Cx. bailyi, Cx. fragilis, Cx. khazani, Cx. bengalensis, Cx. cinctellus, Cx. mammilifer, Cx. peytoni, Cx. quadripalpis, Cx. variatus, Hz. reidi, Hz. chandi, Mi. luzonensis, Ur. novobscura, Ur(Psc.) dibrugarhensis, Ve.(Neo.) assamensis, Oc.(Muc.) laniger and Ae.(Stg.) krombini.* However, 40 mosquito species encountered earlier which were not found by us are *An. aitkenii, An. ahomi, An. annandalei, An. burbumbrosus, An. bengalensis, An. crawfordi, An. gigas var. baileyi, An. insulaeflorum, An. interruptes, An. umbrosus, An. jeyporiensis var.candidiensis, An. lindesayi, An. sinensis, An. pallidus, An. pseudojamesii, An. subpictus, An. theobaldi, An. willmorei, Aediomyia catasticta, Ae. scatophagoides, Ae. quasiferinus, Ae. iyengari, Ae. alboniveus, Ae. dissimilis, Ae. novonorneus, Ae. poecilus, Orthopodomyia albipes, Tx. gravelyi, Tp. indicus, Ur. edwardsi, Ur. micans, Ur. orientalis, Ar. annulitarsis, Ar. flavus, Ar. magnus, Cx. comutus, Cx. minor, Ml. jacobsoni, Mi. intermedia* and *Mi. chamberlaini.*

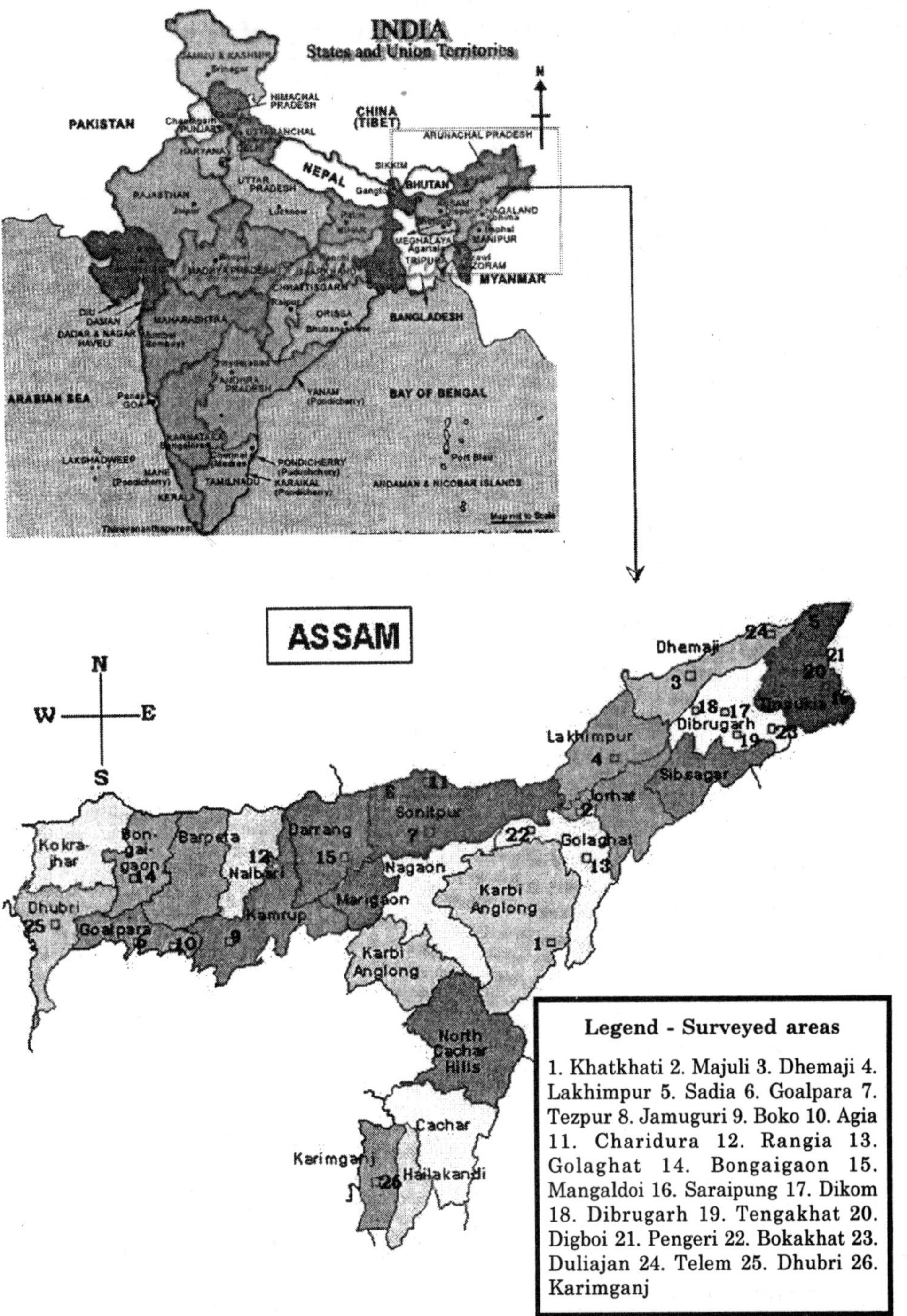

Figure 14.1. Map of Assam, India showing the surveyed areas.

Table 14.1
Different mosquito genera and number of species recorded

Genus	*Monsoon No.of species*	*Post-monsoon No.of species*	*Total No. of species*
Anopheles	14	16	22
Culex	12	16	18
Aedes	8	8	9
Armigeres	2	3	3
Mansonia	4	2	4
Coquillettidia	1	1	2
Toxorhynchites	1	2	2
Malaya	0	1	1
Uranotaenia	0	1	1
Tripteroides	1	2	2
Ochlerotatus	1	4	4
Total: 11	44	56	68

Table 14.2
Names of different Mosquito species recorded in Assam during two different seasons

ANOPHELES

Species	*Monsoon*	*Post-monsoon*
An (Cel) aconitus	•	•
An (Cel) annularis	•	•
An (Cel) culicifacies	•	
An (Cel) dirus	•	
An (Cel) fluviatilis		•
An (Cel) jamesii		•
An (Cel) jeyporiensis	•	•
An (Cel) karwari	•	•
An (Cel) kochi	•	•
An (Cel) maculatus	•	
An (Cel) majidi		•
An (Cel) minimus	•	•
An (Cel) splendidus		•
An (Cel) tessellatus		•
An (Cel) vagus	•	•
An (Cel) varuna		•
An (Cel) philippinensis	•	
An (Cel) nivipes	•	
An (Ano) barbirostris	•	•
An (Ano) gigas	•	
An (Ano) nigerrimus	•	•
An (Ano) peditaeniatus	•	•

(Contd...)

CULEX

Species	Monsoon	Post-monsoon
Cx (Cux) bitaeniorhynchus	•	•
Cx (Cux) epidesmus		•
Cx (Cux) fuscocephala	•	•
Cx (Cux) gelidus	•	•
Cx (Cux) mimeticus	•	•
Cx (Cux) mimulus		•
Cx (Cux) murrelli	•	
Cx (Cux) pseudovishnui	•	•
Cx (Cux) quinquefasciatus	•	•
Cx (Cux) sinensis		•
Cx (Cux) tritaeniorhynchus	•	•
Cx (Cux) whitmorei	•	•
Cx (Cux) vishnui	•	•
Cx (Cui) pallidothorax	•	•
Cx (Lut) fuscanus	•	
Cx (Lut) halifaxi	•	
Cx (Eum) brevipalpis		•
Cx (Eum) malayi		•

AEDES

Species	Monsoon	Post-monsoon
Ae (Stg) aegypti	•	•
Ae (Stg) albopictus	•	•
Ae (Stg) annandalei	•	•
Ae (Stg) vittatus	•	•
Ae (Adm) caecus	•	
Ae (Adm) nigrostriatus		•
Ae (Adm) vexans	•	•
Ae (Neo) lineatopennis	•	•
Ae (Adm) pipersalatus	•	•

OCHLEROTATUS

Species	Monsoon	Post-monsoon
Oc. (Fin) chrysolineatus		•
Oc. (Fin) pseudotaeniatus	•	•
Oc. (Fin) albotaeniatus		•
Oc. (Fin) albolateralis		•

(Contd...)

ARMIGERES

Species	*Monsoon*	*Post-monsoon*
Ar (Arm) kuchingensis	•	•
Ar (Arm) durhami		•
Ar (Arm) subalbatus	•	•

MANSONIA

Species	*Monsoon*	*Post-monsoon*
Ma (Man) indiana	•	•
Ma (Man) dives	•	•
Ma (Man) uniformis	•	•
Ma (Man) annulifera	•	

COQUILLETTIDIA

Species	*Monsoon*	*Post-monsoon*
Cq (Coq) crassipes		•
Cq (Coq) novochracea	•	

TOXORHYNCHITES

Species	*Monsoon*	*Post-monsoon*
Tx (Tox) splendens	•	•
Tx (Tox) minimus		•

MALAYA

Species	*Monsoon*	*Post-monsoon*
Ml (Mal) genurostris		•

URANOTAENIA

Species	*Monsoon*	*Post-monsoon*
Ur (Ura) campestris		•

TRIPTEROIDES

Species	*Monsoon*	*Post-monsoon*
Tp (Trp) tarsalis	•	•
Tp (Trp) aranoides		•

Table 14.3
Status of occurrence of different mosquito species from Assam: an updated checklist

Sr. No.	*Mosquito species*	*Status*
1	*An. (Ano.) aitkenii* (James), 1903	♦
2	*An. (Ano.) ahomi* (Chowdhury), 1929	♦
3	*An. (Ano.) annandalei* (Baini Prashad), 1918	♦
4	*An. (Ano.) barbirostris* (Van der Wulp), 1884	•♦
5	*An. (Ano.) barbumbrosus* (Strickland and Chowdhury),1927	♦
6	*An. (Ano.) bengalensis* (Puri), 1930	♦
7	*An. (Ano.) crawfordi* (Reid), 1953	♦
8	*An. (Ano.) gigas* (Giles), 1901	•♦
9	*An. (Ano.) gigas var. baileyi* (Edwards), 1928	♦
10	*An.(Ano.) hodgkini* (Reid), 1962	•
11	*An.(Ano.) insulaeflorum* (Swellengrebel & Swellengrebel De Graft), 1920	♦
12	*An.(Ano.) interruptes* (Puri), 1929	♦
13	*An.(Ano.) lasteri paraliae* (Sandosham), 1959	•
14	*An.(Ano.) lindesayi* (Giles), 1900	♦
15	*An.(Ano.) nigerrimus* (Giles), 1900	•♦
16	*An.(Ano.) peditaeniatus* (Leicester), 1908	•♦
17	*An.(Ano.) sinensis* (Wiedmann), 1828	♦
18	*An.(Ano.) umbrosus* group (Theobalad), 1903	♦
19	*An.(Cel.) aconitus* (Doenitz), 1902	•♦
20	*An.(Cel.) annularis* (Van der Wulp), 1884	•♦
21	*An.(Cel.) culicifacies* (Gilles), 1901	•♦
22	*An.(Cel.) dirus* (Peyton & Harrison),1979	•♦
23	*An.(Cel.) fluviatilis* (James), 1902	•♦
24	*An.(Cel.) jamesii* (Theobald), 1901	•♦
25	*An.(Cel.) jeyporiensis* (James), 1902	•♦
26	*An.(Cel.) jeyporiensis var. candidiensis* (Koidzumi),1924	♦
27	*An.(Cel.) karwari* (James), 1902	•♦
28	*An.(Cel.) kochi* (Doenitz), 1901	•♦
29	*An.(Cel.) maculatus* (Theobald), 1901	•♦
30	*An.(Cel.) majidi* (Young & Majid), 1928	•♦
31	*An.(Cel.) minimus* (Theobald), 1901	•♦
32	*An.(Cel.) nivipes* (Theobald), 1901	•
33	*An.(Cel.) pallidus* (Theobald), 1901	♦
34	*An.(Cel.) philippinensis* (Ludlow), 1902	•♦
35	*An.(Cel.) pseudojamesii* (Strickland & Chowdhuri),1927	♦
36	*An. (Cel.) splendidus* (Koidzumi), 1920	•♦
37	*An. (Cel.) subpictus* (Grassi), 1899	♦
38	*An. (Cel.) tessellatus* (Theobald), 1901	•♦

(Contd...)

Sr. No.	*Mosquito species*	*Status*
39	*An. (Cel.) theobaldi* Giles, 1901	♦
40	*An. (Cel.) vagus* Doenitz, 1902	•♦
41	*An. (Cel.) varuna* Iyengar, 1924	•♦
42	*An.(Cel.) willmorei* (James), 1903	•
43	*Aediomyia (Ady.) catasticta* (Knab), 1909	♦
44	*Ae. (Adm.) caecus* (Theobald), 1901	•♦
45	*Ae. (Adm.) nigrostriatus* (Barraud), 1927	•♦
46	*Ae. (Adm.) pipersalatus* (Giles), 1902	•
47	*Ae. (Adm.) vexans* (Meigen), 1830	•♦
48	*Ae. (Dic.) iyengari* Edwards, 1923	♦
49	*Ae. (Dic.) nummatus* Edwards, 1923	♦
50	*Oc. (Fin.) albolateralis* (Theobald), 1908	•♦
51	*Oc. (Fin.) alboniveus* Barraud, 1934	♦
52	*Oc. (Fin.) albotaeniatus* (Leicester), 1904	•
53	*Oc. (Fin.) assamensis* (Theobald), 1908	♦
54	*Oc. (Fin.) cacharanus* (Barraud), 1923	♦
55	*Oc. (Fin.) chrysolineatus* (Theobald), 1907	•♦
56	*Oc. (Fin.) dissimilis* (Leicester), 1908	♦
57	*Oc. (Fin.) greenii* (Theobald), 1903	♦
58	*Oc. (Fin.) harveyi* (Barraud), 1923	•
59	*Oc. (Fin.) khazani* Edwards, 1922	♦
60	*Oc. (Fin.) novoniveus* Barraud, 1934	♦
61	*Oc. (Fin.) pallirostris* Edward, 1922	♦
62	*Oc. (Fin.) poicilius* (Theobald), 1903	♦
63	*Oc. (Fin.) prominens* (Barraud), 1923	•
64	*Oc. (Fin.) pseudotaeniatus* (Giles), 1901	•♦
65	*Oc. (Fin.) pulchriventer* (Giles), 1901	♦
66	*Oc. (Fin.) saxicola* Edwards, 1922	♦
67	*Oc. (Muc.) scatophagoides* (Theobald), 1901	♦
68	*Oc. (Muc.) quasiferinus* Mattingly,1961	♦
69	*Oc. (Muc.) laniger*(Wiedmann)	•
70	*Ae. (Neo.) lineatopennis* (Ludlow), 1905	•♦
71	*Ae. (Par.) ostentatio* (Leicester), 1908	•
72	*Ae. (Stg.) aegypti* (Linnaeus), 1762	•♦
73	*Ae. (Stg.) albolineatus* (Theobald), 1904	♦
74	*Ae. (Stg.) albopictus* (Skuse), 1894	•♦
75	*Ae. (Stg.) annandalei* (Theobald), 1910	•♦
76	*Ae. (Stg.) desmotes* (Giles), 1904	♦
77	*Ae. (Stg.) flavopictus* Yamada, 1921	♦
78	*Ae. (Stg.) pseudoalbopictus* (Borel), 1928	•
79	*Ae. (Stg.) vittatus* (Bigot), 1861	•♦

(Contd...)

Sr. No.	*Mosquito species*	*Status*
80	*Ae. (Stg.)krombini*	•
81	*Ve. (Nma.) andamanensis* (Edwards), 1922	♦
82	*Ve. (Nma.) atrius* (Barraud), 1928	♦
83	*Ve. (Nma.) cinca* (Barraud), 1928	♦
84	*Ve. (Nma.) rami* (Barraud), 1928	♦
85	*Ve. (Nma.) vallistris* (Barraud), 1928	♦
86	*Ve. (Nma.) assamensis*	•
87	*Ar. (Arm.) durhami* (Edwards), 1917	•♦
88	*Ar. (Arm.) joloensis* (Ludlow), 1904	•
89	*Ar. (Arm.) kuchingensis* Edwards, 1915	•♦
90	*Ar. (Arm.) subalbatus* (Coquillett), 1898	•♦
91	*Ar. (Lei.) annulitarsis* (Liecester), 1906	♦
92	*Ar. (Lei.) digitatus* (Edwards), 1914	•
93	*Ar. (Lei.) flavus* (Leicester), 1908	♦
94	*Ar. (Lei.) magnus* (Theobald), 1908	♦
95	*Cx. (Cux.) alienus* Colless, 1957	•
96	*Cx. (Cux.) bitaeniorhynchus* Giles, 1901	•♦
97	*Cx. (Cux.) cornutus* Edwards, 1922	♦
98	*Cx. (Cux.) epidesmus* (Theobald), 1910	•♦
99	*Cx. (Cux.) fuscocephala* (Theobald), 1907	•♦
100	*Cx. (Cux.) gelidus* (Theobald), 1901	•♦
101	*Cx. (Cux.) hutchinsoni* (Barraud), 1924	♦
102	*Cx. (Cux.) mimeticus* (Noe), 1899	•♦
103	*Cx. (Cux.) mimulus* (Edwards), 1915	•♦
104	*Cx. (Cux.) murrelli* (Lien), 1968	•♦
105	*Cx. (Cux.) pseudovishnui* (Colless), 1957	•♦
106	*Cx. (Cux.) quinquefasciatus* (Say), 1823	•♦
107	*Cx. (Cux.) sinensis* (Theobald), 1903	•♦
108	*Cx. (Cux.) tritaeniorhynchus* (Giles), 1901	•♦
109	*Cx. (Cux.) vishnui* (Theobald), 1901	•♦
110	*Cx. (Cux.) whitei* (Barraud), 1923	♦
111	*Cx. (Cux.) whitmorei* (Giles), 1904	•♦
112	*Cx. (Cui.) bailyi* (Barraud), 1934	•
113	*Cx. (Cui.) fragilis* (Ludlow), 1903	•
114	*Cx. (Cui.) pallidothorax* (Theobald), 1905	•♦
115	*Cx. (Eum.) brevipalpis* (Giles), 1902	•♦
116	*Cx. (Eum.) khazani* (Edwards), 1922	•
117	*Cx. (Eum.) malayi* (Leicester), 1908	•♦
118	*Cx. (Lop.) bengalensis* (Barraud), 1934	•
119	*Cx. (Lop.) cinctellus* (Edwards), 1922	•
120	*Cx. (Lop.) fraudatrix* (Theobald), 1905	♦

(Contd...)

Sr. No.	*Mosquito species*	*Status*
121	*Cx. (Lop.) mammilifer* (Leicester), 1908	•
122	*Cx. (Lop.) minor* (Leicester), 1908	♦
123	*Cx. (Lop.) peytoni* (Bram & Rattanarithikul), 1967	•
124	*Cx. (Lop.) quadripalpis* (Edwards), 1914	•
125	*Cx. (Lop.) variatus* (Leicester), 1908	•
126	*Cx. (Lut.) fuscanus* (Wiedemann), 1928	•♦
127	*Cx. (Lut.) halifaxi* (Theobald), 1903	•
128	*Cq. (Coq.) crassipes* (Van der Wulp), 1881	•♦
129	*Cq. (Coq.) novochracea* (Barraud), 1927	•♦
130	*Cq. (Coq.) ochracea* (Theobald), 1903	♦
131	*Hz. (Hez.) chandi* (Edwards), 1922	•
132	*Hz. (Hez.) reidi* (Mattingly), 1957	•
133	*Hz. (Mat.) tripunctata* (Theobald), 1908	♦
134	*Ml .(Mao.) genurostris* (Leicester), 1908	•♦
135	*Ml .(Mao.) jacobsoni* (Edwards), 1930	♦
136	*Ma. (Man.) annulifera* (Theobald), 1901	•♦
137	*Ma. (Man.) dives* (Schiner), 1868	•♦
138	*Ma. (Man.) indiana* (Edwards), 1930	•♦
139	*Ma. (Man.) uniformis* (Theobald), 1901	•♦
140	*Mi. (Eto.) luzonensis* (Ludlow), 1905	•
141	*Mi. (Mim.) chamberlaini* (Ludlow), 1904	♦
142	*Mi. (Mim.) intermedia* (Barraud), 1929	♦
143	*Orthopodomyia albipes* (Leicester), 1904	♦
144	*Orthopodomyia anopheloides* (Giles), 1903	♦
145	*Tx. (Tox.) gravelyi* (Edwards), 1921	♦
146	*Tx. (Tox.) splendens* (Wiedemann), 1819	•♦
147	*Tx. (Tox.) minimus* (Theobald), 1905	•♦
148	*Tp. (Rah.) aranoides* (Theobald), 1901	•♦
149	*Tp. (Trp.) indicus* (Barraud), 1929	♦
150	*Tp. (Trp.) tarsalis* (Delfinado and Hodges), 1968	•
151	*Ur. (Psc.) maxima* (Leicester), 1908	♦
152	*Ur. (Psc.) novobscura* (Barraud), 1934	•
153	*Ur. (Psc.) dibrugarhensis*	•
154	*Ur. (Ura.) annandalei* (Barraud), 1926	♦
155	*Ur. (Ura.) campestris* (Leicester), 1908	•♦
156	*Ur. (Ura.) edwardsi* (Barraud), 1926	♦
157	*Ur. (Ura.) hebes* (Barraud), 1931	♦
158	*Ur. (Ura.) longirostris* (Leicester), 1908	♦
159	*Ur. (Ura.) micans* (Leicester), 1908	♦
160	*Ur. (Ura.) orientalis* (Barraud), 1926	♦

• Recorded in RMRC study

♦ Recorded in earlier studies

Moreover, Barraud (1934) reported 90 mosquito species from Assam out of which 26 species *viz, Cx. fraudatrix, Cx. whiti, Cx. hutchinsoni, Ae. assamensis, Ae. cacharanus, Ae. khazani, Ae. greeni, Ae. saxicola, Ae. pulchriventer, Ae. desmotes, Ae. flavopictus, Ae. albolineatus, Ae. nummatus, Ae. pallirostris, Ve. andamenensis, Ve. vallistris, Ve. cinca (hirsutipleura), Ve. atrius, Ve. rami, Hz. tripunctata, Ur. longirostris, Ur. annandalai, Ur. maxima, Ur. hebes, Or. anopheloides* and *Cq. ochracea* were not detected in the present study. Thus, with the addition of these 66 (40+26) species, the total mosquito fauna of the state goes up to 160 (Figure 14.3).

The major vectors of malaria active in Northeast India, *viz. An. dirus and An. minimus* have both been recorded from this state. Potential Japanese encephalitis vectors recorded were *An. barbirostris, An.'hyrcanus' group, Cx. vishnui, Cx. pseudovishnui, Cx. tritaeniorhynchus, Cx. epidesmus, Cx. gelidus, Cx. whitmorei, Cx. bitaeniorhynchus, Cx. fuscocephala* and *Ma. annulifera*. With regards to dengue, both the vectors *viz. Aedes aegypti* and *Ae. albopictus* were found.

[**Acknowledgement:** Authors are grateful to the state health authority, Assam and Forest department for their kind help and cooperation in conducting entomological surveys. Excellent Technical assistance received from Mr. C.K. Sharma, Technical Officer, Mr. P.K. Doloi, Laboratry Technician, Mr. N.K. Baruah, Laboratory Technician is also gratefully acknowledged.]

REFERENCES

Barraud, P.J. 1923. A revision of the culicine mosquitoes of India-VII. The larvae of some species of Stegomyia, Christophersiomyia, Mimomyia and Aedomyia. *Indian. J. Med. Res.* II: 495-505.

Barraud, P.J. 1924. A revision of the culicine mosquitoes of India part XIII. Further descriptions of Indian species of Culex L. including three new species. *Indian J. Med.* Res. II. : 1259-1274.

Barraud, P.J. 1934. *The fauna of British India- Including Ceylon and Burma. Diptera Vol. V. Family Culicidae, Tribes Megarhinini and Culicini.* Taylor and Francis, London; 463 pp.

Barua H.C and Mahanta, J. 1996. Serological evidence of Den-2 activity in Assam and Nagaland. *J Commun Dis*, 28: 56-8.

Baruah, H.C., Biswas, D., Patgiri, D.K. and Mahanta, J. 2002. Clinical outcome and neurological sequelae in serologically confirmed cases of Japanese encephalitis patients in assam,India. *Indian Pediatrics*, 39:1143-1148.

Bentley, C.A. 1902. *Anopheles* mosquitoes in Tezpur, Assam. *Indian Med. Gaz.*, 37: 15-16.

Bhattacharyya, D.R., Prakash, A. Mohapatra, P.K. and Mahanta, J. 2005. Occurrence of *Ochlerotatus (Mucidus) laniger* (Wiedemann) (Diptera: Culicidea) in Assam. India. *Entomo.*, 30: 105-107.

Bhattacharyya, D.R., Prakash,n A., Mohapatra, P.K., Sarma, D.R. and Mahanta, J. 2008. Occurrence of *Aedes (Stegomyia) Krombini* (Diptera: Culicidea) in Assam, India. *Entomo.*, 33(4): 275-276.

Bhattacharyya, D.R., Tewari, S.C., Dutta, P. and Mahanta, J. 1995. Description of larva and pupa of *Aedes (Aedimorphus) nigrostriatus* Barraud, 1927. *Mosq.Syst.*, 27: 191-196.

Bhattacharyya D.R., Prakash, A., Mohapatra, P.K. and Mahanta, J. 1998. A note on the occurrence of *Culex (lophoceraomyia) peytoni* and *Heizmannia (Heizmannia) reidi* (Diptera: Culicidae) in Assam, India. *J Am Mosq. Cont. Assoc.*, 14: 108.

Bhattacharyya D.R., Prakash, A., Tewari, S.C., Mohapatra, P.K. and Mahanta, J. 2000. *Armigeres zoloensis* (Diptera: Culicidae), A rare mosquito in Upper Assam: First report from India. *Entomon.*, 25:63-65

Bhattacharyya, D.R., Prakash, A., Mohapatra, P.K. and Mahanta, J. 2003. Occurrence of *Culex (Lophoceraomyia) quadripalpis, Culex (Lophoceraomyia) mammilifer* and *Uranotaenia (Pseudoficalbia) novobscura* in Assam, India. *Am. Mosq. Contrl. Assoc.*, 19: 13-14.

Bhattacharyya, D.R., Prakash, A., Mohapatra, P.K. and Mahanta, J. 2002. A note on the occurrence of *Aedes (Parades) ostentatio (Leicester),* 1908 in Dibrugarh district, Assam, India (Diptera: Culicidae). *Oriental Insects,* 36: 47-49.

Bhattacharyya, D.R., Prakash, A., Mohapatra, P.K., Mahanta, J. 2004a. *Uranotaenia dibrugarhensis,* a new species in subgenus *Pseudoficallia* from Assam, India. *J. Am. Mosq. Cont. Assoc.,* 20: 1-5.

Bhattacharyya, D.R., Tewari, S.C., Prakash, A., Mohapatra, P.K., and Mahanta, J. 2004b. *Verrallina (Neomacleaya)assamensis.* A new species in Assam, India. *J. Am. Mosq. Cont. Assoc.,* 20: 115-120.

Challam, B.S. 1923. Some notes on malaria investigation in sugarcane estate in Kamrup, Assam. *Indian Med. Gaz.,* 58: 476-479.

Christophers, S.R. 1933. *The fauna of British India- including Ceylon and Burma, Diptera vol. IV-family Culicidae, tribe Anophelini.* Taylor and Francis, London, 371pp.

Das, S.C., Sarkar, P.K. and Das, N.G. 1979. Studies on the biting habits of the haematophagous insects/leeches and evaluation of protective measures against them. Part-II. Adult mosquito survey in certain places of Darrang district, Assam. Technical Report No. FLT/5/79, Mimeographed document.

Das, S.C., Sarkar, P.K., Das, N.G., Hazarika, S. and Malhotra, P.R. 1984. Note on the collection of mosquitoes from animal sheds and human habitations in Assam, India. *Geobios. New Reports,* 3: 142-144.

Dhanpal. J. and Naik, V.A. 1986. Mosquito of Goa. *Indian J. Malariol.,* 23, 39-42.

Dutta, P., Bhattacharyya, D.R. and Dutta, L.P. 1989. Incrimination of *Anopheles dirus* as a vector of malaria in Dibrugarh district, Assam. *Indian J. Malarial.,* 26 : 149-152.

Dutta P, Bhattacharyya, D.R. and Dutta, L.P. 1991. Epidemiological observation on malaria in some parts of Tengakhat Primary Health Centre under Dibrugarh district of Assam. *Indian J. Malario.,* 28: 121-128.

Dutta, P., Bhattacharyya, D.R., Khan, S.A., Sharma, C.K. and Goswami, B.K. 1994. Some observations on malaria in Boko PHC of Kamrup district, Assam. *J. Commun. Dis.,* 26: 52-55.

Dutta P. and Mahanta, J. 1995. Incrimination of *Anopheles minimus* as a vector of malaria in Karbi Anglong district of Assam. *Indian J. Malariol.,* 32: 129-131.

Dutta, P, Gogoi, B.K., Chelleng, P.K., Bhattacharyya, D.R., Khan, S.A., Goswami, B.K. and Mahanta, J. 1995. Filariasis in the labour population of a tea estate in Upper Assam. *Indian J. Med. Res.,* 101: 245-246.

Dutta P, Khan, S.A., Sharma, C.K. and Mahanta, J. 1997. A report on entomological study and malaria vector incrimination in Goalpara district of Assam. *Indian J. Malariol.,* 34: 204-207.

Dutta, P., Khan, S.A., Sharma, C.K., Doloi, P.K. and Mahanta, J. 1999. Medically important mosquitoes of the World's largest river island, Majuli, Assam. *Entomon.,* 24: 33-39.

Gupta, P. and Majumdar, N.R. 1932. A malaria survey of Kachugaon, Goalpara district, Assam. *Rec. Mal. Surv. Ind.,* 3: 253-268.

Kareem, M.A., Singh, Y.K., Bhatnagar, V.N. and Das, M. 1983. A preliminary report on entomological studies under Pf CP in Zone-1. *J. Commun. Dis.*, 15(3): 207-208.

Khan, S.A., Handique, R., Tewari, S.C., Dutta, P., Narain, K. and Mahanta, J. 1998. Larval ecology and Mosquito fauna of Upper Brahmaputra valley, Northeast India. *Indian J. Malarial.*, 35: 131-145.

Khame, J.S. and Khaliwal, M. B. 1998. Mosquitoes of Daman. *Indian J. Malariol.*, 25, 109-111.

Knight, K.L. and Stone, A. 1977. *Catalogue of the mosquitoes of the world (Diptera: Culicidae).* The Thomas Say Foundation, Entomological society of America, 611 pp.

Macdonald, G and Chowdhury, K.L. 1931. Report on a malaria survey of the tea gardens in the Mariani Medical association, Assam. *Rec. Mal. Surv. Ind.*, 2: 111-156.

Mahanta,J, Baruah, U., Sharma Bordoloi, J.N. and Baruah, H.C. 1996. Japanese encephalitis virus antibody among normal individuals of Dibrugarh area, Assam. *J. Commun. Dis.*, 28: 181-184.

Mishra, B.G. 1956. Malaria in North-East Frontier Agency (India). *Indian J. Malariol.*, 10: 331-347.

Mohapatra, P.K., Prakash, A., Bhattacharyya, D.R. and Mahanta, J. 1998. Malaria situation in Northeastern Region of India. *ICMR Bulletin*, 28(3): 21-30.

Nagpal, B.N. and Sharma, V.P. 1987. Survey of mosquito fauna of Northeastern Region of India. *Indian J. Malariol.*, 24: 143-149.

Nandi, J., Mishra, S.P., Rajagopal, R. and Narasimham, M.V.V.L. 1992. Present perspectives of malaria transmission in Boko area of Assam. *J. Commun. Dis.*, 25: 266.

Prakash, A, Mohapatra, P.K. and Srivastava, V.K. 1996. Vector incrimination in Tamulpur Primary health centre, district Nalbari, lower Assam during malaria outbreak 1995.*Indian J, Med. Res.*, 103: 146.

Prakash, A, Bhattacharyya, D.R., Mohapatra, P.K. and Mahanta, J. 1998. Mosquito fauna in a broken forest ecosystem of Dibrugarh district with uptodate systemic list of mosquitoes recorded in Assam. *Entomon*, 23: 47-54.

Puri, I.M. 1931. Larvae of Anopheline mosquitoes with full descriptions of the Indian species. *Ind. Med. Res. Mem.*, 21: 227pp.

Rajavel, A. R., Munirathinam, A., Natrajan, R. and Vaidyanathan, K. 2001. Species Diversity of Mosquitoes (Diptera: Culicidae) in Mangrove Ecosystem in South India. *Entomon.*, 26(3 & 4): 271-277.

Ramsay, G.C. 1930. Some findings and observations in an anopheline malaria infectivity survey carried out in the Cachar district of Assam. *Indian J. Med. Res.*, 18: 533-552.

Rao Ramachandra, T. 1984. *The Anophelines of India.* Indian Council of Medical Research, New Delhi, India, 594pp.

Reinert, John F. 1975. Mosquito generic and subgeneric abbreviations (Diptera: Culicidae). *Mosquito Syst.*, 7(2): 105-110.

Reinert, J.F. 2001. Revised list of abbreviations for genera and subgenera of Culicidae (Diptera)and notes on generic and subgeneric changes. *J. Am. Mosq. Control Assoc.*, 17: 51-55.

Rice, E.M. and Savage, J. de la M. 1932. Malaria survey of two teastates in Upper Assam. *Rec. Mal. Surv. Ind.*, 3: 219-252.

Rice, E.M. and Mohan, B.N. 1936. *A. minimus* in Assam, its cold weather bionomics and their relationship to antilarval control. *Rec. Mal. Surv. Ind.*, 6: 557-594.

Sarkar, P.K., Das, N.G. and Rao, K.M. 1981. Mosquito fauna of certain places of Dibrugarh district, Assam. *Indian J. Med. Res.*, 73: 331-334.

Sarkar, P.K., Rao, K.M., Nath, D.R. Bhuyan, M. and Chowdhury, B.C. 1984. Seasonal incidence of mosquito breeding in plant axils at Tezpur (Assam). *Indian J. Publ. Hlth.*, 28(3): 152-158.

Strickland, C. 1929. The relative malaria infectivity of some species of anophelines in Cachar (Assam). *Indian J. Med. Res.*, 17: 174-182.

Tandon. 1998. Modern trends in Research of vectors of Medical importance. *Adv. Med. Entomol. & Human welfare*, p. 29-37.

Vishwanathan, D.K., Das, S. and Oomen, A.V. 1941. Malaria carrying anophelines in Assam with special reference to the results of twelve months' dissections. *J. Mal. Inst. Ind.*, 4: 297-306.

Wajihullah, J. B. and Sharma, V.P. 1992. *Anopheles minimus* in Assam. *Current Science*. 63(1): 7-9.

Wattal, B.L. and Kalra, N. L. 1951. Region wise pictorial keys to the female Indian Anophelines. *Bullt. Nat. Soc. Ind. Mal. (Mosq.)*, Delhi., 85-137.

15

Wetland Ecology Along the Dhansiri River Channel, Assam, India

—*P. Kotoky*[1], *M.K. Dutta*[2], & *G.C. Borah*[3]

[The Dhansiri river is an extremely meandered important south bank tributary of the mighty Brahmaputra River. The location of the river in a varied geomorphic setup in a valley filled with recent alluvium, morphological adjustment of the channel have developed many low lying areas along the channel. However, associated land use pattern have also influence the development of many wetland areas. Significant changes on the low-lying areas have been observed during the period 1914 to 2000 in the studied stretch of the river. The highly meandered nature of the river channel and consequent migration of meander bends have contributed significantly towards generation of higher amount of ox-bow type swamps followed by linear type. Comparatively higher rate of wetlands are developed on the western side of the river channel. Swampy land or wetlands have represented potentially very rich areas with geomorphologic and ecologic significance. These low lying areas act as storage basins during flood, reducing the flood impact and minimising the intensity of erosion. These wetlands are also helpful in maintaining the rich bio-diversity of the region and can be used as an important eco-potential resource.

Key words: Wetland ecology, Dhansiri river, Assam]

Introduction

Rivers warrant study for three main reasons: first, because of their existence in physical landscape and their significance for producing fluvial landforms; second, because of their importance in relation to many other geomorphologic processes in fluvial dominated landscapes; and third, because of their significance for human use. For these reasons the drainage basin has been acknowledged as the fundamental geomorphic unit.

The global extent of wetlands is estimated to be from 7 to 8 millions km^2 and, compared to the other ecosystems, wetlands are an extremely productive part of the landscape with average annual production above $1000 gcm^{-2}yr^{-1}$ in terms of nutrient recycling and storage, plant and animal harvest, and species conservation (Dugan, 1993: Mitsch *et al.*, 1994). The biological interactions and necessary functions

[1,3] North East Institute of Science and Technology (NEIST-CSIR), Jorhat 785 006, Assam, India.
2 Jorhat Engineering College (JEC), Jorhat–785007, Assam, India.
*Corresponding author: probhatk@yahoo.com

between the elements of soil, water, plants and animals together with their biological diversity and the cultural values, the wetlands are marked invaluable to the people all over the world.

The low-lying areas of the floodplain beyond the natural levee are known as backswamps or flood basins over which the finer materials are deposited during overbank floods. These low-lying areas are naturally waterlogged and marshy areas, serving as regulators of hydrologic regimes, maintaining environmental quality, and resources of diverse aquatic fauna and flora. These areas in local parlance are known as 'Jalah', 'Doloni', 'Pitoni', 'Hola', 'Doba' etc. The reclamation of these disconnected flood basins for cultivation has brought about a general change in their morphology, however, these can still be identified on toposheets and aerial photographs. The perennially waterlogged areas are termed as 'Bils' in Assam. These water bodies are of different sizes and shapes and generally connected to adjacent river systems. There are approximately 1392 bils in the Brahmaputra Valley, Assam, of which 423(30.39 %) are registered fisheries while the remaining 969(69.61%) are unregistered (Deka and Goswami, 1992). Bils originated as a consequence of many naturally interrelated processes and factors. However, the development of bils in this region is strongly related to the geomorphological development of the river basin, neotectonic activities and other allogenic processes.

The rationale behind such study is to generate proper prior attention for the maintenance of natural wetlands in relation to different wetland management projects and their expected positive consequences towards development of rural economy. Wetlands are also beneficial for partial flood storage by temporarily storing the runoff. By storing rainwater and releasing runoff, wetlands can diminish the destructive onslaught of floods downstream. Preserving natural storage can also help to avoid the costly construction of dams and reservoirs. Recharge in wetlands in turn supports more stable biological diversity. Floodplain wetlands in NE region of India are under extreme pressure from farm expansion, water control and human habitation. Apart from the numerous aquatic flora and fauna, wetlands in this region are also the winter home for large number of birds and species.

Most of the swampy lands in Assam are formed as ox-bow lakes or abandoned channel scattered all over the active flood plains of the mighty Brahmaputra River system. A few of these water bodies, however, owe their origin to depressions caused as a result of earthquakes. A swamp or bil is a complex natural system whose origin and development depends on a variety of interacting natural as well as biotic/ anthropogenic features. Any attempt to understand such a system will require consideration of several aspects related to various geological, hydrological, biological and socio-ecological parameters.

The rivers of the NE region of India manifest frequent changes in morphologic behaviour with resultant changes in associated fluviosedimentological attributes. It is therefore felt that the characterisation and quantification of different types of swamp along the stretch of the river Dhansiri within a specific geomorphic setup may throw an insight into their origin and development. Five different types of swamp are identified and characterised as i) compact(C), ii) linear(L), iii) irregular(I), (iv) discrete (D) and (v) ox-bow(O) and the respective populations of each were counted. The areas of the different bills were measured and the generated data are used for the present study. A number of field checks have also been conducted to understand the characteristics and comparative evaluation of the data resource as evident from toposheets and imageries.

Brinson *et al.* (1994) classified wetlands into four broad geomorphic groups: riverine, depressional, fringe (coastal) and extensive peatlands and known as Brinson's Hydrogeomorphic Classification (HGM). Wetlands are also classified according to the source of water (precipitation, overland flow and ground water) and hydrodynamics (the strength and direction of flow). In the present study the wetlands are classified following Deka and Goswami (1992), Sharma and Goswami (1993) and Bezbaruah *et al.* (2003) in relation to localised basinal characteristics, flow behaviour, soil type, and strength and associated geomorphologic attributes.

Methodology

The Survey of India toposheets of 1914 (Scale 1: 63,360) and 1975 (Scale 1: 50,000) and Indian Remote Sensing (IRS) satellite imageries of 1990, 1995 and 2000 were utilized for spatio-temporal change analysis.) The area for the study (Figure 15.1) from Dhansirimukh to Nowakota Kachari ($93^0 30^1$E - $94^0 0^1$E Long. and $25^0 58^1$N- $26^0 45^1$N Lat.) covered by the Survey of India toposheets numbers 83 F/10, F/14, F/15, F/16, G/9 and G/13 was delineated for preparation of geomorphological maps. The Survey of India toposheets and Indian Remote Sensing satellite imageries covering the period from 1914 to 2000 are utilised for the present investigation. The Survey of India toposheets of 1975 (Scale 1:50,000) were scanned, georeferenced, mosaiced and used as base map. The Survey of India toposheets of 1914 (Scale 1:63,360) and the IRS-IB LISS-II, black and white (B/W) imageries of 1990 and 1995 on band four (wave length 0.77 to 0.86ìm) at 1:50,000 scale were also scanned, registered and resampled to keep the uniformity in data set. IRS-IC LISS-III digital image of January 2000 on 1:50,000 scale of path 112, Row 53 were used to map the changes during 2000. The satellite images and SOI toposheets of 1914 were registered to the base map using a set of Ground Control Points (GCPs) in ERDAS IMAGINE 8.5 Software. Thematic maps of different periods were prepared on 1:50.000 scale and were integrated using Arc View GIS. The entire length of the channel have been divided into six sectors and utilized for the present study.

The nature/type of low-lying areas was evaluated both by qualitative and quantitative manner following Nanson and Croke (1992), Deka and Goswami (1992), Dugan (1993), Mitsch (1994) and Bezbaruah et al. (2003). The various types of wetlands were characterized, quantified and areas involved were measured to generate database (Table 15.1 15.2) for the present exploratory study.

Dhansiri River Basin

The Dhansiri river, an important south bank tributary of the mighty Brahmaputra River is originated from Thingtubum Peak of the Barail range at an altitude of 1868 m ($25^0 36^1$N and $93^0 47^1$E). The other name of the Dhansiri is "Dima"-

a word from the BODO Language means big. The important tributaries of the river are Kaliyani, Daigrung, Diphu, Chongajan, Deopani and Nambar. The tributary Dayang is much bigger than the main Dhansiri river. The catchment's area of the river is about 12,548km^2. the total length of the channel is 368.13km., out of which 200 km falls within the plains of Assam. The river attains a maximum breadth of 419m near Bhalaguri-Napamua and average depth is about 6.20m. The hydrological station at Numaligarh, indicate its danger level as 77.42m and maximum, minimum and average water discharges as 2091.85m^3 sec^{-1}, 4.88 m^{3}sec^{-1} and 513 m^{3}sec^{-1}. The annual discharge of sediments as recorded in the station was reported as 6.28lakhs ha.m. The Dhansiri river was reported as one of the highly meandered rivers of the world.

The Dhansiri river basin is situated in a varied geomorphic setup. From the regional geomorphic viewpoints two major geomorphic units can be distinguished for the entire Dhansiri river basin as shown in Figure 15.2. Regional slope, geological formations, structure and topography are the criteria applied to divide the units. Plain and Hills are the major geomorphic units of the area which can further be divided into different subgroups depending on their geochronology status, nature of occurrence and associated structures. The plains and hills can be differentiated into flood plain, younger alluvial plain, valley fills and older alluvial plain and denudation hills, inselberg and structural hills respectively.

Results and Discussion

The different types of wetlands (Figure 15.3) are not uniformly distributed through out the Dhansiri river basin. The total number of swamps with dominant oxbow type as evident from the study was 75 in 1914, 127 in 1975, 111 in 1990, 114 in 1995 and 122 in 2000 (Figure 15.4-15.9). Moreover, total areas covered by the wetlands as observed were 10.24km^2 in 1914, 13.61km^2 in 1975 and 13.23km^2 in 1990, 11.26km^2 in 1995 and 13.06km^2 in 2000. The number of wetlands and their corresponding areas for five different periods of study are projected through Figures 15.4 to 15.9. The geomorphic map of 1914 exhibited dominant association of linear variety of swamps (26.67%) and represented mainly by the paleochannels of the earlier river. The association of different varieties of swamps can be represented as L(26.67% >C(24%) >O(22.67%) >D(20%) >I(6.67%). Out of six sectors the Sector I represent highest development of wetlands in numbers with equal proportion (36.36%) of compact and linear varieties. However, oxbow type swamps have played a major role in development of wetlands under study. The abundance of oxbow type swamps has clearly been attributed their origin from the earlier meandered channels. The maximum number of oxbow type swamp was observed in the Sector III. Maximum land area was also covered by these wetlands in the same sector.

The geomorphic map of 1975 has demonstrated almost an opposite picture of 1914 observations. The distribution of total number of wetlands can be represented as O(45.67%) >L(25.98%)>C(18.11%)>I (7.87%) >D(2.36%). In the 1975 geomorphic map maximum number (47) of wetland was observed in the Sector III however, maximum land area (5.07km^2) of wetlands was observed in the Sector II. The maximum developed sector i.e., Sector III can be represented as O(46.80%)>C(31.91%)>L(8.51 %)>D(6.38%) = I(6.38%). The effect of the 'Great Assam Earthquake' of 1950 with attendant historic flood might have played a key role in changing the morphology of the region. The imbalance in sediment production and distribution associated with the existing hydrologic regime of the area has initiated the development of many low lying areas adjacent to the river channel. The channel experienced a differential rate of geomorphological change as the basic characteristics with the associated soil types along the channel is not identical. The extensive development of swamps as represented by 1975 geomorphic map, compared to 1914 data set can be explained as the intense effect of neotectonics on the nature of the basin. Pascoe (1964) and Geological Survey of India (1977) advocate, it as due to gradual subsidence of the whole or part of the basin, while others advocate that there is both submergence as well as uplift but with dominance of submergence (Prasad *et al.* 1981). Many workers (Pascoe, 1964: GSI, 1977: Goswami *et al.*, 1999: Sarma, 1999) have argued that the valley floor is subsiding due to tectonic movement. As the area is a part of seismically active NE-India situated between the Himalayan and the Arakan-Yoma mountain arches the reasoning seems to be plausible. Subsidence thus helps to transform every former channel into a low-lying area (G.S.I.1977). However, the activity is not uniform throughout the basin and as a result abandoned channels become isolated into cut-off meanders or ox-bow lakes. Moreover, construction of anti flood and some other anthropogenic measures also helps to create many big low lying areas adjacent to active channels. The generation of linear and ox-bow lakes is generally related with migration and morphometric adjustment of stream channels at different level of flow conditions. It is difficult to answer why and how the discrete nature of swamps connected by small segments of channels and irregular low lying structures are generated. However, it can be presumed that the existing lithology of the area, channel network, and associated autogenic processes might have played the key role in the geomorphologic changes of swamps.

Almost similar observation has also been observed in the geomorphic map of 1990. It has clearly been attributed from the study that during the period 1975 to 1990 no significant geomorphic changes took place within the stretch under study. The maximum land area (5.21km^2) covered by the wetlands was observed in Sector II and was followed by Sectors III(3.49km^2) and I(2.76km^2). The distribution of wetlands during this period can be represented as O(35.14%)>L(34.23%)> C(27.03%)>I(3.60%). The rate of geomorphological change took place during the period 1990 to 1995 was appeared to be somewhat different than the earlier periods. The total number of wetlands developed can be represented as L(38.60%) >O(35.97%)> C(22.81%)> D(1.75%) >I(0.88%). Sector II has represented association of both maximum number (35) and land area (3.763km^2) of swamps with dominant linear type within the studied stretch for the period under study. The dominant linear nature of the swamps testifies the transient nature of most of the former stream channels of the basin, which on subsidence generate swamps. The period up to 2000 witnessed a picture with dominance of oxbow type wetlands. Maximum number (38) of swamps was observed in the Sector III. The total number of

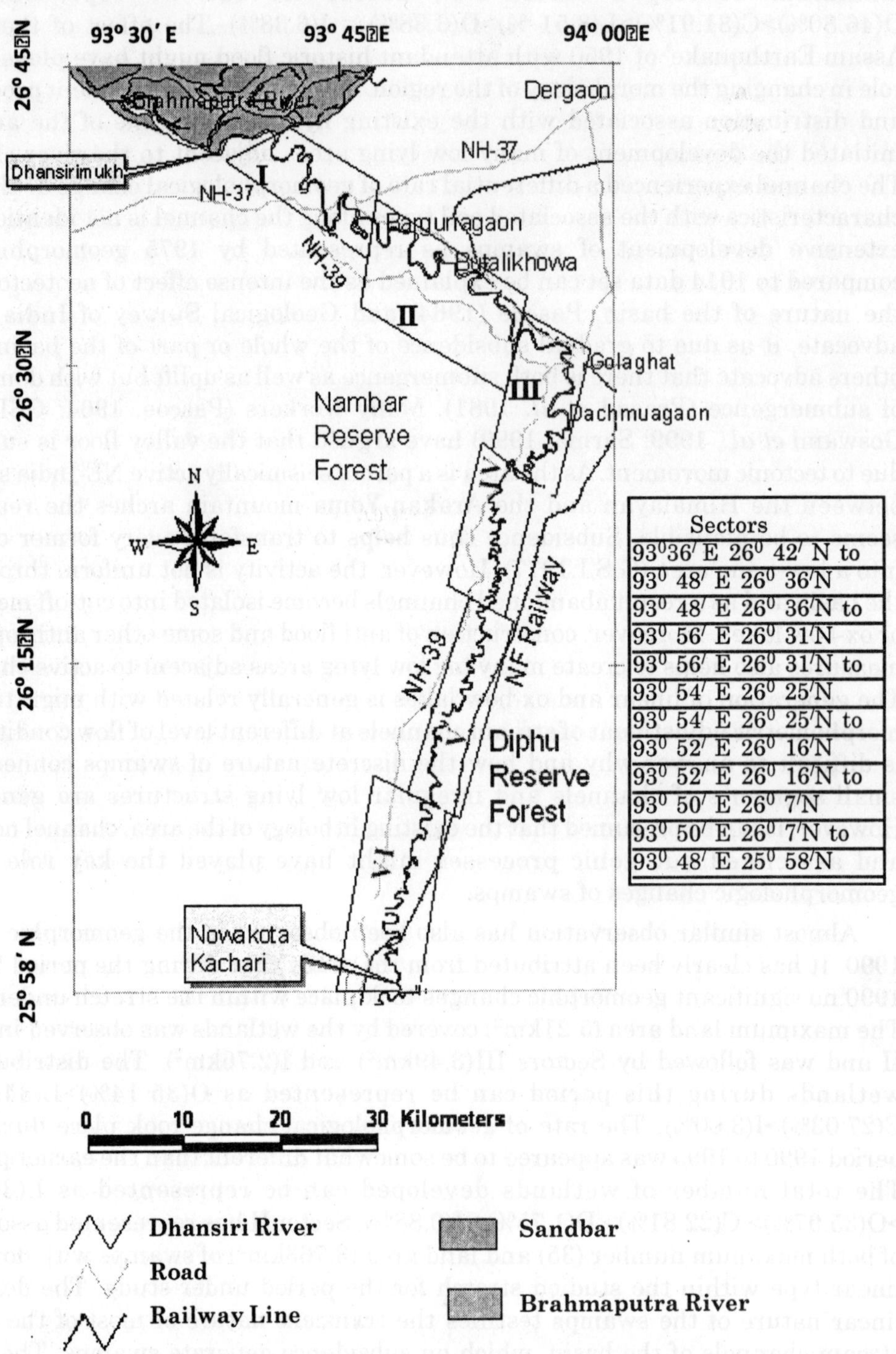

Figure 15.1. Location map of the study area.

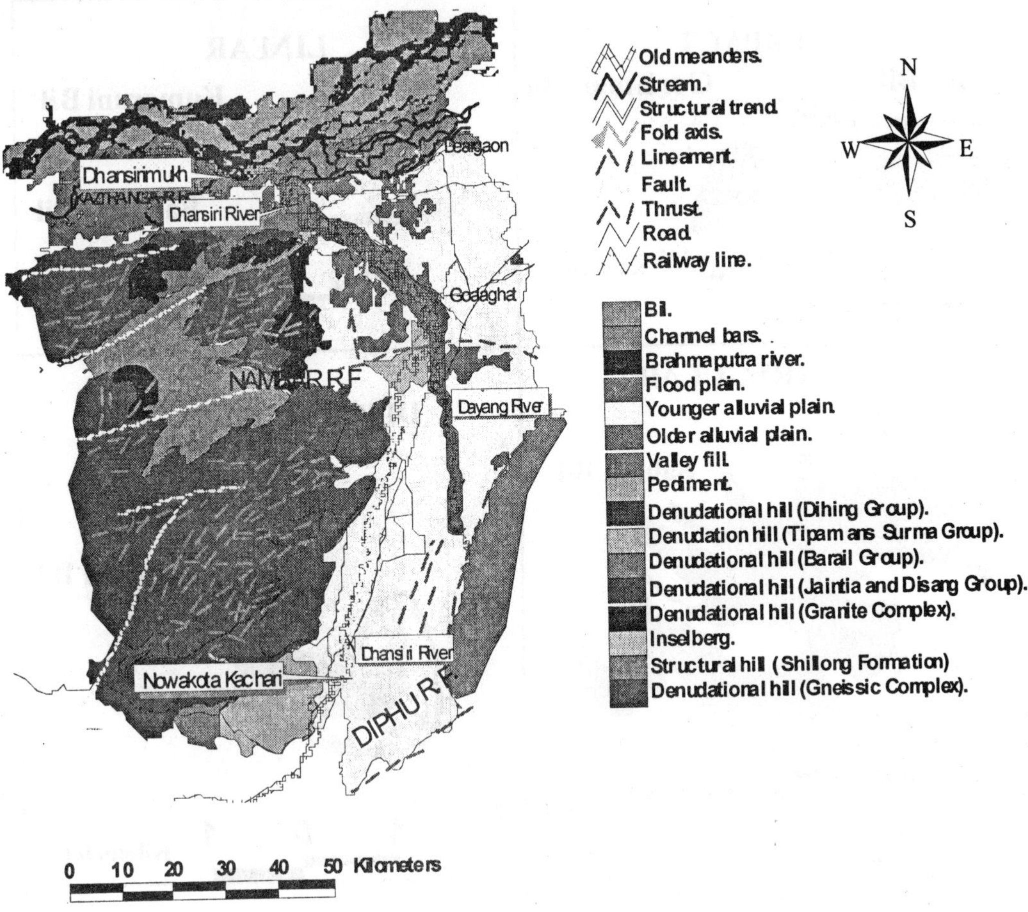

Figure 15.2. Geomorphologic map of the area under study.

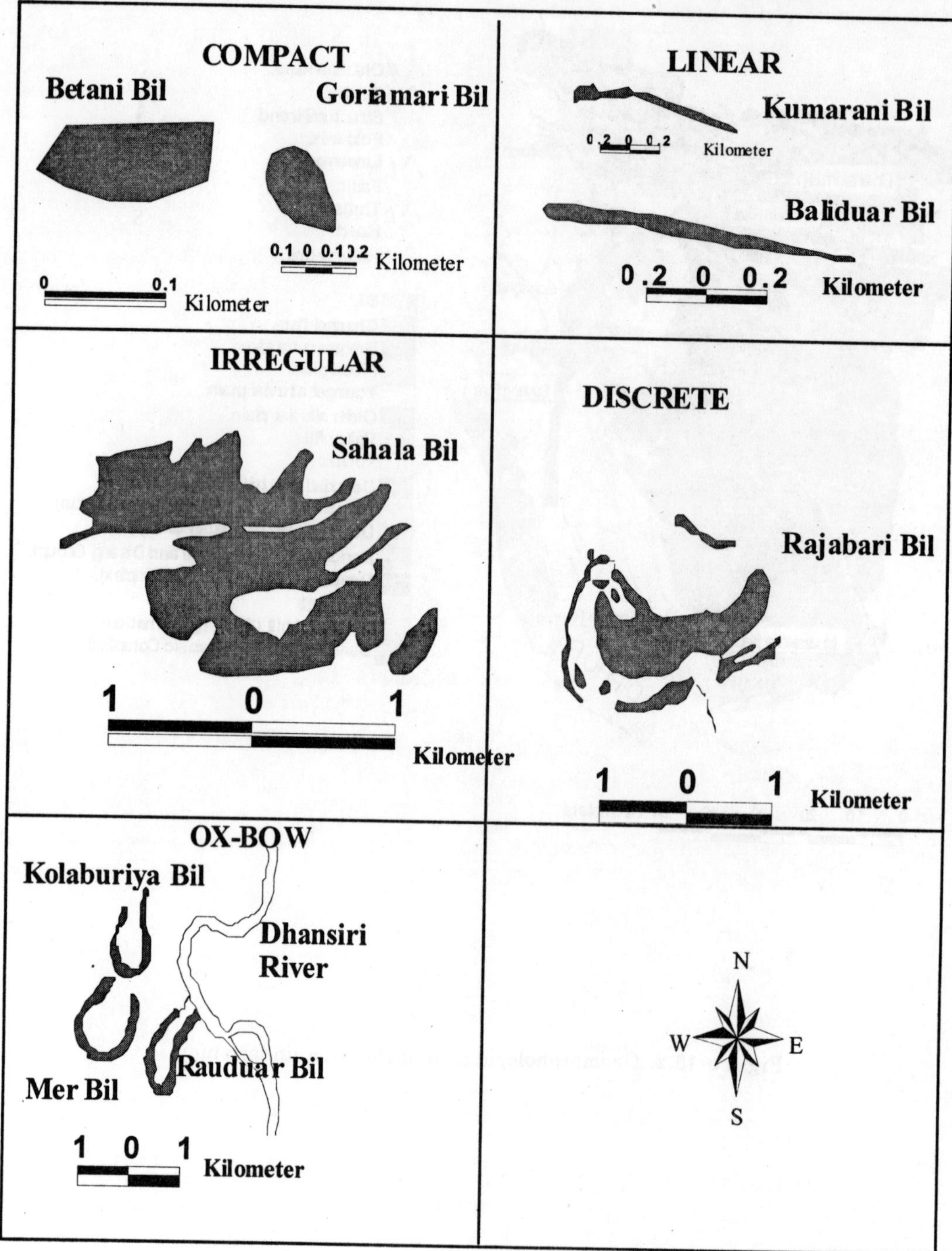

Figure 15.3. Different types of wetlands (Swamps).

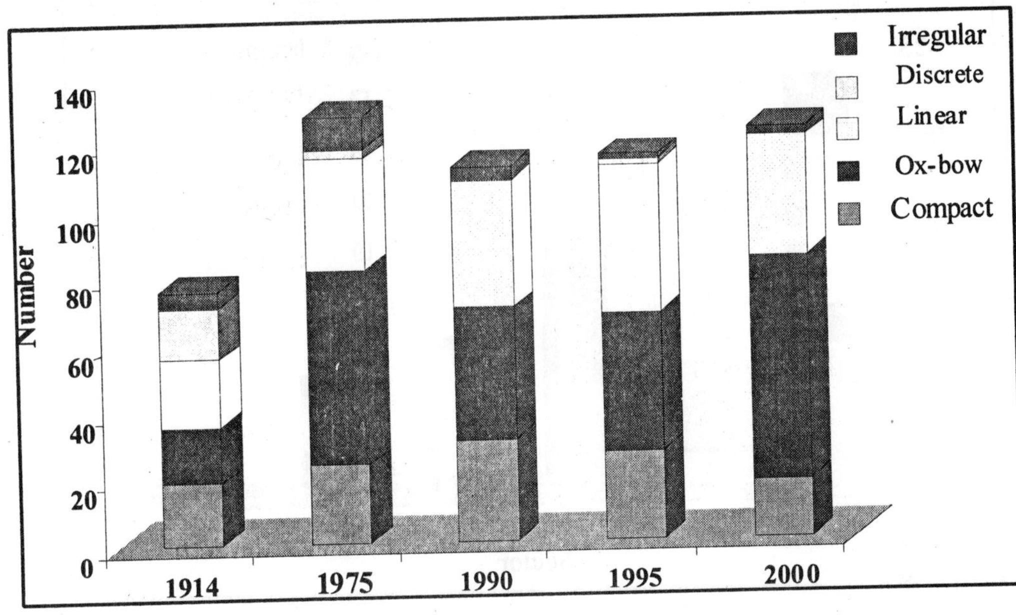

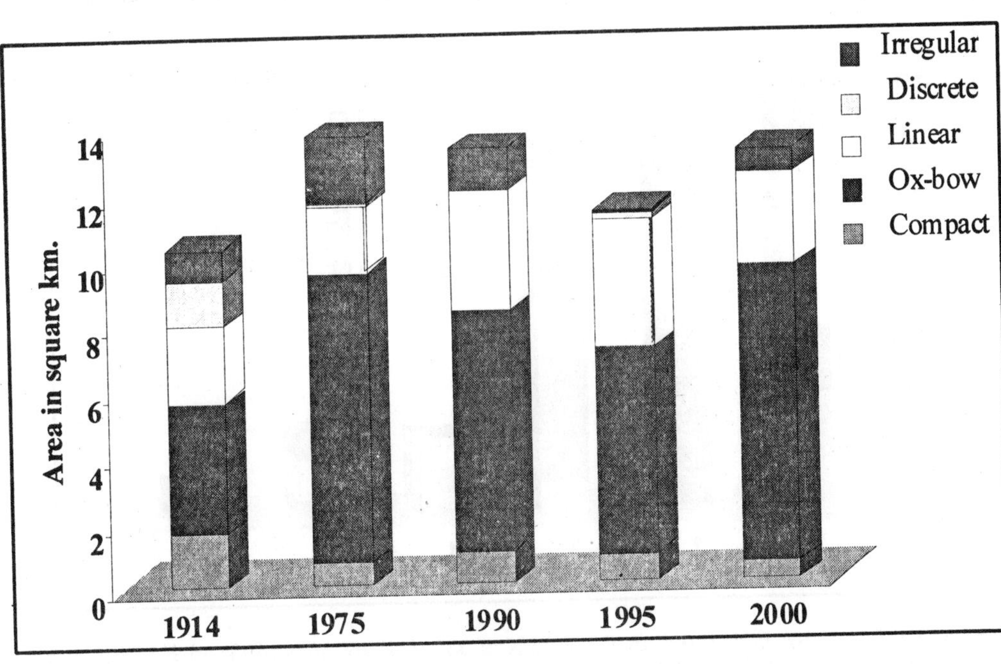

Figure 15.4. Development of swamps in different periods.

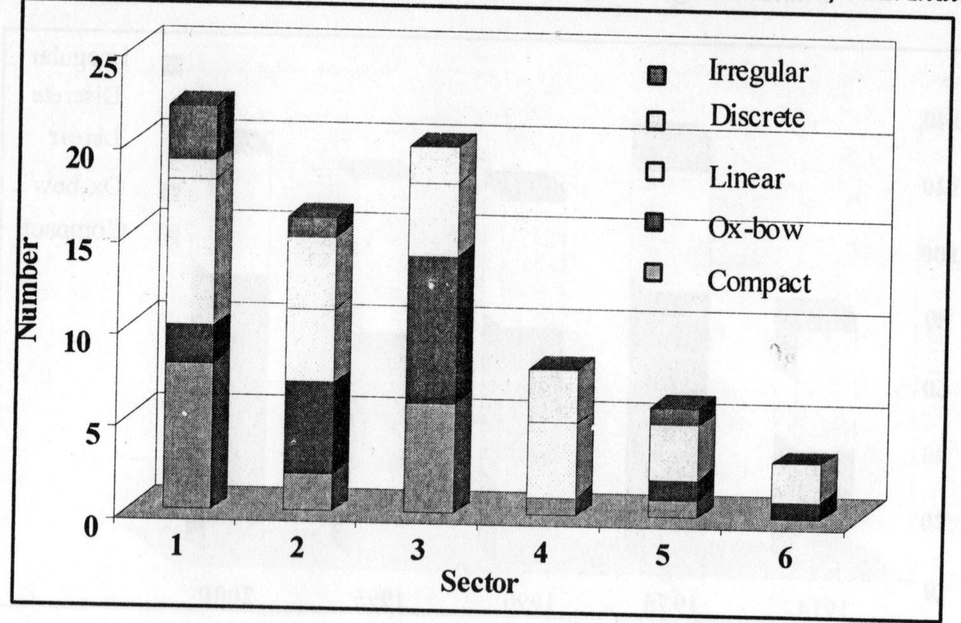

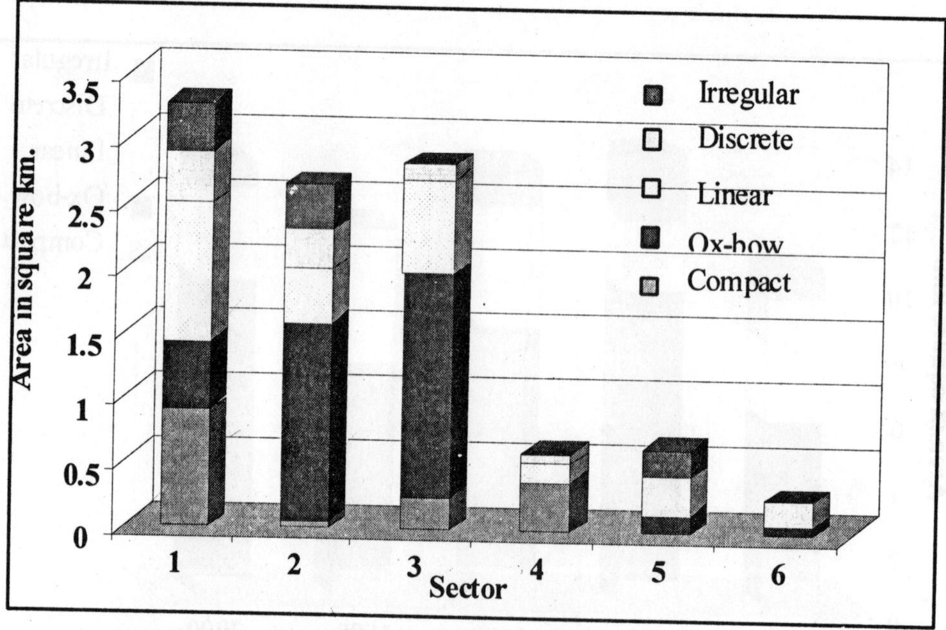

Sectors: **1**-93°36′E 26° 42′ N to93° 48′ E 26° 36′N; **2**-93° 48′ E 26° 36′N to93° 56′ E 26° 31′N;**3**-93° 56′ E 26° 31′N to93° 54′ E 26° 25′N; **4**-93° 54′ E 26° 25′N to 93° 52′ E 26° 16′N; **5**-93° 52′ E 26° 16′N to93° 50′ E 26° 7′N ; **6**-93° 50′ E 26° 7′N to93° 48′ E 25° 58′N

Figure 15.5. Spatial distribution of swamps along banks of the Dhansiri river within the studied reach till 1914.

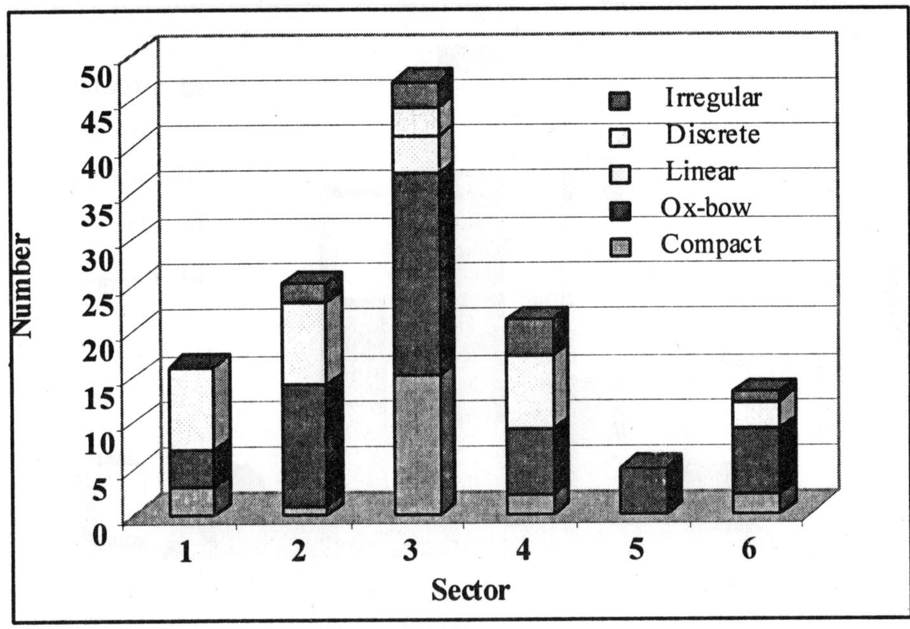

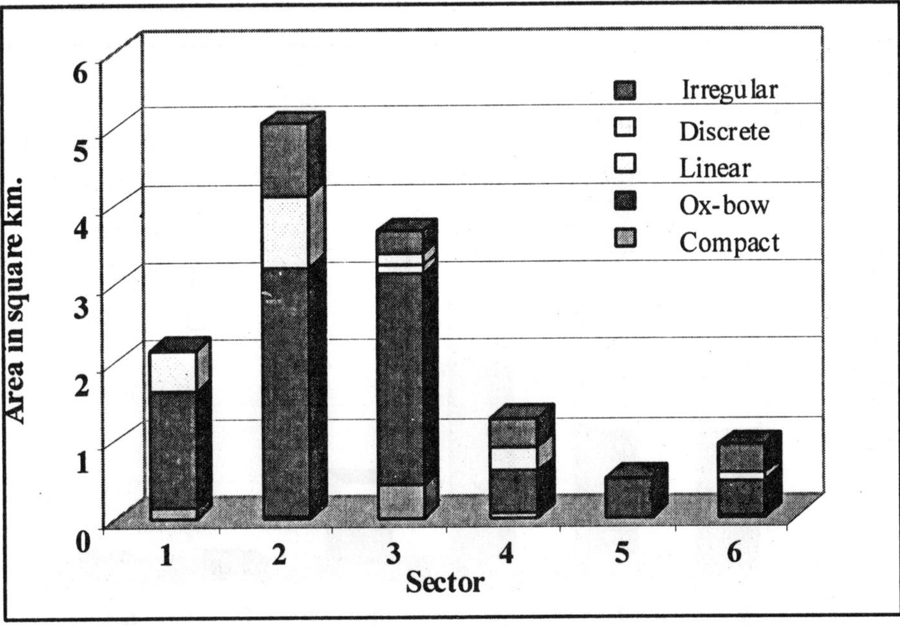

Sectors: 1-93°36′E 26° 42′ N to93° 48′ E 26° 36′N; 2-93° 48′ E 26° 36′N to93° 56′ E 26° 31′N;3-93° 56′ E 26° 31′N to93° 54′ E 26° 25′N; 4-93° 54′ E 26° 25′N to 93° 52′ E 26° 16′N; 5-93° 52′ E 26° 16′N to93° 50′ E 26° 7′N ; 6-93° 50′ E 26° 7′N to93° 48′ E 25° 58′N

Figure 15.6. Spatial distribution of swamps along banks of the Dhansiri river within the studied reach till 1975.

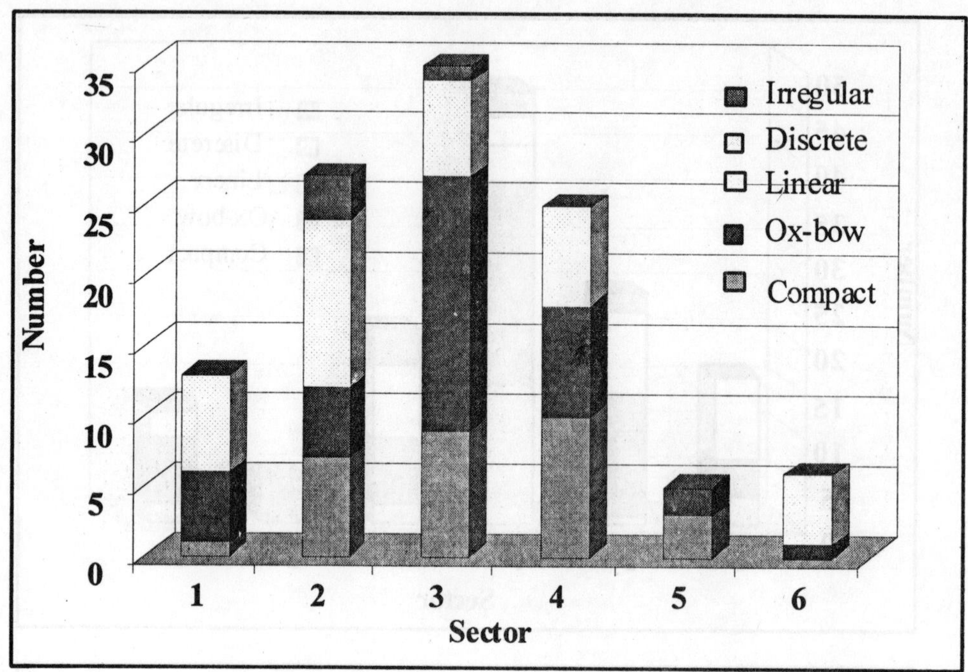

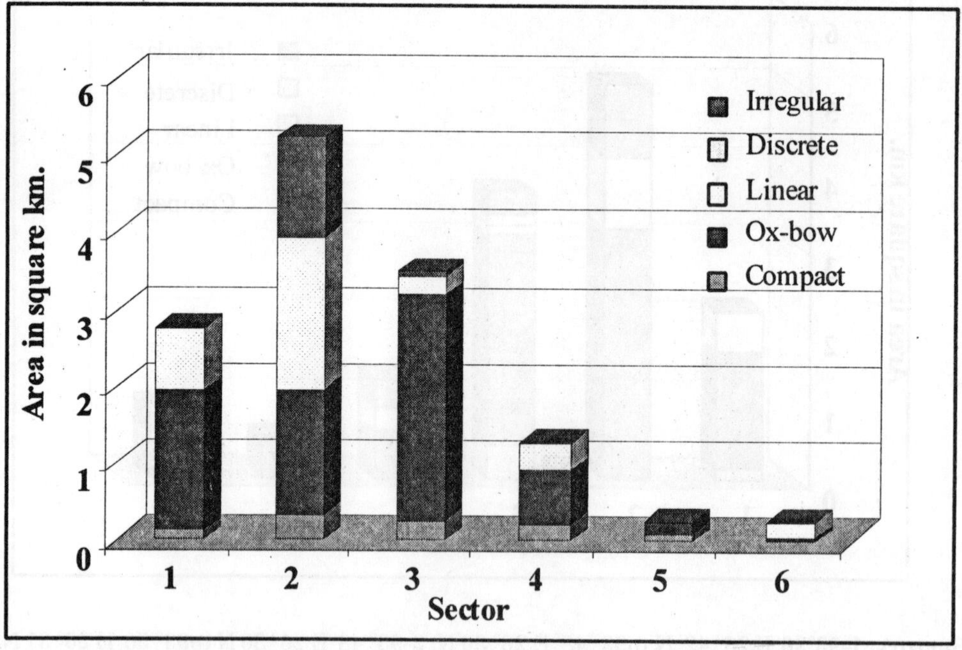

Sectors: 1-93°36′ E 26° 42′ N to93° 48′ E 26° 36′N; 2-93° 48′ E 26° 36′N to93° 56′ E 26° 31′N;3-93° 56′ E 26° 31′N to93° 54′ E 26° 25′N; 4-93° 54′ E 26° 25′N to 93° 52′ E 26° 16′N; 5-93° 52′ E 26° 16′N to93° 50′ E 26° 7′N ; 6-93° 50′ E 26° 7′N to93° 48′ E 25° 58′N

Figure 15.7. Spatial distribution of swamps along banks of the Dhansiri river within the studied reach till 1990.

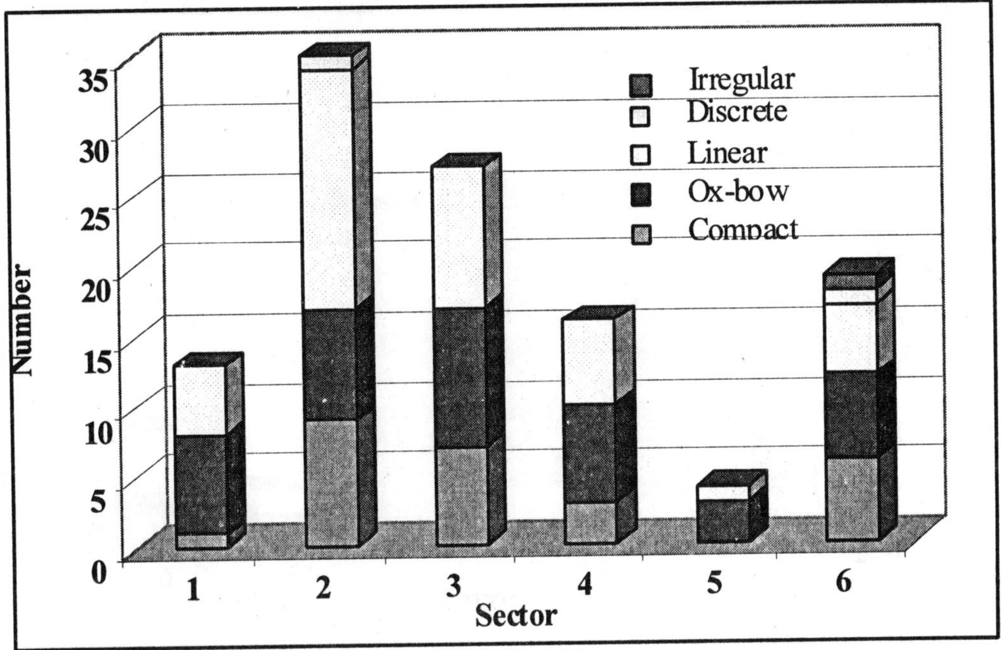

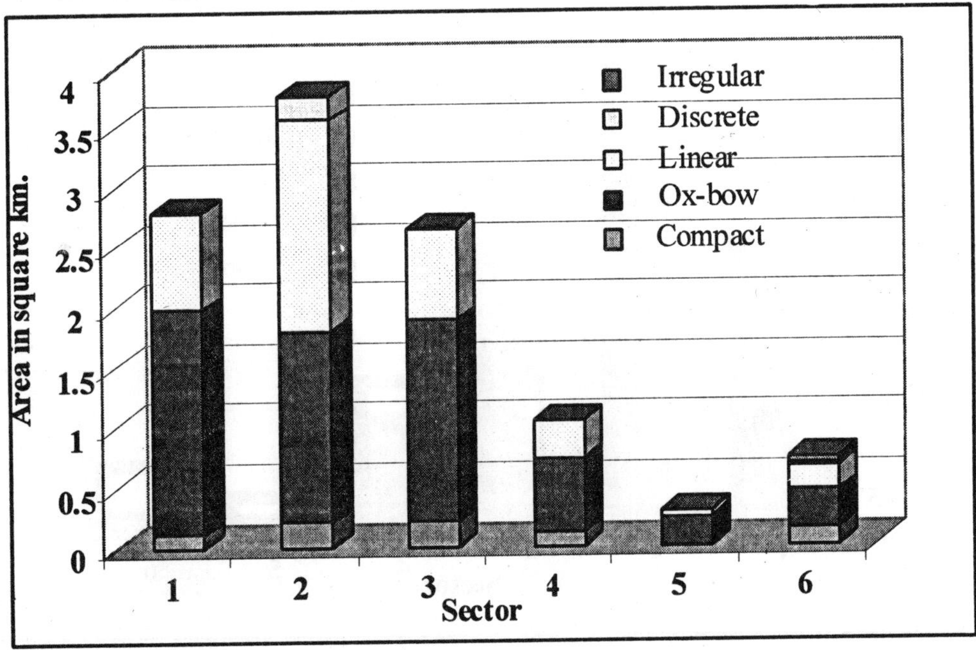

Sectors: 1-93°36′E 26° 42′ N to93° 48′ E 26° 36′N; 2-93° 48′ E 26° 36′N to93° 56′ E 26° 31′N;3-93° 56′ E 26° 31′N to93° 54′ E 26° 25′N; 4-93° 54′ E 26° 25′N to 93° 52′ E 26° 16′N; 5-93° 52′ E 26° 16′N to93° 50′ E 26° 7′N ; 6-93° 50′ E 26° 7′N to93° 48′ E 25° 58′N

Figure 15.8. Spatial distribution of swamps along banks of the Dhansiri river within the studied reach till 1995.

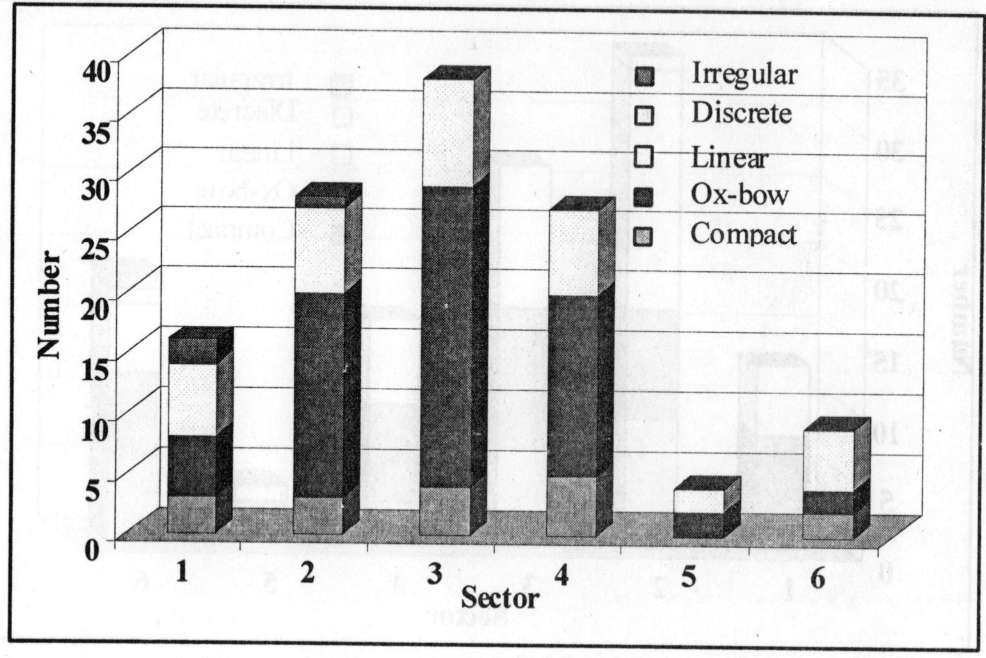

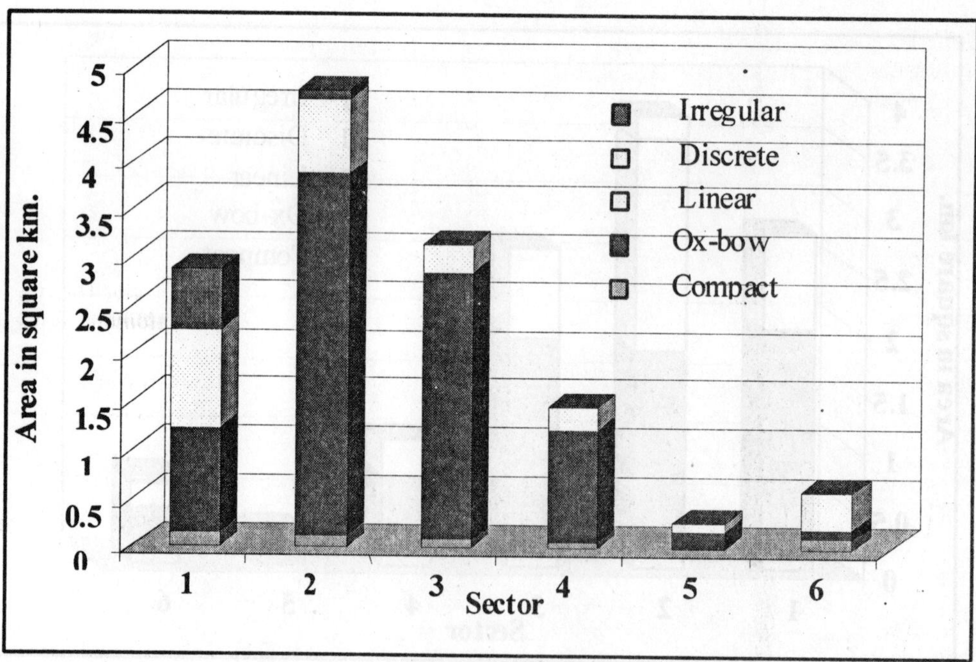

Sectors: 1-93°36′E 26° 42′ N to93° 48′ E 26° 36′N; 2-93° 48′ E 26° 36′N to93° 56′ E 26° 31′N; 3-93° 56′ E 26° 31′N to93° 54′ E 26° 25′N; 4-93° 54′ E 26° 25′N to 93° 52′ E 26° 16′N; 5-93° 52′ E 26° 16′N to93° 50′ E 26° 7′N; 6-93° 50′ E 26° 7′N to93° 48′ E 25° 58′N

Figure 15.9. Spatial distribution of swamps along banks of the Dhansiri river within the studied reach till 2000.

Table 15.1
Different types of wetlands along the Dhansiri river channel with time and space

1914	*I*	*II*	*III*	*IV*	*V*	*VI*	*TOTAL*
COMPACT	8	2	6	1	1	0	**18**
OX-BOW	2	5	8	0	1	1	**17**
LINEAR	8	4	4	4	0	0	**20**
DISCRETE	1	4	2	3	3	2	**15**
IRREGULAR	3	1	0	0	1	0	**5**
TOTAL	**22**	**16**	**20**	**8**	**6**	**3**	**75**
1975	*I*	*II*	*III*	*IV*	*V*	*VI*	*TOTAL*
COMPACT	3	1	15	2	0	2	**23**
OX-BOW	4	13	22	7	5	7	**58**
LINEAR	9	9	4	8	0	3	**33**
DISCRETE	0	0	3	0	0	0	**3**
IRREGULAR	0	2	3	4	0	1	**10**
TOTAL	**16**	**25**	**47**	**21**	**5**	**13**	**127**
1990	*I*	*II*	*III*	*IV*	*V*	*VI*	*TOTAL*
COMPACT	1	7	9	10	3	0	**30**
OX-BOW	5	5	18	8	2	1	**39**
LINEAR	7	12	7	7	0	5	**38**
DISCRETE	0	0	0	0	0	0	**0**
IRREGULAR	0	3	1	0	0	0	**4**
TOTAL	**13**	**27**	**35**	**25**	**5**	**6**	**111**
1995	*I*	*II*	*III*	*IV*	*V*	*VI*	*TOTAL*
COMPACT	1	9	7	3	0	6	**26**
OX-BOW	7	8	10	7	3	6	**41**
LINEAR	5	17	10	6	1	5	**44**
DISCRETE	0	1	0	0	0	1	**2**
IRREGULAR	0	0	0	0	0	1	**1**
TOTAL	**13**	**35**	**27**	**16**	**4**	**19**	**114**
2000	*I*	*II*	*III*	*IV*	*V*	*VI*	*TOTAL*
COMPACT	3	3	4	5	0	2	**17**
OX-BOW	5	17	25	15	2	2	**66**
LINEAR	6	7	9	7	2	5	**36**
DISCRETE	0	0	0	0	0	0	**0**
IRREGULAR	2	1	0	0	0	0	**3**
TOTAL	**16**	**28**	**38**	**27**	**4**	**9**	**122**

Table 15.2
Areas of different wetlands along the Dhansiri River channel at different time and space

1914	*I*	*II*	*III*	*IV*	*V*	*VI*	*TOTAL*
COMPACT	0.894	0.05	0.251	0.375	0.0125	0	**1.5825**
OX-BOW	0.525	1.526	1.725	0	0.125	0.063	**3.964**
LINEAR	1.076	0.425	0.725	0.145	0	0	**2.371**
DISCRETE	0.4	0.3	0.125	0.075	0.3	0.194	**1.394**
IRREGULAR	0.376	0.35	0	0	0.2	0	**0.926**
TOTAL	**3.271**	**2.651**	**2.826**	**0.595**	**0.6375**	**0.257**	**10.2375**
1975	*I*	*II*	*III*	*IV*	*V*	*VI*	*TOTAL*
COMPACT	0.13125	0.025	0.425	0.05	0	0.0375	**0.66875**
OX-BOW	1.4875	3.1875	2.6875	0.5375	0.5125	0.43125	**8.84375**
LINEAR	0.5375	0.90625	0.1125	0.325	0	0.09375	**1.975**
DISCRETE	0	0	0.15	0	0	0	**0.15**
IRREGULAR	0	0.95	0.2875	0.3625	0	0.375	**1.975**
TOTAL	**2.15625**	**5.06875**	**3.6625**	**1.275**	**0.5125**	**0.9375**	**13.6125**
1990	*I*	*II*	*III*	*IV*	*V*	*VI*	*TOTAL*
COMPACT	0.1125	0.3125	0.25	0.2125	0.0875	0	**0.975**
OX-BOW	1.825	1.6375	2.9375	0.7	0.1625	0.025	**7.2875**
LINEAR	0.825	1.9875	0.25625	0.35625	0	0.2125	**3.6375**
DISCRETE	0	0	0	0	0	0	**0**
IRREGULAR	0	1.275	0.05	0	0	0	**1.325**
TOTAL	**2.7625**	**5.2125**	**3.49375**	**1.26875**	**0.25**	**0.2375**	**13.225**
1995	*I*	*II*	*III*	*IV*	*V*	*VI*	*TOTAL*
COMPACT	0.1125	0.206	0.206	0.125	0	0.15	**0.7995**
OX-BOW	1.87	1.589	1.674	0.6	0.25	0.3	**6.283**
LINEAR	0.805	1.768	0.77	0.325	0.0375	0.2	**3.9055**
DISCRETE	0	0.2	0	0	0	0.025	**0.225**
IRREGULAR	0	0	0	0	0	0.05	**0.05**
TOTAL	**2.7875**	**3.763**	**2.65**	**1.05**	**0.2875**	**0.725**	**11.263**
2000	*I*	*II*	*III*	*IV*	*V*	*VI*	*TOTAL*
COMPACT	0.155	0.1125	0.075	0.05625	0	0.1036	**0.50235**
OX-BOW	1.0625	3.75	2.75	1.175	0.1875	0.1036	**9.0286**
LINEAR	1.018	0.7875	0.3125	0.2375	0.075	0.3818	**2.8123**
DISCRETE	0	0	0	0	0	0	**0**
IRREGULAR	0.6325	0.0875	0	0	0	0	**0.72**
TOTAL	**2.868**	**4.7375**	**3.1375**	**1.46875**	**0.2625**	**0.5890**	**13.06325**

distribution of wetlands can be represented as O(54.10%)>L(29.51%)>C (13.93%)>I(2.46%). However, maximum land area covered by the wetlands (4.74km^2) was associated with the Sector II with dominant oxbow type (3.75km^2) wetland.

The number of wetlands during the period of observation in six different sectors is presented in Table 15.3. Significant rise in number of wetlands on the west bank of the Dhansiri River can clearly be observed. This has clearly attributed that the west bank of the river suffers relatively significant geomorphic changes in relation to neotectonic activities of the region. However, the occurrence of highly meandered river channel in the Brahmaputra valley alluvium associated with migration and erosion-deposition processes also have activated the development of wetlands along the river channel.

Table 15.3
Number of wetlands along the Dhansiri river channel

PeriodsSectors	*1914*		*1975*		*1990*		*1995*		*2000*	
	East Bank	*West Bank*	*East Bank*	*West Bank*	*East Bank*	*West Bank*	*East Bank*	*West Bank*	*East Bank*	*West Bank*
I	7	15	5	11	4	9	4	9	6	10
II	7	9	10	15	13	14	15	20	14	14
III	8	12	17	30	15	20	12	15	15	23
IV	3	5	9	12	11	14	7	9	11	16
V	3	3	2	3	2	3	2	2	2	2
VI	1	2	5	8	2	4	8	11	4	5
Total	**29**	**46**	**48**	**79**	**47**	**64**	**48**	**66**	**52**	**70**

Conclusion

The geomorphic, tectonic and hydrologic behaviour of the Dhansiri River basin is intimately connected with the morphologic features and genesis of the structures. Morphological adjustment of stream channels over a relatively short period of time and location of the basin in a tectonically active region have played dominant roles in changing the nature of swampy lands within the basin. Nature of channel and its location in a valley filled with recent alluvium together with associated active geo-tectonics have played a crucial role in generation/development of low lying areas near the Dhansiri River channel. However, associated land use pattern have also influence the development of many wetland areas. Significant changes on the low-lying areas have been observed during the period 1914 to 2000 in the studied stretch of the river. Comparatively higher rate of wetlands are developed on the western side of the river channel.

Swampy land or wetlands have represented potentially very rich areas with geomorphologic and ecologic significance. These low lying areas act as storage basins during flood, reducing the flood impact and minimising the intensity of erosion. These wetlands are also helpful in maintaining the rich bio-diversity of the region and can be used as an important eco-potential resource.

[**Acknowledgement:** We express our deep sense of gratitude to Dr PG Rao, Director, North East Institute of Science and Technology (CSIR), Jorhat, Assam, India, for his constant inspiration and kind permission to publish the work.]

REFERENCES

Bezbaruah, D., Kotoky, P., Baruah, J. and Sarma, J. N. 2003. Geomorphological explanation of swamps along the Brahmaputra River channel, Assam., *Jour. Geol. Soc. India*, 62: 605-613.

Brinson, M.M., Kruczynaki, L.C., Nutter, W.L., Smith, R.D. and Wigham, D.F. 1994. Developing an approach for assessing the functions of wetlands, pp. 615-624. In: *Global Wetlands: Old world and New* (ed. W.J. Mitsch), Elsevier.

Deka, S.K. and Goswami, D.C. 1992. Hydrology, Sediment characteristics and Depositional Environment of Wetlands: A case study of Deepar Bil, Assam. *J. Assam Science Soc.*, 34(2): 62-84.

Dugan, P. 1993. *Wetlands in Danger*. Pub. In association with IUCN. 191.

Geological Survey of India. 1977. Contribution of Geomorphology and Geohydrology of the Brahmaputra Valley, *Geol. Sur. India*, Misc. Pub. 32: 105 – 110.

Goswani, U., Sarma, J.N. and Patgiri, A.D. 1999. River channel changes of the Subansiri in Assam, India. *Geomorphology*, 30: 227-244.

Mitsch, W. J. 1994. *Global wetlands: Old world and new*, Elsevier, pp949.

Mitsch, W. J., Mitsch, R. H. and Turner, R. E. 1994. Wetlands of the old and new worlds: Ecology and management, pp. 3-56. In: *Global wetlands: Old world and New* (ed. W. J. Mitsch), Elsevier.

Nanson, G. C., and Croke, J. C. 1992. A genetic classification of floodplains, *Geomorphology*, 4, 459-486.

Pascoe, E. H. 1964. *A Manual of the Geology of India and Burma*, Vol. III, Geol. Sur. India, Calcutta, 1345 – 2130.

Prasad, K. C., Kakoti, R. and Burman, S. K. 1981. Geology and Geomorphology of a part of the Subansiri River Valley. Geol. Sur. India, Misc. Pub. 46: 1 – 10.

Sarma, J. N. 1999. Flood and bank erosion problems of the river Brahmaputra, Assam. *Plavan*, 14-18.

Sharma, P. and Goswami, D. C. 1993. Geoenvironmental study on wetlands (bils) in the Naogaon and Morigaon districts of Assam using satellite data. *Proc. Nat. Symp. On Remote Sensing Application for resource management with special emphasis on northeastern region*, Guwahati (Nov. 25-27), 232-240.

16

Water Chemistry of the Rivers Around Jorhat, Assam, India

—*P. Kotoky*[1], *R. K. Bordoloi*[2], *N. K. Baruah*[3], *K.K.Borthakur*[4] & *G.C. Borah*[5]

[Jorhat, a thickly populated developing city, is situated in the Upper Assam. The waste water from households, small and medium industries, commercial and business establishments and different medical and health related establishments both under private and public sectors within the city flows to the main Bhogdoi River through a system of mutually independent drains. The Tarajan and Toklai are the two independent natural arteries, which carries the huge amount of untreated wastewater from the surrounding settlements. During high rainfall period the wastewater inundates the low-lying areas and created a significant environmental problem. The present approach has been made to evaluate certain basic parameters on the water system around Jorhat city, Assam. Apart from the occupational and accidental exposures of environmentally unfriendly substances to human beings, contaminants also can obviously influence water, soil and air quality which may ultimately pose a potential hazard to plant and animals. If the situation remains unaltered, adsorption/ association of environmentally unfriendly substances in the associated sediment/ soil system will definitely result a serious eco-societal problem with lesser water availability and increasingly expensive pure water supply treatment. It is expected that the present study will enlighten certain basic requirements towards life sustaining processes for any future studies for the already existing settlements.

Key words: Jorhat, Water, Bhogdoi, Toklai and Tarajan.]

Introduction

Environment in its basic sense mean the physical conditions such as air, water and land in which people, animals and plants live. Environmental degradation caused by the unattended environmental consequences is potentially one of the most important problems in recent years as contamination may impair the basic conditions for survival of life sustaining processes, which may have serious impact on the future years. It is an established fact that the present practice adopted for disposal of wastes in different cities and towns are not at all satisfactory. These are either simply dumped unscientifically without any treatment into low-lying wastelands, ponds and ditches, river banks, streams etc. Such disposals contribute

[1,2,3,4,5] North East Institute of Science and Technology (NEIST), (Formerly RRL-Jorhat, CSIR) Jorhat 785 006, Assam, India

enormously towards degradation of important natural habitat. The putrescible materials present in such disposals lead to irreparable damage to the land and water environment and subsequently damage the aesthetic quality of life. Run-off during monsoon spreads the waste and at the same time percolation of leachates often causes ground water contamination. As a result of growth and expansion of urban areas, there has been deterioration in the living standard of our urban population. The existing processes involved in collection and scientific disposal of household, commercial and industrial wastes needs execution of multi-step environment friendly procedures to resist our environmental degradation. Along with the development of industrial complexes and urbanization, it is also equally essential and prime requisite towards development of environment around. Industrialization has led in the last few decades to an increased mobilization of trace metals either as a direct action or as indirect consequences. Thus, in recent years, the fluxes of many trace elements from terrestrial and atmospheric sources to the aquatic environment have increased (Forstner and Wittman, 1983). As a result of complex physical, chemical and biological processes, a major fraction of these trace metals is found to be associated with the bottom sediments. The processes and factors that control the scavenging of trace metals by the overlying water when the environmental conditions are changed must be understood to assess their impact on the environment.

The last few decades witnessed an increasing emphasis on studies with environmental concern through multidisciplinary approaches all over the world. These types of studies not only provide an introduction to the dynamics of aquatic chemistry but also identify those aspects that jeopardize the resources of both the marine and fluvial domains. Under the ambit of industrialization and population growth, a continuous monitoring on renewable hydrosphere towards its eco friendly relationship becomes a vital key. The river is the most active component of the hydrological cycle involving mobility of water as well as erosion, transportation and deposition of material carried by the different modes. Considering the fact that, more than 90% of the weathering products are transported to the ocean by the rivers. Knowledge of river water chemistry and the elemental flux through them to the soil/water systems is imperative for a comprehensive understanding of exogenic cycles of elements. The complex interaction of rock/ water/ air/life systems represent a very wide subject of critical debate within the framework of environmental geochemistry, which in turn is related with the eutrophication and natural water resources.

The environmental geochemical studies on soil/water systems in relation to technological development and population growth have not yet been received proper attention. The characterization of natural attributes through baseline studies and anthropogenic inputs to them can be considered as a frontier area of study for this naturally rich part of India.

The System Under Study

Jorhat, a thickly populated developing city, is situated in the Upper Assam. The waste water from households, small and medium industries, commercial and business establishments and different medical and health related establishments both under private and public sectors within the city flows to the main Bhogdoi river through a system of mutually independent drains. The Tarajan and Toklai are the two independent natural arteries, which carries the huge amount of untreated wastewater from the surrounding settlements (Figure 16.1). During high rainfall period the wastewater inundates the low-lying areas and created a significant environmental problem. The severity and depth of problem have not yet been looked through and analysed in a scientific way. The City Municipal Corporation, Town and Country Planning Dept., and Public Health Organisation along with other Nongovernmental Organisations must take it seriously with a thrust towards its ecosocietal implications to create a general awareness among the common masses. Most of the wastewater enters the Bhogdoi River which is one of the main sources of drinking water for the people of Jorhat. Such dumping, if continues and consequent absorption/association of environmentally unfriendly substances in the associated systems will definitely result a serious ecosocietal problem with lesser water availability and increasingly expensive pure water supply treatment.

The present approach has been made to evaluate certain basic parameters on the water system around Jorhat city, Assam. Apart from the occupational and accidental exposures of environmentally unfriendly substances to human beings, contaminants also can obviously influence water, soil and air quality which may ultimately pose a potential hazard to plant and animals.

Methodology

For the detailed study, the toposheets of the area is enlarged to the requisite scale. For the location of the sampling points traverses were made through the area of study. Sampling points were selected at six different places giving due consideration to its representative nature and accessibility. The sampling points along with the locations of background sampling points for the Bhogdoi, the Tarajan and the Toklai rivers in and around Jorhat City are presented in Figure 16.1. Samples were collected once in a month covering both monsoon and non-monsoon periods.

Sample collection

For the determination of total metals, water samples were collected in polyethylene bottles, which were previously soaked and washed with 8M Nitric acid HNO_3) and rinsed finally with deionised water. The collected water samples were immediately acidified with conc. HNO_3 to reduce the pH of the sample below 2.

For the determination of other physicochemical parameters, separate water samples in different quantities were collected in polyethylene containers using different preservatives for stabilization (Berman, 1980; De, 1990; Forstner and Wittmann, 1983; Bernas, 1968; APHA, 1995; Faniran and Aerola, 1978; Fresenious *et al.*, 1988).

Investigation of physicochemical parameters on sites

Water quality parameters such as pH, Electrical Conductivity (EC), Dissolved

Oxygen (DO), Total Dissolved Solids (TDS) and temperature of the water samples were measured on the spot with the help of a portable analytical kit [Model NPC-361D, Naina Electronics Ltd., Punjab, India]. DO was also measured iodometrically in the laboratory following modified method of Winkler for cross checking. Before actual measurements the electrodes were calibrated as per instrumental specifications. For every parameter three readings of each sample was taken and averaged out. Total Suspended Matter (TSM) and Biological Oxygen Demand (BOD) were measured in the laboratory as per standard procedure.

Investigations of parameters in the laboratory

In order to overcome the errors due to microbial activities and redox reactions, the nonconservative parameters like Chemical Oxygen Demand (COD), N-NH_3 were measured within 48 hours by standard methods (Fesenius et al., 1988, APHA, 1995). On the other hand, conservative parameters like Cl^-, $SO4^{-2}$, and HCO_3^- were measured within one week. Various methods which were used for the measurement of these physicochemical parameters are summarized in Table 16.1.

Table 16.1
Methods used for determining physicochemical parameters

Physicochemical parameters	*Methods used for analysis*
TSM	By weight difference method
COD	Titrimetry by volumetric dichromate method
Cl-Cl^-	Titrimetry by Mohr's method
SO_4^{-2}	Gravimetry - as $BaSO_4$
N-NH_3	Absorption spectrophotometry (Model:UV-160A, SHIMADZU, JAPAN) by Nessler's method after distillation with NaOH
HCO_3^- alkalinity as ppm $CaCO_3$	By potentiometric titration method
Free CO_2	Acidimetry method

Source: APHA (1995).

Preservation of sample

For accurate analytical results it is essential to protect the water samples form changes in composition and deterioration from aging due to various interactions. The preservation techniques used in this study are summarized in Table 16.2.

Table 16.2
Preservation method for samples

Parameters	*Minimum sample volume (mL)*	*Container*	*Preservation*
COD	500	Polyethylene	pH adjusted to 2 by adding H_2SO_4 and refrigerated
N-NH_3	500	Polyethylene	0.8 ml conc. H_2SO_4 L^{-1} is used
Metals	2500	Polyethylene	5 ml conc. HNO_3 L^{-1} and refrigerated

Source: APHA (1995).

Laboratory investigation

An attempt has been made to summarize the different steps and procedures adopted during the course of investigations in a systematic way and were described in the following paragraphs.

Sample pretreatment

The concentrations of metal ions were evaluated according to APHA (1989). Preliminary digestion of the water samples was necessary to release the metals associated with suspended matter and colloidal organic matter. A preconcentration step through chelation, extraction and subsequent mineralization was also adopted in the case of metals Cd, Cu, Ni, Co, Pb and Cr to separate the analyte from interfering matrices and also bring their concentration within optimum measuring ranges.

Total metal in water column

For the analysis of total metals in water column, initially 500 ml of water sample was used and through a series of analytical steps following standard methodology was reduced to 50 ml. The concentrations of the metals were determined using Perkin-Elmer (2380) Atomic Absoprtion Spectrophotometer.

Results and Discussion

The typical uses of water in an urban household are confined to drinking, bathing, toilet flushing, kitchen and laundry uses, utensil cleaning, lawn watering, automobile washing etc. Industrial and business establishments in the municipal area also use a very large amount of water for various purposes. Water quality depends upon it's source or history. The history of water has generally signified the terrain through which water is flowing, its origin and most important, the extent to which it is contaminated on it's way by anthropogenic means. Physicomechanical and physicochemical characteristics of the terrain through which it flows determines its nature. The extent of dissolved solid materials is a function of the characteristics of the geology of the terrain, the environmental impact on it and the interaction of the systems existed. To ascertain suitability for specific utilization, examination of water quality in a particular river system is very much essential. In cities like Jorhat, no proper sewage dispersal system also exists. The municipal wastewater, the rain waters run-off and the industrial effluent is also carried by some mutually independent, mostly open drains. The major drains converge to small tributaries-Toklai and Tarajan, which carries the wastewater to the Bhogdoi river directly.

Physico-chemical characteristics of river water

Temperature: In thermal pollution, there occurs the discharge of hot effluent in the natural water. This causes thermal change in the water body. Temperature is regarded as an important factor because it is having its effect on chemical and biological reactions in water. Increase in temperature of natural water tends to accelerate chemical reactions, lowers solubility of gases, amplifies taste and odour and increases metabolic activities of microorganisms.

In Toklai river temperature varies in monsoon seasons from 34.8 to35.6^{0}C and 23.7-26.6^{0}C respectively (Table 16.3a). In case of Tarajan river, it varies from 34.5 to 35.4^{0}C for monsoon and 29.9 to 33.5^{0}C for nonmonsoon periods (Table 16.3b). The temperature of water of the river Bhogdoi varies from 33.8 to 35^{0}C for monsoon and 23.2 and 28.8^{0}C for nonmonsoon periods respectively (Table 16.3c). No significant wide variation in temperature was observed throughout the period of study and can be well correlated with the existing nature of climatological changes over time.

pH: pH refers intensity in relation to acidity or alkalinity of a system. It gets increased during daytime largely because of photosynthetic activity (consumption of CO_2). The pH values of Toklai river in monsoon and nonmonsoon periods lie in between 6.53 –7.59 and 6.76-7.40 respectively (Table 16.3a). The pH values observed in between 6.39-7.16 and 6.5-7.05 in monsoon and nonmonsoon periods respectively for the Tarajan river (Table 16.3b). The pH varies from 6.94 to 7.51 in monsoon and from 7.24 to 7.75 in nonmonsoon period for Bhogdoi river (Table 16.3c). No specific changes in the pH values is observed throughout the course of the study, however, in monsoon period a slight decrease in pH value is observed (Figure 16.2).

Table 16.3a

Variation of physicochemical parameters in the Toklai river water

S. pts.	*Temp*	*pH*	*EC*	*DO*	*BOD*	*COD*	*TDS*	*TSM*	*TS*	*NH_3-N*
Monsoon										
TK1	34.5	7.15	84.65	7.88	1.95	5.35	55.02	25.50	80.52	0.33
TK2	35.2	6.53	60.25	6.48	1.85	2.25	33.14	28.32	61.46	0.41
TK3	34.8	7.59	61.65	6.26	2.29	22.56	40.07	30.15	70.22	0.36
TK4	35.1	6.72	89.34	2.37	2.15	12.03	58.07	31.16	89.23	0.39
TK5	35.4	6.80	81.63	6.91	2.06	3.00	53.06	32.15	85.21	0.26
TK6	35.6	6.89	72.54	7.23	1.85	6.32	47.15	25.25	72.40	0.35
Nonmonsoon										
TK1	26.6	6.90	101.34	6.88	1.73	8.64	65.87	20.10	85.97	0.43
TK2	24.6	6.76	96.81	6.26	2.06	10.36	62.93	23.25	86.18	0.72
TK3	24.4	7.40	72.01	3.88	2.12	19.00	46.81	28.65	75.46	0.58
TK4	23.7	7.24	173.57	2.16	1.95	15.55	112.8	30.15	142.9	0.63
TK5	24.2	6.89	125.05	3.24	2.18	8.64	81.28	31.10	112.3	0.32
TK6	24.9	6.90	108.58	5.18	1.87	5.91	70.58	32.30	102.8	0.70

Table 16.3b

Variation of physicochemical parameters in the Tarajan river water

S. pts.	*Temp*	*pH*	*EC*	*DO*	*BOD*	*COD*	*TDS*	*TSM*	*TS*	*NH_3-N*
Monsoon										
TJ1	34.5	6.39	151.89	3.04	1.22	5.15	88.06	32.05	117.1	0.52
TJ2	35.2	6.75	141.55	4.71	1.67	7.36	87.76	35.66	123.4	0.43
TJ3	34.5	6.94	158.05	3.49	1.67	8.09	93.25	36.70	129.9	0.58
TJ4	35.4	7.02	175.45	4.25	1.52	5.88	96.50	42.25	138.7	0.70

(Contd...)

S. pts.	Temp	pH	EC	DO	BOD	COD	TDS	TSM	TS	NH_3-N
TJ5	34.5	7.16	160.17	6.86	1.78	4.32	94.25	41.42	135.6	0.63
TJ6	35.1	7.06	166.52	8.50	1.87	4.50	98.25	45.58	143.8	0.74
Nonmonsoon										
TJ1	31.7	6.15	100.55	2.25	1.12	7.72	65.36	25.62	90.98	0.64
TJ2	30.5	6.67	176.71	3.46	1.37	9.46	98.76	28.61	127.3	0.55
TJ3	29.9	6.86	199.21	2.10	1.32	10.22	111.5	28.98	140.5	0.79
TJ4	32.5	6.98	208.32	2.16	1.43	7.89	116.6	34.16	150.7	0.95
TJ5	32.3	7.05	194.25	3.18	1.55	8.43	108.8	31.61	140.4	0.83
TJ6	33.5	7.01	208.26	5.67	1.65	7.89	116.6	35.68	152.3	0.85

Table 16.3c

Variation of physicochemical parameters in the Bhogdoi River water

Monsoon

S. pts.	Temp	pH	EC	DO	BOD	COD	TDS	TSM	TS	NH_3-N
BH1	33.8	6.78	151.38	6.53	1.52	4.41	98.4	296.1	394.5	0.46
BH2	34.2	6.94	171.07	6.38	1.67	7.36	111.2	286.9	398.1	0.67
BH3	34.8	7.51	185.23	6.08	1.52	8.09	120.4	263.6	384.0	0.86
BH4	34.5	7.16	197.66	6.38	1.52	3.68	118.6	277.6	396.2	0.71
BH5	35.1	7.27	224.57	8.80	2.14	4.32	130.3	313.3	443.2	0.69
BH6	35.2	7.06	226.66	8.60	2.92	5.32	142.8	358.0	500.8	0.54
Nonmonsoon										
BH1	28.8	7.24	148.30	6.46	1.30	3.32	103.2	15.3	118.5	0.75
BH2	25.1	7.31	145.30	6.18	1.25	5.32	101.2	22.6	123.8	0.76
BH3	23.8	7.71	218.90	6.26	1.30	4.85	123.2	32.0	135.2	0.98
BH4	23.2	7.27	138.30	6.20	1.28	2.72	95.5	34.1	129.6	0.99
BH5	27.7	7.75	128.60	6.90	1.90	3.22	94.5	21.0	115.5	0.83
BH6	28.8	7.46	173.40	6.80	1.85	3.93	112.8	44.5	165.3	0.82

Note: All parameters are in ppm except Temp and pH.

This slight depletion in pH value in monsoon period may be attributed as the dilution effect by rain water, which is acidic in nature irrespective of basin characteristics. The average pH values do not differ much and they are found to be within the range of Indian average. The average pH values for the monsoon and nonmonsoon periods, varies from 6.94 to 7.01, 6.78 to 6.88, 7.12 and 7.45 for Toklai, Tarajan and Bhogdoi rivers respectively, which do not favour higher concentration of dissolved metals in water column. Cu^{+2}, Cd^{+2}, Zn^{+2}, Pb^{+2}, Ni^{+2}, Co^{+2}, Cr^{+3} or Cr^{+6}, Fe^{+2} or Fe^{+3} and Mn^{+2} etc. are expected to be hydrolysed under such conditions. Stumm and Morgan (1981) suggested that all chemical reactions associated with the formation, alteration and dissolution of minerals in natural water are controlled by pH. Subramanian (1979) observed active additional pH buffering mechanisms in the Indian River systems resulted higher HCO_3^- concentration, which in turn related to comparatively higher theoretical values than the observed ones. More closer look on pH values indicate higher values around sampling points for the Toklai river at TK1 and TK3 for monsoon, TK3 and TK4 in nonmonsoon periods;

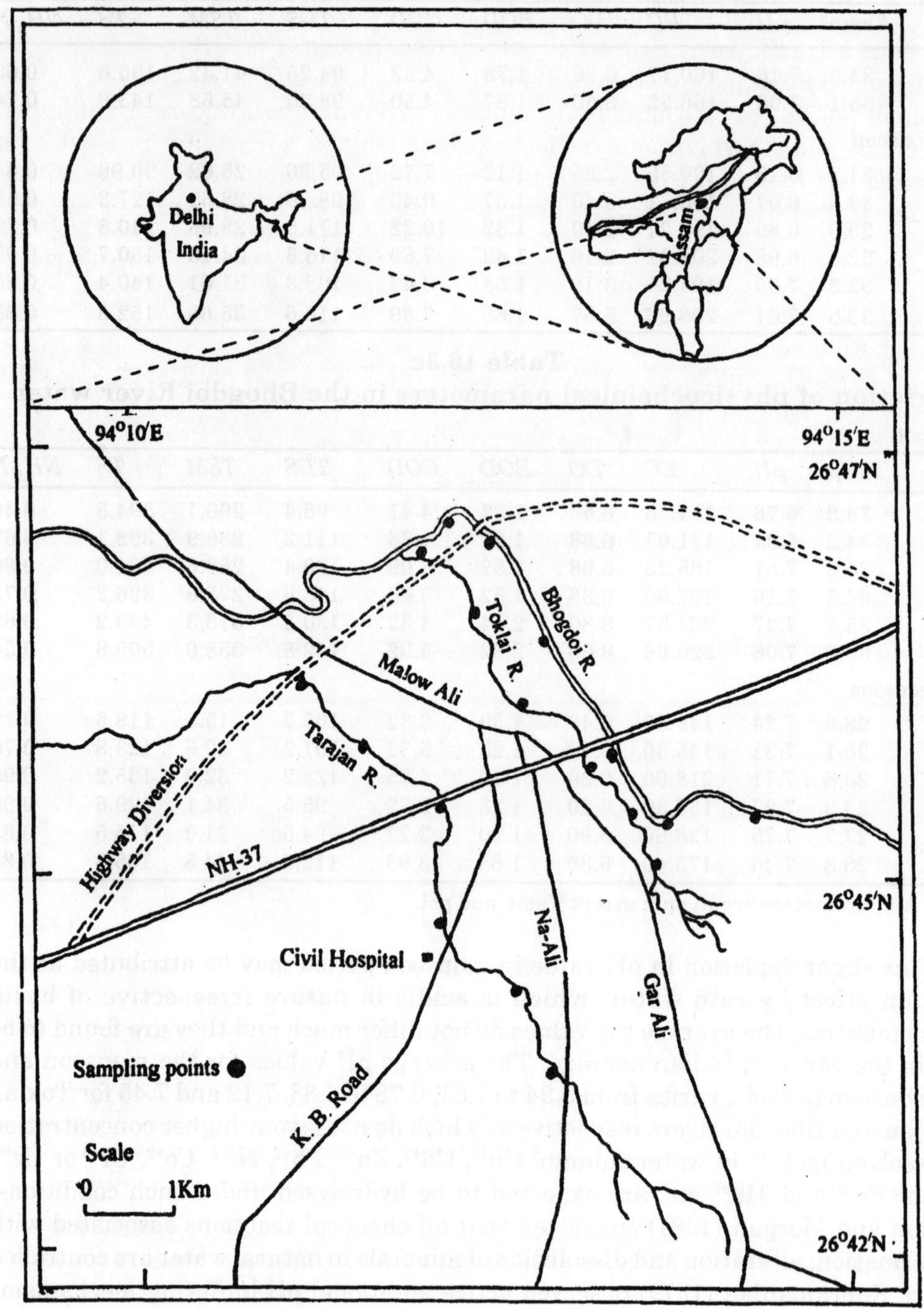

Figure 16.1. Location map of the study area

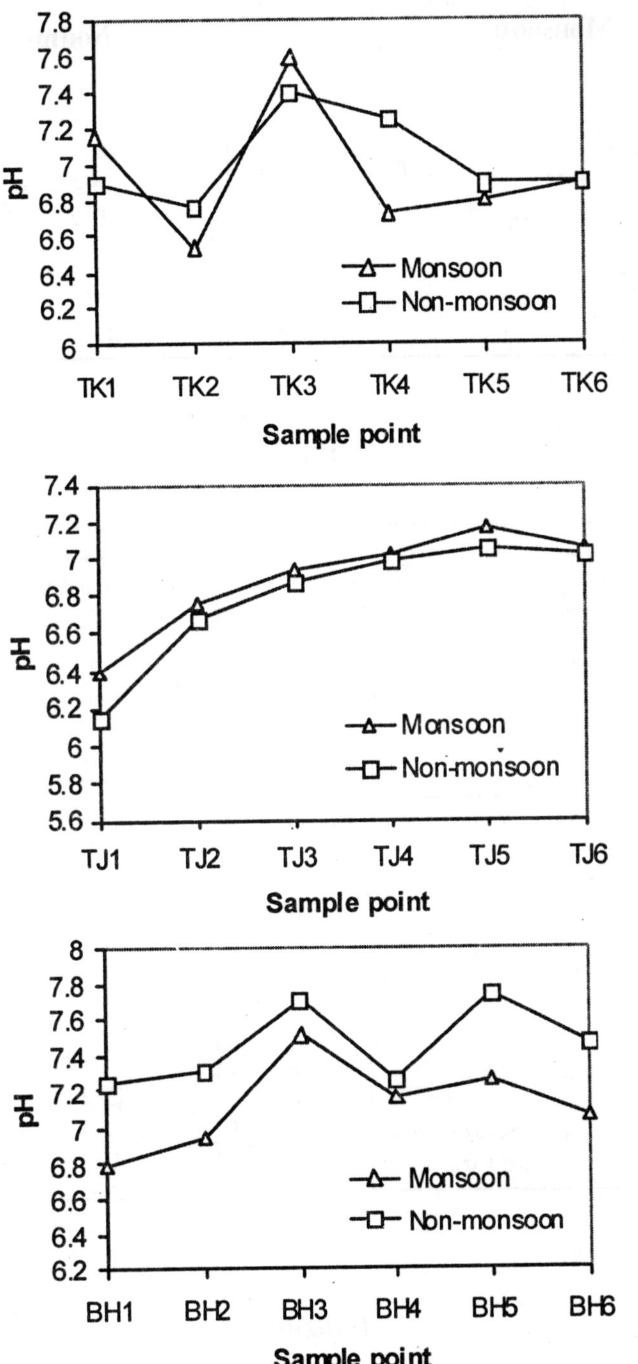

Figure 16.2. Variation of pH in monsoon and nonmonsoon period for the Toklai(TK), Tarajan(TJ) and Bhogdoi (BH) rivers.

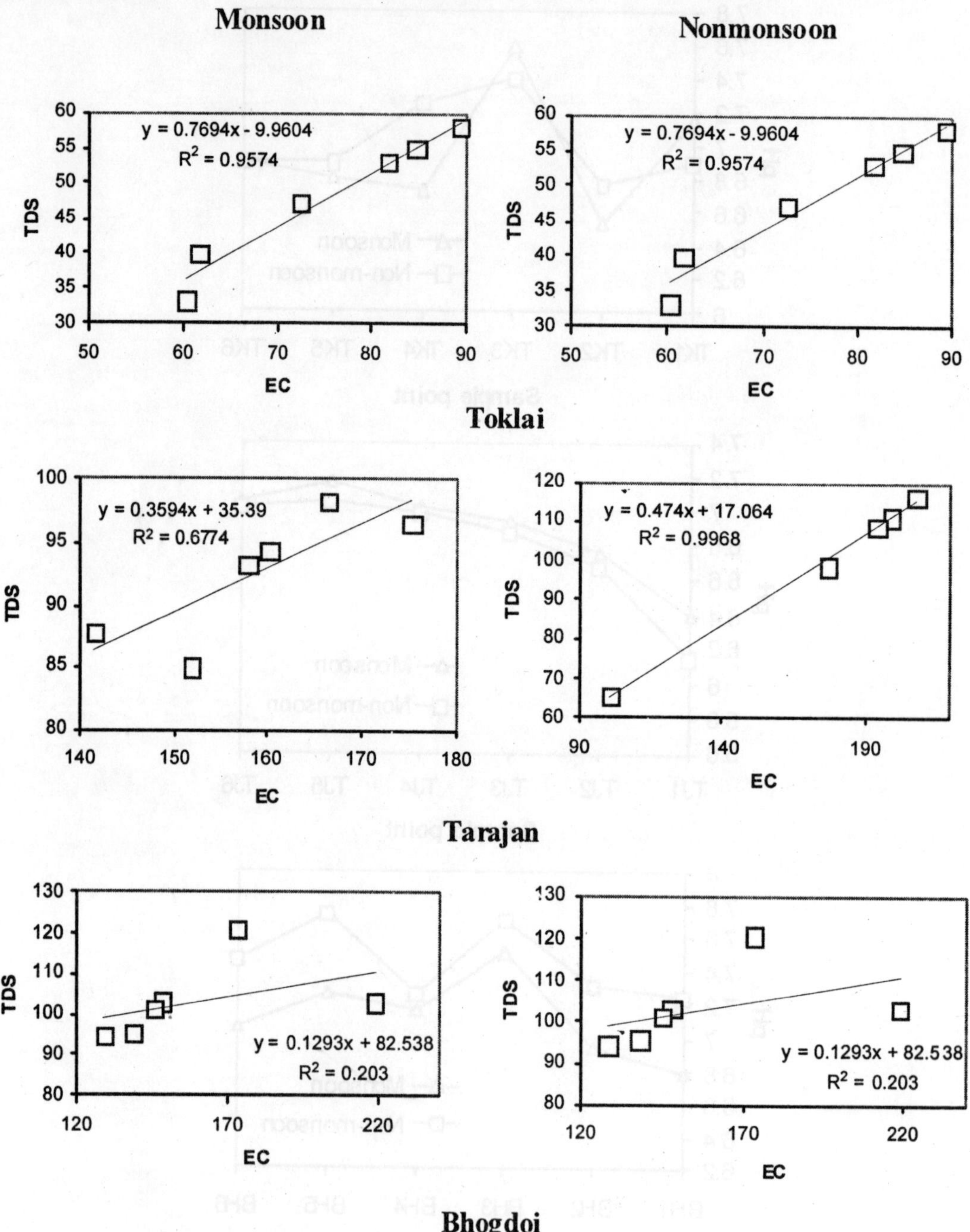

Figure 16.3. Plot of EC vs. TDS in monsoon and nonmonsoon periods of the Toklai, Tarajan and Bhogdoi rivers.

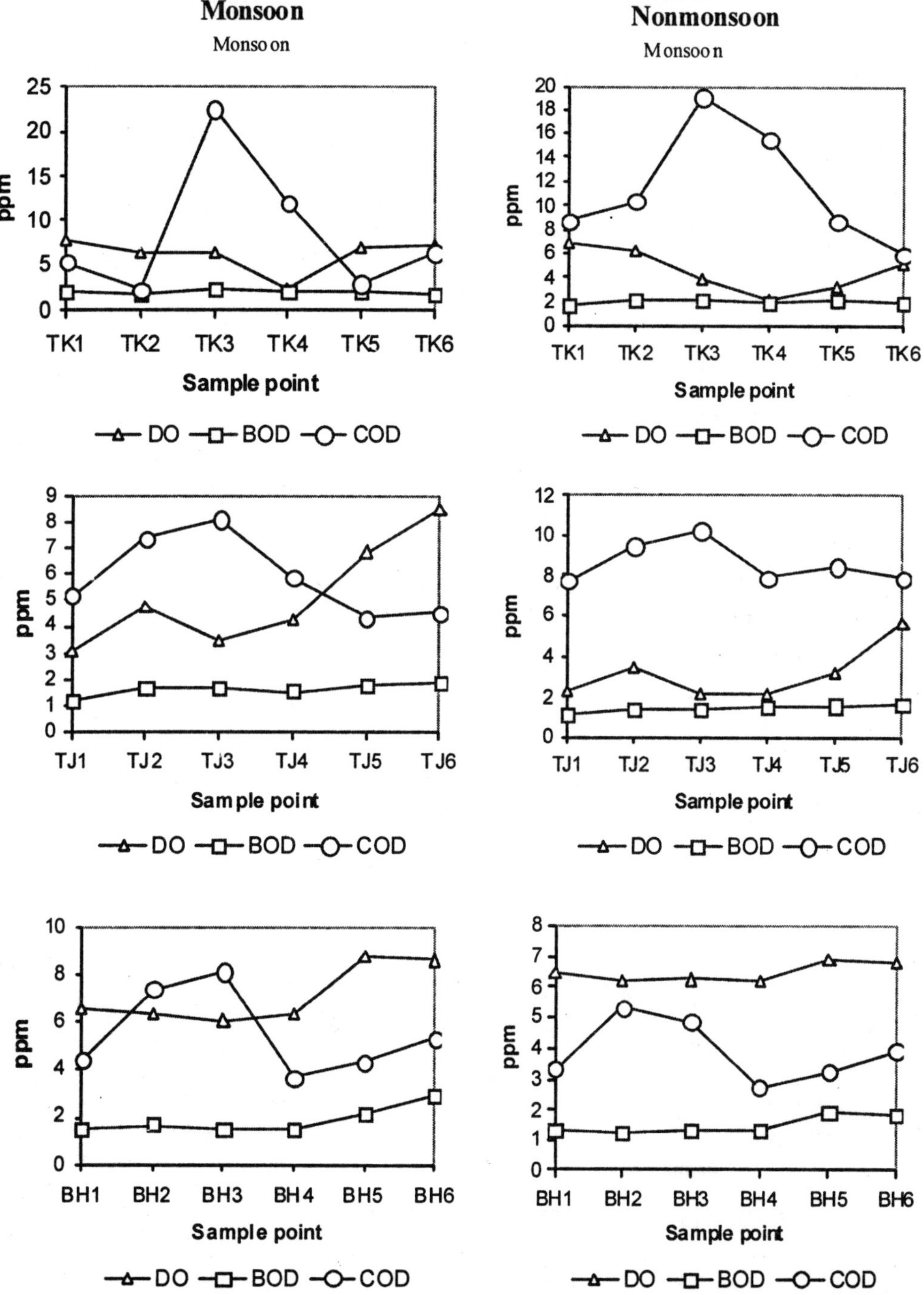

Figure 16.4. Variation of DO, BOD, and COD for the Toklai(TK), Tarajan(TJ) and Bhogdoi (BH) rivers water.

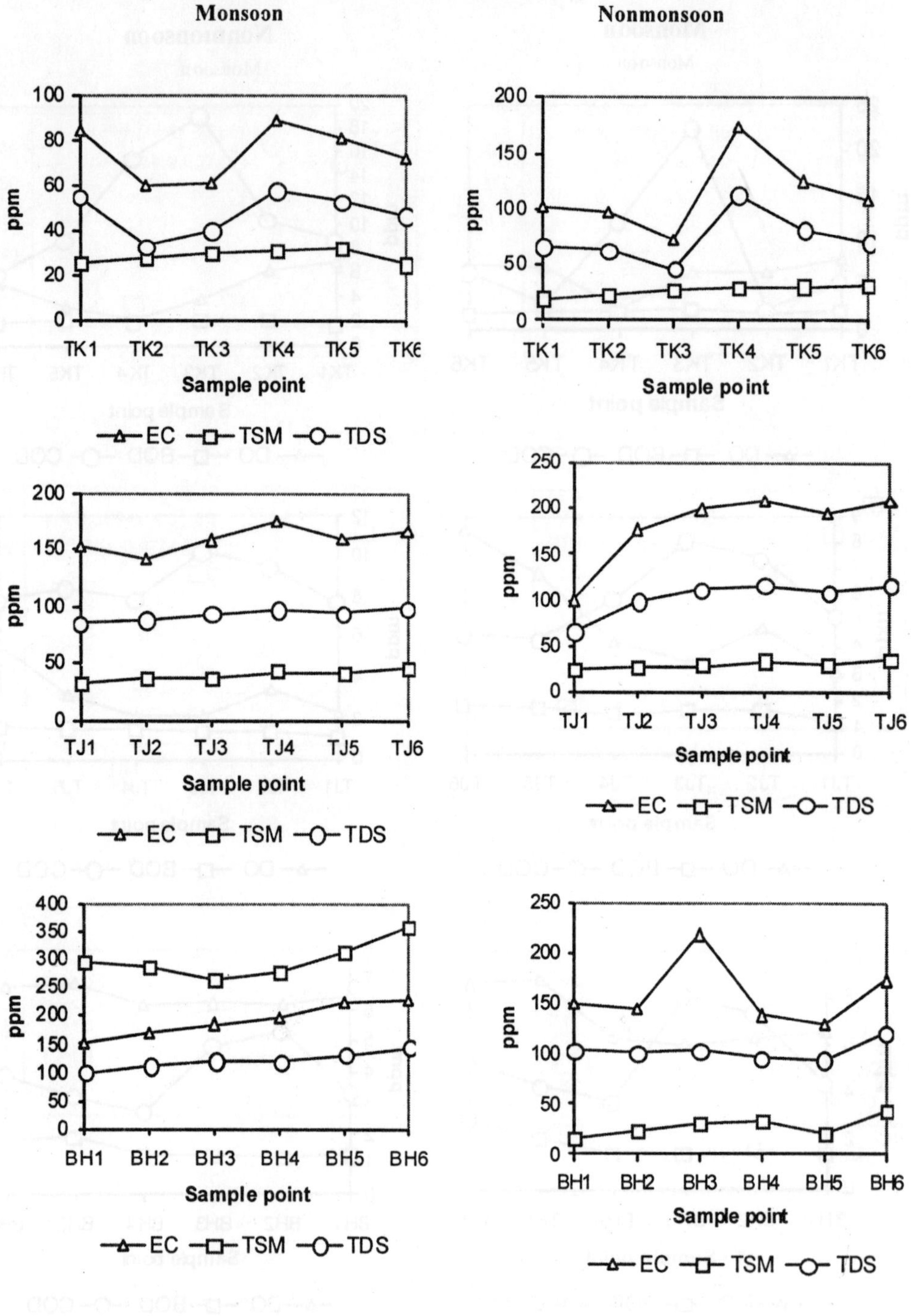

Monsoon **Nonmonsoon**

ppm

TK1 TK2 TK3 TK4 TK5 TK6

Sample point

EC TSM TDS

TJ1 TJ2 TJ3 TJ4 TJ5 TJ6

Sample point

EC TSM TDS

BH1 BH2 BH3 BH4 BH5 BH6

Sample point

EC TSM TDS

Figure 16.5. Variation of EC, TSM and TDS in monsoon and nonmonsoon periods for the Toklai (TK), Tarajan (TJ) and Bhogdoi (BH) rivers.

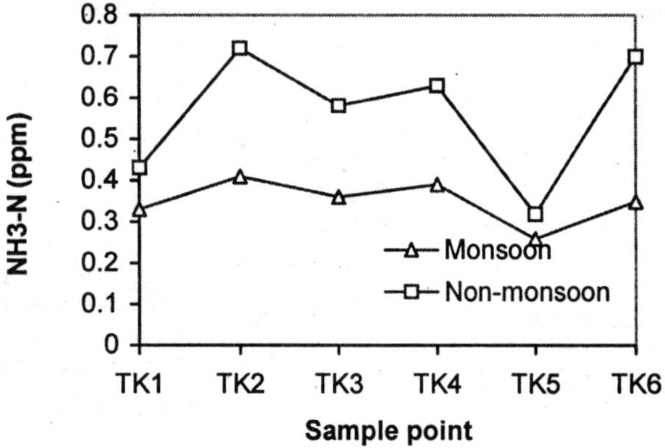

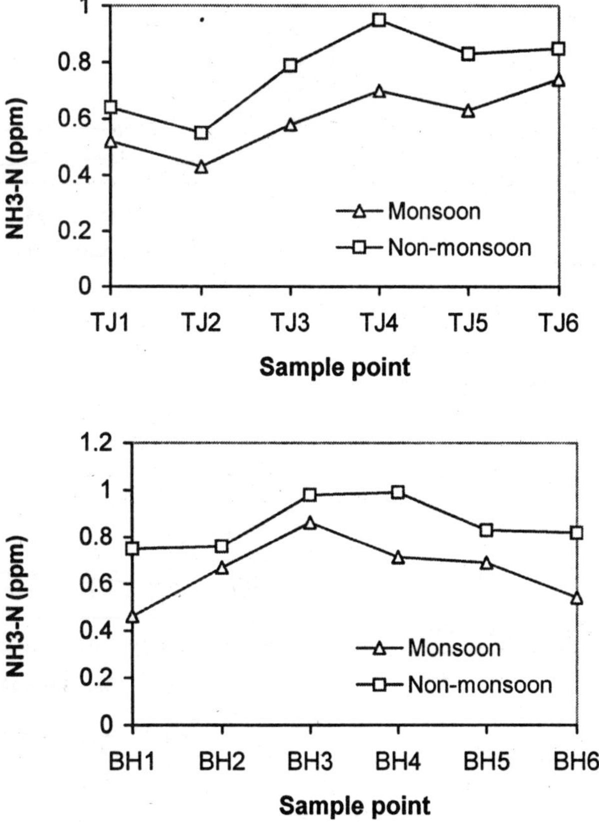

Figure 16.6. Variation of NH_3-N in Toklai (TK), Tarajan (TJ) and Bhogdoi rivers (BH).

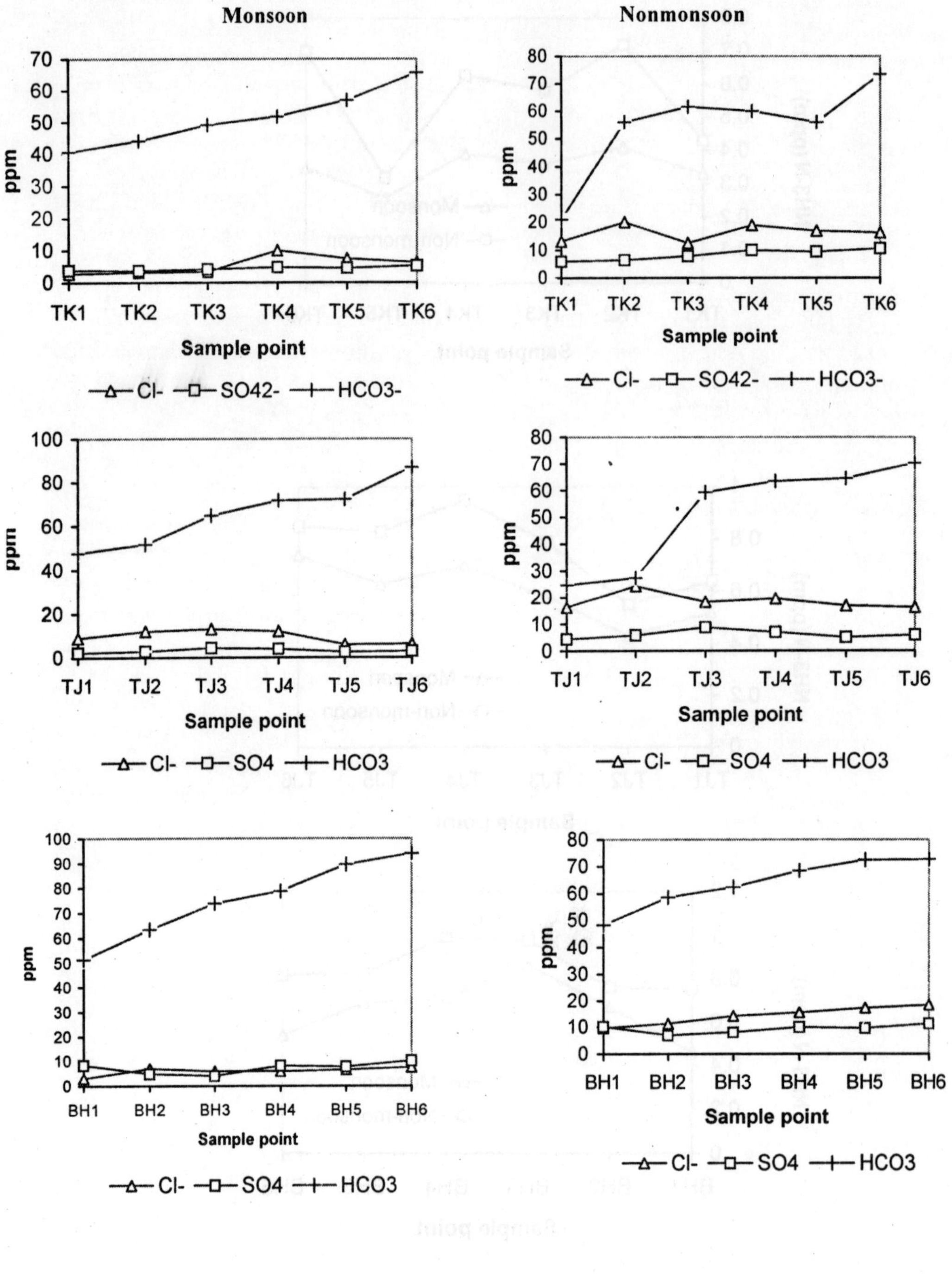

Monsoon **Nonmonsoon**

Figure 16.7. Variation of major anions in monsoon and nonmonsoon periods for the Toklai (TK), Tarajan (TJ) and Bhogdoi (BH) rivers.

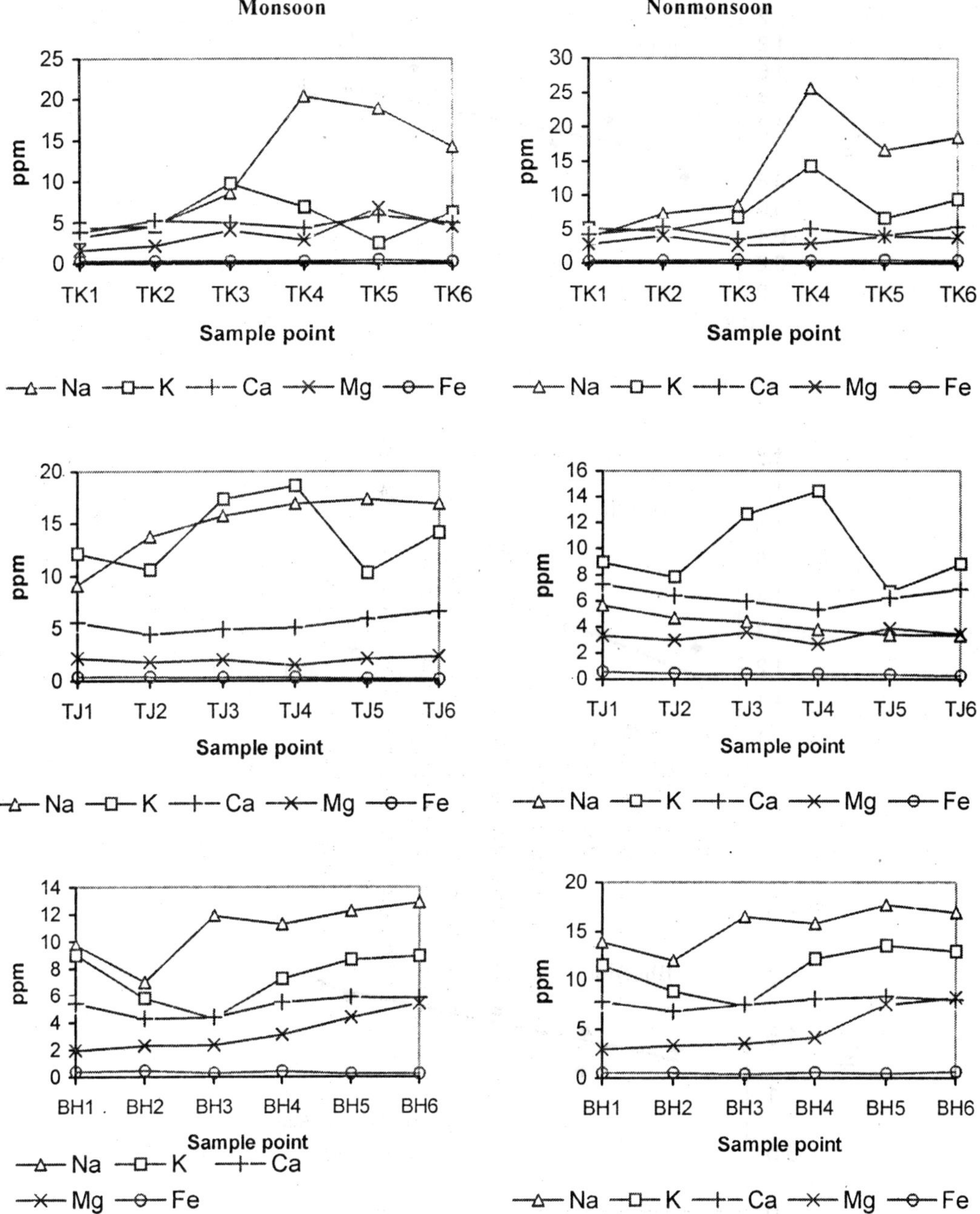

Figure 16.8. Variation of major cations in Toklai (TK), Tarajan (TJ) and Bhogdoi (BH) rivers.

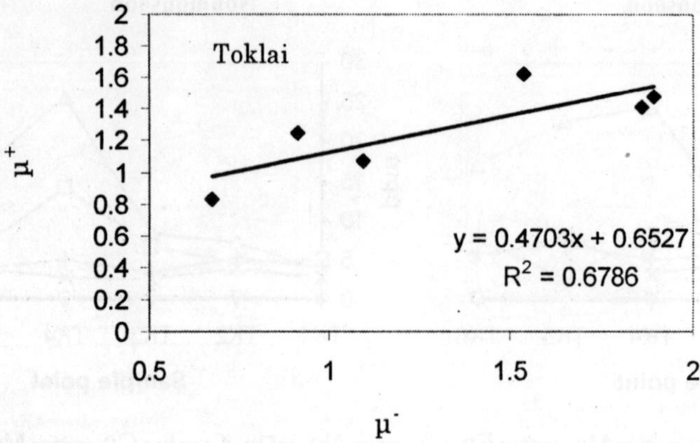

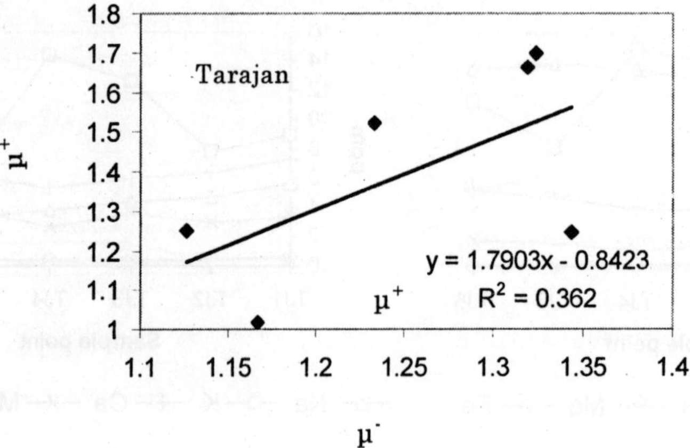

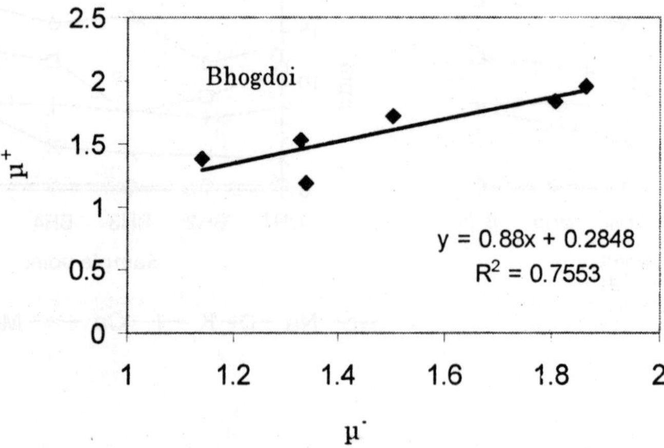

Figure 16.9. Correlation coefficient between μ^+ and μ^- for Toklai, Tarajn and Bhogdoi rivers.

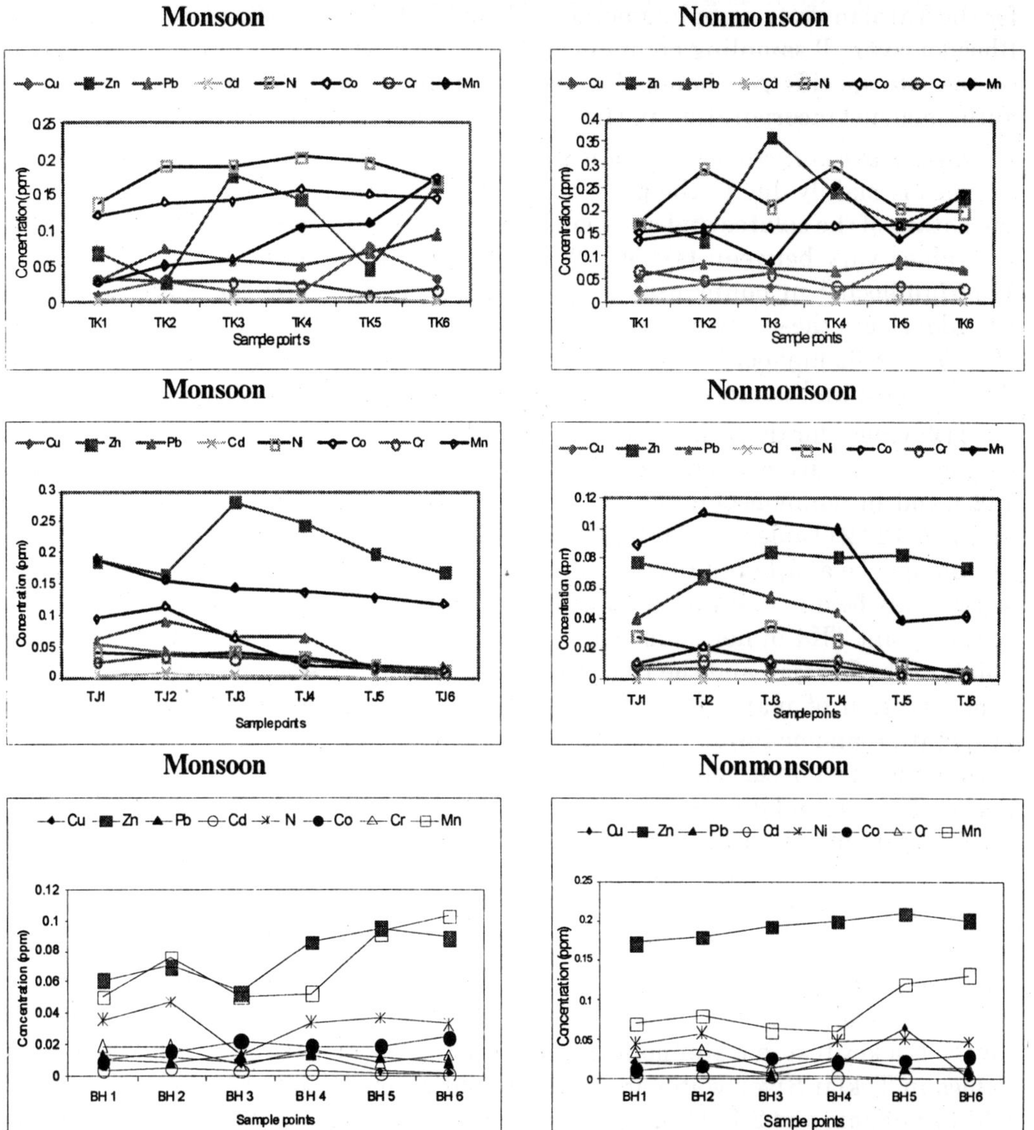

Figure 16.10. Variations of heavy metal concentrations in monsoon and nonmonsoon periods of Toklai (TK), Tarajan(TJ) and Bhogdoi (BH) rivers water.

for the Tarajan river sampling points TJ5 and TJ6 for both the periods and for the Bhogdoi river all sampling points in both periods which might have a close bearing on the site specific geochemical conditions around them. Moreover, the affect of anthropogenic incorporation in these sites also cannot be ruled out.

Electrical conductivity and total dissolved solids: Electrical conductivity of water samples is mainly due to the presence of charged solutes and to a lesser extent in the nature of the solutes and associated temperature. In the case of the Toklai river EC values vary between 60.25 - 89.34 $\mu mhocm^{-1}$ in monsoon season and 72.01 - 173.57$\mu mhocm^{-1}$ in nonmonsoon season (Table 16.3.a). For the Tarajan and the Bhogdoi rivers these variations are observed between 141.55-166.52 $\mu mhocm^{-1}$ and 100.55 - 208.32 $\mu mhocm^{-1}$ and 151.38 - 226.66 $\mu mhocm^{-1}$ and 128.6 - 218.9 $\mu mhocm^{-1}$ for monsoon and nonmonsoon seasons (Tables 16.3b and 16.3c).

TDS values for the Toklai, Tarajan and Bhogdoi rivers are as follows, in monsoon season the variation is observed in between 33.14-58.070; 88.06-98.25 and 98.4-142.8 and in nonmonsoon period is found in between 46.81-112.82, 65.36-116.63 and 94.5-123.2 (Tables. 16.3a, 16.3b and 16.3c). The high TDS values in nonmonsoon season may be attributed to concentration effect while low TDS values in monsoon season may be due to dilution effect of river water. The high conductivity value in sample point TK4 in non-monsoon period of the Toklai river and all the sampling points of the Tarajan and the Bhogdoi Rivers in both the seasons indicate increased concentration of major ions. This increased concentration may be due to the evaporation and/or influence by the municipal waste. The correlation coefficients between EC and observed TDS values are 0.95 in monsoon and 1.0 in nonmonsoon period. The interrelationship between TDS and EC are also observed in the case of the Tarajan and the Bhogdoi rivers through correlation coefficient values. These values for the Tarajan river are 0.82 and 0.99 in monsoon and nonmonsoon periods respectively. Again these observations are 0.95 and 0.45 in monsoon and nonmonsoon periods respectively of the Bhogdoi river (Figure 16.3).

Dissolved oxygen: Dissolved oxygen (DO) level in natural and wastewater depend on the physical, chemical and biochemical activities in the water body. The analysis for DO is a key test in water pollution and waste treatment process control. The DO values in Toklai River water in monsoon and nonmonsoon seasons vary between 2.37 to 7.88ppm and 2.16 to 6.88ppm respectively (Table 16.3a). The DO observation in the Tarajan River in monsoon period is in between 3.04 to 8.5 ppm and in nonmonsoon period it is in between 2.1 to 5.67ppm (Table 16.3b). Similar observation is also found in case of the Bhogdoi River. In this case the variation is found in between 6.08 to 8.8ppm and 6.18 to 6.9ppm for monsoon and nonmonsoon periods respectively (Table 16.3c). The values are slightly higher in monsoon period , this might be due to high flow rate and turbulence of the river by which atmospheric reoxygeneration raises the DO level. The DO value is very low at sampling point TK3 of the Toklai river, 2.37ppm in monsoon and 2.16ppm in nonmonsoon period respectively. This is due to the presence of oxygen demanding pollutants (like organic wastes) as these causes rapid depletion of dissolved oxygen from water. Oxydisable inorganic substances such as hydrogen sulphide, ammonia, nitrites, ferrous iron etc. also bring about a wide decrease in the DO. The variation of DO in two different seasons is shown in the Figure 16.4.

Biological oxygen demand: Biological oxygen demand (BOD) may be defined as the rate of removal (i.e. consumption) of oxygen by microorganisms in aerobic degradation of the dissolved or even particulate oxygen matter in water. It has been used as an index of organic pollution in water. Thus BOD gives us an idea about the biodegradable organic load of receiving bodies of water. In the Toklai River BOD varies between 1.85 - 2.29ppm in monsoon season and 1.75 - 2.18ppm in nonmonsoon seasons (Table 16.3a). This variation is observed in between 1.22 - 1.87ppm in monsoon and 1.12-1.65ppm in nonmonsoon season for Tarajan (Table 16.3b). In case of Bhogdoi, the BOD varies between 1.52 - 1.92ppm in monsoon and 1.25 - 1.9 in nonmonsoon periods (Table 16.3c). In both the seasons, the BOD values are well below the WHO guideline (maximum 6.0ppm). The variation of BOD is shown in the Figure 16.4.

Chemical oxygen demand: Chemical Oxygen Demand is the measure of oxygen, which is required in oxidizing the organic compounds present in the water by means of chemical reactions involving oxidizing substances. The estimation of COD has been of great importance for waters, which are having unfavorable conditions for the growth of microorganisms. In case of river the Toklai, COD values in monsoon and nonmonsoon seasons vary from 2.25 - 22.56ppm and 5.91 - 19.00ppm respectively (Table 16.3a). For the Tarajan river the variation is observed in between 4.32-8.09ppm in monsoon and 7.72-10.22ppm in nonmonsoon season (Table 10.3.b). Again in case of Bhogdoi river water this variation is observed in between 3.68 - 8.09 and 2.72 - 5.32ppm in monsoon and nonmonsoon seasons respectively (Table 16.3c). The high COD value at sampling points TK3 of the Toklai river, 22.56 ppm in monsoon and 19.00ppm in nonmonsoon season may be due to the dumping of municipal wastes and small scale industries, Nursing Homes located in the sample site. The COD values of other sampling points are within acceptable limits. The variation of COD values in two different seasons is shown in the Figure 16.5.

Total suspended matter: The TSM values for the Toklai river water, 25.25 - 32.15ppm in monsoon period and 20.1-32.3ppm in nonmonsoon period are observed (Table 16.3a). The variation of TSM of the Tarajan river in monsoon period is observed in between 32.05 - 45.58 ppm and in nonmonsoon period it is in between 25.62 - 35.68 ppm (Table 16.3b). The TSM varies from 263.6 to 358 ppm in monsoon and from 15.3 to 44.5 ppm in nonmonsoon periods in the Bhogdoi river (Table 16.3c). A higher TSM value is observed in monsoon period, it may be due to heavy downpour as a result of which the stream carries large amount of debris. The seasonal variation of TSM is shown in the Figure 16.5.

N-NH_3: Nitrogen, forms a major constituent (80%) of atmosphere, however it occurs in small amounts in water due to low solubility of molecular nitrogen in water. Nitrogen is also found in small amounts in water as bound forms, such as NH_3, NO_3, NO_2 and organic nitrogen (viz., urea, amino acids etc.).The nitrogen ice water may undergo oxidation to form nitrate or nitrite and also in some cases undergo reduction to form ammonia. The nitrogen also originates from vegetable matter which may come from agriculture load run off N-NH_3 concentration in the Toklai river varies in between 0.26 - 0.41pm in monsoon period and 0.32-0.72ppm in nonmonsoon period (Table 16.3a). In case of the Tarajan River N-NH_3 concentration is observed as 0.43 -0.74ppm in monsoon period and 0.55-0.95ppm in nonmonsoon period (Table 16.3b). The N-NH_3 concentration period. This might be due to the increase in the oxidising environment in the river system which may felicitate the conversion of some of the N-NH_3 to N-NH_2 or NO_3. However in both the seasons the N-NH_3 concentrations are within the tolerance limit. The variation of N-NH_3 is shown in the Figure 16.6.

Major ion chemistry

The major ion chemistry of river waters reveals the nature of weathering on a basin and helps us to understand the exogenic cycles of the elements in the continent-river-ocean-system. The chemistry of river water is largely dominated by the weathering process in the drainage basin. The chemical composition of inland water results from different environmental factors operating simultaneously but with different influences and efficiency. The factors are soil and rock composition, (climatic behaviour, solubility and secondary factors), climatic conditions (rainfall and temperature), morphology, fauna and flora, the time factor and to an increasing extent the anthropogenic influences. The salt content in rivers varies between 10mgl^{-1} and approx. 2000mgl^{-1}, levels between 100mgl^{-1} and 200mgl^{-1} are about average. In river waters with an extremely low salt content, the distributions of major cations correspond exactly to the average rain water composition. With increasing salinity, the cation ratio changes in the direction of the average chemical properties, displayed by sedimentary rocks which is on one way the results of intensive interaction between water and underlying formations. The natural occurrences and development of major ion contents in fresh water is also increasingly affected by anthropogenic attributes. The major ions are introduced into the water systems by both natural as well as anthropogenic sources.

Major anions

Chloride: The main source of cation in river water is atmospheric precipitation. Inland natural waters are generally having low chloride concentration often less than that of bicarbonates and sulphates. However, inland saline waters, coastal, estuarine and seawater are having moderate to heavy chloride content. In natural fresh water high concentration of chloride is regarded as an indicator of pollution because of organic wastes of animal origin (animal excreta is having high quantity of chlorides along with nitrogenous wastes). Industrial effluents also increase the chloride content in natural waters. Except areas, where heavy industrial pollution input dominates, chloride concentration by evaporation should not exceed 20ppm in rivers with rainfall more than 20cm per year (Holland, 1978). A substantial amount of non-atmospheric chloride in the rivers is derived from sedimentary rocks, released possibly by the reaction with sub surface connate brines.

The chloride content of the Toklai water varies from 2.8 ppm to 10 ppm in monsoon period and 12.5 to 20 ppm in nonmonsoon period (Table 16.4a). In case of Tarajan river this variation is observed in between 9 and 13 ppm in monsoon period and in between 16.2 and 24.2ppm in nonmonsoon period (Table 16.4b). For the Bhogdoi river the chloride concentration varies from 3 to 7.6 ppm in monsoon period and from 9.6 to 18.3 ppm in nonmonsoon period (Table 16.4c). The concentration is generally observed as low and hence no significant seasonal trend of variation is distinguishable (Figure 16.7).

Sulphate: Sulphates occur in appreciable quantity in all natural waters, especially high in arid and semi-arid regions where natural waters in general are having high salt content. Domestic sewage and industrial effluents, besides biological oxidation of reduced sulphur species, may add to sulphate content of water. In a region where atmospheric sulphur content is very high due to industrial and automobile emission, the rainwater is having high sulphur content. Anthropogenic sulphate is mostly produced during fossil fuel burning. As indicated by Garrels *et al.* (1975), anthropogenic contribution of the total sulphate in a river system can be as high as 45%.

The sulphate content of Toklai water 3.8 - 5.26ppm in monsoon and 6.4 - 10.58 ppm in nonmonsoon periods are observed (Table 16.4a). The variation of sulphate concentration is observed in between 2.18 and 3.99 ppm in monsoon period and in between 4.36 and 8.8ppm in nonmonsoon period (Table 16.4b) in the Tarajan river water. Again, in case of the Bhogdoi river water the sulphate concentration varies from 4.03 to 10.24 ppm in monsoon period and from 6.97 to 11.33 ppm in nonmonsoon period (Table 16.4c). From the observations it is found that in nonmonsoon period the sulphate concentration is slightly higher than monsoon period. This may be due to high dilution factor brought about by the large amount of water in the monsoon period.

Bicarbonate: The bicarbonate ion (HCO_3^-) is the most abundant among the various dissolved constituents in river water. Carbonates and bicarbonates are formed due to dissolution of CO_2 in water at lower pH (bicarbonate) or at higher pH (carbonate). The other possible sources of bicarbonate are dissolution of limestone and dolomites, weathering of primary and secondary silicate minerals and organic activities. (Garrels and Mackenzie, 1971). However, these types of sources are not directly involved in our present area of study.

Table 16.4a

Variation of major ion composition in the Toklai River water

Monsoon

S. pts.	Cl^-	SO_4^{-2}	HCO_3^-	*Na*	*K*	*Ca*	*Mg*	*Fe*
TK1	2.8	3.8	40.6	3.21	4.32	3.84	1.60	0.300
TK2	3.2	3.7	44.1	4.62	4.54	5.26	2.14	0.272
TK3	3.4	4.2	49.1	8.60	9.82	4.95	4.06	0.288
TK4	10.0	4.8	51.7	20.4	6.96	4.33	2.85	0.296

(Contd...)

S. pts.	Cl^-	SO_4^{-2}	HCO_3^-	*Na*	*K*	*Ca*	*Mg*	*Fe*
TK5	7.5	4.5	56.9	18.9	2.54	5.86	6.84	0.500
TK6	6.3	5.3	65.6	14.3	6.32	4.96	4.50	0.315
Nonmonsoon								
TK1	13.2	5.9	21.0	3.8	5.1	4.21	2.76	0.373
TK2	20.0	6.4	56.0	7.3	4.8	5.31	4.00	0.408
TK3	12.5	7.9	61.5	8.5	6.7	3.56	2.57	0.536
TK4	18.9	10.0	60.0	25.6	14.3	5.04	2.85	0.372
TK5	16.9	9.5	56.0	16.6	6.6	4.15	3.93	0.544
TK6	16.3	10.6	73.5	18.5	9.4	5.32	3.78	0.398

Table 16.4b

Variation of major ion composition in the Tarajan River water

Monsoon

S. pts.	Cl^-	SO_4^{-2}	HCO_3^-	*Na*	*K*	*Ca*	*Mg*	*Fe*
TJ1	9.0	2.18	47.68	9.08	12.16	5.56	2.16	0.401
TJ2	12.0	2.96	51.77	13.74	10.6	4.43	1.79	0.390
TJ3	13.0	4.40	64.78	15.74	17.36	4.90	1.98	0.307
TJ4	12.0	3.99	71.66	16.92	18.62	5.02	1.50	0.319
TJ5	5.9	2.52	72.20	17.32	10.32	5.86	2.08	0.213
TJ6	6.2	2.96	86.90	16.92	14.15	6.60	2.35	0.185
Nonmonsoon								
TJ1	16.25	4.36	24.63	5.67	8.97	7.28	3.32	0.573
TJ2	24.25	5.82	27.38	4.65	7.78	6.38	2.97	0.451
TJ3	18.25	8.80	59.20	4.39	12.65	5.93	3.54	0.410
TJ4	19.56	6.98	63.40	3.76	14.4	5.32	2.65	0.393
TJ5	16.9	5.04	64.46	3.38	6.69	6.21	3.86	0.340
TJ6	16.28	5.93	70.25	3.31	8.80	6.87	3.42	0.285

Table 16.4c

Variation of major ion composition in the Bhogdoi River water

Monsoon

S. pts.	Cl^-	SO_4^{-2}	HCO_3^-	*Na*	*K*	*Ca*	*Mg*	*Fe*
BH1	3.0	8.35	51.40	9.72	9.02	5.4	1.98	0.40
BH2	7.0	5.02	63.39	7.02	5.80	4.3	2.30	0.46
BH3	6.0	4.03	73.98	11.90	4.32	4.4	2.36	0.32
BH4	6.0	8.23	78.92	11.26	7.24	5.5	3.10	0.45
BH5	6.6	7.90	89.56	12.27	8.65	5.9	4.38	0.26
BH6	7.6	10.24	93.98	12.87	8.92	5.9	5.42	0.27
Nonmonsoon								
BH1	9.6	10.25	48.3	13.92	11.52	7.8	2.98	0.52
BH2	11.3	6.97	58.6	12.02	8.86	6.8	3.30	0.56
BH3	14.2	8.20	62.3	16.51	7.32	7.5	3.50	0.43
BH4	15.6	9.96	68.5	15.77	12.21	8.1	4.12	0.58
BH5	17.2	9.70	72.4	17.70	13.55	8.4	7.50	0.45
BH6	18.3	11.33	72.6	16.93	12.93	8.0	8.20	0.67

For the Toklai water it is observed that the concentration of HCO_3^- varies in between 40.6-65.6ppm in monsoon season and 21.0-73.5ppm in nonmonsoon period (Table 16.4a). The variation of HCO_3^- concentration is observed in between 47.66 and 86.9ppm in monsoon period and in between 24.63 and 70.25ppm in non-monsoon period in Tarajan River (Table 16.4b). In case of the Bhogdoi river, the HCO_3^- concentration varies from 51.4 to 93.98 ppm in monsoon period and 48.3 to 72.6 ppm in nonmonsoon period (Table 16.4c). For the Toklai water it is observed that the concentration of HCO_3^- varies in between 40.6-65.6 ppm in monsoon season and 21.0 - 73.5ppm in nonmonsoon period. The seasonal variations of major anions are shown in the Figure 16.7.

Solubility of CO_2: Most natural waters contain dissolved carbonate species due to the solubility of CO_2 gas in water. The main source of CO_2 is the earth's atmosphere with an average volume percentage of 0.03 (the partial pressure of CO_2) acquired from decay of plants and organic matter in the soil zone. Even these sources can be ultimately traced to the atmosphere through the processes like photosynthesis. However, pCO_2 in the soil zone is relatively high because of higher total pressure and large amount of CO_2.

Dissolution of CO_2 in water contributes carbonate species through the following steps:

(1) Formation of carbonic acid, $CO_2 + H_2O \xrightarrow{K} H_2CO_3$

(2) First dissociation, $H_2CO_3 \xrightarrow{K_1} H^+ + HCO_3^-$

(3) Second dissociation $HCO_3^- \xrightarrow{K_2} H^+ + CO_3^{2-}$

The equilibrium constants of these three reactions at an average river water temperature of 25°C can be expressed as :

$$K = aH2CO3/pCO_3 = 10^{-1.5}$$
$$K_1 = aH^+.\ a\ HCO_3^-/aH_2CO_3 = 10^{-6.4}$$
$$K_2 = aH^+.\ a\ CO_3^{-2}/aHCO_3^- = 10^{-10.3}$$

Where, aH^+, $aHCO_3^-$ etc. are activities of the dissolved species. The nature of K_2 indicates that river water pH values being much less than 10.3, CO_3^{-2} is relatively negligible compared to HCO_3^-. The expressions for K and K_1 can also be combined to write:

$$a.H^+ .\ a\ HCO_3^- = 10^{-7.9} \times pCO_2$$
$$\text{or, } \log aHCO_3^- - pH = \log pCO_2 - 7.9$$
$$\text{or, } \log pCO_2 = 7.9 + \log mHCO_3^- - pH$$

where, $mHCO_3^-$ is bicarbonate concentrations in moles L^{-1}, which is practically equal to $aHCO_3^-$ in dilute river waters. This relationship provides a method to estimate "Effective CO_2 pressure" in a river from its pH value and bicarbonate concentration. On this basis, the pCO_2 of Bhogdoi, Tarajan and Toklai rivers in monsoon and nonmonsoon seasons were calculated and are found to be inline of other Indian rivers (Table 16.5). This is a global trend, indicating that the rivers are commonly out of equilibrium with the atmosphere. This trend is probably due to (i) river waters may contain a significant fraction of high CO_2 contained ground water (ii) the rate of reequlibration with the atmosphere by releasing the excess of CO_2 is relatively slow (Stumm and Morgan, 1970; Holland, 1978).

Effective CO_2 pressure in Indian rivers along with the present study calculated at an average temperature 25°C from the compilation by Subramanian (1979).

Table 16.5
Effective CO_2 pressure in Indian rivers with the present study

River	*Location*	*pH*	*HCO_3 ppm*	*Log $mHCO_3$*	*Log pCO_2*
Ganga	Farakka	7.3	170.5	-2.56	-1.96
Brahmaptra	Pandu	7.1	94.1	-2.81	-2.01
Jhelum	Srinagar	7.7	64.0	-2.98	-2.78
Godavari	Rajamundry	8.1	131.5	-2.76	-2.87
Mahanadi	Cuttack	7.5	61.2	-3.00	-2.60
Narmada	Broach	8.2	108.4	-2.75	-3.05
Tapti	Surat	8.3	173.1	-2.55	-2.95
Krishna	Vijaywada	7.7	143.7	-2.63	-2.43
Cauvery	Trichy	7.8	176.8	-2.54	-2.44
Jhanji	Assam	7.4	66.3	-2.96	-2.46
Bhogdoi	Jorhat	7.28	69.49	-2.94	-2.32
Tarajan	Jorhat	6.83	58.69	-3.01	-1.94
Toklai	Jorhat	6.98	53.0	-3.06	-2.14

Source: Raymahashay (1986); Baruah *et al.* (1997).

Chemical weathering of rocks: The weakly acidic H_2CO_3 solution produced by dissolution of CO_2 in natural water reacts with rock forming minerals in a variety of ways Chemical weathering reactions can be grouped into two principal types:

(i) Congruent dissolution of carbonate minerals e.g.
$CaCO_3 + CO_2 + H_2O = Ca^{+2} + 2HCO_3^-$ and

(ii) Incongruent weathering of silicates to clays e.g.
Feldspar + CO_2 + H_2O = Kaolinite + cations, $SiO_2 + HCO_3^-$

However, there are important differences in stoichiometry of the two types of reactions. For weathering of carbonate minerals like, calcite, dolomite and aragonite the HCO_3^- ion concentration is balanced by cations like Ca^{+2}, Mg^{+2} on an equivalent basis. On the other hand, kaolinisations of silicate minerals result in silica in solution which is related to the HCO_3^- ion concentration through the composition of primary silicate (Garrels, 1967).

In a general situation, the total bicarbonate in river water has two fractions viz., (I) Bicarbonate derived from primary carbonates as well as calcareous cements.

(II) Bicarbonate derived from silicates.

Similarly, the total Ca and Mg also have the same two sources.

$$(HCO_3)_{tot} = (HCO_3)_C + (HCO_3)_{Si}$$
$$(HCO_3)_C = (Ca)_C + (Mg)_C$$
$$(Ca)_{tot} = (Ca)_C + (Ca)_{Si}$$
$$(Mg)_{tot} = (Mg)_C + (Mg)_{Si}$$

where, all concentrations are in milliequivalent litre^{-1}. Holland (1978) after a comprehensive review of river water chemistry and composition of rocks in drainage basins arrived at the conclusion that 74±10 percent of Ca and 40±20 percent of Mg in river water are derived from solution of carbonate minerals and remainder largely from silicate minerals. We can therefore, set up a working model as follows:

$$(HCO_3)_C = 0.74\ (Ca)_{tot} + 0.4(Mg)_{tot} \rightarrow (1)$$

$$(HCO_3)_{Si} = (HCO_3)_{tot} - (HCO_3)_C \rightarrow (2)$$

With the help of equations (1) and (2), the $(HCO_3)_C$ and $(HCO_3)_{Si}$ were calculated for the Toklai, Tarajan and Bhogdoi rivers and the results are tabulated along with the bicarbonates derived from rock weathering in other Indian Rivers (Table 10.5) for a comparative study (Figure 16.11). From the table it is evident that in most of the rivers bicarbonate derived from weathering of carbonate minerals is around 1.0 meq l^{-1} (61ppm). On the other hand, it is evident that the variation in the total bicarbonate (alkalinity) in these rivers is mainly due to the variation in bicarbonate derived from weathering of silicate minerals. In our present study, 67% bicarbonate comes from silicate weathering and 33% comes from carbonate weathering for the Bhogdoi river and for the Toklai and Tarajan rivers the percentages are 66.6% and 33.4% and 68.4% and 31.6% respectively (Table 16.6), which agrees with Raymahashay's (1986) findings. According to Subramanian (1979), kaolinite is a common product of silicate weathering in most of the Indian River basins. XRD reports of bed sediments of these three rivers also show the presence of kaolinite, illite and chlorite. The study of seven component phase diagram (K_2O-Na_2O-MgO-CaO-Al_2O_3-SiO_2-H_2O) on Jhanji river water by Baruah (1996) also supports our contention. The study indicates that Albite, Kaolinite, Chlorite and Ca-feldspar and to a lesser extent K-mica/K-feldspar have made significant contribution of major cations (viz. Na^+,K^+, Ca^{+2}, Mg^{+2}) to the rivers of this region.

Major Cations

The principal source of the major cations viz., Na^+, K^+, Ca^{+2}, Mg^{+2}, Fe^{+3} in our study area is the result of chemical weathering of clay minerals originated from rock system of the basin. The nature of parent rock and the proton source determines the relative concentration of cations released to the solution. The other important sources are the industrial and human activities; e.g. dumping of waste materials through several drains leading to these rivers can also contribute these ions to the river system.

Table 16.6

Calculated values for bicarbonate of the Toklai, Tarajan and Bhogdoi Rivers

Toklai water				
Ca	Mg	$(HCO_3^-)_{tot}$	$(HCO_3^-)_C$	$(HCO_3^-)_{Si}$
0.236	0.290	0.868	0.290(33.4%)	0.578(66.6%)
Tarajan water				
Ca	Mg	$(HCO_3^-)_{tot}$	$(HCO_3^-)_C$	$(HCO_3^-)_{Si}$
0.293	0.219	0.962	0.304(31.6%)	0.658(68.4%)
Bhogdoi water				
Ca	Mg	$(HCO_3^-)_{tot}$	$(HCO_3^-)_C$	$(HCO_3^-)_{Si}$
324	0.341	1.139	0.376(33%)	0.763(67%)

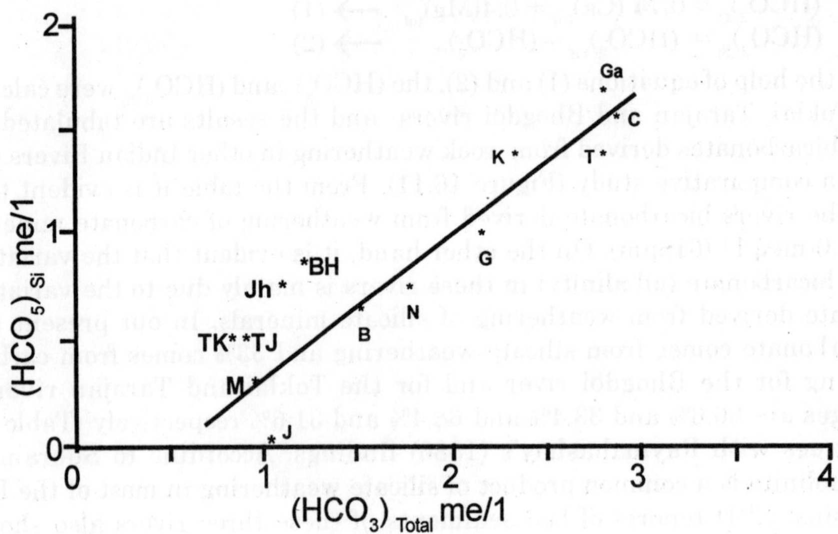

(All the values are expressed in m eqv l^{-1})

Figure 16.11. Variation of total bicarbonate and silicate derived bicarbonate in Indian rivers along with the present study. [J=Jhelum, M=Mahanadi, Jh= Jhanji, B=Brahmaputra, N=Narmada, Go=Godavari, K=Krishna, T=Tapti, G=Ganga, C=Cauvary].

Sodium and Potassium: The average concentration of Na^+ of the Toklai river water in monsoon season varies from 3.21 to 20.4 ppm and in non-monsoon season it varies from 3.8 to 25.6 ppm (Table 16.4a). In the Tarajan river the Na+ concentration varies in between 9.08 to 17.32 ppm in monsoon period and in between 3.31 to 5.67 ppm in nonmonsoon period (Table 16.4b). Again, in the Bhogdoi river water this variation is observed as 7.02 - 12.87 ppm in monsoon period and between 12.02 to 17.7 ppm in nonmonsoon period (Table 16.4c). The low concentration of Na^+ in monsoon season (Figure 16.8) may be due to the high dilution factor brought about by enormous amount of water discharge whereas, in nonmonsoon season dilution factor is reduced significantly which may lead to slight increase in concentration of Na^+. Higher concentration of Na^+ ion at sampling points TK4 and TK5 of the Toklai River may be attributed to the dumping of municipal waste through several drains near these points. The main source of Na^+ in the river water can be considered as the plagioclase feldspar minerals e.g. weathering of albite may produce Na^+ to the river water as follows:

$$2H_2CO_3 + 9H_2O + 2NaAlSi_3O_8 \text{ — } \underset{\text{Kaolinite}}{Al_2Si_2O_5(OH)_4} + 4H_4SiO_4 + 2Na^+ + 2HCO_3^-$$

A substantial part of potassium ion comes from terrestrial and anthropogenic sources. Potassium feldspar is chemically inert in case of weathering and a little dissolved material is derived from the alteration of potassium feldspar and most of the K^+ ion comes from the dark silicate minerals like biotite, hornblende etc. According to Berner and Berner (1987) only 15% of the river transported K^+ is in dissolved load while the rest is in particulate form.

In the Toklai river water K^+ ion varies from 2.54 to 9.82 ppm in monsoon season and in non-monsoon season it is found as 4.8 to 14.3 ppm (Table 16.4a). The K^+ ion concentration of the Tarajan river water varies from 10.32 to 18.62ppm in monsoon season and from 7.78 to 14.4 ppm in nonmonsoon season (Table 16.4b). Again, this variation in the Bhogdoi River water is observed in between 4.32 and 9.02 ppm in monsoon period and in between 12.02 and 17.7 ppm in non-monsoon period (Table 16.4c). The K^+ ion concentration in monsoon season is significantly reduced than that of nonmonsoon. It is directly related to climate dominance monsoonal factor which has helped in dilution of highly mobile major cations.

In terms of percentage contribution towards TDS, Na^+ and K^+ ions together contributed 36.42% in monsoon and 25.26% in nonmonsoon periods of the Toklai river. Millot (1970) from his studies of hydrosphere concluded that in the case of chemical weathering Na^+ turns out to be much more mobile than K^+ and dominates the latter in natural system. In case of the Toklai river the contribution of Na^+ towards TDS in monsoon and nonmonsoon periods are 23.72% and 14.74% respectively. But K^+ contributed 12.70% and 10.62% towards TDS in monsoon and nonmonsoon periods respectively. From the study it can be attributed that Na^+ contributed a significant proportion towards TDS followed by K^+. Similar observations have been made in case of Tarajan river, where Na^+ contributed 15.99% and 4.40% in monsoon and nonmonsoon periods. On the other hand, in case of Bhogdoi river the contribution of Na^+ and K^+ is proportionately higher in nonmonsoon period than monsoon period. The contributions of Na^+ and K^+ in monsoon period are 7.47% and 6.16% and in nonmonsoon period these are 14.79% and 10.74% respectively (Table 16.7). From this observation it can easily be concluded that Na^+ and K^+ ions are mainly contributed by the chemical weathering of feldspar group of minerals and associated clay fraction in the Bhogdoi river. The increase in Na^+ and K^+ content along with the increasing concentration of $SO4^{-2}$ and Cl^- indicates that during lean flow period there is significant contribution of major ions from soil salt/ground waters (Sarin *et al.*, 1989).

Calcium and Magnesium: Calcium occurs in great abundance in all natural waters and are derived mainly from the weathering of carbonates, sulphates and silicates. Calcite, aragonite, dolomites are the important carbonates and gypsum, anhydrites are the important sulphates and plagioclase feldspars, biotites are the important silicates

Table 16.7
Percentage contribution of major cations in relation to TDS

Toklai

S. points	*Na(%)*		*K(%)*		*Ca(%)*		*Mg(%)*	
	M	*NM*	*M*	*NM*	*M*	*NM*	*M*	*NM*
TK1	5.83	5.76	7.85	7.74	6.97	6.39	2.90	4.19
TK2	13.94	11.60	13.69	7.62	15.87	8.43	6.45	6.35
TK3	21.46	18.15	24.50	14.31	12.35	7.60	10.13	5.49
TK4	35.13	22.69	11.98	12.67	7.45	4.47	4.90	2.52
TK5	35.65	20.42	4.78	8.06	11.04	5.10	12.89	4.83
TK6	30.32	26.21	13.40	13.31	10.51	7.53	9.54	5.35
Average	23.72	14.74	12.7	10.62	10.69	6.58	7.80	4.79

Tarajan

S. points	*Na(%)*		*K(%)*		*Ca(%)*		*Mg(%)*	
	M	*NM*	*M*	*NM*	*M*	*NM*	*M*	*NM*
TJ1	10.31	8.67	13.80	13.72	6.31	11.13	2.45	5.07
TJ2	15.65	4.70	12.07	7.87	5.04	6.46	2.03	3.01
TJ3	16.87	3.93	18.61	11.34	5.25	5.31	2.12	3.17
TJ4	17.53	3.22	19.29	12.34	5.20	4.56	1.55	2.27
TJ5	18.37	3.10	10.95	6.14	6.21	5.70	2.20	3.54
TJ6	17.22	2.83	14.40	7.54	6.71	5.89	2.39	2.93
Average	15.99	4.40	14.85	9.83	5.79	6.51	2.12	3.33

Bhogdoi

S. points	*Na(%)*		*K(%)*		*Ca(%)*		*Mg(%)*	
	M	*NM*	*M*	*NM*	*M*	*NM*	*M*	*NM*
BH1	9.87	13.48	9.16	11.16	5.48	7.56	2.01	2.88
BH2	6.31	11.87	5.21	8.75	3.85	6.71	2.07	3.26
BH3	9.88	13.40	3.58	5.94	3.64	6.09	1.96	2.89
BH4	9.49	16.51	6.10	12.78	4.63	8.48	2.61	4.31
BH5	9.42	18.73	6.64	14.33	4.53	8.86	3.36	7.93
BH6	9.01	14.74	6.25	11.46	4.09	7.08	3.79	7.27
Average	7.47	14.79	6.16	10.74	4.37	7.46	2.63	4.75

contributing Ca^{+2} ion into the river water. Ca^{+2} and Mg^{+2} ions are considered to be the two ions with similar behaviours and this is true in many phenomena of the hydrosphere. Its source has been in the rock from which it gets leached. Its concentration has been found to vary greatly in natural waters depending upon the nature of the basin. In an aquatic environment, Ca is an important micro nutrient. As it is an important contributor to hardness in water, it is considered as effective parameters towards the utility of water for domestic use. In the Toklai river water concentration of Ca^{+2} is found in between 3.84 - 5.86 ppm in monsoon season and in the case of nonmonsoon season it varies from 3.56 - 5.32 ppm (Table 16.4a).

The concentration of Ca^{+2} ions in the Tarajan river varies in between 4.4 and 6.6 ppm in monsoon period and in nonmonsoon period it varies in between 5.32 and 7.28 ppm (Table 16.4b). The variation of Ca^{+2} ion in monsoon period is between 4.29 - 5.9ppm and in nonmonsoon period this is between 6.79 - 8.37 ppm for the Bhogdoi river water (Table 16.4c).

Like calcium, magnesium also occurs in all natural waters. It has been somewhat in low concentration than calcium. Its high content lowers the utility of water for domestic use, whereas a concentration above 500 mgl^{-1} gives an unpleasant taste and makes it unfit for drinking purpose. Mg^{+2} ion varies from 1.60 - 6.84 ppm in monsoon season and 2.57 - 4.00ppm in nonmonsoon season of the Toklai water (Table 16.4a). In the Tarajan river this variation is observed in between 1.5 - 2.35 ppm in monsoon period and in between 2.65 - 3.86 ppm in nonmonsoon period (Table 16.4b). In the Bhogdoi river this variation is observed in between 1.98 - 5.42 ppm in monsoon period and in nonmonsoon period it is in between 2.98 - 8.9 ppm (Table 16.4c).

In case of percentage contribution towards TDS, Ca^{+2} and Mg^{+2} ions contributed 10.69% and 7.8% in monsoon period and 6.58% and 4.76% in nonmonsoon period to the Toklai river water, while in the Tarajan river Ca^{+2} and Mg^{+2} in monsoon period contributed 5.79% and 2.12% and in nonmonsoon period these contributions are 6.51% and 3.33% respectively . The contributions of Ca^{+2} and Mg^{+2} towards TDS in case of the Bhogdoi River are observed to be much lower than the Na^{+} and K^{+} ions. The contribution of Ca^{+2} and Mg^{+2} in monsoon period are 4.37% and 2.63%, while they contributed 7.46% and 4.75% in nonmonsoon period (Table 16.7).

Iron: In nature iron exists in the Fe (II) and Fe (III) oxidation states. The reactions of iron in weathering processes depend on the Eh-pH system of the environment and on the oxidation state of its compounds. Significant amounts of iron are also released into the environment as a result of coal mining. In aqueous solution, ferric and ferrous iron hydrolyses various species, forming complexes with both inorganic and organic materials. Iron occurs as particulate ferric hydroxide or in the form of organic complexes in natural system. In river water near the bottom may be depleted of oxygen and attain a low Eh, such water may contain dissolved ferrous (reduced) iron concentration up to several mgl^{-1}. Since iron compounds are extensively used in water treatment, a guideline value of 0.3ppm has been recommended by WHO. For fresh water aquatic life, the criterion is 1ppm. In the presence of dissolved oxygen, iron in water is precipitated as $Fe(OH)_3$, which in increasing quantity is detrimental to physiological function of life sustaining processes.

The study of Fe^{+3} contents in the Toklai water reveals that in monsoon season the concentration varies from 0.272 - 0.315 ppm which is almost within the WHO's limit with an exception in the sampling point TK5 (0.50ppm). But in case of nonmonsoon season Fe^{+3} concentration varies between 0.372 - 0.544ppm which is well above the WHO's guidelines (Table 16.4a). In the case of the Tarajan river water the concentration of iron varies in between 0.185-0.401ppm in monsoon season and in between 0.285 - 0.573ppm in nonmonsoon season (Table 16.4b). Again in the Bhogdoi river water this variation is observed in between 0.27 - 0.46ppm in monsoon season and in between 0.43 - 0.67ppm in nonmonsoon season (Table 16.4c). The Fe concentration both in the Bhogdoi river water and in the Tarajan River water is above the guideline value. The high concentration of Fe observed in nonmonsoon season may be the result of release of this metal from the breakdown of planktonic matter or through sedimental regeneration into the overlying waters because of existing environmental conditions i.e. low overflow and calm condition in the river. High concentration throughout the year is observed in some points in both the rivers, which may be attributed that the organic pollutants released from agricultural run off will chelate iron to form soluble complexes (Shyama Sundar *et al.*, 1994). The low concentration of Fe^{+3} in monsoon period over the nonmonsoon period for the three rivers attributed dilution/dispersion of associated iron contents in monsoon period.

The observed charge balance between major cations (Σ^+ in milliequivalent l^{-1}) and major anions (Σ^- in milliequivalent l^{-1}) for the three rivers are presented in the Tables 8.a, b and c. Statistical analysis indicates the coefficients of correlation between Σ^+ and Σ^- for the Bhogdoi river R = 0.87, for the Toklai River R = 0.83 and for the Tarajan river R = 0.62 (Figure 16.9). The observed charge balance provides an added proof on the precision and accuracy of generated data set. The relationships indicate good correlation in case of the Bhogdoi and the Toklai river, where as it shows a fair accuracy in case of the Tarajan river. This proved that the cations (mainly Na^+, K^+, Ca^{+2}, Mg^{+2} and Fe^{+3}) and anions (SO_4^{-2}, Cl^- and HCO_3^-) considered in the present study might have significant contributions in case of the Bhogdoi and the Toklai rivers, whereas in case of the Tarajan river, there might have some other ionic components which were not measured in the present investigation played a significant role in maintaining the charge balance over the basin. Statistical analysis of data between the Σ^+ and Σ^- are coupled by the relation —

Bhogdoi: $\Sigma^+_{(\text{meqv}\, l^{-1})} = 0.88\ \Sigma^-_{(\text{meqv}\, l^{-1})} + 0.2848$

Tarajan: $\Sigma^+_{(\text{meqv}\, l^{-1})} = 1.790\ \Sigma^-_{(\text{meqv}\, l^{-1})} - 0.8423$

Toklai: $\Sigma^+_{(\text{meqv}\, l^{-1})} = 0.4703\ \Sigma^-_{(\text{meqv}\, l^{-1})} + 0.6527$

Tables 16.8a,b,c

Major ion equivalence of the Toklai(TK), Tarajan(TJ) and Bhogdoi(BH) Rivers water

(a)Monsoon (Toklai)

S. pts	*Cl⁻*	*SO_4^{-2}*	*HCO_3^-*	*Na*	*K*	*Ca*	*Mg*	*Fe*	*μ^+*	*μ^-*
TK1	0.078	0.079	0.665	0.139	0.111	0.192	0.133	0.016	0.591	0.822
TK2	0.090	0.077	0.723	0.200	0.116	0.263	0.178	0.015	0.772	0.890
TK3	0.095	0.087	0.805	0.373	0.252	0.247	0.338	0.015	1.225	0.987

(Contd...)

TK4	0.282	0.100	0.847	0.887	0.178	0.216	0.237	0.016	1.534	1.229
TK5	0.211	0.093	0.933	0.823	0.065	0.293	0.570	0.027	2.265	1.237
TK6	0.177	0.109	1.075	0.622	0.162	0.248	0.375	0.017	1.424	1.361
Nonmonsoon										
TK1	0.371	0.123	0.344	0.165	0.130	0.222	0.230	0.020	0.756	0.838
TK2	0.563	0.133	0.918	0.317	0.123	0.265	0.333	0.022	1.060	1.614
TK3	0.352	0.164	1.008	0.369	0.172	0.178	0.214	0.029	0.963	1.152
TK4	0.532	0.208	0.984	1.113	0.366	0.253	0.237	0.020	2.246	1.724
TK5	0.476	0.197	0.918	0.722	0.169	0.207	0.327	0.029	1.454	1.591
TK6	0.459	0.220	1.205	0.804	0.241	0.266	0.315	0.021	1.647	1.884

(b)Monsoon (Tarajan)

S. pts	Cl^-	SO_4^{-2}	HCO_3^-	*Na*	*K*	*Ca*	*Mg*	*Fe*	μ^+	μ^-
TJ1	0.253	0.045	0.781	0.394	0.312	0.278	0.180	0.021	1.185	1.079
TJ2	0.338	0.061	0.849	0.597	0.272	0.221	0.149	0.021	1.260	1.248
TJ3	0.366	0.091	1.063	0.684	0.445	0.245	0.165	0.017	1.556	1.519
TJ4	0.338	0.083	1.175	0.735	0.471	0.251	0.125	0.017	1.599	1.596
TJ5	0.166	0.053	1.184	0.753	0.265	0.293	0.173	0.012	1.496	1.403
TJ6	0.175	0.061	1.425	0.735	0.363	0.330	0.196	0.010	1.634	1.661
Nonmonsoon										
TJ1	0.457	0.091	0.404	0.247	0.230	0.364	0.277	0.031	1.149	0.952
TJ2	0.683	0.121	0.449	0.202	0.199	0.319	0.248	0.024	0.992	1.253
TJ3	0.514	0.183	0.970	0.191	0.324	0.297	0.295	0.022	1.129	1.667
TJ4	0.550	0.145	1.039	0.163	0.269	0.266	0.221	0.021	1.040	1.734
TJ5	0.477	0.105	1.057	0.147	0.172	0.311	0.322	0.018	0.970	1.639
TJ6	0.457	0.124	1.152	0.144	0.226	0.344	0.285	0.015	1.014	1.733

(c)Monsoon (Bhogdoi)

S. pts	Cl^-	SO_4^{-2}	HCO_3^-	*Na*	*K*	*Ca*	*Mg*	*Fe*	μ^+	μ^-
BH1	0.084	0.174	0.843	0.423	0.231	0.270	0.165	0.022	1.111	1.101
BH2	0.197	0.105	1.039	0.305	0.149	0.214	0.192	0.025	0.885	1.341
BH3	0.169	0.084	1.211	0.517	0.111	0.219	0.197	0.017	1.061	1.464
BH4	0.169	0.171	1.294	0.489	0.186	0.275	0.258	0.024	1.232	1.634
BH5	0.186	0.164	1.468	0.533	0.222	0.295	0.365	0.014	1.429	1.818
BH6	0.214	0.213	1.541	0.559	0.223	0.293	0.452	0.015	1.542	1.968
Nonmonsoon										
BH1	0.270	0.214	0.791	0.605	1.295	0.390	0.248	0.028	1.566	1.275
BH2	0.318	0.145	0.961	0.523	0.227	0.340	0.275	0.030	1.395	1.424
BH3	0.400	0.171	1.021	0.718	0.188	0.376	0.292	0.023	1.597	1.592
BH4	0.439	0.208	1.122	0.686	0.313	0.405	0.243	0.031	1.778	1.769
BH5	0.484	0.202	1.186	0.769	0.347	0.418	0.265	0.024	2.163	1.872
BH6	0.515	0.236	1.190	0.736	0.332	0.399	0.683	0.036	2.186	1.941

Heavy metal distribution in water column

The discharge of heavy metals into river system may result in numerous physical, chemical and biological responses. These can be separated into two broad categories: (1) Effect of environment on the metal and (2) Effect of the metal on the environment. The first category emphasis's that conditions in receiving bodies may lead to a change in the speciation and toxicity of the metals. Such conditions include differential input of anthropogenic and geochemical material, concentration of chelators and suspended solids. Biological responses under the second category are often equally diverse. Depending on environmental conditions, there may be a change in density, diversity, community structure and species composition of populations. The nature and extent of change depends largely on the concentration of heavy metal species in the water and sediment. Hence, physicochemical processes within the natural water have a major but indirect effect on biological responses.

Distribution and transport of heavy metals in natural aquatic environments are primarily controlled by the sediment and water column respectively. The physicochemical composition of the water column determines the mode (s) of transport of metals, such as particulate, colloidal, dissolved ionic and dissolved complex forms. Geological and anthropogenic inputs determine the character of the active surface sediments and the carrier compartment-the water column. Thus, the natural water of the world is bound to vary in their effect on metal distribution and transport. In this present investigation eight heavy metals such as Zinc (Zn), Nickel (Ni), Cobalt (Co), Copper (Cu), Cadmium (Cd), Lead (Pb), Chromium (Cr) and Manganese (Mn) are used considering the subject of concern towards environmental destabilization and human health. Heavy metal distribution of Toklai, Tarajan and Bhogdoi rivers are tabulated in Tables 16.9a, b and c. and the seasonal variation is shown in the Figure 16.10.

Zinc: Zinc is the most abundant essential trace element in the human body. Zn is essential for the normal activity of DNA polymerises and for protein synthesis and thus plays a vital role in the healthy development of many life forms. Excessive amount of Zn, however, may be toxic, especially to the aquatic biota (Nriagu and Coker, 1980). Three types of toxic reactions of Zn to human beings have been reported by Prasad (1976). Most rocks in the earth's crust contain Zn in varying concentrations depending upon the type of rock. It's content is highest in basic rocks (basalt, gabbro) = 70-130ppm average in acid rocks (granite, Rhyolite) = 50 - 60ppm, intermediate in metamorphic (schists), average 80ppm and low in loessic loams and glacial clays (= 30-40ppm) as well as in carbonate rocks and in sandstones (16-20ppm). Zn is found in high concentrations in soils (50ppm on an average) (Aubert and Pinta, 1980). Moderately increased zinc concentrations in water may affect metabolism especially children and patients already suffering from irregular metabolism. The WHO guideline value of Zn concentration for drinking water is 5ppm.

Tables 16.9a, b and c
Heavy metal distribution in the Toklai (TJ), Tarajan(TJ) and Bhogdoi(BH) rivers

(a)Monsoon

S. pts	*Cu*	*Zn*	*Pb*	*Cd*	*Ni*	*Co*	*Cr*	*Mn*
TK1	0.010	0.069	0.031	0.0028	0.138	0.121	0.032	0.027
TK2	0.028	0.027	0.074	0.0050	0.189	0.139	0.031	0.051
TK3	0.014	0.178	0.056	0.0040	0.191	0.141	0.029	0.059
TK4	0.014	0.143	0.050	0.0050	0.203	0.157	0.025	0.104
TK5	0.076	0.047	0.070	0.0064	0.195	0.151	0.010	0.110
TK6	0.032	0.163	0.095	0.0022	0.168	0.145	0.017	0.173
Nonmonsoon								
TK1	0.023	0.175	0.056	0.0066	0.175	0.151	0.068	0.136
TK2	0.042	0.134	0.082	0.0066	0.292	0.163	0.045	0.153
TK3	0.032	0.361	0.072	0.0054	0.209	0.163	0.062	0.086
TK4	0.015	0.240	0.068	0.0020	0.297	0.166	0.035	0.255
TK5	0.091	0.172	0.084	0.0056	0.205	0.169	0.035	0.139
TK6	0.070	0.232	0.072	0.0025	0.197	0.162	0.032	0.236

(b)Monsoon

S. pts	*Cu*	*Zn*	*Pb*	*Cd*	*Ni*	*Co*	*Cr*	*Mn*
TJ1	0.0066	0.077	0.040	0.0002	0.029	0.011	0.009	0.089
TJ2	0.0064	0.068	0.066	0.0006	0.020	0.021	0.012	0.110
TJ3	0.0062	0.084	0.054	0.0012	0.035	0.013	0.012	0.105
TJ4	0.0048	0.081	0.044	0.0038	0.026	0.009	0.012	0.099
TJ5	0.0031	0.082	0.008	0.0002	0.012	0.003	0.004	0.039
TJ6	0.0026	0.073	0.007	0.0003	0.004	0.002	0.002	0.042
Nonmonsoon								
TJ1	0.055	0.187	0.060	0.005	0.041	0.095	0.024	0.189
TJ2	0.043	0.165	0.090	0.007	0.035	0.114	0.036	0.156
TJ3	0.038	0.283	0.068	0.004	0.043	0.063	0.031	0.144
TJ4	0.029	0.246	0.064	0.004	0.033	0.021	0.028	0.136
TJ5	0.019	0.197	0.013	0.001	0.021	0.016	0.016	0.128
TJ6	0.015	0.168	0.009	0.002	0.012	0.013	0.010	0.118

(c)Monsoon

S. pts	*Cu*	*Zn*	*Pb*	*Cd*	*Ni*	*Co*	*Cr*	*Mn*
BH1	0.0142	0.062	0.010	0.0040	0.036	0.010	0.019	0.051
BH2	0.0124	0.070	0.009	0.0054	0.047	0.016	0.019	0.075
BH3	0.0082	0.053	0.013	0.0036	0.014	0.023	0.007	0.051
BH4	0.0142	0.086	0.015	0.0028	0.034	0.019	0.016	0.053
BH5	0.0032	0.095	0.012	00018	0.037	0.019	0.009	0.092
BH6	0.0026	0.089	0.009	0.0012	0.033	0.025	0.014	0.103
Nonmonsoon								
BH1	0.022	0.172	0.020	0.0045	0.046	0.013	0.034	0.070
BH2	0.018	0.180	0.020	0.0053	0.059	0.018	0.038	0.081
BH3	0.009	0.193	0.006	0.0046	0.022	0.027	0.015	0.063
BH4	0.019	0.199	0.024	0.0035	0.049	0.023	0.028	0.060
BH5	0.065	0.210	0.015	0.0029	0.052	0.024	0.014	0.120
BH6	0.003	0.201	0.011	0.0015	0.048	0.030	0.015	0.132

In the Bhogdoi River, the concentration of Zn^{+2} varies from 0.05 to 0.095ppm in monsoon period and 0.17 to 0.21ppm in nonmonsoon period (Table 9.c). The surface depletion of dissolved zinc may be due to uptake of zinc by organisms. The low value of zinc in non-monsoon period may be attributed to transportation of zinc in particulate form. On the other hand, during the monsoon period organic matter released from agricultural run off from surrounding areas is likely to chelate Zn^{+2} to form soluble complexes in the favourable environmental conditions resulting enhancement of zinc concentration. But in monsoon period, river carries enormous amount of water and suspended load. Due to fine grain nature and presence of comparatively high organic component in suspended sediment, zinc is sorbed on it and as a result concentration of dissolved zinc in the overlying lying water column decreases.

A similar trend is observed in case of the Tarajan and Toklai rivers also. For the Tarajan river zinc concentration varies from 0.06 to 0.08 ppm and from 0.16 ppm to 0.28 ppm in monsoon and nonmonsoon seasons respectively (Table 16.9b). Again for the Toklai river zinc concentration varies from 0.02 to 0.17ppm in monsoon season and from 0.13 to 0.36 ppm in nonmonsoon season. (Table 16.9a).

Nickel: Nickel is a beneficial heavy metal occurring in natural water in trace quantities. It is considered as a borderline element between hard and soft acid acceptors in chemical interactions towards donor atom. This is reflected in it's abundance in the earth's crust as oxides, carbonates, silicates, with iron, magnesium and sulfides arsenides and tellurides. Nickel is one of the major elemental constituents of the earth, constituting about 2% by weight. However, it is a minor constituent (0.01%) of the earth's crust. Nickel salts are soluble and can occur as a leachate from nickel bearing rocks.

Nickel has a central function in metabolism, but under practical conditions, situations of deficiency in humans are unlikely to occur. Nickel interacts with only five essential elements, Ca, Co, Cu, Fe and Zn in plants and microorganisms. Competitive interactions are more prevalent with Cu than Zn. Ni, when administered in large excess, interferes with detoxification activities and drug metabolism of the liver by altering both the synthesis and degradation of cellular heme and the liver glutathione levels. Ni powder and it's compounds cause a variety of cancers in rodents and are listed as possible causative agencies for occupational and environmental cancer in man. Inhalation and ingestion through diet are the major route of intake. Percutaneous absorption is important in its desmatopathological effects on human. Thus, Ni appears to posses a dual status with respect to it's role in environmental health. WHO guideline for Ni in drinking water is 0.05ppm. In our present study, the concentration of Ni varies from 0.013 to 0.047ppm in monsoon season and 0.022 to 0.059ppm in nonmonsoon season in the case of the Bhogdoi River (Table 9.c). For the Tarajan River it is found as 0.004 to 0.035ppm in monsoon period and 0.012 to 0.043ppm in nonmonsoon period (Table 16.9b). Again, in case of Toklai river the concentration of Ni varies from 0.138 to 0.203 ppm in monsoon period and from 0.175 to 0.297ppm in nonmonsoon period (Table 16.9a). In both the periods the concentration of Ni is found to be well below the WHO guideline value in the

Bhogdoi and the Tarajan river water. But in case of the Toklai river the concentration of Ni is slightly higher than the WHO guideline values. Again, slightly higher values of Ni concentration are observed in nonmonsoon period, which might be due to the formation of soluble complex with various organic and inorganic ligands present in water column and also to abiogenic scavenging process followed by resuspension from the bottom sediments. But in monsoon period, Ni gets sorbed on fine grains of suspended sediments and also due to absorption to the Fe and Zn hydroxides as suggested by Jenne and Luoma (1977), for which a low value of Ni concentration is observed in this period. The apparent increase in concentration in nonmonsoon period in relation to volume of water present and dilution/dispersion during the monsoon period resulting decrease in concentration also plays a major role in its distribution.

Cobalt: Cobalt is an essential trace metal as human body requires vitamin B_{12} which is a cobalt complex. Having the ability to occupy low symmetry sites in enzymes, Co (II) is an enzyme activator (Forstner and Wittmann, 1983).

Co is found in all the rocks of the earth's crust, the content varying with the nature of the rocks. The concentration of Co in the soil of the different climatic zones may vary from 0.05 to 200 – 300 ppm (Aubert and Pinta, 1980). Co has recognised toxic effects on human thyroid, the heart and possibly the kidneys, in addition to the occupational lung disease seen in cement and carbide industries and also gives allergic manifestations. WHO has not suggested any guide line value of Co for drinking water.

In the present study, Co concentration is observed in between 0.01 to 0.025ppm in monsoon period and in between 0.02 to 0.059ppm in nonmonsoon period for the Bhogdoi river (Table 16.9c). Incase of Tarajan river, the concentration of Co varies from 0.002 to 0.021ppm and from 0.013 to 0.11ppm in monsoon and nonmonsoon period respectively (Table 16.9b). A variation of Co concentration ranging from 0.012 to 0.157 ppm for monsoon period and from 0.151 to 0.169 ppm for nonmonsoon period is observed in case of the Toklai river (Table 16.9a). Co has shown almost a similar behaviour with that of Ni in our present study.

Copper: Copper is an essential element required to human metabolism and is generally considered to be non toxic for man at the levels encountered in the drinking water. Several Cu containing proteins have been identified in biological systems. Ceruloplasmin, a blue protein in mammalian serum, accounts for over 95% of the circulating Cu. The levels of ceruloplasmic is highly sensitive to a variety of pathological changes and reflects the disorder of Co metabolism; the reduction of both serum Cu and ceruloplasmin Cu are widely accepted as an important diagnostic tool. Cu and Fe are involved in the natural selection of aerobic cells and evolution of metaloproteins and metallo enzymes. The evolution resulted in the development of Cu-Zn enzymes, heme enzymes, Fe-Cu enzymes and oxygen carrying proteins. This adaptation demands elaborate storage and transport proteins, exclusively Cu and Fe. In all three proteins Cu is coordinated to N, O and/or S ligands. Hence, the dominant role of Cu in biological systems is its ability to stabilize sulfur radicals.

Copper is present in all the rocks of the earth's crust, but its concentration varies according to the nature of the rocks. (basic rocks – 100 – 200 ppm, metamorphic rock – 30 - 40ppm and acid rocks (granite, rhyolite) – 10 – 20 ppm. At levels above 5ppm it imparts colour and an undesirable taste to water. Staining of laundry and plumbing fixtures occur at copper concentration above 1 ppm. WHO (1984) recommended a guideline value of 1ppm for copper in drinking water. Copper concentration varies from 0.01 to 0.076 ppm in monsoon period and 0.015 to 0.091ppm in nonmonsoon period for the Toklai river (Table 16.9a). This variation is observed for the Tarajan river in between 0.0026 and 0.066ppm in monsoon period; and 0.015 to 0.055 ppm in nonmonsoon period (Table 16.9b). Again in the Bhogdoi River water a variation in between 0.0026 and 0.014 ppm in monsoon period and in between 0.003 and 0.065ppm in nonmonsoon period is observed (Table 16.9c). In all the rivers a comparatively low concentration of copper is observed. This might be due to the formation of water insoluble complexes of copper with fulvic acid at pH range 6 - 8. The high value of copper concentration in nonmonsoon period might be due to the concentration effect. On the other hand, a low concentration of copper in monsoon period might be due to the sorption of copper on the organic matter present in the fine grain suspended sediment.

Cadmium: Cadmium is non-essential, non-beneficial element and recognized to be of high toxic potential. The stable state of cadmium in the natural environment is Cd^{+2}. Once ingested, cadmium is transported to all parts of the body through the blood stream. Almost all organs probably absorb cadmium, the highest concentrations are invariably found in the liver and kidney through its strong binding with cystine reduces of metellothionein, with somewhat lower concentration in pancreas and spleen. Cadmium is a cumulative poison; ingestion of small amount over the years may lead to the accumulation of chronically or even acutely toxic levels of cadmium in the body (Laws, 1993). Extreme cases of Cd poisoning are associated with a softening of the bones apparently caused by disruption of calcium phosphorous balance in the renal tube.

Cd is most commonly found in nature associated with Zn in carbonate and sulfide ores. It is an oxyphilic and sulphophilic element. It undergoes multiple hydrolysis at pH values encountered in the environment. In the absence of any precipitating anions cadmium exists as Cd^{+2} up to pH 8 in a natural aquatic environment. However, Cd begins to hydrolyses at pH 8 forming Cd $(OH)^+$ species. Higher hydroxy species of Cd are not relevant at the pH values commonly found in the environment. In the absence of any precipitating anion Cd^{+2} will be available for sorption on to the suspended solids and organic complexes with a variety of organic components. Being a soft acid acceptor, it prefers to coordinate with soft donor atoms like sulfur, nitrogen etc. The organic matter found in natural water, including amino acids, hydroxy and carboxylic acids of aliphatic and aromatic nature, contains suitable donor atoms for complex formation with cadmium. Considering its acute toxicity, WHO (1984) has recommended a guide line value of 0.005 ppm for Cd in drinking water. In the present environmental studies, the concentration of Cd varies, in the Toklai river, from 0.0022 to 0.0064 ppm in the monsoon period and from 0.002 to 0.0066 ppm in nonmonsoon period (Table 16.9a). In case of the Tarajan river this variation occurs in between 0.0002 to 0.0038 ppm in monsoon and in between 0.001 to 0.007ppm in nonmomsoon period (Table 16.9b). In the Bhogdoi river water, a variation of 0.0012-0.0054 ppm in monsoon period and 0.0015 to 0.0053 ppm in nonmonsoon period is observed (Table 16.9c). A slightly high concentration of Cd in non-monsoon period might be due to the formation of soluble complexes with the ligands present in the system; the low concentration of Cd in non-monsoon period might be due to sorption on the fine-grained suspended sediments. Except a few points average Cd concentration is found to be well below the WHO limit. Also owing to the dilution of river water, Cd concentration gradually decreases as the river spreads towards downstream direction.

Lead: Lead is a non-essential toxic heavy metal. It enters the human body primarily via inhalation and through the ingestion of contaminated food and water. As in the case of Cd, the lungs retain Pb more efficiently than the gastro-intestinal tract. Once absorbed into the blood stream, it is transported to all parts of the body primarily by the red blood cells; although, its incorporation into tissues apparently occur through the blood plasma. Pb begins to appear in liver and kidney within a few hours after absorption but ultimately about 90% of the inorganic lead in the body is deposited in the bones where it replaces the calcium. Once incorporated in the bone structure, lead released back to the blood stream at a steady but a very slow rate. In contrast to inorganic lead, organic lead does not accumulate in the bones but tends to concentrate in the lipid tissues including those of central nervous system. The highest concentration of lead is often found in the brain and liver. Of all the environmental neuro-toxins, lead is considered one of the most serious because, of its widespread subtle and insidious impact on man. Even at low levels it causes increased tension, anger, depression and impaired memory. Also, evidences have been adduced that environmental neuro-toxins like lead are a significant factor in promoting criminal behaviour. The behaviour of lead in natural water is a combination of precipitation equilibria and complexation equlibria with inorganic and organic ligands. The degree of mobility of Pb depends on the physic-chemical state of the complexes formed. Hydrolysis of precipitates above pH 6, solubilises Pb as $Pb(OH)^+$ (Moore and Rammorthy, 1984). Insoluble Pb $(OH)_2$ is not formed till pH 10. The mobility of Pb may alter after interaction with OH^- ions present in water. Pb forms moderately strong chelates with organic ligand containing S, N and O donor atoms. Depending on the composition of water, a significant portion of Pb in drinking water is bound to colloids, either hydrous iron oxides or organic macromolecules.

Due to its acute toxicity, WHO has recommended a guideline value of 0.5 ppm in drinking water and ISI has recommended a value of 0.1 ppm (IS: 3025-1964). The concentration of Pb in the Toklai river water varies from 0.31 to 0.095ppm in monsoon period and from 0.056 to 0.084 ppm in nonmonsoon period (Table 16.9a). In case of the Tarajan river this variation occurs in between 0.007 to 0.066 ppm in monsoon period and 0.009 to 0.09ppm in nonmonsoon period (Table 16.9b). This is observed in case of the Bhogdoi River between 0.009 and 0.015 ppm in monsoon;

0.006 to 0.024 ppm in nonmonsoon period (Table 16.9c). No seasonal trend of variation is observed in all the three rivers. However, Pb concentration in all the three rivers is well below the WHO value and ISI guide line values.

Chromium: Chromium has been identified both as an essential micronutrient and as a chemical carcinogen. These paradoxical effects are dependent on the oxidation state and solubility. In the earth crust Cr ranks 4th in abundance among 29 elements of biological importance from the essential and toxic perspectives. Cr is more abundant than Cd, Co, Cu, Pb, Ni or Zn. The average crustal concentration of Cr is 100mgkg^{-1}. Cr is present in food crops and other vegetation at concentrations in the 50mgkg^{-1} range.

In general Cr^{3+} is less toxic than Cr^{6+} with respect to acute, chronic and oral toxicity, dermal irritancy and allergy, systematic effects, cytotoxicity and genotoxicity as well as carcinogenicity. Cr^{3+} has been established as an essential nutrient which helps in maintaining normal metabolism of glucose, cholesterol and fat while bronchionic carcinomas have been associated with the inhalation of slightly soluble and insoluble hexavalent Cr compounds. Renal failure from nacrosis can result from ingestion of Cr^{6+}.

Since Cr is a naturally occurring element of abundance, hence river water is considered as one of the main source, through which Cr makes way into the body. Considering its environmental impact, WHO has recommended as guideline value of 0.05ppm for total Cr.

The present study of water chemistry of the river Toklai indicates that the total Cr concentration varies from 0.01 to 0.032 ppm in monsoon period and from 0.032 and 0.068 ppm in nonmonsoon period (Table 16.9a). In case of the Tarajan river this variation occurs in between 0.002 and 0.012 ppm in monsoon period and between 0.01 and 0.036ppm in nonmonsoon period (Table 16.9b). This is observed in the Bhogdoi river water in between 0.007 to 0.019 ppm in monsoon period and in between 0.014 to 0.038 ppm in nonmonsoon period (Table 16.9c). Cr concentration in monsoon period is slightly lower than the nonmonsoon period. Concentration effects may predominate in non-monsoon period for which a slightly higher concentration of Cr is observed. On the other hand, a low concentration of Cr in monsoon period might due to the sorption of chromium on the fine grained suspended sediment present within the system. However, irrespective of both the periods, the average Cr concentration in all three rivers is well below the WHO guideline value.

Manganese: Mn is the second most abundant essential metal in nature (exceeded only by Fe). Mn is involved in glucose utilisation (Forstner and Wittmann, 1983). At levels exceeding 0.15ppm, Mn in water supplies stains plumbing fixtures and laundry. At higher concentration, it causes an undesirable taste to beverages. In combination with Fe, it's presence in drinking water may lead to accumulation of deposits on the distribution system. A guideline value of <0.05ppm (WHO, 1984) for drinking water is based on its staining property.

Variation in Mn^{+2} concentrations in monsoon and non-monsoon seasons of all the three rivers were observed in the present study. In the Bhogdoi river Mn^{+2}

concentration varies from 0.05 to 0.103 ppm in monsoon period and in nonmonsoon period it is observed in between 0.06 to 0.132ppm (Table 16.9c). In the Tarajan river, variation of Mn^{+2} concentration is observed as: 0.039 - 0.11ppm in monsoon and 0.118 – 0.189ppm in non-monsoon periods (Table 16.9b). In case of the Toklai river a similar trend is observed. In monsoon season the concentration is observed in between 0.05 – 0.17 ppm and in nonmonsoon season it varies between 0.08 – 0.25 ppm (Table 16.9a). The depletion of Mn^{+2} concentrations in monsoon and increase in non-monsoon seasons reflects its distribution as abiogenic scavenging process followed by resuspension from the bottom sediments. It is found that in both seasons the concentration of Mn^{+2} is slightly higher than the guideline value for drinking water.

Conclusion

Considering the geographical location, population growth and nature of development, an attempt has been made to monitor the chemistry of water system around Jorhat city. One of the major south bank tributary of the mighty river Brahmaputra, the Bhogdoi river and two small streamlets, the Tarajan river and the Toklai river passing through the study area are considered for the present investigation. These two mutually independent natural arteries carry almost all the waste water from the city and debauch them to the Bhogdoi river. During rainfall the wastewater inundates low-lying areas and created significant problems. If the situation remains unaltered, adsorption/association of environmentally unfriendly substances in the associated sediment/soil system will definitely result a serious eco-societal problem with lesser water availability and increasingly expensive pure water supply treatment. It is expected that the present study will form a strong database for any future studies for the already existing settlements.

The present study has contributed a significant database towards geoenvironmental attributes and can be summarised under the following heads:

1. Like all other Indian rivers, the water of the Bhogdoi, the Tarajan and the Toklai rivers also manifests almost neutral to alkaline in nature. Without any significant variation, a slight depletion in pH value was observed in monsoon period, which can be corroborated with the dilution by rainwater. The pH value as observed do not favour higher concentration of dissolved metal in water column.
2. EC, TDS, TSM, DO, BOD, COD, Cl^-, SO_4^{-2}, N-NH_3^- and HCO_3^- values are in general within the guide line values of WHO (1984) and IS(1964). However, at points TK3 and TK4 of the Toklai river a deviation in COD and DO values are observed and can be correlated with the anthropogenic introduction from nearby areas.
3. The study on observed charge balance between major cations ("+) and major anions ("-) for the three rivers indicates significant precision and accuracy in data generation in the present study.
4. The major elements (Na^+, K^+, Ca^{+2}, Mg^{+2}, Fe^{+2}) and heavy metals (Pb^{+2}, Cd^{+2}, Cr^{+3}, Zn^{+2}, Co^{+2}, Ni^{+2}, Cu^{+2} Mn^{+2}) concentrations are evaluated for the three

rivers. The major cations Na^+ and K^+ indicate significant impact of silicate weathering within the basin with a little impact on carbonate weathering. The reduction in concentration during the monsoon period is directly related to climate dominant monsoonal factor, which has helped in dilution of highly mobile major cations. The Ca^{+2} and Mg^{+2} contents also exhibit their higher concentration than the guide line value proposed by WHO (1984), may be the result of sedimentological regeneration and anthropogenic influx from the nearby areas. The major cations (Na^+, K^+, Ca^{+2}, Mg^{+2}) contributed a significant proportion (up to a maximum of 54.91%) towards total dissolved solid fraction.

5. The heavy metal content in water (Pb^{+2}, Cd^{+2} , Cr^{+3} , Cu^{+2}, Zn^{+2} , Ni^{+2}, Co^{+2}) of the Bhogdoi and the Tarajan rivers are within the tolerance limit and established guide line values. However, in case of the Toklai river higher concentration Pb^{+2} and Ni^{+2} are observed within the urbanised part of the channel indicating incorporation from the anthropogenic source.

The present study will provide a better understanding towards existing environmental situation over there and necessity to implement a continuous monitoring system for this developing township coupled with increasing population growth. The situation may not be alarming at present but it may create environmentally unfriendly scenario in future years to come.

[**Acknowledgements:** The authors are grateful to Dr. P. G. Rao, Director, North East Institute of Science and Technology (CSIR), Jorhat, Assam, India, for his kind permission to publish the work.]

REFERENCES

A.P.H.A. 1995. Standard *methods for the examination of water and waste water.* 19th. Edition. Amercan Public Health Association.

Aubert, H. and Pinta, M. 1980. Trace elements in siols. Dev. *In soil Sciences.* Series 7, 2nd Ed. Elsevier Pub. Con.

Baruah, N.K., Kotoky, P. Bhattacharyya, K.G. Borah, G.C. 1996. Metal speciation in Jhanji River sediments. *The Sci. of the Tot. Environment,* 193: 1-12.

Baruah, N.K., Kotoky, P., Baruah, J. and Bhattacharyya. K. G. 1997 Correlation coefficients of some physicochemical parameters of the river Jhanji, Assam. *Ind. Jour. Env. Prot.,* 2: 81-85.

Berman, E. 1980. *Toxic metals and their analysis.* Heyden and Son., London, 289 pp.

Bernas, B.1968 A new method for decomposition and comprehensive analysis of silicates by AAS *Anal. Chem.* , 40: 1682 - 1686.

Berner, E.K. and Berner, R.A. 1987. *The global water cycle.* Prentice-Hall, Englewood Cliffs, NJ.

Chakraborty, D. , Adams, F., Van Mol, W. and Ingolic, K. 1987. Determination of trace metals in natural waters at nanogram per liter level by electrothermal atomic adsorption spectrometry after extraction with NaDDc. *Anal. Chim. Acta.,* 196: 23-31.

De, A.K. 1990. *Environmental Chemistry.* Wiley Eastern Ltd., New Delhi, 263.

Faniran, A. and Aerola, O. 1978. *Essential of Soil Study.* Heinemann, London, 273 pp.

Fresenious, W., Quentin, K.E. and Schnider, W. 1988. *Water Analysis.* Springer Verlag, Berlin, 779 pp.

Forstner, U. and Wittmann, G.T.W. 1983. *Metal pollution in the aquatic environment,* 2nd Ed. Springer-Verlag, 486 pp.

Garrels, R.M.1967. *Genesis of some ground waters from igneous rocks, in researches in Geochemistry,* VOL.2. edited by P.H.Abelson, 405-420. John Wiley, New York., 2: 405-420.

Garrels, R.M., and Mackenzie, F.T. 1971. *Evolution of sedimentary rock.* Norton and Co. Inc. New York, 397 pp.

Garrels, R.M., Mackenzie, F.T. and Hunt, C. 1975. *Chemical cycles and the global environment - Assessing Human influences.* C.A. William, Kaufman Co., 206 pp.

Holland, H.D., 1978. *The chemistry of the atmosphere and oceans.* Willey Interscience, New York, 351 pp.

Jenne, E.A. and Luoma, S. 1977. Forms of trace elements in soil, sediment and associated waters: An over view of their determination and biological availability. In: *Biological Implications of Metals in the Environment* (eds. R.E. Wikling and H. Druken), 110-143 NTIS, Springfield, VA,

Laws, E.A. 1993. *Aquatic Pollution: An Introductory Text,* 2nd Ed. John Wiley and sons, New York, 345 pp.

Millot, G. 1970. *Geology of Clays.* Springer-Verlag, Paris, 305 pp.

Moore, J.W. and Ramamoorthy, S. 1984. *Heavy metal in natural waters.* Springer-Verlag, New York, 263 pp.

Nriagu, J.O. and Coker, R.D. 1980. Trace metals in humic and fulvic acids from Lake Ontario Sediments. *Environ. Sci. and Tech.* 4: 443-446.

Prasad, A.S. (ed.) 1976. *Deficiency of zinc in man and its toxicity in trace elements in human health and disease* Vol. 1. Academic Press New York.

Raymahashay, B.C. 1986. Geochemistry of bicarbonate in river water. *J. Geol. Soc. India.,* 27: 119-120.

Sarin, M.M., Krishnaswamy, S., Dilli, K., Somayajulu, B.L.K. and Moore, W.S. 1989. Major ion chemistry of the Ganga - Brahmaputra river systems. Weathering processes and fluxes to the bay of Bengal. *Geochim. Cosmochim. Acta.,* 53: 997-1009.

Stumm, W. and Morgan, J.J., 1970. *Aquatic Chemistry.* Wiley Interscience. New York, 583 pp.

Stumm, W. and Morgan, J. J. 1981. *Aquatic Chemistry: An Introduction Emphasizing Chemical Equilibria in Natural Waters.* J. Wiley and Sons.

Subramanian, V. 1979. Chemical and suspended sediment characteristics of river of India. *Jour. Hydr.,* 44, 37 - 55.

Shyama Sundar, B., Madhu, G., Srinivas Murty, K. and Mangathayaramma, V. 1994. River Krishna - estimation of trace metals and their distribution. *Ind. J. Env. Prot.,* 14(9): 654-663.

World Health Organisation (WHO). 1984. *Guidelines for Drinking Water Quality,* Vol. 1, *Recommendations.* W.H.O. Geneva.

17

Fish Fauna and Hydrological Environment of Deopani Stream, Arunachal Pradesh

—*Rajdeep Dutta*[1], *Debajit Sarma*[2] *& B.C. Tyagi*[3]

[The Deopani river, originates in Lower Dibang Valley district of Arunachal Pradesh, India. In the present study, an attempt has been made to study the fish fauna and hydrological condition of Deopani river from October, 2008 to September, 2009 at an interval of two moths. The major fish groups recorded during the study were Catfishes (*Glyptothorax horai*), Mahseers (*Tor putitora, Neolissocheilus hexagonolepis*) and Barils (*Barilius bendelensis, B. tileo*). Among all these fish species, *Glyptothorax* spp. were to found to be the most common species recorded in all the seasons. The rocky substratum of the river favors the occurrence of this species. During the study, the physico-chemical regime of the river was also investigated. River water with fast water current showed water velocity (0.53-5.88 msec^{-1}), temperature (16.3-2.54^{0}C), pH (7.0-8.0), Alkalinity (32.0-65.0 mg/liter), DO (3.2-14.8 mg/liter) and carbon dioxide (0.32-12.0 mg/liter.) The study reflected that the physico-chemical condition of the river was in general conducive for aquatic habitat. The investigation also revealed unpolluted and pristine status of Deopani river.

Key words: Fish fauna, hydrology, Deopani river, Arunachal Pradesh.]

Introduction

The state of Arunachal Pradesh is known for its rich riverine resources. The state has a potential reverine length of about 2000 km, including the high altitude cold water streams. The fish fauna of these streams is a combination of both torrential and plain water forms occupying diverse ecological regime in their distributional range.

Deopani stream of Lower Dibang Valley district of Arunachal Pradesh is located at a longitude of 95^{0}51^{1}12.96$^°$ E and latitude of 28^{0}09^{1}31.58$^°$ N. The stream originates at the confluence point of two small streams Eze and Eme at a longitude 95^{0}51^{1}16.00$^°$E and latitude of 28^{0}09^{1}34.60$^°$N having altitude of 475m msl. This perennial stream with rocky substratum traverses a course of 12 km before it

[1] Krishi Vigyan Kendra, Tirap district, PO–Deomali, Pin–786 629, Arunachal Pradesh
[2,3] Directorate of Coldwater Fisheries Research, ICAR, Bhimtal, Nainital , Pin–263 136, Uttarakhand
Email: rajdeep18j@rediffimail.com

debouches into the Dibang river at a longitude of $95^{0}44^{'}14.83^{\circ}$ E and latitude of $28^{0}10^{'}26.60^{\circ}$N having altitude of 217m msl.

Materials and Methods

The study was conducted for one year starting from October 2008 to September 2009. During this period fish and water samples were collected from the stream at an interval of two months. The samples were collected from four different stations of the stream selected based on their altitude and fishing status. Station 1 is the point of origin of Deopani stream, the confluence point of two small streams Eze (Ezengo in local language) and Eme located at a longitude of $95^{0}51^{'}16.00''$E and latitude of $28^{0}09^{'}34.60''$N at an altitude of 475m msl, Station 2 is located near Roing (district headquarter of Lower Dibang valley) and Anini (district headquarter of Dibang valley) connecting bridge at a distance of about 1.14 km from station 1 having longitude of $95^{0}50^{'}34.27^{\circ}$E and latitude of $28^{0}09^{'}35.50^{\circ}$N at an altitude of 431m msl. Station 3 is located about 600m distance from station 2 at a longitude of $95^{0}50^{'}12.35^{\circ}$ E and latitude of $28^{0}09^{'}42.98^{\circ}$N having altitude of 413m msl. The fourth station was fixed at a distance of about 1.6km from Station 3 at a longitude of $95^{0}49^{'}24.93^{\circ}$E and latitude of $28^{0}09^{'}49.08^{\circ}$N having altitude of 379m msl.

Water samples were collected bimonthly from the four stations during the period under study. Air temperature, water temperature, pH and water velocity were recorded at the spot. Water samples were collected with specific precaution and brought to the laboratory for the analysis of dissolved oxygen, total alkalinity and free carbon dioxide. The analytical procedure for the above mentioned parameters was followed as described by APHA (1989). Fish samples were caught using a cast net of mesh size 17mm. The net was operated for forty times in every station during each sampling and the length and weight of the fishes were recorded. The fish samples were preserved using alcohol packed in plastic containers and labeled properly. The collected fishes were identified following Nath and Dey (2000).

Results and Discussion

The river substratum is mainly composed of stones, rock girders and large boulders. Gravelly sand is deposited where the water flows serenely in sheltered bays in the rocks and corner reaches of the river. Riverbed is marked by the conspicuous absence of many-rooted plants. The seasonal variation of different physico-chemical parameters of the stream water in Stations 1, 2, 3 and 4 are presented in Table 17.1-17.4.

Water Temperature

Water temperature ranged between 16.3 and 24.5 ^{0}C, minimum during winter and maximum in the monsoon indicating clear-cut seasonal variations (Figure 17.1). Variation in water temperature in relation to seasonal changes is a common phenomenon in aquatic ecosystem and is also observed by many workers in other major rivers of India (Laal *et al.*, 1986; Pathak *et al.*, 2001b and Nath, 2001). Choudhury (2002) observed the fluctuation of water temperature (7.0-22.0^{0}C) in relation to seasonal changes in the rivers of Arunachal Pradesh during pre-monsoon, retreating monsoon and winter. The fluctuation of water temperature in different seasons is corroborated with the earlier findings of Nath and Dey (2000) in the rivers Kameng (13.0 -23.0 ^{0}C), Dikrang (16.0- 24.0 ^{0}C), Ranga (9.0-24.0^{o}C) and Subansiri (10.0-16.0^{o}C).

pH

No marked seasonal variation in the pH of the stream water was observed during the study period. pH was found to be neutral to slightly (pH 7.0-8.0) alkaline in all the stations. Weak alkaline reaction (pH 7.0-8.0) is considered most suitable for fish production due to availability of various nutrient elements in the aquatic environment and physiological activity of fish remains optimum at this point. The present finding revealed the productive nature of Deopani stream. This agrees with the similar observation by Nath and Dey (2000) in four rivers of Arunachal Pradesh ranged (6.2-7.2) in Kameng, (6.2-7.5) in Dikrang, (6.2-7.0) in Ranga and (6.5-7.5) in Subansiri. Different pH values observed by Choudhury (2002) in the rivers of Arunachal Pradesh also ranged from 6.8-8.1.

Water Current

River water current was ranging between 0.53-5.88 m.sec^{-1} (Figure 17.2). In all the stations higher velocity was recorded during monsoon (June to September) and retreating monsoon (October- November) seasons whereas lower values were recorded during winter season (December- February). It is to be mentioned here that Nath and Dey (2000) in different streams of Arunachal Pradesh reported water velocity range of (0.9-1.3 m.sec^{-1}) in Kameng, (0.6-1.5 m.sec^{-1}) in Dikrang, (0.4-1.0 m.sec^{-1}) in Ranga and (0.6-0.9 m.sec^{-1}) in Subansiri.

Dissolved Oxygen

Rich dissolved oxygen was observed in Deopani river and the value fluctuated from 3.2 to 14.8 mgl^{-1} (Figure 17.3). The dissolved oxygen content may be considered as conducive for aquatic habitat. High dissolved oxygen content is a characteristic feature of hilly streams. This agrees with the phenomenon reported by Pathak *et al.* (2000) having high DO values from Siang (10 mgl^{-1}), Dibang (9.02 mgl^{-1}) and Lohit (9.73 mgl^{-1}) rivers of Arunachal Pradesh. Choudhury (2000) also reported high DO values (7.6-15.2 mgl^{-1}) in the rivers of Arunachal Pardesh.

Alkalinity

Alkalinity values were found to be low, the maxima was recorded during retreating monsoon (October- November, 65 mgl^{-1}) while the minima was observed in monsoon (June to September, 32 mgl^{-1}) (Figure 17.4). Lower alkalinity value of hilly streams was also reported by earlier workers. Gurumayum *et al.* (2000) observed a range of 15- 82.60 mgl^{-1} in some selected rivers of Meghalaya. Pathak *et al.* (2001a) recorded an average alkalinity of 62.67 mgl^{-1} and 35.72 mgl^{-1} in rivers Siang and Dibang of Arunchal Pradesh respectively. An alkalinity range of 33.98-66.7 mgl^{-1} was recorded in the river Subansiri by Daimari *et al.* (2005). According to Unni (2003), Indian rivers with an alkalinity range of 50-100 mgl^{-1} can be considered as low or unpolluted. Keeping in mind the alkalinity values, the present finding revealed unpolluted and pristine status of Deopani stream.

Table 17.1
Seasonal variation of different physico-chemical parameters at Station 1

Sl. No.	*Parameters*	*First Sampling (Oct'08)*	*Second Sampling (Nov'08)*	*Third Sampling (Jan'09)*	*Fourth Sampling (Apr'09)*	*Fifth Sampling (Jun'09)*	*Sixth Sampling (Sep'09)*
1.	Air Temperature (°C)	34	28	27	28	30	28
2.	Water Temperatures(°C)	24	19	17	20.5	21	21.5
3.	Water pH	7.0	7.0	7.0	7.0	7.0	7.5
4.	Water velocity (m/sec)	5.88	0.97	0.68	1.79	0.96	2.33
5.	Dissolved oxygen (mg/l)	9.89	6.4	9.6	9.2	6.8	12.4
6.	Total Alkalinity (mg/l)	30	60	44	50	40	20
7.	Carbondioxide as CO_2 (mg/l)	3.52	4	4	8	12	5

Table 17.2
Seasonal variation of different physico-chemical parameters at Station 2

Sl. No.	*Parameters*	*First Sampling (Oct'08)*	*Second Sampling (Nov'08)*	*Third Sampling (Jan'09)*	*Fourth Sampling (Apr'09)*	*Fifth Sampling (Jun'09)*	*Sixth Sampling (Sep'09)*
1.	Air Temperature (°C)	34	27.5	27	27	30	31.5
2.	Water Temperatures(°C)	24	19.5	17	19	21	22
3.	Water pH	7.0	7.0	7.0	7.0	7.0	8.0
4.	Water velocity (m/sec)	5.66	0.89	0.60	1.49	0.81	1.17
5.	Dissolved oxygen (mg/l)	9.66	6.8	9.4	10.8	5.2	6.8
6.	Total Alkalinity (mg/l)	30	65	46	55	30	20
7.	Carbondioxide as CO_2 (mg/l)	2.64	3	4	7	9	4

Table 17.3
Seasonal variation of different physico-chemical parameters at Station 3

Sl. No.	*Parameters*	*First Sampling (Oct'08)*	*Second Sampling (Nov'08)*	*Third Sampling (Jan'09)*	*Fourth Sampling (Apr'09)*	*Fifth Sampling (Jun'09)*	*Sixth Sampling (Sep'09)*
1.	Air Temperature (°C)	34	27.5	28	31	32	32.5
2.	Water Temperatures(°C)	24	19	17	20	22	22.5
3.	Water pH	7.0	7.0	7.0	7.0	7.0	7.5
4.	Water velocity (m/sec)	4.28	0.83	0.56	2.01	0.82	0.69
5.	Dissolved oxygen (mg/l)	10.12	4.4	9.4	8.4	5.6	10.8
6.	Total Alkalinity (mg/l)	32	60	46	55	30	40
7.	Carbondioxide as CO_2 (mg/l)	2.66	1	3	0.35	0.55	4

Table 17.4
Seasonal variation of different physico-chemical parameters at Station 4

Sl. No.	*Parameters*	*First Sampling (Oct'08)*	*Second Sampling (Nov'08)*	*Third Sampling (Jan'09)*	*Fourth Sampling (Apr'09)*	*Fifth Sampling (Jun'09)*	*Sixth Sampling (Sep'09)*
1.	Air Temperature (°C)	34	27.5	28	29	32	31
2.	Water Temperatures(°C)	24	19.5	17.5	20	22	22.5
3.	Water pH	7.0	7.0	7.0	7.0	7.0	7.0
4.	Water velocity (m/sec)	3.75	0.78	0.53	0.82	0.64	2.04
5.	Dissolved oxygen (mg/l)	10.35	6.8	10.2	8.4	3.2	14.8
6.	Total Alkalinity (mg/l)	32	60	46	60	50	30
7.	Carbondioxide as CO_2 (mg/l)	4.40	3	3	8	6	6

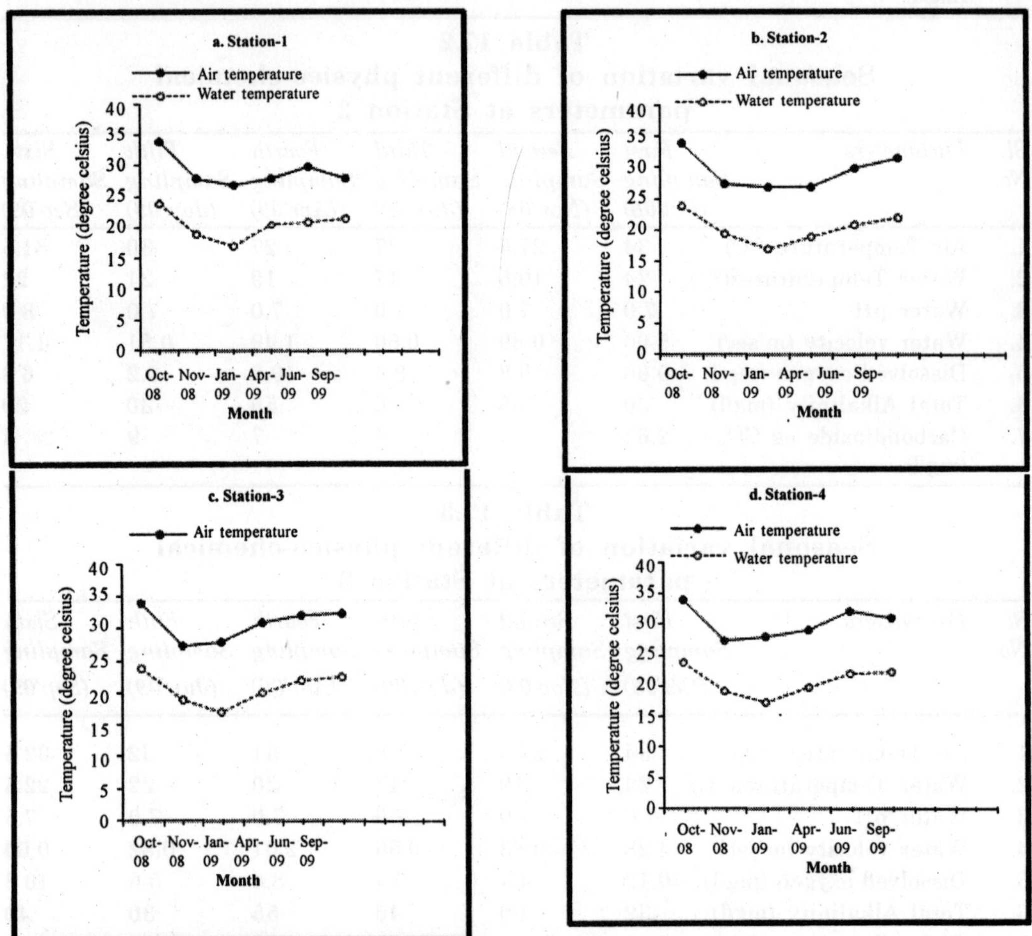

Figure 17.1. Monthly variations in air and surface water temperature of Deopani stream at Stations 1 to 4

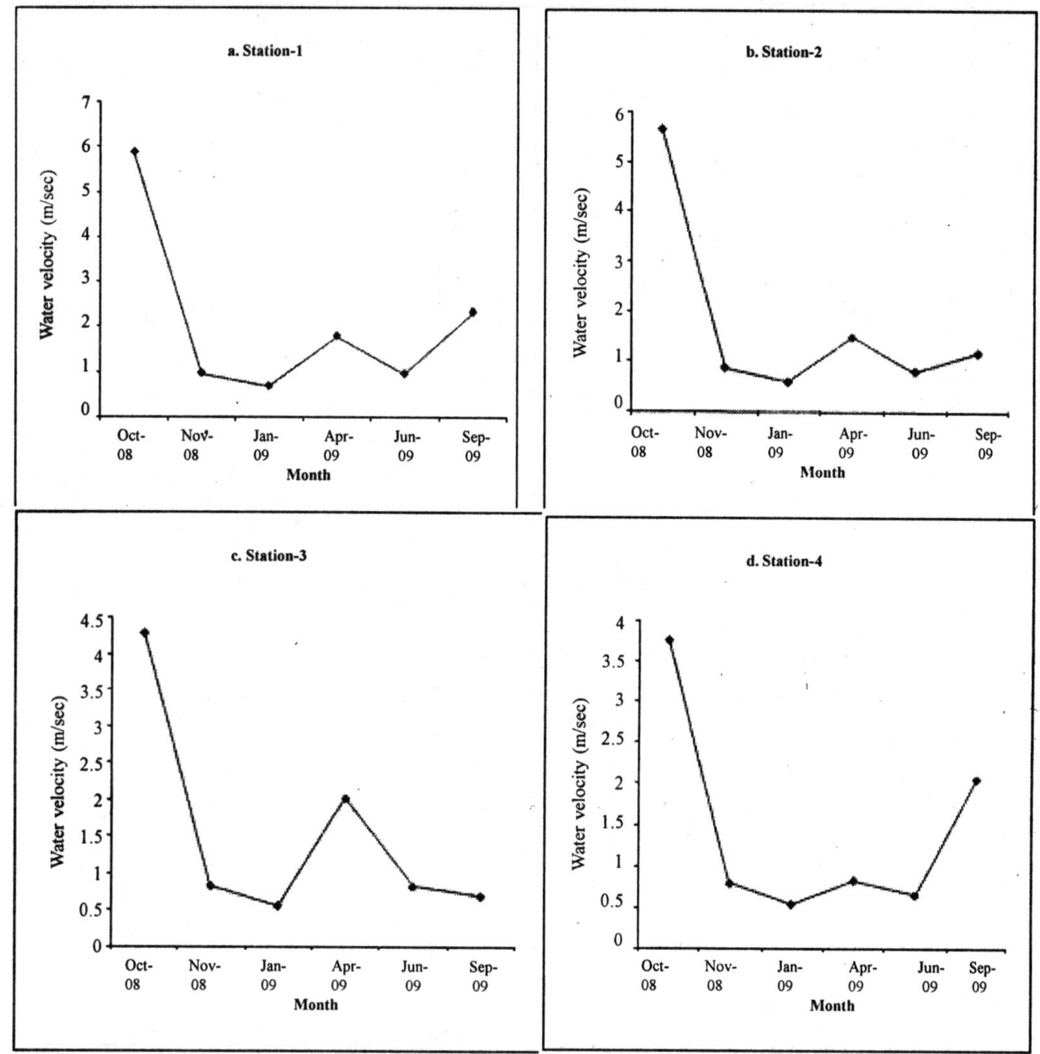

Figure 17.2. Monthly variations in water velocity of Deopani stream at Stations 1 to 4

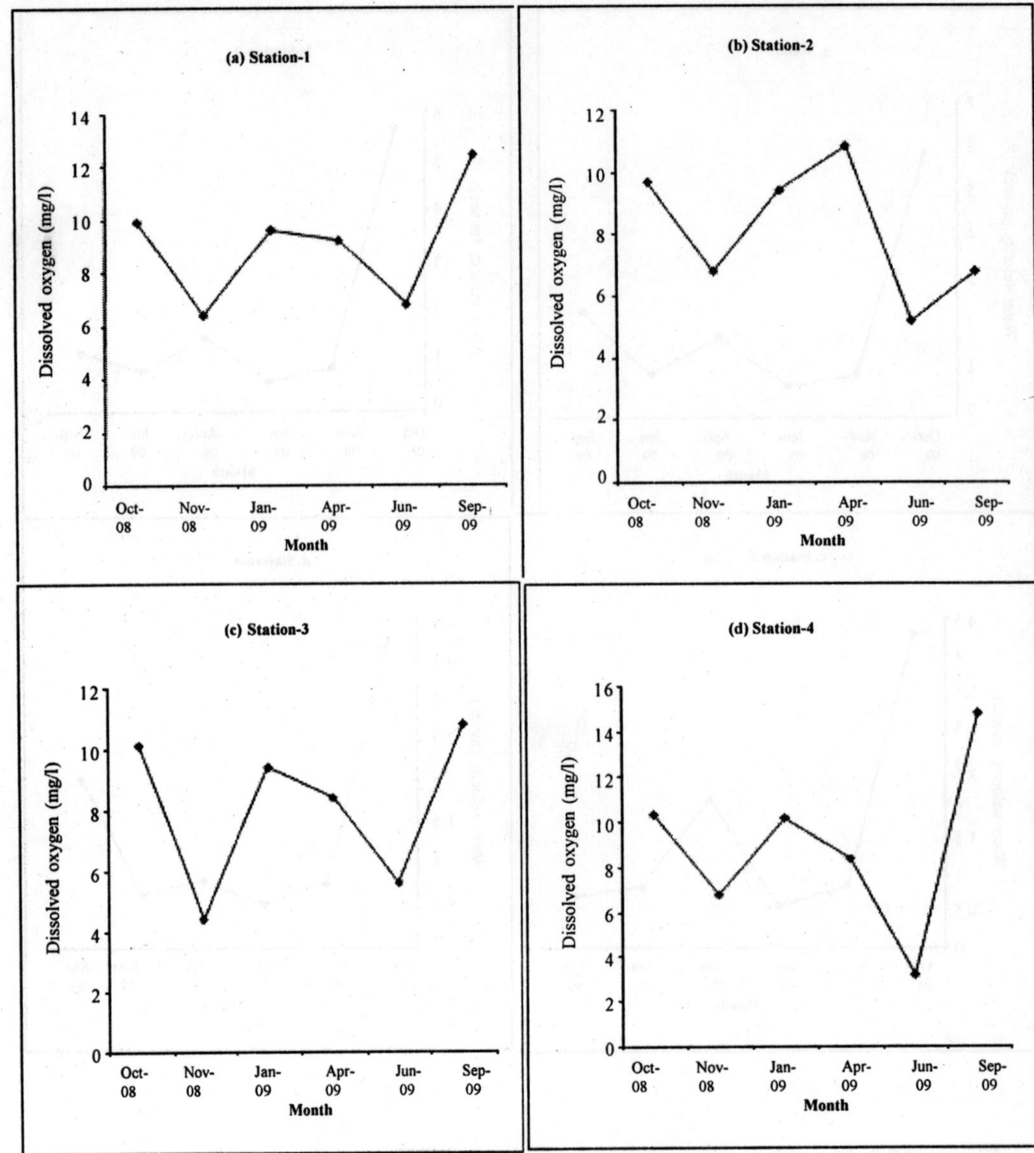

Figure 17.3. Monthly variations in Dissolved Oxygen of Deopani stream at Stations 1 to 4

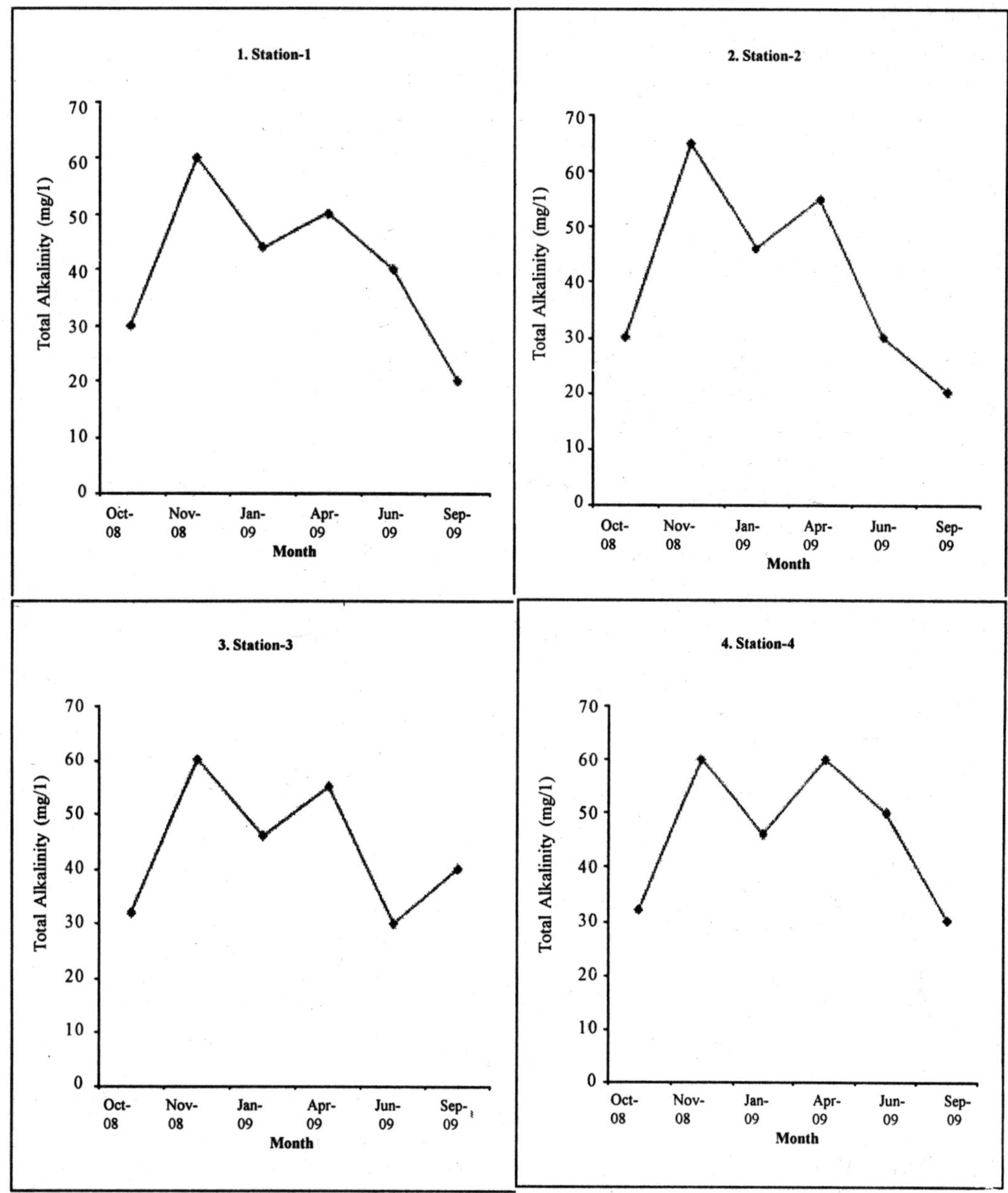

Figure 17.4. Monthly variations in Total alkalinity of Deopani stream at Stations 1 to 4

Free CO_2

Free CO_2 content did not show any distinct trend in seasonal variation in all the stations. The values ranged between 0.38 to 12 mgl^{-1}. Choudhury (2002) observed CO_2 values within the range of (0.9-3.8 mgl^{-1}) in the rivers of Arunachal Pradesh during premonsoon, retreating monsoon and winter season.

Fish Fauna

The Eastern Himalayan region encompassing the Northeast is considered as one of the hot spots of freshwater fish biodiversity in the world (Kottelat and Whitten, 1996). The mixing of drainages and their fish fauna in the geological past has rendered the Eastern Himalayas very important from faunistic point of view. The region shares its fish genetic resources with that of the Indo-Gangetic plains and to a lesser extent with the Myanmar and South Chinese fauna. So far, 267 fish species belonging to 114 genera under 38 families and 10 orders has been recorded and reported from the region. This is approximately 33.13% of total Indian freshwater fishes (Sen, 2000). During the present study, the major fish groups recorded were Catfishes (*Glyptothorax horai*), Mahseers (*Tor putitora, Neolissocheilus hexagonolepis*) and Barils (*Barilius bendelensis, B. tileo*). Among all these fish species, *Glyptothorax horai* were to found to be the most common species recorded in all the seasons. The rocky substratum of the river favours the occurrence of this species.

It may be mentioned here that Nath and Dey (2000) have reported the occurrence of 132 fish species belonging to 10 orders and 27 families. Although the ichthyofaunal diversity of Deopani river is less however, the river harbors very important fishes like Catfishes, Mahseer and Barils and considered an important fishery for the local people for their livelihood. It is therefore, necessary to protect the pristine water resources of Deopani river and its important faunistic diversity.

[**Acknowledgement:** The authors are thankful to Dr. P.C. Mahanta, Director, Directorate of Coldwater Fisheries Research, ICAR, Bhimtal for providing financial assistance and Dr. Narendra Kumar, Programme Coordinator, Krishi Vigyan Kendra, Tirap district, Arunachal Pradesh for providing necessary facilities to carry out the study.]

REFERENCES

APHA, 1989. *Standard methods for the examination of water and wastewater.* American Public Health Association Washigngton D.C., 1134pp.

Choudhury, M. 2002. Ecological status and fishery potential of six northeastern states Arunachal Pradesh, Manipur, Meghalaya, Mizoram, Nagaland and Tripura. Final report NERC, CIFRI, Guwahati.

Daimari, P., Choudhury, M. and Dutta, A. 2005. Ecology and Fishery of River Subansiri (Arunachal Pradesh). *Environment & Ecology,* 23 (1): 49-54.

Gurumayum, S.D., Daimari, P., Goswami, B.S., Sarkar, A. and Choudhury, M. 2000. Physico-chemical qualities of water and plankton of selected rivers in Meghalaya. *J. Inland Fish. Soc. India,* 34(2): 50-54.

Nath, D. 2001. Water and soil characteristics of the Narmada estuary before commissioning of Sardar Sarobar dam. *J. Inland Fish Soc. India,* 2: 64-74.

Nath, P. and Dey, S.C. 2000. *Fish and fisheries of North Eastern India (Arunachal Pradesh).* Narendra Publishing House, Delhi. 217pp.

Pathak, V., Sarkar, A., Mahavar, L.R. and Bhattacharjya, B.K. 2001a. Ecological status and Fish production potential of Siang, Dibang and Lohit- the three forerunners of river Brahmaputra. *J. Inland Fish. Soc. India,* 33(2): 23-28.

Pathak, V., Mahavar, L.R. and Sarkar, A. 2001b. Ecological Status and production dynamics of sretch of river Mahanadi. *J. Inland Fish. Soc. India,* 33(1): 25-31.

Pathak, V., Choudhury, M., Laal, A.K., Bhattacharjee, B.K., Sarkar, A. and Mahavar, L.R. 2000. Ecology and production dynamics of river Brahmaputra with special emphasis on its tributaries. *Central Inland Fisheries Research Institute, Barrackpore, India,* Bull. No. 97.

Kottelat, M. and Whitten, T. 1996. Freshwater biodiversity in Asia with special reference to fish. World Bank Tech. Paper No. 343. The World Bank, Washington, D.C.

Laal, A.K., Sarkar, S.K. and Sarkar, A. 1986. Ecology and fisheries of river Ganga at Bhagalpur (Bihar). *Procc. Natt. Symp. Fish and Fisheries Env.* : 51-55.

Sen, N, 2000. Occurrence, distribution and status of diversified fish fauna of North East India, pp 31-48. In *Fish Biodiversity of North East India* (ed. A.G. Ponniah and U.K. Sarkar), NATP Publ.2, NBFGR, Lucknow, India.

Unni, K.S. 2003. *Ecology of Indian Rivers.* International Book Distributors, Dehra Dun-248001, 312pp.

18

Pattern of Microdistribution and Physical Habitat Preferences of *Neolissochilus hexagonolepis* (McClelland) in the Rivers in Arunachal Pradesh

—Boni Amin Laskar[1] & Debangshu Narayan Das[2]*

[Several investigators have conducted the studies on fish diversity in various drainage systems of Arunachal Pradesh, India. The present study was intended precisely to inspect the distribution of one of the important riverine fish species *Neolissochilus hexagonolepis* and to map the habitat features of the fish only within drainages of Lower Dibang Valley District of the state. Besides, the assessment of species composition of various other fishes in those habitats and the contribution of *Neolissochilus hexagonolepis* in the catches were also evaluated during the study. The study revealed that *N. hexagonolepis* is abundantly available in the study area, and therefore it readily dispels the notion of present vulnerability of the species. The study further revealed that the fish occupies various types of microhabitats in lotics and the physical features of such microhabitats were found distinctly diverse in respect to their size groups. A total of 65 fish species were identified which included commercially important fish such as *Neolissochilus hexagonolepis, Tor putitora, Schizothorax* sp., *Labeo dyocheilus, Labeo* sp., *Semiplotus semiplotus, Barilius bendelisis, B. vagra, B. barna, Garra annandalei, G. gotyla, Amblyceps* sp. The collection also included ornamental fishes like *Aborichthys* sp., *Amblyceps* sp., *Badis* sp., *Chanda* sp., *Channa* sp., *Colisa* sp., *Danio* sp., *Garra* sp., *Hara* sp., *Olyra* sp., *Parambassis* sp., *Puntius* sp., etc.,. *Neolissochilus hexagonolepis* was found in both the summer and winter seasons in almost all the rivers and streams having the altitude ranging from 130 m to 460m asl within the study area.

Key words: *Neolissochilus hexagonolepis,* distribution, lotic system, habitat characters, fish diversity.]

Introduction

Fishes occur in diverse habitats like lakes, streams, estuaries and oceans

[1,2] Department of Zoology, Rajiv Gandhi University, Rono Hills, Doimukh 791112, Arunachal Pradesh
*Corresponding author's Email: aminboni@rediffmail.com, boniamin.laskar@gmail.com

throughout the world. However, their distribution pattern within their habitats also varies in spatial and temporal scales from global to local levels and is governed by different limiting factors within a particular habitat. At global level some species have small ranges of distribution and few have large ranges within a continent. Species that have wide geographic ranges of distribution is the most common species and such widespread or generalist species are often adapted to use broad ranges of variations in food and habitat resources in and around their inhabitating ecosystem. On the contrary, specialist species often have narrow geographic ranges of capability for resource use limited by their strict preferences either on food items or on physical shelter (Krebs, 2008). Such a specialist species mostly show either endemicity in distribution pattern in certain geographical area or restricted availability in particular segment of any habitat. Several investigators (Hora and Mukerji, 1935; Hora, 1937; Ghosh and Lipton, 1982; Sen, 1985; Sen, 2000; Kottelat, 1990; Talwar and Jhingran, 1991; Nath and Dey, 1997; 2000; Jayaram 1999; Menon, 1999; Ng and Rainboth 2001; Ng, 2005; 2006; Tamang *et al.* 2006; 2007; Vishwanath and Darshan, 2007; Nebeshwar *et al.* 2007) have conducted the studies on fish diversity in various drainage systems of Arunachal Pradesh, India in different times. The studies were mostly concentrated on systematic enumeration of the collected fishes from various locations aiming towards reporting of available fish species, identification and naming of new candidates from drainage systems of the state. Nath and Dey (2000) had made quite an extensive documentation of fishes of Arunachal Pradesh with description of two new species. Recently, Bagra *et al.* (2009) have updated and compiled a checklist of ichthyofaunal diversity covering their exploration activities in many rivers and streams along with all the acceptable previous reports on fish diversity of the state. However, none the previous studies contributed hardly any knowledge either on pattern of micro distribution any species or habitat preferences of any of the commonly available fish species in different lotic system of the state. On back ground of conservation of any fish species *in-situ*, the prior knowledge on species specific micro distribution pattern and habitat preference within the ecosystem is inevitable prerequisite for proper strategy development. Therefore, present study was intended to explore the distribution of *Neolissochilus hexagonolepis* commonly known as Chocolate mahseer in the Lower Dibang Valley District of the state. The survey of literature found that only Dibang river has so far been studied for the fish diversity in the district and it is only Nath and Dey (2000) who have reported the occurrence of 43 number of fish species from the Dibang river. There are many rivers and streams flowing through the district. But under this present study attempts have been made to explore the left bank tributaries of Dibang river; located in the two administrative blocks of the district. The study also encompasses (i) the assessment of various fish species composition in the lotic systems, (ii) the contribution of *Neolissochilus hexagonolepis* in the catch, and most importantly (iii) the physical habitat mapping of *N. hexagonolepis* in the lotic systems in relation to its distribution pattern.

Study area

Physiography of the area: The Lower Dibang Valley district with an area of

3900 sq. km and total population of 50,438 (as per Census 2001) forms a part of eastern Arunachal Pradesh. The district derives its name from the mighty Dibang river, which flows through the heart of district and finally joins into the river Brahmaputra. The district is bounded on the North by the Dibang Valley district, on the East the Lohit district and McMahon Line, on the West the East Siang and Upper Siang districts of Arunachal Pradesh and on the South by the Tinsukia district of Assam. Major portions of the district are the part of the Siwalik and the foothills, the southern part especially the Paglam, Kornu and Roing circles have some flat land while the other three circles viz. Hunli, Desali and Dambuk are completely hilly. The topography of the district is extremely variable; mountains are the most imposing features with the highest Snow-capped peak of Mayudia at a height of 2655 meters from mean sea level. Slopes of the hills are very steep. The district is well known for its largest cover of thick green forest with almost 80% of the area in the form of reserved forest, wildlife sanctuaries or unclassified state forests. The hills are traversed by a large number of rivers and rivulets, to name some of the big rivers are the Dibang, Ithun, Ason, Chipu, Aba, Epi, Sisiri, Datung, Deopani and Kundil. The altitude in the Lower Dibang Valley ranges between 110m to 2700m.

Climate: Three very important meteorological data namely, air temperature, humidity and rainfall have been recorded from the reliable source, for six consecutive years from 2003 to 2008. The data have been analysed taking the average during a month in all the six years. However, the total rainfall for a month in every year and the maximum and minimum temperature in a day during all the years also analysed. The seasonal variations in temperature, humidity and rainfall are shown in Figure 18.1.

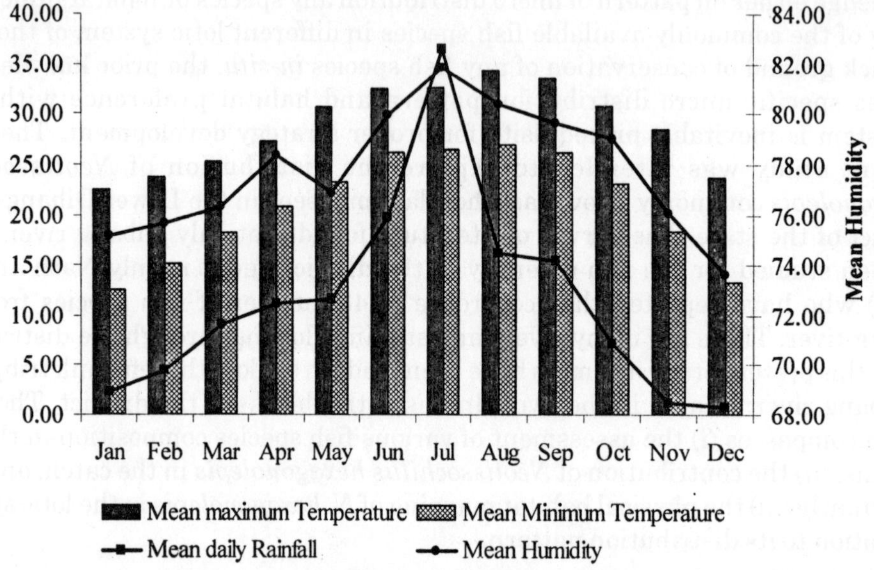

Figure 18.1. Seasonal variations in maximum temperature, minimum temperature, daily rainfall and humidity (2003-2008)

Methods

Selection of sampling site

One of the major constraints usually faced in selecting the study sites in a mountain stream is its inaccessibility up to a desired point due to hilly terrain and even if some stretches are accessible during spring and summers, it may prove to be an uphill task during wet seasons. The toposheets published by the survey of India was used to obtain information on the origin and topography of the rivers, tributaries, catchments area, etc. The segment of river which represents characteristics of a particular riverine zone was considered ideal for sampling. During site selection, due attention was paid to riffles, pools, rapids, sandy bottom, etc which facilitates easy and exclusive sampling.

Fish sampling, preservation and identification

Eight streams/ rivers were examined in the water-course of the study area. The fish sampling periods have been categorized in to two seasons depending on the frequency of rain. It has been revealed from the meteorological records of past six years (Figure 18.1) that the period from October to January is most dry season, while the period from April to July is the heavily raining season. Hence the sampling was conducted in the winter season (October to January) and summer season (April to July).

Fish specimens were collected from river sources through experimental fishing with cast net, gill nets and electro-fishing. The size of the cast net was 5 meter diameter, 2.25 meter length and 10-15 mm mesh. The sizes of gill nets were 15 to 25 m length and 35 to 60 mm mesh. The electro-fishing was done with diesel engine generator fixed on a small country boat. Cast netting was done along the river towards upstream, mostly at the riffle sites with pebble substratum and at the confluence of streams. Gill nets were fixed along the river preferably at the sites with less flow rate of water and the pools having apparently no flow. During sampling the gill nets were fixed standing at the suitable sites and cast netting continued up to certain length of the stream. While return the fishes caught in the gill nets were collected. The electro-fishing was done only when the water level in the rivers was not too high.

Fishes were preserved in 10% formalin. All the varieties of fishes caught in the nets from a site were preserved together and the specimens of Chocolate Mahseer were segregated in the laboratory and preserved separately with proper labeling of the date, time and sites of collection. All the fish specimens were identified following Jayaram (1981, 1999), Menon (1974, 1999), Talwar and Jhingran (1991), Sen (1985) and Nath and Dey (2000). Also other literatures have been consulted as and when required.

Study of Microhabitats of Neolissochilus hexagonolepis

Fishes by virtue of their existence in the water medium remain unseen from air medium. Unlike other terrestrial organisms fishes are more difficult to locate them. So the *in-situ* study of preferences of habitats for feeding, breeding and hiding of fishes is troublesome. The conventional methods of fish sampling with cast netting and other fishing gears are somewhat a trial and error method for such kinds of study. As the fishes move locally for a certain extend due to social gathering, so fish may sometime be sampled from habitat of occasional passages or corridors. Conversely during operation of fishing gears the fish occurring in a particular habitat may pass aside due to the disturbances. The present study has been carried out by diving under water with the help of self-designed snorkel musk. Macro habitat types were identified on the basis of availability of the fish and the physical study was carried out following the NBFGR manual (NBFGR, 2001), methods described by Overton *et al.* (1997) and Armontrout (1990). Many important fish habitats where the *N. hexagonolepis* was not encountered during conventional sampling were overlooked for the detail study of microhabitat.

Results and Discussion

In India the distribution of *Neolissochilus hexagonolepis* is said to have been restricted only in the North-Eastern states and Eastern Himalaya; where it has been reported from the rivers and lakes of Assam (Jayaram 1982), Arunachal Pradesh, Manipur, Meghalaya (Yazdani, 1977), Mizoram (Laskar *et al.*, 2002), Nagaland, Sikkim and North Bengal (Lepcha *et al.*, 2004). Kar and Sen (2007) reported the distribution of the species in the rivers of Mizoram, Tripura and Barak drainage of Northeastern India, at altitude ranging from 14.45 m MSL to 464 m MSL. The name of the rivers (altitudes given in parenthesis) in Mizoram where the species reported to has been occurred are river Tuirial (161.5 m and 338 m), Kolodyne (158.84 m), Karnafuli (51.2 m), Mat (464 m), Tlawng (74.08 m), Tuirini (266.46 m) and river Serlui (34.14 m). In the state Tripura the Monu river (68.29 m), Khowai (54.2 m), Gomati (14.45 m) and Feni river (39.63 m); while in Barak Valley of Assam, the Barak river (31.40 m), Jatinga (26.2 m), Sonai (28.04 m) and Dhaleswari river (34.75 m) are in the list (Kar and Sen, ibid).

In the state Nagaland the species is widely distributed in the river Doyang, Tizu, Intangki, Arachu, Lanyi, Tesuru and river Likhimro (Govt. of Nagaland, 2005). In Arunachal Pradesh the species was previously recorded from Yembung division (Chaudhuri 1913) and Tirap division (Srivastava 1966). It is also found in the river Kameng, Subansiri, Dikrong, Pachin, Dibang, Ranga, Siang, Lohit, Noadihing, Buridihing and river Tirap (Nath and Dey 2000). But the literature survey revealed that the species widely distributed in the lotics in North East India. *Neolissochilus hexagonolepis* was also reported from Bangladesh, Bhutan, China: Yunnan, Indonesia, Malaysia, Myanmar, Pakistan, Thailand and Vietnam (Jayaram, 1981). The fish is abundant in most of the big rivers, lakes and reservoirs of Nepal from 250m to 1500m altitude, having preferences for water temperature of 10^0 to 30^0 C (Rai and Swar, 1989). The species was also reported from Tadi,

Trisuli, Kali Gandaki drainage at elevations up to 1440m, from the Kosi tributaries–mainly from Sun Kosi, Indravati, Tamar and Arun rivers at elevations up to 1220m, and from the river Bagmati at elevations up to 1350m. *N. hexagonolepis* is dominant fish species in Indrasarobar reservoir, where in 1990 it had contributed 77.2% of the catch (Rai, 1990). It is one of the abundant fish species in the lakes of the Pokhra valley in Nepal (Ferro, 1980, Ferro and Badagami, 1980, Shrestha, 1981). So the literature survey revealed that the species is distributed in wide range of altitudinal topography of very low to as high as 1500 m asl.

Neolissochilus hexagonolepis and Co-existing Fish Species in the Study Area

General Ichthyodiversity: A list of the fish collected from the study area along with their distribution and seasonal occurrences is furnished in Table 18.1. In the present study altogether 65 (Sixty five) fish species were identified from the study area. Menon (1962) reported 218 fish species from the Himalayas. The Eastern Himalaya drained by the river Brahmaputra has great diversity of cold water fish than the Western Himalayan drainage. The ichthyofaunal composition of the study area comprised of economically important food fishes like *Neolissochilus hexagonolepis, Tor putitora, Schizothorax* sp., *Labeo dyocheilus, Labeo* sp., *Semiplotus semiplotus, Barilius bendelisis, B.vagra, B. barna, Garra annandalei, G. gotyla, Amblyceps* sp. etc. The collection also included ornamental fishes like *Aborichthys* sp., *Amblyceps* sp., *Badis* sp., *Chanda* sp., *Channa sp., Colisa* sp., *Danio* sp., *Garra* sp., *Hara* sp., *Olyra* sp., *Parambassis* sp., *Puntius* sp., etc. However, some of the fishes were seen to have been occurring in the summer rainy season and similarly few species of the winter dry season were absent in the other season. The variation may be discussed as during summer the sampling in some of the river sites were not possible because of unapproachable terrain in the study area. In the study area, 43 species of fishes were recorded from river Dibang by Nath and Dey (2000); in which the fish resources in the tributaries of river Dibang probably not have been considered. In this circumstance the present study gives an enlisted document of the resource wealth of the freshwater fish species in the study area. However, there is only a meager contribution of economically viable food fishes harbouring in the natural water bodies in the study area. Also there is neither any organized fishing in the region nor any specific fisher community. The bulk of the resources is under unorganized and unregulated capture fishery and hence these fish resources will be facing threat from indiscriminate and unauthorized destructive fishing.

The relative contribution of *Neolissochilus hexagonolepis* and other fish species is compared among seasons and site and presented in Table 18.2. The population dynamics of the species is compared between sample from lentic and lotic habitats, and presented in Figure 18.2.

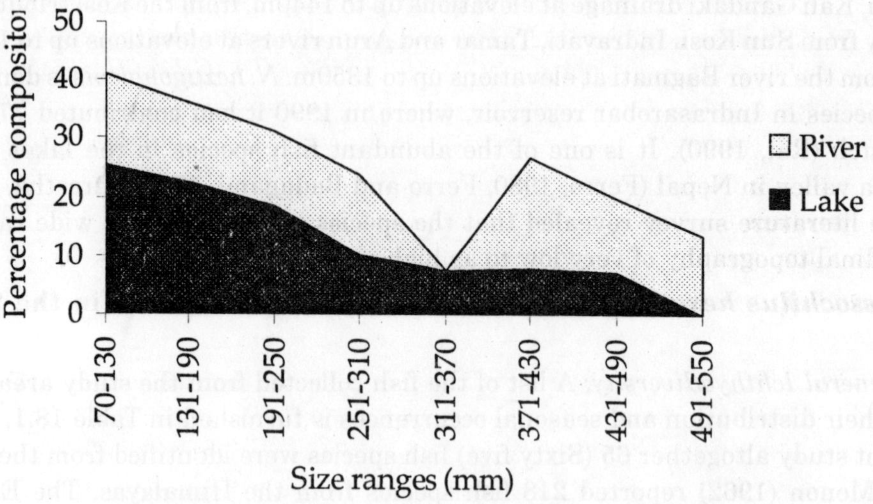

Figure 18.2. Relative catch composition (in percentage) among different size ranges of *Neolissochilus hexagonolepis* in different ecosystems in a year.

In the lotic habitats the species was seen to has been colonized in almost all the studied rivers and streams located between the altitude of 130m and 460m asl. The study of relative composition of the species among its various size groups from the lotic habitats is carried out from the data recorded from market survey and experimental fishing in both rivers and lakes. The fishes appearing in the market are virtually afforded from various sites and composite of different fishing gears. Also the present study in lotics is not sufficient to justify the nature of the population structure. Yet the population in rivers/streams do comprise of early fries to adults and brooders, hence the population can be said expanding in rivers too. The fry of the species in the size range of 19 – 55 mm had been collected during the month of early November, when the water levels in the rivers are at low. Density of fry collected from the various streams ranged between 25 – 175 fry/m^2.

This study revealed that the species is abundantly available in the study area in both lakes and rivers. An additional effort was made to explore the population status of the species in a lake called Sally Lake in the study area; where it has been found that the species is thriving well with the other exotics and Indian major carps. The lake provides habitat congenial for the species to spawn naturally which was revealed from the observation of occurrence of early fries of the species in the lake. Experimental fishing with the standard cast net conducted in the Sally lake revealed that higher relative composition of the species (25.55%) is contributed by fishes in the size range of 70-130 mm, followed by 22.63% in 131-190 mm and 18.61% in the size range of 191-250 mm. This indicates that the population of Chocolate mahseer in the lake is established with greater numbers in the young group. However, due to regular harvesting and poaching, bigger sizes of the species are not available and it is assumed that the population will be declining in the future. The lake has potential for quality seed and capacity to sustain mahseer

Table 18.1
Distribution of various Ichthyospecies in the study area

Sl. No.	*Fish species*	*DEO*	*EPI*	*DIB*	*KUN*	*CHI*	*CHA*	*ZIN*	*KAL*
1	*Aborichthys elongatus*								
2	*Aborichthys kempi*								
3	*Acanthocobitis* (unknown species)								
4	*Acanthocobitis botia*								
5	*Amblyceps (unknown species)*								
6	*Anguila bengalensis*								
7	*Aspidoparia* (unknown species)								
8	*Aspidoparia mora*								
9	*Badis assamensis*								
10	*Badis* (unkonwn species)								
11	*Bagarius bagarius*								
12	*Barilius* (unknown species)								
13	*Barilius* (unknown species)								
14	*Barilius barna*								
15	*Barilius bendelisis*								
16	*Barilius vagra*								
17	*Chaca chaca*								
18	*Chagunius chagunio*								
19	*Chanda nama*								
20	*Channa* (unknown species1)								
21	*Channa* (unknown species2)								
22	*Channa orientalis*								
23	*Channa punctata*								
24	*Chrossocheilus* (unknown species)								
25	*Colisa fasciata*								
26	*Danio* (unknown species)								
27	*Danio dangila*								
28	*Devario aequipinnatus*								
29	*Esomus danricus*								
30	*Garra* (unknown species)								
31	*Garra annandalei*								
32	*Garra gotyla*								
33	*Glyptothorax cavia*								
34	*Glyptothorax* (unknown species)								
35	*Hara* (unknown species)								
36	*Labeo* (unknown species)								
37	*Labeo dyocheilus*								
38	*Lepidocephalicthys guntea*								
39	*Macrognathus* (unknown species)								

(Contd...)

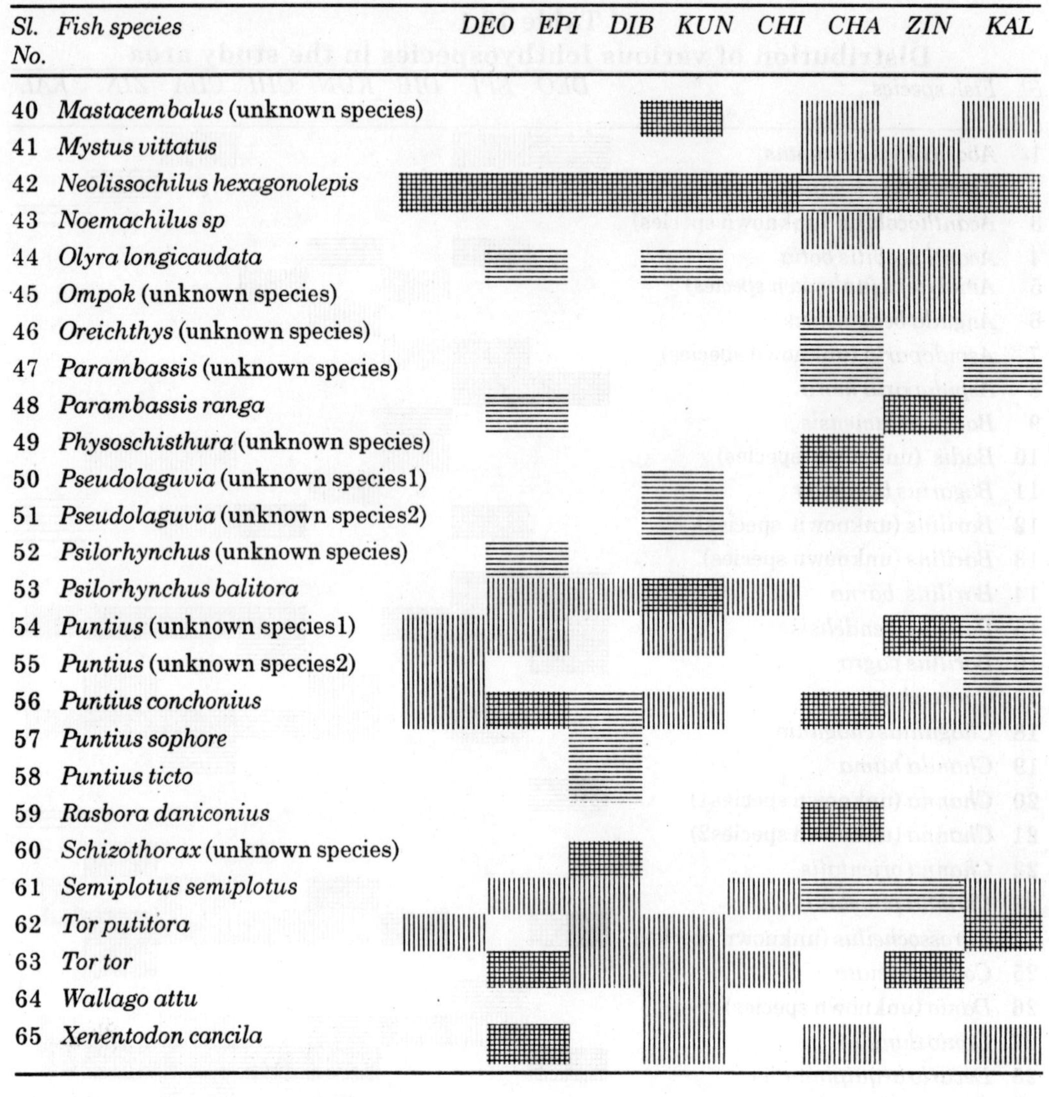

Sl. No.	Fish species	DEO	EPI	DIB	KUN	CHI	CHA	ZIN	KAL
40	*Mastacembalus* (unknown species)				Both		Summer		Summer
41	*Mystus vittatus*						Summer	Summer	
42	*Neolissochilus hexagonolepis*	Both	Both	Both	Both	Both	Winter	Both	Both
43	*Noemachilus sp*						Summer		
44	*Olyra longicaudata*		Winter		Winter			Summer	
45	*Ompok* (unknown species)						Summer	Summer	
46	*Oreichthys* (unknown species)						Winter		
47	*Parambassis* (unknown species)						Winter		Winter
48	*Parambassis ranga*		Winter					Both	
49	*Physoschisthura* (unknown species)						Both		
50	*Pseudolaguvia* (unknown species1)				Winter		Both		
51	*Pseudolaguvia* (unknown species2)				Winter				
52	*Psilorhynchus* (unknown species)		Winter						
53	*Psilorhynchus balitora*		Summer	Summer	Both	Summer			
54	*Puntius* (unknown species1)	Summer	Summer		Summer			Both	Winter
55	*Puntius* (unknown species2)	Summer							Winter
56	*Puntius conchonius*	Summer	Both	Winter	Summer		Both	Summer	Summer
57	*Puntius sophore*			Winter					
58	*Puntius ticto*			Winter					
59	*Rasbora daniconius*						Both		
60	*Schizothorax* (unknown species)			Both					
61	*Semiplotus semiplotus*		Summer	Summer		Summer	Winter	Winter	Summer
62	*Tor putitora*	Summer		Summer	Summer				Both
63	*Tor tor*		Both	Summer	Summer	Summer		Both	
64	*Wallago attu*				Summer				
65	*Xenentodon cancila*		Both		Summer		Summer		Summer

Present during winter season

Present during summer season

Present during both winter and summer seasons

The abbreviations used are: DEO= River Deopani, EPI= River Epipani, DIB= River Dibang, KUN= River Kundil, CHI= Chipu nalah, CHA= Chaya nalah, ZIN= Zingu nalah and KAL= kaliya nalah.

Table 18.2

Fish species composition (%) in the study area during winter season and summer season.

Sl. No.	Fish species	DEO W/S	EPI W/S	DIB W/S	KUN W/S	CHI W/S	CHA W/S	MAJ W/S	KAL W/S
	1	*2*	*3*	*4*	*5*	*6*	*7*	*8*	*9*
1	*Aborichthys elongatus*	30.43/0	0/1.39				0/5.26		
2	*Aborichthys kempi*							3.18/0	
3	*Acanthocobitis* sp.								8.05/0
4	*Acanthocobitis botia*		1.32/2.08		6.36/0				
5	*Amblyceps* sp.			0/1.63		0/2.33			
6	*Anguila bengalensis*				0/2.56				
7	*Aspidoparia* sp.		0.99/4.17	0/0.81					
8	*Aspidoparia mora*	1.30/4.04	0.99/5.56						
9	*Badis assamensis*			1.27/0					
10	*Badis* sp.			0/1.63			0/1.75	0/2.35	
11	*Bagarius bagarius*				0/1.28				
12	*Barilius* sp. 1		4.95/4.86			7.83/11.63	0/10.53		17.45/0
13	*Barilius* sp. 2								13.42/0
14	*Barilius barna*	4.34/7.07	5.94/5.56						
15	*Barilius bendelisis*	10.86/10.10	24.09/8.33	18.35/9.76		31.32/17.44	15.44/7.89	25.47/5.88	0/3.95
16	*Barilius vagra*	6.08/2.53				16.87/0			
17	*Chaca chaca*				0/0.64			0/2.35	
18	*Chagunius chagunio*		2.973.477		0/10.90	12.65/0	8.08/14.04	8.91/5.88	0/22.37
19	*Chanda nama*						2.21/0		
20	*Channa* sp. 1	0.87/0							
21	*Channa* sp. 2								2.01/0
22	*Channa orientalis*							0/2.35	
23	*Channa punctata*				4.54/1.28		0/5.26		0/2.63
24	*Chrossocheilus* sp.	2.60/9.09		14.55/8.13	0/8.97	0/13.95	0/4.39	14.01/1.18	5.36/7.89
25	*Colisa fasciata*							0/8.24	
26	*Danio* sp.		2.31/0						
27	*Danio dangila*		0/6.94	4.43/5.69	10.00/0		0/1.75		0/7.89
28	*Devario aequipinnatus*	0.86/5.05			2.27/7.69	9.04/0	13.2310.53	11.46/8.24	0/5.26
29	*Esomus danricus*						2.21/0		

(Contd...)

	1	*2*	*3*	*4*	*5*	*6*	*7*	*8*	*9*
30	*Garra* sp.	20/22.73	6.60/0	28.48/4.07		9.03/11.63			2.68/0
31	*Garra annandalei*		0/13.89			0/10.47	7.35/0	8.92/0	
32	*Garra gotyla*			0/4.88					
33	*Glyptothorax cavia*	13.91/0		11.39/7.32	6.36/0				
34	*Glyptothorax* sp.			0/4.07	1.82/0	0/12.79			
35	*Hara* sp.				7.27/0		5.88/2.63		2.68/2.63
36	*Labeo* sp.		1.98/0		0/1.28			0/2.35	4.02/6.58
37	*Labeo dyocheilus*	0/1.52	0/0.69	0/3.25					0/2.63
38	*Lepidocephalicthys guntea*						1.47/0		
39	*Macrognathus* sp.		0.33/0		5.45/0			0/4.71	
40	*Mastacembalus* sp.				3.63/1.92		0/1.75		0/3.95
41	*Mystus vittatus*						0/3.51	0/3.53	
42	*Neolissochilus hexagonolepis*	8.69/11.62	7.59/9.72	3.79/4.88	20.90/16.03	13.25/9.30	23.53/0	10.19/3.53	9.39/15.79
43	*Noemachilus* sp.						0/2.63		
44	*Olyra longicaudata*		1.65/0		1.36/0			0/4.71	
45	*Ompok* sp.						0/0.88	0/7.06	
46	*Oreichthys* sp.						5.88/0		
47	*Parambassis* sp.						1.47/0		2.68/0
48	*Parambassis ranga*		1.32/0					1.91/5.88	
49	*Physoschisthura* sp.						1.47/4.39		
50	*Pseudolaguvia* sp. 1				0.91/0		1.47/2.63		
51	*Pseudolaguvia* sp. 2				3.64/0				
52	*Psilorhynchus* sp.		7.92/0						
53	*Psilorhynchus balitora*		0/2.78	0/13.01	25.45/5.77	0/2.33			
54	*Puntius* sp. 1	0/5.05	0/6.94		0/3.85			9.55/11.76	17.45/0
55	*Puntius* sp. 2	0/4.04							9.40/0
56	*Puntius conchonius*	0/16.16	22.44/14.58	7.59/0	0/26.92		4.41/8.77	0/14.12	0/3.95
57	*Puntius sophore*			0.63/0					
58	*Puntius ticto*			1.90/0					
59	*Rasbora daniconius*						1.47/8.77		
60	*Schizothorax* sp.			7.59/4.88					
61	*Semiplotus semiplotus*		0/4.17	0/8.13		0/6.98	4.41/0	3.18/0	0/6.58
62	*Tor putitora*	0/1.01		0/14.63	0/3.85				5.36/2.63
63	*Tor tor*		2.31/2.78	0/3.25	0/1.28	0/1.16		3.18/5.88	
64	*Wallago attu*				0/0.64				
65	*Xenentodon cancila*		4.29/2.08		0/5.13		0/2.63		0/5.26

W=winter dry season (October to January), S=summer/rainy season

population, hence the study suggests for the imposition of regulation to sustain ably manage the resource of valued sport fish in the lake. The study recorded a maximum size of 460mm length of *N. hexagonolepis* from Sally lake and 550mm (1720gm) from the riverine ecosystem. This study may readily dispels the notion of present vulnerability of the species. However, threats are many and will be increasing due to natural and anthropogenic distortions.

Microhabitats of *Neolissochilus hexagonolepis*

The capture of juveniles in peripheral shallow pool habitat along the stream confirms that this habitat is being used as rearing habitat. The presence of eggs or fry in a given habitat type verifies that it provides spawning habitat (Figure 18.3). Collection sites of seeds of *Neolissochilus hexagonolepis* in different streams/lakes in the district were identified. Deopani river (Lat. 28^0 09′ 34″ N; Long. 95^0 50′ 33.6″E; Alt. 421 m asl) and Epipani river (Lat. $28^0$10′ 21″ N; Long. $95^0$49′ 58″ E; Alt. 388m asl) have been identified as the potential sites for seeds of *Neolissochilus hexagonolepis*. These sites can be exploited for restoration of the stocks of *Neolissochilus hexagonolepis* in depleted waters. The fry were found to be inhabited in the braids of rivers, and small pockets formed amongst boulders and stones on the banks having no flow to slow flow of water of spring nature. The seeds of other species of genus *Barilius, Danio, Garra, Puntius* etc. were also seen to have been coexisting with *N. hexagonolepis.*

Table 18.3
Structural variables of habitats inhabited by the fry of *Neolissochilus hexagonolepis*

	Habitat structures			
Parameters	*1. Underneath big boulder*	*2. Small pockets formed by cobbles at rivers side*	*3. Remaining of water and disconnected from main river*	*4.Reservoir at hydro-electric power generation station*
Depth (cm)	10-25	5-10	5-10	10-150
Flow	No flow	Moving ridge of water	No flow	Moderate
Substratum	Sand & gravels	Gravels	Gravels	Cobbles, gravels & occasional boulders
Approximate numbers of seeds	20-30	5-10	10-20	More than 1000
Number of seeds collected per effort	5-20	5-8	10-20	300-500

Menon (1954) related the distribution of Himalayan fish to morphological characteristics which enable them to inhabit the torrential streams. He recognized that *Tor* species dwell in shallow, clear cold water in the foot hills without any striking modification to current. The need for shelter from the current has lead to territoriality. Mahseer and Schizothoracines chase intruders to defend the limited food sources and available shelter. Mahseers have wider tolerance and even survive at water temperature over 25°C. During winter when the water level remains at its lowest and water is highly transparent, all size groups of Mahseer and Schizothoracines are present in pools of rivers (Sehgal, 1988).

The study revealed various microhabitats inhabited by *N. haxagonolepis*. The physical features of such microhabitats are found different with respect to size group of fish. All such physical characters of each microhabitat are described below along with the numbers of fish caught through different fishing methods.

1. *Pool habitat:-* A portion of the stream with reduced current velocity, often with water deeper than the surrounding areas. Two types viz. Scour pool and Dammed pool were found inhabiting by the fish.

A) *Scour pool 1:-* Pool formed by scouring action when flowing water infringes against and is diverted by a stream bank or channel obstruction (**rooted & woody debris**). Following three kinds of micro habitats were identified to be inhabited by the species.

i) *1st microhabitat* is underneath the network of overhanging roots of riparian trees

Fish size:- only fingerlings.

Fishing method:- no standard fishing methods except underwater arrow shooting is possible.

Fish behaviour:- stand stationary by slowly undulating the fins for grasping the food (debris) coming along with the water.

ii) *2nd microhabitat* is underneath the dense network of overhanging shrubs

Fish size:- only adults.

Fishing method:- no standard fishing methods except underwater arrow shooting is possible.

Fish behaviour:- stand stationary by slowly undulating the fins for grasping the food (debris) coming along with the water.

iii) *3rd microhabitat* is at the bottom of deeper pools

Fish size:- only adults.

Fishing method:- only cast netting and underwater arrow shooting are possible.

Fish behaviour:- move slowly from point to point and scrap bottom for food.

B) *Scour pool 2:-* Pool formed by scouring action when flowing water infringes against and is diverted by a stream bank or channel obstruction (**boulder and bedrock**). Microhabitat found is only one at about 75% to the bottom at the deeper portion.

Fish size:- only adults.

Fishing method:- no standard fishing methods except cast netting and electro-fishing are possible.

Fish behaviour:- stand stationary by slowly undulating the fins and scrap the boulder and cobble substratum for food.

C) *Dammed pool:* - Pool formed by down stream damming action (water reservoir made along the side of the main river for the purpose of hydro electric power generation. Three kinds of microhabitat were found in a dammed pool to be inhabited by the fish.

i) *1st microhabitat* is at the periphery of the pool at the shallower water.

Fish size:- only spawns and fries.

Fishing method:- only drag netting is possible.

Fish behaviour:- not specific

ii) *2nd microhabitat* is at about 75% to the bottom at the deeper portion.

Fish size:- only adults.

Fishing method:- no standard fishing methods except underwater arrow shooting is possible.

Fish behaviour:- stand stationary by slowly undulating the fins and scrap the boulder and cobble substratum for food.

iii) *3rd microhabitat* is in between the above two

Fish size:- only fingerlings.

Fishing method:- only drag netting is possible.

Fish behaviour:- if a small stone is thrown in to the water at about 1 to 2 m from the periphery the fingerlings appear near the surface and congregate at the point where stone was thrown. This suggests that the fingerlings are attracted thinking that a fruit is dropped or some kind of food is dropped in to the water.

2. *Run habitat:*- A habitat type that is deep and fast with a defined thalweg (the line followed by majority of the stream flow) and little surface agitation.

i) *1st microhabitat* is underneath or behind the submerged single large wood debris which is a relatively stable woody material located within the bankfull channel and appearing to influence the bankfull flows.

Fish size:- adults and brooders.

Fishing method:- no standard fishing methods except electro-fishing with boat is possible.

Fish behaviour:- not observed

ii) *2nd microhabitat* is at the river side and underneath a dense network of riparian vegetation of *Sacrum spontianum* and *Alpinia galanga* (overhanging and emergent).

Fish size:- only adults.

Fishing method:- no standard fishing methods except electro-fishing with boat is possible.

Fish behaviour:- not observed

iii) *3rd microhabitat* is behind the submerged aggregates woody debris and underneath the root masses attached to trunk less than 3m and at least 0.1m in diameter, sometimes the root mass is visible mostly at the main flow of the channel

Fish size:- adults and brooders.

Fishing method:- only electro-fishing is possible.

Fish behaviour:- not observed

3. *Riffle habitat:-* Shallow rapids where the water flows swiftly over completely or partially submerged obstruction to produce surface agitation.

i) *1st microhabitat:* is along the lateral flow of the riffles

Fish size:- fingerlings and young.

Fishing method:- only cast netting is possible.

Fish behaviour:- running against the flow from column to surface and moving very fast in all direction.

ii) *2nd microhabitat* is confluence of small stream and main river

Fish size:- fingerlings and very rarely adults.

Fishing method:- only cast netting is possible.

Fish behaviour:- not observed.

REFERENCES

Bhaumik, U., Das, A. K. and Paria, T. 2006. Impact of environmental hazards in decreasing fish productions of Indian reservoirs. Pp. 273-293. In *Advances in fish research* Vol-IV, (eds. J. J. Munshi Dutta and H. R. Singh), Narendra Publishing House, Delhi.

Chaudhary, K. K. 2002. Fish farming and its prospects in the hilly areas of Arunachal Pradesh. Pp. 269-274. In K. K. Vass and H. S. Raina (eds.), Highland fisheries and aquatic resource management. India. National Research Centre on Coldwater Fisheries (ICAR), Bhimtal.

Chaudhuri, B. L. 1913. Zoological results of the Abor expedition (1911-1912). *Records of Indian Museum* 8: 243-258.

Choudhury, S. and Sen, N. 1977. On a collection of fish from Arunachal Pradesh with some new records. *Newsletter Zoological Survey of India* 3(4): 217-223.

Conway, K. W. and Kottelat, M. 2007. A new species of *Psilorhynchus* (Teleostei: Psilorhynchidae) from the Ataran River Basin, Myanmar, with comments on the generic name *Psilorhynchoides.* Zootaxa 1663: 47–57.

Dutta. A. K. and Sen, T. K. 1977. *Schizopygopsis stoliczkae* Steindachner- First record from Arunachal Pradesh, India, with observation on the extension on the geographical range. *Newsletter Zoological Survey of India* 3(4): 143-144.

Eschmeyer, B and Fricke, R. 2008. Catalog of Fishes, Online Version. Electronic Database accessible at http://research.calacademy.org/research/Ichthyology/catalog/fishcatsearch.html. California Academy of Sciences, San Francisco, USA. Captured on 10 December 2008.

Ghosh, S. K. and Lipton, A. P. 1982. Ichthyofauna of N.E.H. region with special reference to their economic importance. ICAR Special Bulletin (1): 119-126.

Hora, S. L. 1921. On some new record and rare species of fish from the Eastern Himalayas. *Records of Indian Museum* 22(5): 731-744.

Hora, S. L. 1937. Notes on fishes in the Indian Museum. XXXV. A further note on Hamilton's *Cyprinus (Garra) lamta. Records of Indian Museum* 29: 344-348.

Hora, S. L. and Mukerji, D. D. 1935. Fish of the Naga Hills, Assam. *Records of Indian Museum* 37: 381-404.

Jayaram, K. C. 1963. A new species of sisorid from the Kameng frontier division (NEFA). *Journal of Zoological Society of India* 15(1): 85-87.

Jayaram, K. C. 1999. *The freshwater fishes of the Indian region.* Delhi. Narendra Publishing House. 551 p.

Jayaram, K. C. and Mazumdar, N. 1964. On a collection of fish from Kameng frontier division, NEFA. *Journal of the Bombay Natural History Society* 61(2): 264-280.

Jhingran, V. G. and Sehgal, K. L. 1978. *Cold water fisheries of India.* Barrackpore India. Inland Fishery Society. 239 p.

Kottelat, M. 1990. *Indochinese nemacheilines. A revision of nemacheiline loaches (Pisces: Cypriniformes) of Thailand, Burma, Laos, Cambodia and southern Viet Nam.* München. Verlag Dr. Friedrich Pfeil. 262 p.

Kottelat, M. 2001. *Fishes of Laos. Cambodia.* Wildlife Heritage Trust Publication. 198 p.

McClealland, J. 1839. Indian Cyprinidae. Calcutta. *Asiatic Researches,* 19(2): 217-417.

Menon, A. G. K. 1964. Monograph of the cyprinid fishes of the genus *Garra,* Hamilton. *Memoirs of the Indian Museum,* 14(4): 173-260.

Menon, A. G. K. 1999. Check list - fresh water fishes of India. *Records of Zoological Survey of India,* Miscellaneous Publication, Occasional Paper No. 175: 1-366.

Nath, P. and Dey, S. C. 1997. *Fish and fisheries of North Eastern India Vol-I. Arunachal Pradesh.* New Delhi. Narendra Publishing House. 201 p.

Nath, P. and Dey, S. C. 2000. *Fish and fisheries of North Eastern India (Arunachal Pradesh).* New Delhi. Narendra Publishing House. 217 p.

Nebeshwar, K., Bagra, K. and Das, D. N. 2007. A new species of the Cyprinoid genus *Psilorhynchoides* Yazdani *et al.* from Arunachal Pradesh, India (Cypriniformes: Psilorhynchidae). *Zoos' Print Journal,* 22(3): 2632-2636.

Ng, H. H. 2005. *Glyptothorax botius* (Hamilton, 1822), a valid species of catfish (Teleostei: Sisoridae) from northeast India, with notes on the identity of *G. telchitta* (Hamilton, 1822). *Zootaxa,* 930: 1-19.

Ng, H. H. 2006. The identity of *Batasio tengana* (Hamilton, 1822), with the description of two new species of *Batasio* from north-eastern India (Teleostei: Bagridae). *Journal of Fish Biology,* 68: 101-118.

Ng, H. H. and Rainboth, W. J. 2001. A review of the sisorid catfish genus Oreoglanis (Siluriformes: Sisoridae) with descriptions of four new species. *Occasional Papers of the Museum of Zoology,* The University of Michigan 732: 1-32.

Sen, N. 2000. Occurrence, distribution and status of diversified fish fauna of North East India. Pp. 31-48. In *Fish Biodiversity of North East India,* (eds. A. G. Ponniah and U. K. Sarkar),. NBFGR. NATP Publication 2.

Sen, T. K. 1985. The fish fauna of Assam and the neighbouring Northeastern States of India. *Records of Zoological Survey of India* Occasional paper No. 64: 1-216.

Talwar, P. K. and Jhingran, A. G. 1991. *Inland fishes of the India and adjacent countries.* Oxford and IBH Publishing Co., New Delhi, 541 pp.

Tamang, L., Chaudhry, S. and Choudhury, D. 2006. On a new record of freshwater fish, *Pseudolaguvia shawi* (Hora) from Arunachal Pradesh, India (Teleostomi: Erethistidae). *Zoos' Print Journal,* 21(11): 2443-2446.

Tamang, L., Chaudhry, S. and Choudhury, D. 2007. Icthyofaunal Contribution to the state and comparison of habitat contiguity on taxonomic-diversity in senkhi stream, Arunachal Pradesh, India. *Journal of the Bombay Natural History Society,* 104(2): 172-179.

Vass, K. K. 2002. Trout farming concept and its potential in India. Pp. 223-241. In *Highland fisheries and aquatic resource management, India* (eds. K. K. Vass and H. S. Raina), National Research Centre on Coldwater Fisheries (ICAR), Bhimtal.

Vass, K. K. and Gopakumar, K. 2002. Coldwater fisheries and research status in India. Pp. 3-29. In *Highland fisheries and aquatic resource management, India* (eds. K. K. Vass and H. S. Raina), National Research Centre on Coldwater Fisheries (ICAR), Bhimtal.

Vishwanath, W. and Darshan, A. 2007. Two new catfish species of the genus *Pseudecheneis* Blyth (Teleostei: Siluriformes) from Northeastern India. *Zoos' Print Journal,* 22(3): 2627-2631.

Vishwanath, W. and Linthoingambi, I. 2007. Redescription of catfishes *Amblyceps arunachalensis* Nath and Dey and *Amblyceps apangi* Nath and Dey (Teleostei: Amblycipitidae). *Zoos' Print Journal,* 22(4): 2662-2664.

Yazdani, G. M., Singh, D. F. and Rao, M. B. 1990. *Psilorhynchoides.* A new genus for the cyprinid fish, *Psilorhynchoides homoloptera* Hora and Mukerji and *P. pseudecheneis* Menon and Dutta, with a definition of the subfamily Psilorhyninae (Cyprinidae). *Matsya,* 15-16: 1-7.

19

Percid Fishes (Teleostomi: Perciformes) of Manipur

—*L. Arunkumar**

[17 species belonging to 10 genera, 5 subfamilies of percid fishes and their distribution in the three river systems in Manipur are reported in this paper. Barak basin of Manipur has the greatest number of percid fishes in Manipur. *Channa stewartii* is the distinctive fishes of Barak basin of Manipur.

Key words: Diversity, Perciformes, Manipur, India.]

Introduction

Manipur lies in one of the hottest biodiversities of the world. The study of the fish fauna of this region is of much importance considering its high degree of biodiversity. There are few useful books of the identification and description of fishes of different countries, out of which Day (1878, 1889) is still the last complete reference to Indian and Burmese fishes. However, there have been great changes in nomenclature, classification and systematic status of many fish groups. Talwar and Jhingran (1991) and Jayaram (1999) are present more recent compilation for fish of India and Myanmar are concerned, there is no up-to-date report. The workers had access only materials collected before 1940. The monograph of fishes of North-east India made by Sen (1985) needs thorough revision in the light of fresh collections and many fishes reported after his work.

Ichthyological research in the Indian region started with the pioneering works of Hamilton-Buchanan (1822), followed by Day (1878). Hora made tremendous contribution to the ichthyology of the region till the middle of 20th centuary. However, works on the Burmese forms became almost standstill after Hora's (1941) report on Chindwin Hopwood Expedition. In the later part of 20th century and in the beginning of 21st century, ichthyologists from all over the world have focussed their research on ichthyological exploration in South-East Asia. Many new species have been discovered, many taxa have been reviewed and worked out the phylogeny (Allen and Burgess, 1990; Datta and Chaudhuri, 1993; Roberts, 1994, 1995; Anderson and Heemstra, 2003, Kottelat, 2003; Sasaki, 1989, 1990, 1992, 1994; Sasaki and Amaoka, 1989; Sasaki and Kailola,

*Department of Zoology, Mayai Lambi College, Yumnam Huidrom 795 009, Manipur

1991; Schwarzhans, 1993; Kullander and Britz, 2002; Ruber *et al*, 2004; Vishwanath and Shanta, 2004; Chakrabarty *et al.*, 2006; Ng *et al.*, 1996; Rema Devi, 1992; Pezold, 1993; Hoese and Allen, 1990; Derist, 1997a, b; Paepke, 1997; Ng and Lim, 1990, 1991; Lee and Ng, 1994; Ng *et al*, 1995, 1999; Musikasinthorn, 1998; Musikasinthorn and Taki, 2001; Zhang *et al.*, 2002; Vierke, 1991; Murray, 2006 and Britz, 2007.

Manipur, a hill state in the Northeast corner of India is a distinct geographical entity. It lies between 23.5^0 N and 25.4^0 N latitude and 93.4^0 E and 94.3^0 E longitudes. It is bounded by Nagaland on the north, Mizoram and Chin hills (Myanmar) on the south, the Somra tract and upper Chindwin district of Myanmar on the east and Cachar district of Assam on the west. The western, the central and the southeastern water bodies of this state drained by the Barak, the Manipur and the Yu river systems. The records of Perciform fishes of this state are very poor except Hora, 1921, Menon, 1952, 1954, Arunkumar 2000, Arunkumar and Tombi Singh, 1997, 1999, Selim and Vishwanath, 1997; Datta and Laishram, 1984; Vishwanath and Shanta, 2004. A detailed survey of the Freshwater fishes (Percomporpha: Perciformes) of Manipur, India was carried out in respect of taxonomic status and species diversity.

Thus, the present study has been undertaken to study the fauna of Perciform fishes of Manipur and their distribution. Systematic indexes of Percid fishes are provided.

Materials and Methods

Fishes are collected from different markets of Manipur *viz.*, Khwairamband Bazar, Singjamei market, Pishum market, Jiri market and Moreh market. Care was taken to keep the external morphology intact for taxonomic studies details of the location, date of collection, local name, colour of fish etc. for each specimen were also recorded. Photographs were taken the fish was in fresh conditions in order to keep the original coloour of fish. Collection from different size was packed into different packets with labels. Fishes were kept in the Zoology Department, Liberal college, Luwangshangbam, Manipur. Catalogue, Uncatalogue (Uncat.) and unregistered specimens of MUMF (Manipur University Museum of Fishes) were vividly observed.

Fishes were fixed and preserved in 5% to 10% formalin following "Wet Preservation" technique of Ayapan and Satyamurthi (1960). Measurements were made with dial caliper to the nearest 0.1 mm. Head length and measurements of the body parts are given as proportions of standard length (SL). Subunits of head are presented as proportions of head length (HL) Total length=TL.

Morphometric analysis for identifying the fishes was carried out generally with refrence to Talwar and Jhingran (1991) and Jayaram (1999). For further confirmation following literatures were followed: viz: Roberts (1994) for genus *Chanda*; Fraser-Burnner (1955) Roberts (1994), Musikasinthorn (1998), Anderson and Heemstra (2003), and Kotelat (2003) for genus *Parambassis*;

Sasaki (1989), Talwar (1995), Mohan (1981), and Talwar and Jhingran (1991) for Genus *Johnius*; Barlow *et al* (1968), Kullander and Britz (2002) and Vishwanath and Shanta (2002) for genus *Badis*; Ng *et al.* (1996), Liem (1970), and Chakravarty *et al* (2006) for genus *Nandus*, Trewavas (1983), Talwar and Jhingran (1991) for genus *Oreochromis*, Akihito and Meguro (1975), Rema Devi (1992), Pezold (1993), Hamilton-Buchanan (1822), Hoese and Allen (1990), Geevarghese and John (1983), and Allen (1991) for genus *Glossogobius*; Talwar and Jhingran (1991) and Munshi and Srivastava (1988) for genus *Anabas*; Derijst (1991a, b), Paepke (1997) and Vishwanath and Linthoingambi (2004) for genus *Polyacanthus*; and Playfair (1867), Ng and Lim (1989, 1990, 1991), Lee and Ng (1994), Ng *et al.* (1995, 1999), Musikasinthorn (1998, 2000), Vierke (1991), Murray (2006) and Britz (2007) for genus *Channa*.

The sequence of families in phylogenetic, following the classification of Nelson (1994) with a few deviations based on the recent studies. For each species, separate systematic accounts are given under subheadings *viz.*, local name, material examined, diagnosis, description, colour, distribution and remarks.

Results and Discussions

The list of percid fishes (Teleostomi: Perciformes) in Manipur is shown in Table 19.1. It includes 17 species. 17 species of perciform fishes in the three river systems of Manipur are shown in Table 19.2. *Johnius coitor, Nandus nandus Badis tuivaiei* and *Channa stewartii* are the distinctive fish fauna of the Barak river system in Manipur. 17, 12 and 12 species of perciform fishes are distributed in the Barak, the Manipur and the Yu river systems of Manipur respectively.

The comprehensive survey of the fish fauna (Percomorpha: Perciformes) in Manipur yielded 17 species belonging to 10 genera, 5 subfamilies and 9 families. The systematic account is as follows:

Phylum	:	Chordata
Subphylum	:	Vertebrata (=Craniata)
Class	:	Actinopterygii
Division	:	Teleostei
Series	:	Percomorpha
Order	:	Perciformes
Suborder	:	Percoidei
Superfamily	:	Percoidea
Family	:	Chandidae (=Ambassidae)
Genus	:	*Chanda* Hamilton-Buchanan
Species	:	*nama* (Hamilton-Buchanan)
Genus	:	*Parambasis* Bleeker
Species	:	*ranga* (Hamilton-Buchanan)
Family	:	Sciaenidae
Group	:	IV
Subfamily	:	Otolithinae

Tribe	:	Johniini
Genus	:	*Johnius* Bloch
Species	:	*coitor* (Hamilton-Buchanan)
Family	:	Nandidae
Subfamily	:	Nandinae
Tribe	:	Nandini
Genus	:	*Nandus* Valenciennes
Species	:	*nandus* Hamilton-Buchanan
Family	:	Badidae
Subfamily	:	Badinae
Genus	:	*Badis* Bleeker
Species	:	*badis* Hamilton-Buchanan
Species	:	*tuivaiei* Vishwanath and Shanta
Suborder	:	Labroidei
Family	:	Cichlidae
Genus	:	*Oreochromis* Gunther
Species	:	*mossambica* (Peters)
Suborder	:	Gobioidei
Family	:	Gobiidae
Subfamily	:	Gobiinae
Genus	:	*Glossogobius* Gill
Species	:	*giuris* (Hamilton-Buchanan)
Suborder	:	Anabantoidei
Family	:	Anabantidae
Genus	:	*Anabas* Cuvier
Species	:	*testudineus* (Bloch)
Family	:	Belontiidae (= Polycanthidae)
Subfamily	:	Trichogastrinae
Genus	:	*Polyacanthus* Cuvier
Species	:	*fasciatus* (Schneider)
Species	:	*labiosus* (Schneider)
Species	:	*sota* (Hamilton-Buchanan)
Suborder	:	Channoidei
Family	:	Channidae
Genus	:	*Channa* Scopoli
Species	:	*gachua* (Hamilton-Buchanan)
Species	:	*marulius* (Hamilton-Buchanan)
Species	:	*punctatus* (Bloch)
Species	:	*stewartii* (Playfair)
Species	:	*striata* (Bloch)

Order Perciformes

(Spiny ray fishes)

The order Perciformes, the largest group of vestebrates, encompasses 18 suborders of fish, 148 families about 1496 genera and about 9239 species

(Nelson, 1994). The member of this huge group of spiny-rayed fish have a world wide distribution and are especially common in tropical and subtropical seas and are less abundant in temperate waters. They normally inhabit marine inshore areas and although 1922 species (mainly cichlids) occurs only in freshwater. The largest suborder of the perciformes, the percoidei, contains 71 families, about 528 genera and 2860 species (Craig, 2000).

The order Perciformes in the most diversified of all fish orders. The members of the Perciformes are very varied in sized and shape but in general have the following characteristics. They usually have laterally compressed bodies and most have ctenoid scales although cycloid scales are found in some species. The perciformes are physoclists (the duct connecting the swin bladder to the gut is closed). They are usually two dorsal fins but never an adipose fin. Spines and the second support the first dorsal fin by soft rays. Anal fin with spines. Pelvic fin thoracic or Jugular with spines. Mouth protractile. Head and cheeks with musciferous canals, pores. Preopercle entire or serrated. Caudal fin skeleton with 5 or fewer hypurals. No free second ural centrum. Caudal fin with 17 or fewer principal rays. Weberian apparatus is absent.

Suborder: Percoidei

Family: Chandidae (=Ambasidae)

Oblong, short body, compressed, slightly semi transparent. Cycloid scales present. First dorsal fin with 7-8 spines and second, with a spine and 9-10 soft rays. Precopercular, suborbital and lacrymal bones serrated. Lateral line complete with very fin scales (125 in lateral series, Roberts, 1994). A forwardly directed recumbent spine in font of dorsal fin present.

Key to genera

Lower jaw prominent, mouth large scales minute 100 to 107 in lateral line series. Lower jaw with canine teeth...... *Chanda*

Lower jaw not prominent. Mouth moderate. Scales moderate 30-90 in lateral line series. Lower jaw without canine teeth. .... *Parambassis*

Genus *Chanda* Hamilton

Chanda Hamilton-Buchanan, 1822, *Fish Ganges*: 103, 307 (type species: *Chanda nama* Hamilton-Buchanan). -Talwar, 1971, *Bull. Zool. Nomencl*; 1976, *Bull. Brit. Mus. Nat. Hist.*, 29(1): 10-11 (family position). -Roberts, 1994. *Nat Hist. Bull. Siam. Soc.* 42: 265 (revision).

Diagnosis: Body highly compressed and nearly transparent. Abdomen rounded. Preorbital edge with 4 serrae. Lower jaw strongly projecting. Cheek, gill cover, predorsal above the lateral line and body just below base of dorsal fin scaleless. Jaws, palate and tongue with villiform teeth. Lower limb of preopercle with double serrated edge, opercle without a prominent spine. A forwardly directed recumbent spine present in front of the dorsal fin. Anal fin with three spines. Caudal fin forked. Lateral line complete.

***Chanda nama* Hamilton**

Chanda nama Hamilton-Buchanan, 1822, *Fish Ganges*; pp. 109, 3701, pl. XXXIX, fig. 37 (type locality: ponds throughout Bengal).

Ambsssis nama Day, 1878, *Fish. India*; P. 50, pl. XIV, fig. 5.

Chanda nama Jayaram, 1981, *F.W. Fish. India, Handbook*, Calcutta, Z.S.I. pp. 317-318.

Chanda nama Jayaram, 1999, *F.W. Fish. Indian Region*. N.P.H. Delhi, P. 369, fig. 193.

Chanda nama Archis *et al*, 2004, *Copeia*; P. 903 (valid).

Material examined: Uncat. MUMF; 2 exs; Barak river, March 1985; 3 exs; Lokchao stream; July 1985; TL=40-52 mm, Collected by M.G. Sharma. Uncat MUMF., 2 exs; Taret hill stream at Saibol, 15. V. 1987; 2 exs; Jiri river at Jiribam; 17.X.1987; 47-112 mm TL. Collected by L.I. MUMF-535/3,3 48.6-50.0 mm SL; Serov, Manipur river, 65 km South of Imphal, Manipur, Feb. 26, 1994. MUMF 402-4025, 2 exs. 38.0.38.5 mm SL; Barak river, Vanchengphai, Tamenglong district, Manipur, December 10, 1999. Uncat. MUMF., 1 ex; 38.7 mm SL; Iril river Manipur, December 2, 2004, Collected by IL.

Local name: Ngamhai/Nga mingshel/Khihungsha

Diagnosis: A monotypic species of Chanda with 110-114 scales in longitudinal series, mouth particularly large, with very prominent lower jaw.

Description: D.VII+1, 14; P.i.10, V.1, 5, A iii, 15; C. 9+8. Body oval, transparent and strongly compressed. Supraorbital ridge smooth. Mouth large, with conspicuous lower jaw; teeth viliform on jaws, with canines on either side of lower jaw, langue edentate. Scales minute.

Proportional measurements (in%): Body depth 47.4 (46.8-48.1), head length 37.9 (37.7-38.2); predorsal length 46.4 (46.2-46.6), dorsal fin height 22.4 (22.1-22.6); pectoral fin length 24.6. (24.2-24.9), ventral fin length 19.6 (19.5-19.7); anal fin height 22.1 (21.6-22.6); caudal fin length 27.5 (27.3-27.6); caudal peduncle length 16.2 (16.1-16.3); and caudal peduncle height 12.3 (12.2-12.4) of SL. Head width 41.4 (41.4-41.5); 32.8 (32.4-33.1); eye diameter 32.4 (32.4-32.5); interorbital space 27.6 (27.5-27.6); dorsal fin length 123.1 (120.7-125.5), Pectoral fin length 57.6 (51.7-63.5); and anal fin length 94.8 (93.1-96.6) of HL. Caudal peduncle height 75.8 (75.8-75.9) of its length.

Colour: Body silvery-yelowish, sparrely scattered minute black dots on body. Eyes black. Fin yellowish; caudal fin dusky and yellowish, with a pale outer border.

Distribution: Barak, Manipur and Yu river systems of Manipur.

Remarks: This species is widely distributed in both Barak, Manipur and Yu river basins of Manipur. It seldom exceeds 3 to 4 inches as reported by Hamilton-Buchanan (1822). The genus *Chanda* is now restricted to its type species *Chanda nama* Talwar (1971). Vishwanath *et al* (1998) kept it under the genus *Parambassis*. Archis *et al* (2004) worked out the ontogeny of this species.

Genus *Parambassis* (Bleeker)

Parambassis Bleeker, 1874, *Natuurk. Verh. Holland. Maatsch. Wetensch;* 2(2): 86,102 (Type species: *Ambassis apogonoides* (Bleeker). – Guha and Talwar, 1975, *J. Inland Fish. Soc. India,* 8.P. 76 (Status discussed)-Roberts, 1994, *Nat. Histo. Bull. Siam. Soc.,* 42pp. 271-289 (revision). –Roberts, 1989, *Mem. Calif. Acad. Sci.,* 4, P. 161 (Synonymy). Jayaram, 1999. *F.W. Fish. Indian Region.* NPH. Delhi, P. 370.

Diagnosis: Body elongate and compressed. Abdomen rounded. Pointed snout. Mouth large, gape oblique, extending to anterior border of orbit. Jaws straight or slightly upturned. Supraorbital ridge with one or two spines posteriorly. Preorbital serrated on both ridge and edge. Suborbital serrated. Tongue without tooth. Anal fin with three spines. Cheek with four to seven transverse scale rows. Lateral line complete.

Parambassis ranga (Hamilton)

Chanda ranga Hamilton-Buchanan, 1822, *Fish. Ganges.* PP. 113, 371, pl. XVI. Fig. 18.

Ambassis ranga Day, 1878. *Fish. India.,* P.51, pl. XIV, Fig. 6.

Chanda ranga Jayaram, 1981. *F.W. Fish. India. Handbook No.* 2., Calcutta, Z.S.I., PP. 317-318.

Parambassis ranga Jayaram, 1999. *F.W. Fish. Indian Region.* N.P.H. Delhi, P. 370-371, fig. 194.

Material examined: Uncat. MUMF. 2 exs., Jiri River, Feb., 1985; TL=65-90 mm; Collected by M.G. Sharma. Uncat. MUMF. 2 exs; Ikoppat near Khangabok, 29.XII. 2000; 52-58 mm TL; 3 exs., 38.0-40.5 mm SL; Leimatak river, Leimatak, Tamenglong district, March 10, 2000.

Local name: Ngamhai/Ngamingshel.

Diagnosis: A species of *Parambassis* with preopercular hind edge smooth, often with few serrations, lateral line scales 51-52.

Description: D.vii+i, 12; P.i,10; V.i,5 AII, 12-13; C. 9+8. Body stout, deep and compressed, preopercular hind edge smooth, with one or two serrations at angle. Mouth oblique. Scales small; cheek with 5-6 transverse scale rows.

Proportional measurements (in%): Body length 38.1 (36.7-40.0); head length 35.0(34.6-35.0); predorsal length 44.2 (43.5-45.7); dorsal fin height 22.3 (20.8-23.7); dorsal fin length 45.0 (43.5-46.4); pectoral fin length 24.7 (22.7-25.8); ventral fin length 23.9 (20.2-28.1); anal fin height 22.2 (20.8-25.0); caudal fin length 27.7 (26.6-29.6); caudal peduncle length 17.2 (16.5-17.6); and caudal peduncle height 11.9 (11.6012.4) of SL. Head width 39.4 (37.2-42.0); head height at occiput 76.6 (72.4-80.0); snout length 27.4 (25.4-29.0); eye diameter 33.1 (32.4-34.1); pectoral fin length 70.4 (65.7-74.2); anal fin length 90.1 (87.6-91.3) of HL. Caudal peduncle height 69.2 (66.2-71.6) of its length.

Colour: Silvery-yellowish body, transverse lateral light brown them bands on sides of body, a definite dusky spot on shoulder. Fins hyaline; dorsal and caudal fins with blackish edges.

Distribution: Leimatak river, Jiri river.

Remarks: This small glass fish was first reported as *Ambassis ranga* in 1921 by Hora from the biggest freshwater lake in the north eastern India, Loktak lake and sluggish stream. Later, it was reported by Menon (1952) and Menon (1954). After that, it was not reported in the central valley of Manipur. Vishwanath (2000) remarked that the specimens collected for the Jiribam market has 3 blue transverse bands on body, which also confirmed to the present observation. It can prove to be a good aquarium fish. It is a bony not relished much as food. Sen (1985) kept it under the genus *Chanda.*

It is different from *Parambassis pulcinella* Kottelat, 2003 by having less dorsal soft rays (12 VS. 15-16) and anal soft rays (12-13 VS. 16-17).

Family: Sciaenidae

Body elongate and compressed. Head and body covered with ctenoid or cycloid scales. Muciferous system well developed. Dorsal fin long, a notch separating spinous and soft parts. Lateral line scales extending to end of caudal fin. Anal fin with one or two weak spines. Bony upper edge of opercle forked, bony flap present above gill opening in two points. Post temporal with a fimbriate edge. Supramaxillary is absent. Premaxillary and maxillary are present in the upper jaw. Swim bladder is thick walled and single chambered, usually being suspended by the third to fifth vertebrae, and with many chambers. 36 species of sciaenids are recorded in the Indian waters (Mohan, 1981). 8 species and 8 genera of sciaenids are reported from the fresh waters of the Indian region (Jayaram, 1999).

Genus: *Johnius* Bloch

Johnius Bloch, 1793, *Naturges. Ausland. Fische*; 7,P.132 (type species: *Johnius carutta* Bloch).-Talwar and Shetty, 1971, *Proc. Indian Acad. Sci.*, 74(2), PP. 74-80 (generic relationship).

Diagnosis: Elongate body. Abdomen rounded. Prominent snout but swollen posteriorly. Inferior type of mouth. Upper jaw longer. Lower jaw with villiform teeth. Palate endentate. Two weak opercular spines. Two dorsal fins continuous. Anal fin with robust, pointed two spines. Caudal fin wedge shaped. Scales ctenoid on boy and cycloid on percle. Swim bladder hammer headed. 13 species of genus *Johnius* are recorded from Indian waters (Mohan, 1981). 1 species of *Johnius* is recorded from Indian region by Jayaram (1999). 16 species of *Johnius* are reported by Sasaki (1992).

Johnius is mainly distributed in the Indo-west pacific region and with 20 species. Species of *Johnius* found in the Indo-west pacific region are small and highly specialized in swimbladder.

***Johnius coitor* (Hamilton)**

Bola coitor Hamilton-Buchanan, *Fish. Ganges*; PP. 75, 368, pl. 27, fig. 24 (type locality: Ganges river).

Johnius coitor Talwar and Shetty, 1971. *Proc Indian Acad. Sci.* (B) 74: 75, fig 1 and 2 (Redescription).

Johnius coitor Mohan, 1981, *Indian J. Fish*; 28 (1and2) P. 16; fig. 23(a), (b).

Material examined: Uncat. MUMF; 2 exs., Jiri river at Jiribam; 16. X. 1992; Total length = 142-145 mm. MUMF 42, 1 ex., 97.5 mm SL; Jiri river, Jiribam, Manipur- Assam border; December 6, 2000.

Local name: Ngachi, Nga-thumyakpi.

Diagnosis: A species of *Johnius* with prominent and projecting snout; lower gill rackers 11; ctenoid scales on body and head; cycloid on snout.

Description: D. X+i, 28; P.i.15; V.1,5; A.11, 7; C. 17; L.l.51. Body elongate; snout conspicuously prominent and projecting. Mouth inferior; teeth villiform. Gill marks 11 on lower arm of first arch. Dorsal fin deeply notched, dorsal spines moderately weak; second anal spine strong. Caudal fin acutely rhomboid. Scales cycloid on snout.

Proportional measurement (in%): Body depth 24.1; head length 29.2; predoreal length 34.1; dorsal fin height 17.7; dorsal fin length 65.6; pectoral fin length 21.2; ventral fin length 21.5; anal fin height 20.7; caudal fin length 27.2; caudal peduncle length 24.6; caudal peduncle height 7.9; and body width 15.6 of SL. Head with 53.3; snout length 28.8; eye diameter 26.3; interorbital space 26.3; dorsal fin height 60.7; pectoral fin length 72.6; and anal fin length 30.9 of HL. Caudal peduncle height 32.1 of its length.

Colour: Silvery brown on dorsal and white on abdomen, spinous dorsal fin with a dusky edge; soft dorsal fin, anal and caudal fin with a dull grey border.

Remarks: Salted dried fishes of this species are important food in Manipur. Fowler (1933) considered *Corvina grypota* as a synonym of *Johnius coitor* (Hamilton-Buchanan, 1822) whereas Chu *et al* (1963) synonymised it with *J. dussumieri* (Cuvier, 1830). *J. coitor* of Gloerfelt and Kailola, 1984) was changed to *J (J) laevis* by Sasaki and Kailola (1991).

Family: Nandidae

Body oblong, and compressed. Head and body covered with medium sized ctenoid scales. Operculum armed with a spine. Mouth large and strongly protrusible. Dorsal fin with continuous spiny and soft part. Pelvic fin thoracic. Anal fin with spines. Lateral line incomplete or absent. Pelvic fin without axillary process. They are highly cryptic colouration.

Subfamily: Nandinae

Tribe: Nandini

Presence of highly protectile and very large mouth. Maxilla reaching to hind border of eye. Incomplete lateral line. A spine with triangular operculum.

Genus: *Nandus* Valenciennes

Nandus Valenciennes, 1831, in Cuvier and Valenciennes, ***Hist. Nat. Poiss.***, 7,P.481 (type species: *Nadus marmoratus* Cuvier = *Coius nandus* Hamilton-Buchanan).-Liem, 1970, *Fieldiana Zoo.*, 56, PP. 1-166.-Jayaram, 1999, *F.W. Fish. Indian Region,* N.P.H. Delhi, P. 395, fig. 209.

Diagnosis: Body oblong, compressed. Abdomen rounded Head large, compressed. Snout pointed, conical. Mouth terminal, highly protectile, its cleft wide, extending to below posterior border of orbit. Lower jaw longer. Villiform teeth absent on vomer. Opercle with one spine. Opercular bones serrated or entire. Dorsal fin contiguous, soft part longer.Dorsal fins inserted above pectoral fin base. No scaly axillary process. Anal fin with three spines. Caudal fin rounded. Large and simple air bladder.

***Nandus nandus* (Hamilton)**

Gobius nandus Hamilton-Buchanan, 1822, *Fish Ganges,* PP. 96, 370, pl. 30. fig. 32 (type locality: Gangetic provinces).

Nandus marmoratus Day, 1878. *Fish India;* PP. 129-130. Pl. xxxii, fig. 1.

Nandus randus Shaw and Shebbeare, 1937. *J. Roy. Asiat. Soc. Bengal, Sci.* 111, P. 115, fig. 120.

Nandus nandus Jayaram, 1981. *F.W. Fish. India,* Handbook No. 2. Calcutta. Z.S.I., pp. 335-336.

Nandus nandus Jayaram, 1999 *F.W. Fish. Indian Region,* N.P.H., Delhi. P. 395, fig. 209.

Material examined: Uncat. MUMF; 2 exs., Awangkhul, Iyai stream, Feb., 1985; 2 exs., Jiri river August, 1985; Total length = 8.2-9.6 cm; Collected by M.G. Sharma. MUMF 4261-4262, 2 exs; 44.5-47.5 mm SL., Jiri river, Jiri Manipur Assam border, December 6, 2000.

Local name: Nga-kharaobi, Kalapia (Manipuri), Gadguddi and Badvahdei (Assamese), Nagwangae.

Diagnosis: A species of *Nandus* with 14 species and 11 soft rays; operculum triangular with a single prominent spine; preopercle serrate in a continuous band near its angle; lateral line incomplete. Mouth very large and highly protrusible; lower jaw projecting maxilla reaching beyond kind edge of orbit.

Description: D. XIV, 11; P. 15; V.1.5; A. III, 7; C. 7+7. Body oblong and compressed. Snout pointed. Eye large. Dorsal anal fins extend beyond caudal fin base pectoral fin extends far beyond the ventral fin base. Caudal fin slightly rounded. Scales in longitudinal series.

Proportional measurement (in%): Body depth 35.8 (34.8-35.8); head length 45.1 (44.9-45.3); predorsal length 47.6 (46.9-48.3); dorsal fin length 21.9 (21.4-22.3); caudal fin length 26.6 (25.8-27.4); caudal peduncle length 14.1 (13.8-14.3) and caudal peduncle height 11.7 (11.6-11.7) of SL. Head width 34.3 (33.5-

35.0); head height at occiput 64.7 (62.8-66.5); snout length 30.4 (29.3-31.5); eye diameter 27.0 (26.5-27.5); interorbital space 18.9 (17.7-20.0); dorsal fin length 109.9 (104.7-115.0); pectoral fin length 42.7 (41.9-43.5) and anal fin length 31.6 (30.2-33.0) of HL. Caudal peduncle height 32.4 (80.9-83.9) of its length.

Colour: Grayish brown; vertically three broad patchy blotches, a dusky on caudal fin base, some narrow dark bands radiate from eye. Fins grayish; yellowish narrow bands on spots across soft portions of dorsal, anal and caudal fins.

Distribution: Barak river system in Manipur.

Remarks: Barman (1988) reported that the fish is a piscivorous fish preying upon smaller fishes in the paddy field and ditches. In Manipur, at is found in the lower course of the Barak and Yu river systems. *Nandus marmoratus* Valenciennes in Cuvier and Valenciennes, 1931 is an unwarranted replacement name for *N. nandus* by Day (1878) and Eschmeyer, 1990. *N. nandus, N. nebulosus, N. oxyrhynchus*, and *N. prolixus* are well distributed in India and Burma (now Myanmar), Peninsular Malaysia and Sundaland, Mekong, Maeklong and Chao Phraya basins (Laos and Thailand), and Sepilok river drainage in Sabah, northeastern Borneo respectively. Vishwanath *et al* (1998) reported *N. nandus* from the Chatrickong River, the Yu river system, Chindwin basin of Manipur.

Family: Badidae

Subfamily: Badinae

Presence of slightly protrusible and small mouth. Triangular operculum with a spine. Barlow *et al* (1968) proposed a separate family Badidae whereas Nelson (1994) recognized only a subfamily status for this group. Jayaram (1999) obeyed Nelson's (1994) points. Recently, Ruber *et al* (2004) worked out the evolutionary and biogeographic patterns of the family Badidae.

Genus: *Badis* Bleeker

Badis Bleeker, 1853, *Verh. Bat. Genootsch*; 25,P.106 (type species: *Labrus buchanani* Bleeker = *Labrus badis* Hamilton-Buchanan). -Jayaram, 1999. *F.W. Fish. Indian Region.*, N.P.H. Delhi; P. 394, fig. 208.

Diagnosis: Body moderately elongated, compressed, snout bluntly rounded. Abdomen rounded. Mouth small, terminal, slightly upturned and slightly protectile, maxilla not extending behind anterior margin of orbit. Opercle with one sharp spine. Lower jaws longer, Villiform teeth on jaws, vomer and palatine none on tongue. Dorsal fin is single and the spinous portion is of much greater extent than the soft one. Anal fin with 3 spine. Caudal fin rounded of wedge shaped. Lateral line absent or interrupted and tubed. 2-4 dentary foramina. Hypobranchial 3-toothed. Males with short dorsal fin lappets. Ruber *et al* (2004) reported 15 valid species of this genus. Kullander and Britz (2002) described 10 new species of genus *Badis*. They also also erected a new genus *Dario* with three new species viz., *Dario dario, Dario hysginon* and *Dario dayingensis*. They are small badid fishes with adult siz below 25 mm.

***Badis badis* (Hamilton)**

Labrus badis Hamilton-Buchanan, 1822, *Fish. Ganges,* PP. 70, 368, pl. xxv, fig. 23.

Badis buchanani Day, 1878. *Fish. India*. PP. 128-129, pl. xxxi, fig. 6.

Badis badis Shaw and Shebbeare, 1937. *J. Roy. Asiat. Soc. Bengal, Sci.* III. Pp. 112-113 fig. 117.

Badis badis Jayaram, 1981, *F.W. Fish. India*. Handbook No. 2. Calcutta., Z.S.I. pp. 334-335.

Badis badis Jayaram, 1999. *F.W. Fish Indian Region*. N.P.H. Delhi. PP. 394, fig. 208.

Materials examined: Uncat. MUMF; 4 exs; Barak river, Feb., 1985; 2 exs., Jiri river, March 1985, Total length = 6-7.5 cm. Collected by M.G. Sharma. 2 exs; Jiri river at Jiribam 16.10.1992, Total length = 40-55 mm Collected by I.S. MUMF 4204-4206, 2 exs; 39.0-46.0 mm SL.; Iyai river, Noney, Tamenglong district, October 2, 2000.

Local name: Nga-sakthibi, Ookabi-macha or Ching-ookabi (Manipuri), Ungokpamapu (Maring-Naga people of Manipur).

Diagnosis: A species of *Badis* with small protractible mouth; operculum, distinctly triangular, some indistinct vertical bands along lateral side from operculum to caudal base and bluish dark spot behind the gill opening.

Description: D.17/8-9; P.i, 11-12; V.1,5; A. iii, 7; C. 7+7. L.l. 28-30; L.tr. 3/1/7; Predorsal scales 13-14. Body moderately elongate, relatively low and slightly compressed. Abdomen rounded. Snout pointed. Eyes large. Mouth small, upturned, lower jaw longer than upper, lips thin, teeth villiform on jaws. Dorsal fin insertion opposite to pectoral fin, nears to the tip of snout than the base of the caudal fin. Caudal fin rounded. Scales ctenoid. Lateral line interrupted.

Proportional measurements (in%): Body depth 28.3 (27.4-29.4); head length 31.3 (31.1-31.5); predorsal length 33.2 (32.2-33.7); dorsal fin height 66.1 (58.2-72.2); pectoral fin length 21.6 (21.0-22.2); ventral fin length 24.1 (23.1-25.1); anal fin height 33.3 (32.1-34.1); caudal fin length 23.1 (22.2-23.9); caudal peduncle length 19.1)19.5-20.0) and caudal peduncle height 15.7 (14.9-16.2) of SL. Head width 51.4 (49.2-53.6); head height at occiput 65.1 (64.1-65.7); snout length 28.0 (27.1-29.5); eye diameter 24.8 (24.1-26.2); interorbital space 21.0 (20.5-22.7); pectoral fin length 69.2 (67.9-71.4); anal fin length 57.2 (55.7-58.6); and anal fin height 97.9 (82.8-108.6) of HL. Caudal peduncle height 82.1 (79.5-85.9) of its length.

Colour: Body dark brown, indistinct bands on lateral side of body extending from operculum to caudal, the bands round from mid-dorsal to ventral surface. A dark bluish spot behind gill opening and at the base of caudal fin.

Remarks: Hora (1921) did not get a single specimen of it from Manipur but reported from Dhanashori Stream near Dimapur, Nagaland. It is listed in the fish of Manipur (Menon, 1954) it kept under the family Nandidae by Sen (1985).

Kullander and Britz (2002) based on external morphological characters, but mostly on information derived from colour patterns, assigned the species to 5 species group: (1) *Badis assamensis* species group e.g.: *B. assamensis, B. blosyrus,* (2) *B. badis* species group e.g. *B. badis, B. kanabos, B. chittagongsis, B. ferrarisi,* (3) *B. corycaeus* species group e.g.: *B. corycaeus, B. pyema,* (4) *B. rubber* species group e.g.: *B. rubber,B. siamensis, B. khwae* and (5) *B. kyar* species group e.g. *B. kyar.*

Badis tuivaiei Vishwanath and Shanta

Badis tuivaiei Vishwanath and Shanta, 2004. *Zoos' Print J.* P. 1619-1621, fig. 1.

Material examined: Holotype: 9.VII. 2000, Tuivai river, Churachandpur district, Manipur 34.6 mm SL., Coll. K. Shanta, MUMF 5124. Paratypes: 20 exs., 9.VII. 2000, Tuivai river, Churachandpur district, Manipur, 26.4-54.0 mm SL., MUMF 5115/20; 2 exs., 25.VII. 2000, both from Tuivai river, Churachandpur district, Manipur, Coll. Juliana L; MUMF 5116 (49.00 mm SL), MUMF 5118 (30.2 mm SL); 3 exs., 3.II. 2004, Irang R. Tamenglong district, Manipur 53.3-59.4 mm SL., Coll. L. Shakuntala, MUMF 5125-5127.

Local name: Ng-sakthibi, Ookabi-macha, Ching-ookabi (Manipuri).

Diagnostic characters: A *Badis* species with the following combination of characters; a conspicusous black blotch covering the superficial part of the cleithrum above pectoral fin base, a black blotch between third and fourth dorsal spine, a mid basal round black spot on caudal fin; 16-20 circumpeduncular scales, 31 vertebrae; 26-32 scales in lateral line row, interorbital (5.6-6.6% SL) and body depth (25.9-29.2% of SL) respectively.

Description: D.16-18/9; P.13-14; A.iii,,6-10; Ll.20/2-25/4; L.tr.1½/1/7. Gill rakers 6-8. Vertebrae 31. Body elongate, moderately compressed. Predorsal contour straight in small specimens, sloping at some angle as prepelvic contour in some larger specimens, more strongly as the size increases. Eye situated in anterior half of head and at about mid axis of body. Jaws almost equal anteriorly, lower jaw slightly projecting, maxilla reaching to one by third of eye's orbit. Opercular spine slender with a simple sharp tip. Dental pores three, anguloarticular pores two, preopercular pores three. Ctenoid scales on lateral body of head. Caudal and pectoral fins are rounded pectoral fin extending two by third distance to anal fin origin pelvic fin pointed, inner branch of second soft ray the longest, not reaching to vent. Head length (26-31.4%), snout length (6.1-7.6%), orbital diameter (6.1-9.4%), interorbital width (5.6-7.2%), upper jaw length (8.5-10.4%), lower jaw length (10.9-16.4%), body depth (25.9-29.2%), pelvic fin length (20.9-25.8%) and pelvic to anal fin length (29.8-36.7%) in standard length respectively.

Colour: Pale brownish to yellowish with brown to black markings. Preorbital stripe dark grey continued across chin. Postorbital stripe blackish, formed by a single blotch close to orbit. No supraorbital stripe, and opercular blotch. Dark brown cleithral blotch is prominent. Dorsal fin dusky. Caudal fin uniformly brown, and with a small rounded brownish blotch at mid base.

Distribution: Trivai and Irang rivers which both tributaries of the Barak river system of Manipur.

Remarks: A purely hill stream fish of Manipur. It is similar to *B. kanabos* Kullander and Britz (2002) and *B. badis* (Hamilton-Buchanan, 1822) in their colour patterns, particularly the horizontal lines on sides and blotch on caudal peducnle. It differs from *B. kanabos* by its narrower interobital width (5.6-6.6 VS. 7.3-8.6), shallower body (25.9-29.2 Vs. 29.0-35.0), more lateral line scales (26.32 VS. 25.26), more circum-pedunclular scales (20 VS. 16-17), more vertebrae (30-31 VS. 26-28), more dorsal fin rays (XVIII, 9 VS. XVI, 9), lesser body depth (25.9-29.2 VS. 29-35), smaller interorbital width (5.6-66.VS. 7.3-8.6). It differs from *B. chittagongsis* by having more lateral scale rows (26-32 VS. 27-29), more vertebrae (30-31 VS. 28-29), lesser body depth (25.9-29.2 VS. 29-34), lesser gill rakers (6-8 VS. 8-11). It differs from *B. badis* by having lesser body depth (25.9-29.2 Vs 30.7-38.9%SL), lesser introbital width (5.6-6.6 VS. 6.5-8.3%SL) more lateral scale row (26-32 VS. 25-28), more vertebrae (30-31 VS. 26-28). It differs from *B. ferrarisi* in presence of distal extra scapular. *Badis tuivaiei* belongs to *Badis badis* group (Kullander and Britz, 2002).

Suborder: Labroidei

Family: Cichlidae

Body deep, oblong and compressed. Head and body with ctenoid scales. Nostril single on each side. Inferior pharyngeal bones triangular with median longitudinal suture. Teeth on jaws. Palate edentate. Dorsal fin spinous and soft part. Anal fin with spines and a soft part. Lateral line interrupted. Truncate or rounded caudal fin.

Genus: Oreochromis Gunther

Oreochromis Gunther, 1889, *Ann.; Mag. Nat. Hist.* (6) 4:70 (type species: *Oreochromis hunteri* (Gunther). Trewavas, 1983, *Publ. Brit. Mus. Nat. Hist.*, No. 878, pp. 139-373 (revision). – Jayaram 1999. *F.W. Fish. Indian Region* N.P.H. Delhi, P. 400, fig. 213.

Diagnosis: Body elongate and compressed. Abdomen rounded. Head compressed, with concave upper profile. Mouth terminal, large, cleft extending to below anterior border of orbit. Snout rounded. Eyes large, lateral, in middle of head, not visible from below ventral surface. Jaws equal, with 3-5 rows of teeth. Anal fin with three spines, third spine a little longer than the last dorsal. Spinous part of dorsal fin longer than soft part which may be produced into a filament. Scales cycloid. Incomplete lateral line.

***Oreochromis mossambica* (Peters)**

(Figure 19.1)

Chromis (Tilapia) mossambica Peters, 1852 *Monatsb. Akad. Wiss. Berlin* 618.

Tilapia mossambica Sen, 1985, *Rec. Zool. Surv. India, Occ. Paper*. 64. 189, fig. 100.

Oreochromis mossambica Jayaram, 1999. *The Freshwater Fishes of Indian Region* 400, fig. 213.

Material examined: Uncat. MUMF, 2 ex. 112-154 mm SL. 1ex. 100 mm SL.

Local name: Tunghanbi.

Diagnosis: Body elongated, compressed. Mouth terminal, large, cleft extending to below anterior border to orbit. Jaws with 3-5 rows of teeth. Spinous part of dorsal fin longer than soft part which may be produced into a short filament.

Description: D. 15 or 16/10-12, A. iii, 10; C. 19; Ll. 29-30. Body deep and compressed, its depth 3.0-4.0% SL. Single nostril on each side. Mouth terminal. Eyes larged, lateral Lips thin. Jaws equal. Caudal fin truncated or rounded. Scale below eyes in 3 times. Snout rounded scales cycloid, lateral line incomplete, with 30-32 scales.

Colour: Three to four dark blotches along sides. Males deep black during breeding seasons. Dorsal, pectoral and caudal fin edges red.

Distribution: Widely distributed in Manipur.

Remarks: Exotic species introduced in India in 1952 and after 1970 in Manipur. A hardy fish, unsuitable for culture along with Indian major carps because of its adverse effects on the growth and production. However, it is now cultured in Manipur and is becoming popular. A good example for parental eare by mouth i.e. mouthbreeder. It belongs to family Cichlidae not Ciclidae of Vishwanath (2002).

Suborder: Gobioidei

Family: Gobiidae

Body oblong to elongate. Head with mucous canals and open pores. Spinous and soft dorsal separate. United pelvic fins which formed an adhesive disc. Ctenoid or cycloid scales. Lateral line absent;

Subfamily: Gobiinae

No scales between and before eyes on head. Soft part of dorsal fin longer than spinous part. Caudal fin lanceolate, rounded or truncate. Simple teeth present-teeth on lower jaw in several rows.

Genus: *Glossogobius* Gill

Glossogobius Gill, 1859, *Proc. Acad. Nat. Sci. Philad.*, P. 46 (type species: *Gobius platycephalus* Richardson). -Akihito and Meguro, 1975, *Japan. J. Ichthyol*, 22, P. 127-142. -Geevarghese and John, 1983, *J. Fish. Biol.* 22.P.231-240. -Rema Devi, 1992, *Rec. Zool. Surv. India* 90(1-4), P. 174. -Koumans, 1941,

Mem. Indian Mus. 13(3): 205-329. -Hoese and Allen, 1990, *Rec. West. Aust. Mus. Suppl. No.* 34:117-129. -Jayaram, 1999. *F.W. Fish. Indian Region*, N.P.H. Delhi, 413.

Diagnosis: Body elongate, rounded anteriorly and compressed posteriorly. Abdomen round. Head depressed. Operculum and cheek maked. Mouth Small, oblique, not extending to anterior margin of orbit. Eyes large, superior, in the mid of head, not visible from ventral surface. Lips thick. Jaws with villiform teeth in several rows. Tongue bilobate. Gill openings extend far forward of orbit. Pelvic fins oblong united midventrarlly. Scales cetenoid on body and cycloid on head.

***Glossogobius giuris* (Hamilton)**

Gobius giuris Hamilton-Buchanan, 1822, *Fish. Ganges* PP. 51, 366, pl. 33 fig. 15 (type locality: Gangetic provinces).

Korimans 1941, *Mem. Indian Mus.* 3(3): 256 (Valid).

Vishwanath, 2002, *Fish N.E. India.* P. 184; fig. 160.

Talwar and Jhinghran 1999. *Inland Fish. India.* P. 541 (Description).

Material examined: Uncat. MUMF. 3 exs; Thoubal river at Irong, 27. XII. 2000, Total length = 75-85 mm. 2 exs; Jiri river at Jiribam, 18.10.10992, Total length = 78-82 mm. Collected by I.S. unregistered, 2 examples, 79.5-80.6 mm SL; Sugnu, Manipur river, Manipur, August, 20, 1994. MUMF 5017, 3 exs. 59.6-70.1 mm SL., Tuivai river, Singhat, Churachandpur district, Manipur, September 16, 2001. MUMF 6373-6375; 3 exs; 60.3-71.4 mm SL.; Iril river at Bamonkampu, Imphal, September 16, 2002, collected by IL.

Local name: Nylon-nga, Nylon-ngamu (Manipuri), Khaivachar.

Diagnosis: A species of *Glossogobius* with branchiostegal membranes attached to sides of isthmus; body pale without longitudinal lines, iris without a lappet, transparent yellowish grey with 4-6 blotches along the central line, eye place on top of head.

Description: D.Vi+i,9; P.i.17; V.10; A.i.8; C.12. Body rounded, slightly compressed posteriorly. Head depressed. Cheek and operculum naked. Mouth small, oblique, not extending to anterior margin of orbit. Lips thick. Eyes small, superior in the middle of head. Dorsal fins separated; pelvic fins oblong united mid-ventrally, scales ctenoid on body and cycloid on head.

Proportional measurements (in%): Body depth 18.8 (17.3-21.3); head length 30.5 (29.5-31.5); caudal length 26.7 (25.7-27.7), predorsal length 37.6 (36.8-38.6); post dorsal length 61.9 (60.9-62.7); caudal peduncle length 22.7 (18.7-24.8); caudal peduncle height 10.5 (10.3-10.7); anterior dorsal fin base length 15.7 (12.9-18.5); its height 20.7 (20.6-20.8); posterior dorsal fin base length 18.0 (16.3-20.8); its height 17.3 (17.0-17.8) of SL. Height of head at occiput 49.5 (46.6-51.4); head height at eye 47.0 (44.9-50.5); snout length 35.7 (33.2-39.8); eye diameter 19.0 (17.8-20.5); interorbital species 9.6 (7.4-13.6), and maximum head width 66.7 (60.8-72.4) of HL.

Colour: In life, yellowish brown with five dark blotches on sides, side of head with irregular dark spots. All fin dusky.

Distribution: Tuivai river. Iril river, Imphal river, Barak river, Manipur river and Yu river system in Manipur.

Remarks: Rainboth (1996) remarked that the species in mostly found in the lower course of river. In Manipur, the fish is widely distributed in the plains and lower course of the hill-streams. It has interesting features which makes it ornamental. It feeds on small fishes and crustaceans.

Suborder: Anabantoeidei

Family: Anabantidae

Body oblong, compressed. Head and body covered with ctenoid scales. A suprabrachial organ well developed usually border of interopercular serrated. Dorsal fin base longer than anal fin base, pelvic fin thoracic, inserted below base of pectoral fins. Anal fin with spines. Lateral line interrupted. Vertebrae 25 (After Liem, 1963).

Genus: *Anabas* Cuvier

Anabas Cuvier, 1816, *Le Regne Animal.* 2; p. 339 (type species: *Perca scandens* Daldorf).

Diagnosis: Body oblong, compressed. Mouth small, terminal, upper jaw slightly protrusible, jaws with villiform. First epibranchial expanded to form labyrinthine suprabranchial organ for accessory breathing or respiration. Dorsal fin 16-18 spines and 8-10 soft rays. Anal fin with 8-11 spines and 9-11 soft rays. Caudal fin rounded. Ctenoid scales are present. Lateral line in two rows.

***Anabas testudineus* (Bloch)**

(Figure 19.2)

Anthias testudineus Bloch, 1793, *Naturges. Ausland. Fische*, 7, P. 121, pl. 322.

Anabas testudineus: Rainboth 1996: 214 (valid).

Local name: Ukabi, Nga samchet (Manipuri), Tampakhai.

Material examined: Uncat. MUMF, 2 exs., 63.7-67.7 mm SL; Iril river, January 2, 2002, IL.

Diagnosis: A species of *Anabas* with the following characters; body depth 28.6-33.0% SL; snout length 5.1-8.1%SL; mouth fairly large; teeth villiform on jaws; scales large, 21-29 in lateral line series.

Description: D.XVI-XVIII,8-9; P.I, 13-14; V.I,5, A.X-XI, 9-10, C.15-16, Ll. 28. Body elongated and moderately deep. 28.6-33.0%SL. Head moderately compressed, its length much longer than body depth and 35.2-37.8% SL, its width 58.4-66.2, depth at occiput 77.6-82.1; snout bluntly rounded; mouth fairly large, and terminal; eye diameter 22.0-26.3; interorbital space 31.2-33.5% HL. Scales large, lateral line incomplete. Dorsal fin inserted at vertical level of pectoral fin anterior base.

Colour: Greenish to dark grey of dorsal side and flanks, fading pale yellow on belly. A distinct opercular black spot is present.

Remarks: The fish mostly inhabited in pools, swamps, paddy fields etc. It is caught during rainy season, in the month of June on the lower course of rivers. It is widely distributed in Manipur valley.

It was firstly reported from Manipur by Menon (1952) and secondly by Menon (1954). Menon (1974) reported that this type locality is an error as the fish does not occur in Japan. The generic and the specific names of this fish are derived from Latin having the meaning of the climber and the turtle like respectively.

In Manipur, it is highly esteemed for its flavour. Smoked and sun dried fishes are stored in airtight earthen vats for consumption in off season. However, the fish has not been taken up for culture.

Family: Belontidae (=Polycanthidae)

Body short and compressed. Head and body covered with ctenoid scales. A suprabranchial cavity is present. Pelvic fin modified into a filiform ray. Dorsal fin base shorter than anal fin base. Lateral line vestigial, absent or sudimentary. Caudal fin truncate, rounded or emarginate.

Table 19.1
List of fauna of Percid fishes of Manipur

Sl. No.	*Scientific Name*
1	*Chanda nama* (Hamilton-Buchanan)
2	*Parambasis ranga* (Hamilton-Buchanan)
3	*Johnius coitor* (Hamilton-Buchanan)
4	*Nandus nandus* Hamilton-Buchanan
5	*Badis badis* Hamilton-Buchanan
6	*B. tuivaiei* Vishwanath and Shanta
7	*Oreochromis mossambica* (Peters)
8	*Glossogobius giuris* (Hamilton-Buchanan)
9	*Anabas testudineus* (Bloch)
10	*Polyacanthus fasciatus* (Schneider)
11	*P. labiosus* (Schneider)
12	*P. sota* (Hamilton-Buchanan)
13	*Canna gachua* (Hamilton-Buchanan)
14	*C. marulius* (Hamilton-Buchanan)
15	*C. punctatus* (Bloch)
16	*C. stewartii* (Playfair)
17	*C. striata* (Bloch)

Subfamily: Trichogasterinae

Supraorbital crest absent. Lateral line absent.

Genus *Polyacanthus* Cuvier

Polyacanthus Cuvier, 1829, *Hist. Nat. Poiss.*, 3:227 (type, *Trachopodius colisa* Hamilton, 1822; subsequent designation by Jordan., 1919, PP. 173.- *Colisa* Cuvier, 1931, *Hist Nat. Poiss.* 7.p.359 (type species: *Colisa vulgaris* Cuvier = *Trichopodus colisa* Hamilton-Buchanan). -Jayaram, 1999. *F.W. Fish.*

Indian Region. N.P.H., Delhi. P. 441.-Menon, 1999, Checklist F.W. Fish. India. P. 197:366.

Diagnosis: Body compressed and short. Snout blunt. Mouth terminal, upturned, small and highly protectile. Conical teeth. Ventral border of preopercle serrated. Opercle entire. Dorsal fin with 15-22 spines and 10-20 soft rays. Pelvic fin modified into a single elongate filliform ray. Scales are ctenoid. Caudal fin truncate or slightly emarginated. Lateral line absent or interrupt.

Key to species

* 11-13 or more oblique bands from back to belly.
  Non-papillated thin lips.. *P. fasciatus*
* 8-10 oblique bands from back to belly.
  Papillated thin lips.. *P. labiosus*
* Lateral dark longitudinal band from eye to lower half
  to caudal fin. Body without bands......................... *P. sota.*

***Polyacanthus fasciatus* (Schneider)**

(Figure 19.3)

Trichogoster fasciatus Schneider, 1801, *Syst. Ichth.* 164, pl. 36 (type locality: Tranquebar).

- Day, 1878. *Fishes of India*: 374, pl. 78. fig. 6.
- Day, 1889. *Fauna Br. India, Fishes*, 2: 372, fig. 123.
- Jayaram, 1999 *F.W. Fish. India Region*, N.P.H., Delhi, P. 442., fig. 228.
- *Polyacanthus fasciatus* Menon, 1999. *Checklist F.W. Fish India.* P. 298 (Valid).

Material examined: Uncat. MUMF-541/4, 4, 42.8-47.5 mm SL., Serou, 65 km. from Imphal, Manipur Feb., 26, 1994. MUMF L0001/18, 46.0-65.0 mm SL., Mayang Imphal Ponds and Iril river, 5. VI. 2001. Collected by IL. MUMF L0001-L0005, 5 exs; 45.0-63.0 mm SL; Iril river, June 5, 2001.

Local name: Ngabemma, Ngapemma (Maniupri) Khaishim.

Diagnosis: An oval species of *Polyacanthus* with several (11-13) oblique dark-blue bands across flanks. Lips simple and non-papillated but highly protractible. Tip of dorsal and anal fins rounded not produced upto dorsal and ventral base of caudal fin respectively.

Description: D.XV-XVII, 9-14; p.i; V.i,7-9; A. XV-XVII, 14-19; C.13. Body ovate and strongly compressed, head moderate, eyes small, mouth small. Preorbital serrated. Dorsal and anal fin base long, tips of soft portion somewhat rounded or slightly pointed but not reaching base of caudal fin. Pelvic fin modified into a filamentous prolongation, extending upto caudal fin base. Caudal fin truncate. Lateral line interrupted.

Proportional measurements (in%): Body depth 41.3 (37.0-44.4); predorsal length 37.4 (35.1-39.1); head length 32.2 (30.0 (30.0-35.0); dorsal fin base length 59.0 (45.3-64.0); pectoral fin length 31.0 (30.0-34.0); ventral fin length

89.0 (80.0-106.0); anal fin length 62.1 (58.6-68.0); caudal fin length 31.4); head height at eye 18.0 (16.0-21.0); snout length 9.0 (8.0-10.2); eye diameter 9.5 (8.0-10.3; interorbital space 15.3 (14.2-16.4); maximum head width 20.0 (19.0-21.4); and internarial space 10.2 (9.0-13.4) of HL.

Colour: In life, greenish with 11-13 alternate orange and bluish bas descending obliquely. Dorsal, anal and caudal fins with dark-blue and orange spots and bars. Anal fin with red margin. Filamentous pelvic fin is deep orange.

Distribution: Widely distributed in Manipur.

Remarks: Small fish with medium food value in Manipur. Smoked fishes are used to add flavour while cooking. This species is decorated with a beautiful colour and mostly culture in the aquarium in Manipur. It is fairly abundant in Manipur valley during rainy season.

Menon (1952) reported it for the first time from Manipur and secondly followed by Menon (1954). It kept under the family Anabantidae by Sen (1985).

***Polyacanthus labiosus* Day**

Trichogaster labiosus Day, 1878, *Fish. India.* 374 pl. 79. fig. 4 (type locality: Rangoon, Myanmar).

Trichogaster labiosus Day, 1889. *Fauna of Br. India. Fishes.* 2: 372.

Trichogaster labiosus Jayaram, 1999. *F.W. Fish. Indian Region.*

Colisa labiosus Vishwanath and Linthoingambi, 2004. *J. Bombay Nat. Hist. Soc.* 101: 466, Fig. 1.b.

Material examined: Uncat. MUMF. L0019-L0037, 19 exs., 47.0-64.5 mm SL. Iril river, Manipur June 5, 2001. IL. MUMF L0019/18, 40.7-64.5 mm SL, 5. VI. 2001-15. VIII. 2001, Mayang Imphal ponds and Iril river, Collected by IL.

Local name: Ngapemma.

Diagnosis: It is distinct in having following combination of characters: body with 8-10 oblique orange brown bands; lips thick and papillated with villi-like projections, tip of soft dorsal and anal fins produced reaching both dorsal and ventral base of caudal fin respectively.

Description: D. XV-XVIII, 9-11; P.IV-7-9; A. XV-XVIII 15-18 C. 15. Body oblong and compressed, Mouth small, highly producible. Scales of predorsal 8-9 pelvic fin modified into a single filamentous rays extending upto posterior extremity of caudal fin. Anal fin contiguous with caudal fin. Caudal fin slightly emarginated or wedge shaped.

Colour: Greenish, lighter ventrally. Blue horizonal stripe on body. Fin dark, outer edge of anal fin yellowish red.

***Polyacanthus sota* (Hamilton)**

Trichopodius sota Hamilton-Buchanan, 1882. *Fish Ganges.* 120, 373 (type locality: Ganges River).

Trichopodus chuna (Hamilton-Buchanan) Day, 1876. *Fish. India,* 373. pl. 78, fig. 3.,

Trichopodus chuna Day, 1889, *Fauna Br. India, Fishes*. 2: 371.

Material examined: Uncat. MUMF L0037/6, 21.4-24.3 mm SL., Canchipur, 5.X. 2001, IL. MUMF, 3 exs., 21.4-24.3 SL., Khuga river, Churachandpur, October 2001, Collected by KSD.

Description: D. XVII-XVIII, 6-9; P.i., V.i.9; A. XVIII-XXII, 11-13; C.13. Body ovate and compressed. Mouth small, upturned and highly protrusible. Dorsal spine increasing in length posteriorly; its soft portion not produced. Anal fin base scaly. Scales are larged. Interrupted lateral line. Eyes bid, pelvic fin modified into a filamentous ray, reaching posterior extremely of anal fin. Preorbital is serrated.

Proportional measurements (in%): Body depth 36.0-40.0; caudal fin length 28.1-33.0; head length 33.0-35.0; predorsal length 41.6-42.5; dorsal fin base length 57.5-63.0; pectoral fin length 25.0-30.0; ventral fin length 64.5-77.0; anal fin base length 61.0-64.0; preanal length 40.5-43.4 of SL. Height of head at occiput 26.0-28.0; snout length 10.0-12.0; eye diameter 11.2-13.3; interorbital space 14.0-12.0; eye diameter 14.0-16.0; maximum head width 18.0-19.2 of HL.

Colour: Patle white with a longitudinal dark-blue hand extending from eye to base of caudal fin. All fins are hyaline.

Distribution: Widely distributed in Manipur.

Remarks: Small size fish with medium food value. Mostly consumed as dry fish. Widely distributed in the lakes, small streams and rivers of Manipur valley. Menon (1954) reported it for the first time from Loktak lake as *Colisa chuna* which is junior synonym of *C. sota* (Menon, 1974). It kept under the family Anabantidae (Sen, 1985).

Suborder: Channoidei

Family: Channidae

Body elongate with scales. Head with plate-like scales. Suprabranchial accessory respiratory apparatus well developed. A single long dorsal fin, and a similar anal fin, both without spines. Pelvic fin present. 29 species of *Channa* currently considered valid (Britz, 2007).

Genus: *Channa* Scopoli

Channa Scopoli, 1777, *Introd. Hist. Nat*; P. 459 (type species: *Channa orentalis* Bloch and Schneider). – *Ophicephalus* Bloch, 1793, *Naturges. Ausland, Fische*, P. 137 (type species: *Ophicephalus punctatus* Bloch – Musikasinthorn, 1998, *Ichthyol. Res.* 45: 335 (Sittang river basin, Myanmar) – Musikasinthorn, 2000, *Ichthyol Res.* 47: 27-37 (Brahmaputra river basin, India).

Diagnosis: Body elongate, cylindrical anteriorly. Head depressed, large, covered with plate like scales. Mouth opening large and wide. Lower jaw longer than the upper jaw. Teeth on jaws and palate. Suprabranchial chamber with bony lamellae as accessory respiratory organ. Spines an absent on dorsal and anal fins, and both the fins are free from caudal fin. Caudal fin rounded.

Key to species

1. Predorsal scales 12-14. Scales between orbit and angel of preopercle 4 or 5.............................. 2
   Predorsal scales 15-22. Scales between orbit and angle of preopercle 9-17 or 5 3
2. Length of pelvic fin less than half of that of pectoral fin... 4
   Length of pelvic fin more than half of that of pectoral fin. Pectoral fins plain, dorsal fin rays 28-32.. *C. punctatus.*
3. Predorsal scales 18-20.................................. *C. striatus.*
   Predorsal scales 15-17.................................. 5
4. Anal fin with 27 rays, scales with black spots. Dorsal fin rays 38-40.................................. *C. stewartii.*
   Dorsal fin rays 32-37. pectroral fins with bluish vertical bands.. *C. gachua.*
5. Black ocellus with white edges at caudal fin....... *C. marulius.*

Table 19.2

The distributional pattern of freshwater fishes (Percomorpha: Perciformes) in the three river systems of Manipur. Presence and absence of each species in indicated by + and – respectively

Sl. No.	*Scientific Name*	*River systems in Manipur*		
		Barak	*Manipur*	*Yu*
1.	*Chanda nama* (Hamilton-Buchanan)	+	+	+
2.	*Parambasis ranga* (Hamilton-Buchanan)	+	+	-
3.	*Johnius coitor* (Hamilton-Buchanan)	+	-	-
4.	*Nandus nandus* (Hamilton-Buchanan)	+	-	+
5.	*Badis badis* (Hamilton-Buchanan)	+	+	+
6.	*B. tuivaiei* (Vishwanath and Shanta)	+	-	-
7.	*Oreochromis mossambica* (Peters)	+	+	-
8.	*Glossogobius giuris* (Hamilton-Buchanan)	+	+	+
9.	*Anabas testudineus* (Bloch)	+	+	+
10.	*Polyacanthus fasciatus* (Schneider)	+	+	+
11.	*P. labiosus* (Schneider)	+	+	+
12.	*P. sota* (Hamilton-Buchanan)	+	+	+
13.	*Channa gachua* (Hamilton-Buchanan)	+	+	+
14.	*C. marulius* (Hamilton-Buchanan)	+	-	+
15.	*C. punctatus* (Bloch)	+	+	+
16.	*C. stewartii* (Playfair)	+	-	+
17.	*C. striata* (Bloch)	+	+	+

***Channa gachua* (Hamilton)**

Ophiocephalus gachua Hamilton-Buchanan, 1822. *Fish Ganges*. **68, 367** (type locality: ponds and ditches of Bengal).

Material examined: MJMF 5140, 1 ex., 72.0 mm SL; Tuivai river, Churachandpur, August 15, 2001, Collected by KSD. MUMF-343, 118.3 mm SL., Chakpi river (Tributary of Manipur river) at Mombi, South of Imphal, Manipur, December 18, 1992. Uncat. MUMF. 3 exs; 80-152.70 mm SL; Chatrickong river, 150 km. from Imphal, Manipur, April 20, 1996.

Local name: Meitei ngamu (Manipuri), Khaiva.

Diagnosis; A colourful *Channa* species with vertically striped pectoral fins. Scales on head are larged. Separated pelvic fin present. Dark green both on dorsal and flanks but pale ventrally. Caudal fin rounded, complete lateral line.

Description: D. 35-37; P. 13-16; V.5; A.23; C. 12. Ll. 44-45, Ltr. 4/1/8, PDS. 12. Body elongate and cylindrical anteriorly and compressed posteriorly. Anterior part of head depressed, mouth large, eyes moderate. Entire body scale; scales bigger on the head. Anterior nostril tubular. Dorsal and anal fins long and entirely soft rayed. 5 scale rows between preopercular angle and posterior border of orbit. Snout rounded. Circum peduncular scales 24.

Proportional measurements (in%): Body depth 17.0-20.3 caudal fin length 19.9-27.87; head length 30.4-34.0; length of caudal peduncle 7.3; its height 11.7; pre-dorsal length 35.8-37.50; post dorsal length 65.0; dorsal fin base length 60.6; its height 10.9; pectoral fin length 21.0-21.62; pelvic fin length 8.8-115; anal fin base length 38.1 %SL. Height of head at occiput 43.4-46.2; that at eye 33.7; snout length 20.1; eye diameter 15.4-17.7 interorbital space 27.3-29.4 maximum head width 59.0-64.3; inter narial space 17.3% HL.

Colour: In life, dark green both on dorsal and flanks but pale ventrally. Body with dark oblique bands above and below lateral line. Pectoral fin with 6 distinct alternating blue or orange bands and at the base a black spot present. Outer margins of dorsal and caudal fins bright orange.

Remarks: It is important food fish of the Manipuris. Hora (1921) reported *Ophiocephalus harcourtbutleri* Annadale from Loktak lake which is synonymous with *C. orientalis* Schneider (De Witt, 1960); Hora (1936) and Menon, M.A.S. (1952) reported *O. gachua* (Hamilton) which is also synonymous with *C. orientalis* Schneider (Dewitth, 1960); Menon, A.G.K. (1954b). Talwar and Jhingran (1991), Jayaram (1999) and Rainboth (1999) put the species as synonym of *Channa orientalis*. But, Kottelat (1998) remarked that *C. orientalis* is known only from Sri Lanka and South India. The species are easily distinguished from *Channa orientalis* in morphological, ecological and behavioured characters though both the species occur in Sri Lanka (Kottelat, 2001). Ng *et al* (1999) revalidated *C. harcourtbutleri*, a small snake head restricted to the Inle basin, but up till now reported from Inle lake as *Channa gachua*.

Chauna gachua is very similar to *Channa orientalis*. However, the species can be easily distinguished from the later in having ventral fin separated VS. ventromedially fused. Zoogeographically, *C. orientalis* is endemic to Sri Lanka, whereas *C. gachua* is widely distributed in South East Asia. The diploid number of chromosomes for *C. gachua* is 78 from India (Banerjee and Others, 1988) and 112 for specimens from Thailand (Donsakul and Magtoon, 1991), a strong indication that this represents a species complex.

***Channa marulius* (Hamilton)**

Ophiocephalus marulus Hamilton-Buchanan, 1822, *Fish. Ganges*. Pp. 65, 367, pl. 17, fig. 19 (type locality: River Ganges).

Ophiocephalus marulius Day, 1878, *Fish. India* PP. 263, 364, pl. ixxvI, fig. 4 *Channa marulius* Jayaram, 1981. *F.W. Fish. India*, Handbook No. 2. Calcutta. Z.S.I. pp. 306-397.

Channa marulius Pethiyagoda, 1991. *F.W. Fish. Sri Lanka*. P. 279.

Material examined: Uncat. MUMF; 1ex., Jiri river, Jiribam, 18.X.1922. 328 mm in total length. Collected by I.S. 3 exs., Lokchao stream, Lokchao, 6-6-1985; 1 ex., Barak river, Tamenglong, Oct., 1985, total length = 20-30 cm. Unregistred MUMF. 1 ex;, 488.00 mm SL., Barak river, Vanchengphai, Tamenglong district, March 19, 1999.

Local name: Ngamu Porom, Gojar, Ngamu Usoichabi (Manipuri).

Diagnosis: Black grayish in colour with 5-6 rounded black ocellus present below the lateral line and one prominent black ocellus at the upper base of caudal fin.

Description: D.45-55, P.16-18, V.6, A. 28-36. C. 14, Ll. 66 (60-70) Perdorsal scales 15-16. No patch of scales on gular region of head. Lateral line scales drop two rows between 16th and 18th perforated scale. Scale rows between posterior margin of orbit and propercular angle 10. Scales on top of head moderate sized with a rosette of head scales between the oribits, with the froutal head scale in the centre of rosette; two scales between rosette and the basal head scale. 10 scale rows between preopercular angle and posterior border of orbit. Pectoral fin length about half head length. Mouth large, lower jaw with 7-18 canines hehind a single row of villiform teeth that widen to 5-6 at jaw symphysis.

Colour: Pale-edged occllus are present.

Distribution: Barak and Yu river systems of Manipur.

Remarks: Teeth present on prevomer but absent on palatines (Talwar and Jhingran, 1991), and Smith (1945) indicated that canine teeth were absent from both prevomer and palatines. Talwar and Jhingran (1991) provided information on life colours fin both young and adults, commenting that there is a "pale-edged ocellus" toward the uperside of the caudal fin base that "fades wit growth", but provided on information as to size when that happens.

The only other species of *Channa* having an ocellus on the upper lobe of the caudal fin near its base is *C. marulioides*. In contrast to *C. marulius, C.*

marulioides has 55-58 lateral line scales, 13-15 predorsal scales, and 5-7 preopercular scales. The lateral line drops abruptly between the 17th and 20th scale in *C. marulioides*, and there is a two row drop of the lateral line in *C. marulius* may have a series of dark blotches along the sides, margined posteriorly and posterodorsally by the a series of white scales; late juveniles and adults of *C. maruliodes* also have a series of dark blotches on the sides, but posterior and posterodorsal scales of these blotches are dark, usually black, and margined with white. The diploid number of chromosome of *C. marulius* is 44 (Donsakul and Magton, 1991).

***Channa punctatus* (Bloch)**

(Figure 19.4)

Ophiocephalus punctatus Bloch, 1793, *Naturges. Ausland. Fische*, 7, p. 139, pt. 358 (type locality: Coromandel coast).

Material examined: Uncat. MUMF. 4 exs., Ikop pat (lake) at Khangabok, 29. XII. 1990, total length = 108-120 mm. MUMF 5141, 1 ex., 63.0 mm SL., Tuivai river, Churachandpur, August 15, 2001, Collected by KSD.

Local name: Ngamu bogra.

Diagnosis: A species of *Channa* with plain pectoral fin. Dorsal fin rays 28-32 and anal fin rays 20-23. pectoral fin extends to anal fin.

Description: D.28-32, P.15-17, V.6.A.20 C. Body elongate, and fairly round in cross section. Eyes moderate. Head depressed, mouth large, lower jaws with 3-6 canine teeth behind a single raw of villiform teeth snout pointed. Anterior nostril tubular. Caudal peduncle shorter than its depth. Edges of dorsal, pectoral, anal and caudal fin rounded. 4 or 5 scale rows between preopercular angle and posterior border of orbit. 38 (37-40) lateral lie scales pelvic fin more than half the length of pectoral fins, extending to anal fin.

Proportional measurements (in %): Body depth 20.6-22.3; caudal fin length 23.7; head length 36.2-38.4; length of caudal peduncle 6.4; its height 10.6; predorsal length 39.3; post dorsal length 60.4; dorsal fin base length 54.4; its height 11.4; pectoral fin length 21.3; ventral fin length 13.4; anal fin base length 37.7% SL; Height of head at occiput 49.1; that at eye 29.7; interorbital space 22.8; eye diameter 15.4-17.4; snout length 18.1-20.4 HL.

Colour: 4-5 longitudinal rows of spots dorsally. Sideward on the anal (bottom) fin there are 3-4 rows of spots. The pale yellow to dark olive rounded caudal fin has 4-5 dark, transverse bands with a light olive transverse band. Caudal fin spotted. A sell defined black shoulder mark above the pectoral fin and 6-7 dark transverse triangular or saddle bands, descending from base of dorsal fin to lateral line. Here, we can also be viewed two-three pairs of blotches and three pairs of blotches and three-four dots. Shaw and Shebbeare (1937) described a browner bodied fish with 12 rays in caudal fin. Dorsal fin with 3 bands. Pectroal fin base with 1 spot and 6-7 rows of transversely arranged spot on the fin. Anal fin with 5 stripes. Edges of dorsal and anal fins are light yellowish or whitish. Caudal fin with irregular arranged 6-7 bands.

Remarks: Unlike other varieties, the pectoral fin is not spotted or striated and it has a higher than average lateral scale count (37-40). It is also important among the Manipuris. It was firstly recorded from Manipur by Hora (1921).

***Channa stewartii* (Playfair)**

Ophiocephalus stewartii Playfair, 1867. *Proc. Zool. Soc. London*; p. 14, pl. 3. (type locality: Cachar, Assam).

Channa stewartii Talwar and Jhingran, 1991, *Inland Fish. India*. 2: 1021.

Material examined: Uncat. MUMF. 1 ex., Jiri river at Jiribam, 18.X. 1986; Total length 245 mm; Collected by I.S. and Rajen. MUMF. 4223-4225, 3 exs; 106.00-120.00 mm SL; Barak river, Vanchengphai, Tamenglong district, March 18, 1999. Syntypes: BMNH 1867. 2.14. 19-20.

Local name: Gofa, Ngamu, Assamese snakehead, Golden snakehead.

Diagnosis: A short *Channa* species without patch of scales on gular region of head. Many scales with well defined circular black spots.

Description: D.39-40; A.27, Ll. 47-50. Pelvic fin about one-third as long as pectoral. Scale rows between preopercular angle and posteror border of orbit 4-5. Predorsal scales 13. Dorsal fin originates above as pectoral fin.

Colour: Black spots on may body scales. A sireesof indistinct dark bands above the lateral line. Dorsal fin with deep blue iridescence along its base, white or white and orange outer edge. Pectoral fin spotted with dark colour to form light and transverse end.

Remarks: They love maggots. Endemic to Brahmaputra (upper, middle, lower) river basin of India and Bangladesh, and the Ganges river basin from southern Nepal southwards (Sen, 1985, Talwar and Thingran 1991, Musikasinthorn, 2000). In southern Nepal it occurs in the Kamala, Bangmati, Koshi, Gandaki, and Karnali river basins (Shrestha, 1990). The diploid chromosome number of *Channa stewartii* is 104 (Rishi and Haobam, 1984). It is only found in the Barak river system of Manipur, and often found feeding on young shorts of bamboo plants in swampy areas away from rivers.

***Channa striata* (Bloch)**

(Figure 19.5)

Ophiocephalus striatus Bloch, 1793, *Naturg. Ausland Fishche*, 7, pl. 141, pl. 359 (type locality: Malabar).

Material examined: Uncat. MUMF. 3 exs; Kharungpat near Kaching 132-155 mm TL. 2 ex. Singjamei Bazar. 130-150 mm TL.

Local name: Porom, Porong (Manipur).

Diagnosis: Body with chevron-shaped bars pointing forward, the upper part usually indistinct in older individuals, lateral line with 57-59+2 scales, lateral

Figure 19.1. *Oreochromis mossambica*

Figure 19.2. *Anabas testudineus*

Figure 19.3. *Polyacanthus fasciatus*

Figure 19.4. *Channa punctatus*

Figure 19.5. *Channa striata*

line scales 6/1/10½, dorsal fin with 44 rays, anal fin with iii, 24 rays; Circumpeduncular scales 32.

Description: D. 44 (37.46) p.ii, 15-17 V.i.5; A.iii, 24 or (23-29) C.11. Body elongated and cylindrical. Snout rounded. Mouth large, lower jaw longer, maxilla and premaxillary process extending to vertical level of beyond posterior margin of orbit. Eyes large. Lateral line complete with 57-58 scales. Lateral lines transverse 6/10½. Circumpeduncular scales 32. Pectoral fin not reaching anal fin. Pelvic fin origin at same vertical level as dorsal fin origin. Anal fin long, when pressed, touches caudal fin. Vent situated at anal fin base. Caudal fin rounded.

Colour: Body with chevron-shaped bars pointing forward, more distinct at lower part. Dorsal fin with 5 inter-radial bands on the last 7 rays. Pectoral and ventral fin plain. Anal fin with 3 bands on the last 4 rays and 5th –7th rays has 1 band at base.

Remarks: On 23 July 2002, the secretary border of the Interior announced a proposed rule to list of the family Channidae (Snakehead fishes) as injurious wildlife under the Lacey Act (18 U.S.C. 42 (Walter *et al* 2004). It was firstly recorded from Manipur by Menon (1952) and followed by Menon (1954). A good food fish, which grows to large size, occurs in the lakes, rivers and swamps. Jayaram (1999) used two specific names in pages 446 and 447 for this species.

Banerjee and others (1988) recorded the diploid chromosome number of *C. striata* from India as 40, but Donsakul and Magtoon (1991) reported a count of 44 for this species from Thailand, indicating that it represents a species complex. *Channa striata* is an obligate air breather.

REFERENCES

Akihito, P. and Meguro, K. 1975. Description of a new gobiid fish *Glossogobius aureus*, with notes on related species of the genus. *Japan J. Ichthyol.*, 22: 127-142.

Akihoto, P. and Meguro, K. 1975. First record of *Glossogobius celebius* from Japan. *Japan J. Ichthyol.*, 21(4): 227-230.

Alfred, E.R. 1964. *Channa bistriata* (Weber and de Beaufort) the young of the snakehead fish *Channa lucius* (Cuvier). *Bull. Natn. Mus. Singapore*, 32: 155-156.

Alfred, E.R. 1968. Rare and endangered freshwater fishes of Malaya and Singapore. *In Technical session-4-Threatened species. Conference on conservation of nature and natural resources in tropical south-East Asia, Bangkok, Thailand IUCN Pub; New series* 10: 325-331.

Allen, G.R. and Burgess, W.E. 1990. A review of the glassfishes (Chandidae) of Australia and New Guinea. *Rec. West. Aust. Mus. Suppl.*, 34:139-206.

Allen, G.R. 1982. *Parambassis altipinnus*, a new species of freshwater glassfish from western New Guinea (Pisces, Ambassidae). *Bull. Zool. Mus. Univ. Amster.*, 8(20): 165-169.

Allen, G.R. 1991. *Field guide to the freshwater fishes of New Guinea.* Christensen Research Institute, Madang, Papua, New Guinea.

Anderson, M.E. and Heemstra, P.C. 2003: Review of the glassfishes (Perciformes, Ambassidae) of the Western Indian Ocean. *Cybium*, 27(3): 199-210.

Archis, R., Kirk, G. and Winemiller, O. 2004. Ontogeny of scale feeding in the Asian. Glassfish, *Chanda nama* (Ambassidae). *Copeia*, 4: 903-907.

Arunkumar, L. 2000. Threatened fishes of Manipur. *Ecol. Env. and Cons.,* 6(2): 185-191.

Arunkumar, L. and Tombi Singh, H. 1997. On a collection fishes from the head-waters of Yu-River System with four new records in Manipur. *J. Freshwater Biol.*, 9(3-4): 126-133.

Arunkumar, L. and Tombi Singh, H. 1999. Fauna of the Yangoupokpi-Lokchao Wildlife Sanctuary, Manipur, India. *Indian J. Forestry*, 22(4): 304-315.

Banister, K.E. 1994. *Glossogobius ankaranensis,* a new species of blind cave goby Madagascar (Pisces: Gobioidei: Gobiidae). *Aqua, J. Ichthyol. Aqua. Biol.,* 1(3): 25-28.

Barlow, G.W.,Liem, K.F. and Wickler, W. 1989. Badidae a new fish family-behavioural, osteological, and developmental evidence. *J. Zool.*, 156:415-447.

Barman, R.P. 1988. On a collection of fishes from river Gumati, Tripura North East India. *Rec. Zool. Surv. India, Occas. Pap.* 19: 1-86.

Britz, R. 2007. *Channa ornatipinnis* and *C. pulchra,* two new species of dwarf snakeheads from Myanmar (Teleostei: Channidae). I*chthyol. Explor. Freshwater,* 18(4): 335-342.

Chakrabarty, P., Oldfield, G. and Ng, H. H. 2006. *Nandus prolixus*, a new species of leaf fish from northeastern Borneo (Teleostei: Perciformes: Nandidae). *Zootaxa,* 1328: 51-61.

Chaudhuri, B.A. 1911. Contributions to the Fauna of Yunnan based on Collections made by J. Coggin Brown, B. Sc., 1909-1910. Part 2. Fishes. *Rec. Indian Mus.*, 6: 13-24.

Chaudhuri, B. I. 1912. Descriptions of some new species of freshwater fishes from north India. *Rec. Indian Mus.*, 7:437-444.

Chu, K.-Y. 1956. A review of the Sciaenoid fishes of Taiwan. *Rep. Inst. Fish. Boil., Minist. Economic Affairs and natn. Taiwan Univ.*, 1(1): 13-46., pls. 1-5.

Coburn, M.M. and Gaglione, J.I. 1992. A comparative study of percoid scales (Teleostei, Perciformes). *Copeia,* 4: 986-1001.

Craig, John F. 2000. *Percid Fishes systematic, Ecology and Exploitation.* Nigel Balmforth, professional division, Blackwell Science, Osney Mead, Oxford OX2OEL, UK; XVII+352pp.

Datta, N.C. and Chaudhuri, S. 1993. Two new glassfishes from Myanmar. *Indian Biologist.*, 25(1): 1-4.

Datta, N.E. and Laishram, I.S. 1984. Zoogeographical significance of the ichithyofauna of Manipur, India. *Intl. J. Acad. (Proc. IV AISI)*, 5: 83-93.

Day, F. 1878. *The fishes of India; being a natural history of the fishes known to inhabit the sea and fresh waters of India, Burma and Ceylon.* Text and Atlas in 4 parts. London 778pp; 195pls.

Day, F. 1889. *Fauna of British India including Ceylon and Burma.* Fishes Vol.1. 245-302pp. London, Taylor and Francis.

Dutt, S. and Reddy, P.B. 1979. On the Snake-head fishes of Andhra pradesh. *Mem. Soc. Zool. Guntur,* 1: 103-108.

Eschmeyer, W.N. 1990. *Catalog of the genera of recent fishes.* San Francisco, California Academy of Sciences, 697 pp.

Fraser-Brunner, A. 1955. A synopsis of the centropomid fishes of the subfamily Chandidae, with description of a new genus and two new species. *Bull. Raff. Mus. Singapore,* 25:185-213.

Geevarghese, C. and John, P.A. 1983. A new goby, *Glossogobius minutus* (Teleostei: Gobiidae) from the South-west coast of India. *J. Fish. Biol.*, 22: 231-240.

Gill, T. 1862. Revision of the genera of North American Sciaeninae. *Proc. Acad. Nat. Sci. Phila.*, 13:79-89.

Greenwood, P.H. 1976. A revision of the family Centropomidae (Pisces, Perciformes). *Bull. Br. Mus. Nat. Hist. (Zool.),* 29:1-81.

Greenwood, P.H. 1977. A review of the family Centropomidae (Pisces: Perciformes): an appendix. *Bull. Br. Mus. Nat. Hist. (Zool),* 31(6): 297-301.

Greenwood, P.H., Rosen, D.E., Weitzman, S.H. and Meyers, G.S. 1966. Phyletic studies of Teleostean fishes, with a provisional classification of living forms. *Bull. Amer, Mus. Nat. Hist.*, 131(4): 339-455.

Guha, S. and Talwar, P.K. 1981. Generic Status and the type selection for the Indian Glassfish, *Ambassis thomassi* Day (Pisces: Ambassidae), with remarks on its natural distribution. *J. Inland Fish. Soc. India,* 13(2): 50-56.

Gunthar, A. 1861. Catalogue of the Acanthopterygian fishes in the collection of the British Museum. Vol. 3. *Brit. Mus. Trustees, London,* 586pp.

Gupta. S.K. 1984. A note on the Systematics and bionomics of *Chanda* spp. (Centropomidae). *Indian J. Fish.*, 31(3): 352-356.

Hamilton, F. 1822. *An account of the fishes fond in the river Ganges and its branches.* Archibald Constable and Co. Edinburgh and London, 408pp.

Herre, A.W.C. T. and Myers, G.S. 1937. A contribution to the ichthyology of the Malay Peninsula. *Bull. Raffles Mus.*, 13: 5-75.

Hoese, D.F. and Allen, G.R. 1990. Description of two new species of *Glossogobius* (Pisces: Gobiidae) from northern Papaua Guinea. *Rec. West. Aust. Mus. Suppl.*, 34:117-129.

Hora, S.L. and Mukerji, D.D. 1935. Fish of Naga Hills, Assam. *Rec. Indian Mus.*, 36: 571-573.

Hora, S.L. 1921. Fish and fisheries of Manipur with the some observations on those of Naga Hills. *Rec. Indian Mus.*, 22: 165-214.

Jayaram, K.C. 1981. *The freshwater fishes of India, Pakistan, Bangladesh, Burma and Sri Lanka – A handbook.* Edited by the Director, Zoological survey of India, Calcutta, 475pp., XIII pls.

Jayaram, K.C. 1999. *The freshwater fishes of the Indian region.* Narendra Publ. House, New Delhi, 551 pp.

Kottelat, M. 1985. Freshwater fishes of Kampuchea - A provisory annotated Check-list. *Hydrobiologia,* 212: 249-279.

Kottelat, M. 1991. Notes on the taxonomy and distribution of some western Indonesian freshwater fishes, with diagnoses of a new genus and six new species (Pisces: Cyprinidae, Belontiidae, and Chaudhuriidae) *Ichthyol. Explor. Freshwaters,* 2(3): 273-287.

Kottelat, M. 2003. *Parambassis pulcinella,* a new species of glass perch (Teleostei: Ambassidae) from the Ataran River basin (Myanmar), with comments on the family-group names Ambassidae, Chandidae and Bogodidae. *Ichthyol. Explor. Freshwaters,* 14:9-18.

Kottelet, M. 1989. Zoogeography of the fishes from Indochinese inland waters with an annotated check list. *Bull. Zool. Mus. Unv. Amsterdam,* 12:1-55.

Koumans, F.P. 1941. Gobiod fishes of India. *Mem. Indian Mus.* 13(3): 205-329.

Koumans, F.P. 1953. *Gobioidea. In: The fishes of Indo-Australian Archipelago.* X (M.Weber and L.F. De Bearefort eds.). PP. XIII and 423. Leiden E.J. Brill.

Kullander, S.O. and Britz, R. 2002. Revision of the family Badidae (Teleostei: Perciformes), with description of a new genus and ten new species. *Ichthyol. Explor. Freshwaters,* 13(4): 295-372.

Lee, P.G. and Ng, P.K.L. 1994. The systematics and ecology of snakeheads (Pisces: Channidae) in peninsular Malaysia and Singapore. *Hydrobiologia,* 285:59-74.

Liem, K.F. 1963. The comparative osteology and phylogeny of the anabantoidei (Teleostei, Pisces). *Ill. Biol. Monogr. No.* 30,149pp.

Liem, K.F. 1965. The status of the anabantoid fish genera *Ctenops* and *Trichopsis Copiea*, 1965(2): 206-213.

Martin, T.J. and Heemstra, P.C. 1988. Identification of *Ambasisis* species (Pisces: Perciformes, Ambassidae) from South Africa. *S. Afr. J. Zool.*, 23(1): 7-12.

Martin, T.J. and Blaber, S.J.M. 1983. The feeding ecology of Ambassidae (Osteichthyes: Perciformes) in Natal estuaries. *S. Afr. J. Zool.*, 18(4): 353-362.

Martin, T.J. 1989. Niche separation in *Ambassis* (Pesces: Perciformes) species co-existing in estuaries of South-east Africa, *J. Fish. Biol. Supp. A.*, 35:321-328.

Menon, A.G.K. 1954. Further observation on the fish fauna of the Manipur state. *Rec. Indian Mus.*, 52(1): 21-26.

Menon, A.G.K. 1974. A check list of the fishes of the Himalayan and the Indo-Gangetric plains. *Spl. Publ. No. 1. Inland Fisheries Soc. India, Barrackpore*, 136pp.

Menon, A.G.K. 1954. Further observations on the fish fauna of the Manipur state. *Rec. Indian Mus.*, 52:21-26.

Menon, A.G.K. 1999. Checklist of freshwater fishes of India. *Rec. Zool. Surv. India, Occ. Pap.* 175:1-366.

Menon, M.A.S. 1952. On a small collection of fish from Manipur, Assam. *Rec. Indian Mus.*, 50:265-270.

Menon, M.A.S. 1952. On a small collection of fish from Manipur, Assam. *Rec. Indian Mus.*, 50: 265-270.

Mohan, R.S. Lal. 1969. On three new genera of Sciaenid fishes (Pisees: Sciaenidae) from India. *Curr. Sci.*, 38: 295-296.

Mohan, R.S. Lal. 1972. A synopsis of the Indian genera of the fishes of the family Sciaenidae. *Indian J. fish.*, 16(1 and 2): 82-89.

Mohan, R.S. Lal. 1976. Two new species of Sciaenid fishes *Johnius elongatus* and *Johnieops macrorhynus* from India. *Matsya,* 1: 19-25.

Mohan, R.S. Lal. 1981. An illustrated synopsis of the fishes of the family Sciaenidae of India. *Indian J. Fish.*, 28(1 and 2): 1-24.

Mohan, R.S. Lal. 1984. Systematic position of the Sciaenid *Corvina carouna* Cuvier, 1830. *Indian J. Fish.*, 31(3): 374-377.

Munshi, J. S. Data, and Srivastava, M.P. 1988. *Natural History of Fishes and systematic of freshwater fishes of India.* Nerendra Publishing House, New Delhi, 403pp.

Murray, A.M. 2006. A new channid (Teleostei: Channiformes) from the Eocene and Oligocene of Egypt Murray. *J. Paleon.*, 80: 117-1178.

Musikasinthorn, P. and Taki, Y. 2001. *Channa siamensis* (Gunther 1861), a junior synonym of *Channa lucius* (Cuvier in Cuvier and Valenciennes, 1831). *Ichthyol. Res.*, 48(3): 319-324.

Musikasinthorn, P. 1998. *Channa panaw,* a new channid fish from the Irrawaddy and Sittang river basins, Myanmar. *Ichthyol. Res.,* 45(4): 355-362, figs. 1-7.

Musikasinthorn, P. 1998. First record of *Parambassis lala* (Pisces: Ambassidae) from Inle Lake, the Salween river basin, Myanmar. *Nat. Hist. Bull. Siam Soc.*, 46: 203-205.

Musikasinthorn, P. 2000. *Channa aurantimaculata,* a new channid fish from Assam (Brahmaputra River Basin), India, with designation of a neotype for *C. amphibeus* (McClelland, 1845). *Ichthyol. Res.*, 47(1): 27-37.

Nelson, J.S. 1976. *Fishes of the world.* John Wiley and Sons, New York. 416pp.

Nelson, J.S. 1994. *Fishes of the world.* Third Edition, John Wiley and Sons, Inc. XVII+600pp.

Ng, H.H. and Lim, K.K.P. 1990. Snakeheads (Pisces; Channidae): *Natural history, biology and economic importance*. Essays in Zoology. Papers commemorating the 40th Anniversary of the Department of Zoology, National university of Singapore: 127-152.

Ng, H.H., Ng P.K.L. and Britz, R. 1999. *Channa harcourtbutleri* (Annandale, 1918): a valid species of Snakehead (Perciformes, Channidae) from Myanmar. *J. South Asian. Nat. Hist.* 4(2):

Ng, H.H., Vidthayanon, C. and Ng, P.K.L. 1996. *Nandus oxyrhynchus*, a new species of leaf fish (Teleostei: Nandidae) from the Mekong basin. *Raff. Bull. Zool.* 44(1): 11-19.

Ng, H.H., Tan, S.H., Ng, P.K.L. 1995. Revalidation of *Channa baramensis* (Steindachner, 1901), a species of snakehead from northern Borneo (Teleostei: Channidae) *Sarawak Mus. J.*, XLVIII (69): 219-226.

Ng, P.K.L. and Lim, K.K.P. 1990. The Black snakehead, *Channa melasoma* (Bleeker, 1851) (Channidae): First record from Singapore. *Raff. Bull. Zool.*, 38:21-24.

Ng, P.K.L. and Lim, K.K.P. 1991. The identity of *Ophicephalus cyanospilos* Bleeker from Sumatra, and a new record of *Channa bankanensis* (Bleeker) from Peninsular Malaysia (Pisces: Channidae). *Raff. Bull. Zool.*, 39:119-130.

Ng, P.K.L. and Lim, K.K.P. 1989. Rediscovery of the Dwarf Snakehead, *Channa gachua* (Hamilton, 1822) (Channidae) in Singapore. *Raff. Bull. Zool.*, 37:172-174.

Ng, P.K.L. and Kottelat, M. 1994. Revision of the *Betta waseri* species group (Teleostei, Belontiidae). *Raff. Bull. Zool.*, 42(3): 593-612.

Ng. P.K.L., Tay, J.B. and Lim, K.K.P. 1994. Diversity and conservation of blackwater fishes in Peninsular Malaysia, Particularly in the North Selangor peat swamp forest. *Hydrobiologia*, 285: 203-218.

Ng, HH and Jaafar, Z. 2008. A new species of leaf fish, *Nandus andrewi* (Teleostei: Perciformes: Nandidae) from northeastern India. *Zootaxa*, 1731: 24-32.

Pethiyagoda, R. 1991. *Freshwater fishes of Sri Lanka*. Wildlife Heritage Trust. Colombo, Sri Lanka, 362pp.

Pezold, F. 1993. Evidence for a monophyletic Gobiinae. *Copiea*,(3): 634-643.

Playfair R.L. 1867. One the fishes Cachar. *Proc. Zool. Soc. Lond. (Part. 1)*: 14-17, 3pls.

Rainboth, W. 1996. *Fishes of the Cambodian Mekong*. FAO Species identification field guide for fishery purposes. Food and agriculture Organization of the United Nations, Rome, 265pp., 27pls.

Rao, Seshagiri, B.V. 1968. Systematic studies on *Anabas testudineus* (Bl. 1872) and *A. oligolepis* Bleeker, 1855. *Proc. Indian. Acad. Sci.* 57(5): 207-214.

Reddy, W.P.B. 1978. Studies on the taxonomy of Indian species of family Channidae (Pisces: Teleostei) and some aspects of the biology of *Channa punctata* (Bloch, 1993) from Guntur, Andhra Pradesh. *Matsya,* 4:95-96.

Rema Devi, K. 1992. Gobioids of Ennore estuary and vicinity. *Rec. Zool. Surv. India*, 90(1-4): 161-189.

Rishi, K.K. and Haobam, M.S. 1984. Karyotype of three forms of fishes having high chromosome number: Modinagar, India. *Intl. J. Acad. Ichthyol.*, 5 (½): 139-144.

Roberts, T.R. 1978. An ichthyological Survey of the Fly River in Papua New Guinea with description of a new species. *Smithsonian Contributions to Zoology*, 281: 1071.

Roberts, T.R. 1989. The freshwater fishes of western Borneo (Kalimantan Barat, Indonesia). *Mem. Calif. Acad. Sci.*, 14, 210pp.

Roberts, T.R. 1993. The freshwater fishes of Java, as observed by Kuhl and van Hasselt in 1820-23. *Zool. Verhand*, 285:1-94.

Roberts, R.T. 1994. Systematic revision of the Tropical Asian Freshwater glass perches (Ambassidae), with description of the new species. *Nat. Hist. Bull. Siam Soc.*, 42:263-290.

Ruber, L., Britz, R., Kullander, S.O. and Zardoya, R.. 2004. Evolutionary and biogeography patterns of the Badidae (Teleostei: Perciformes) inferred from mitochondrial and nuclear DNA sequence data. *Mol. Phylo. Evol.*, 32(3): 1010-1022.

Sasaki, K. and Amaoka, K. 1989a. *Johnius distinctus* (Tanaka, 1916), a senior synonym of *J. tingi* (Tang, 1937) (Perciformes, Sciaenidae). *Japan J. Ichthyol.*, 35(4): 446-468.

Sasaki, K. and Amaoka, K. 1989b. Records of *Johnius vogleri* (Perciformes, Sciaenidae) from Chinese waters, with notes on its synonym. *Japan J. Ichthyol.*, 36(1): 125-128. figs. 1-2.

Sasaki, K. and Kailola, P.J. 1991. *Johnius* (Johnius) *laevis*, a new sciaenid from Northern Australia and Papua New Guinea. *Japan J. Ichthyol.*, 38(2): 119-123.

Sasaki, K. 1989. Phylogeny of the family Sciaenidae with notes on its zoogeography (Teleostei, Perciformes). *Mem. Fac. Fish; Hokkaido Univ.*, 36(1/2): 1-137.

Sasaki, K. 1990. *Johnius grypotus* (Richardson, 1846), resurrection of a Chinese sciaenid species. *Japan J. Ichthyol.*, 37(3): 224-229.

Sasaki, K. 1992. Two new and two resurrected species of the sciaenid genus *Johnius* (*Johnius*) from the west Pacific. *Japan. J. Ichthyol.*, 39(3): 191-199.

Sasaki, K. 1994a. *Johnius ąneus* Bloch, a senior synonym of *Pennahia macrophthalmus* (Bleeker), with comments on the identity and status of an alleged lectotype of *J. aneus* (Sciaenidae: Perciformes). *Japan J. Icthyol.*, 40(4): 498-499.

Sasaki, K. 1994b. *Johnius mannarensis* Mohan (Sciaenidae); a valid species distinct from *J. macropterus* (Bleeker). *Japan J. Ichthyol.*, 41(1): 65-67.

Selim, K. and Vishwanath, W. 1997. Ichthyological investigation in Chatrickong River of Ukhrul District, Manipur. *J. Freshwater Biol.*, 9(3-4): 134-137.

Sen, T.K. 1985. The fish fauna of Assam and the neighboring north-eastern states of India. *Rec. Zool. Surv. India, Occas. Pap.* 64: 1-216.

Silas, E.G. 1952a. Speciation among the freshwater fishes of Ceylon. Symposium on organic Evolution. *Bull. No. 7. Nat. Inst. Sci. India, New Delhi*, 248-259.

Talwar, P.K. and Jhingran, A.G. 1991. *Inland fishes of India and adjacent countries.* Vol. 2. Oxford –IBH Pub. Ltd., New Delhi, 616pp.

Talwar, P.K. and Shetty, H.P.C. 1971. On the generic relationship of *Bola coitor* Hamilton, 1822 (Pisces: Sciaenidae) with a redescription of the species. *Proc. Indian Acad. Sci.*, 74(2): 74-80.

Talwar, P.K. 1971. *Chanda* Hamilton-Buchanan, 1822 (Pisces: Centropomidae); proposed use of the plenary power to designate *Chanda nama* Hamilton-Buchanan, 1822, as the type species. *Bull. Zool. Nomencl.*, 28(3and4): 104-105.

Talwar, P.K. 1971. *Chanda* Hamilton-Buchanan, 1822 (Pisces: Centropomidae); Proposed use of the Plenary power to designate *Chanda nama* Hamilton-Buchanan, 1822 as the type species. *Bull. Zool. Nomencl.*, 28(3and4): 104-105.

Talwar, P.K. 1995. Fauna of India and the adjacent counters Pisces, Perciformes. Sciaenidae. *Zoo. Surv. India; Calcutta*, 1-144.

Tan, S.H. and Tan, H.H. 1994. The freshwater fishes of Pulau Bintan Riau Archipelango, Sumatra, Indonesia. *Tropical Biodiversity,* 2(3): 351-367.

Trewavas, E. 1961. The sciaenid fishes with a singhle mental barbel. *Copeia,* (1): 107-117.

Trewavas, E. 1962. A basis for classifying the sciaenid fishes of tropical West Africa. *Ann. Mag. Nat. Hist.*, 13(5): 167-176.

Trewavas, E. 1977. The Sciaenid fishes (croakers or drums) of the Indo-west Pacific. *Trans. Zoo. Soc. London*, 33: 253-541.

Tweedie, M.W.F. 1936. A list of fishes in the collection of the Raffles Museum. *Bull. Raff. Mus.*, 12: 16-28.

Tweedie, M.W.F. 1940. Additions to the collection of fishes in the Raffles Museum. *Bull. Raff. Mus.*, 16: 68-82.

Tweedie, M.W.F. 1950. Notes on Malayan Freshwater Fishes. 2. The species of *Channa* Scopoli (=*Ophicephalus*) in the collection of Raffles Museum. *Bull. Raff. Mus.*, 21: 99-105.

Vierke, J. 1991. Ein farbenfroher neuer Schlangenkopffisch aus Assam *Channa bleheri* spec. nov. *Das Aquarium,* 259: 20-24.

Vinod, K., Mahapatra, B.K. and Mandal, B.K. 2006. Fish biodiversity of Nagaland and their conservation status. *Envir. Eco.*, 24 S(1): 203-207.

Vishwanath, W. and Shanta, K. 2004. A new fish species of the Indo-Burmese Genus *Badis* Bleeker (Teleostei: Perciformes) from Manipur, India. *Zoos' Print J.*, 19(9): 1619-1621.

Vishwanath, W. 2000. *Fish fauna of Manipur*. Manipur Association for Science and Society (MASS), 143 pp.

Vishwanath, W. and Geetakumari, Kh. 2009. Diagnosis and interrelationships of fishes of the genus *Channa* Scopoli (Teleostei: Channidae) of northeastern India. *J. Threatened Taxa*, 1(2): 97-105.

Vishwanath, W., Manojkumr, ,Kosygin, L. and Selim, K.S. 1998. Biodiveristy of Freshwater fishes of Manipur, India. *Italian J. Zool.*, 65 Supplementary: 321-324.

Vishwanth, W. and Linthoingambi, I. 2004. Fishes of the genus *Colisa* Cuvier from Manipur and first record of *Colisa labiosus* (Day) from India. *J. Bombay Nat. Hist. Soc.*, 101(3): 466-469.

Weber, M. and de Beaufort, L.F. 1922. *The fishes of the Indo-Australian Archipelago. 4. Heteromi, Solenichthys, Synentognathi, Percesoces, Labyrinthici, Microcyprini*. E.J. Brill, Leiden, 410pp.

Weber, M. and de Beaufort, L.F.1936. *The fishes of the Indo-Australian Archipelago*. Vol. 7. E.J. Brill, Leiden, 607pp.

Website: http://www.Snakeheads.co.UK/Species htm..

Zakaria-Ismail, M. 1987. The fish fauna of the Ulu Endau river system, Johore, Malaysia, *Malay, Nat. J.*, 41: 403-411.

Zakaria-Ismail, M. and Lim, K.K.P. 1995. The fish fauna of Tasik Temengor and its tributaries South of Banding, Hulu Perak, Malaysia. *Malay. Nat. J.*, 48: 319-332.

Zakaria-Ismail, M. 1994. Zoogeography and biodiversity of the freshwater fishes of Southeast Asia. *Hydrobiologia*. 285:41-48.

Zakaria-Ismail, M. 1984. Checklist of fishes of Taman Negara. *Malay. Nat.*, 37:21-26.

Zakaria-Ismil, M, 1993. The fish fauna of the Sungai Teris and Sungai Rengit Krau Game Reserve, Pahang, Malaysia. *Malay. Nat. J.*, 46:201-228.

Zhang, C.G., Musikasinthorn, P. and Watanabe, K. 2002. *Channa nox.*, a new channid fish lacking a pelvic fin from Guangxi, China. *Ichtyol. Res.*, 49(2): 140-146.

20

Fish Breeding Grounds on Lohit River, Arunachal Pradesh: A Case Study

—*Pranab Nath*[1]

[The present study focuses on the fish breeding sites in Lohit river from Tidding confluence to its confluence with river Brahmaputra, Lohit district, Arunachal Pradesh, India. Three categories of fish namely, stenothermic, eurythermic and exclusively warm water fish species have been recorded from the stretch of Lohit river. The study revealed presence of breeding grounds of the above mentioned fishes; however they are under several threats and need appropriate measures to be adopted for better management.

Keywords: Breeding ground, conservation, Lohit river]

Introduction

India is endowed with vast freshwater bodies consisting of 45,000 Km. of rivers, 26,334 Km. of canals, 2.36 million hectares of ponds and tanks and 2.05 million hectares of reservoirs, which present like harbor a rich and diversified fish fauna characterized by many rare and endemic fish species. About 21,730 species of fishes have been recorded in the world; of which, about 11.7% are found in Indian waters. Out of the 2546 species so far listed, 73 (3.32%) belong to the cold freshwater regime, 544 (24.73%) to the warm freshwater domain, 143 (6.50%) to the brackish water and 1440 (65.45%) to the marine ecosystem.

The State of Arunachal Pradesh is rich in fresh water resources in the form of rivulets and lentic water bodies that is reflected by the presence of rich fish diversity. Arunachal Pradesh (AP) state is situated in the North-Eastern part of India between 26° 28' to 29° 30' N and 90° 30' to 97° 30′ E, with 83,743 km2 total area. Almost 95 % of the streams and rivers in the state drain in to the mighty Brahmaputra river and the rest drain into the Chindwin-Irrawaddy drainage. The river Brahmaputra also receives water from the state of Assam and ultimately discharges the contents in to the Bay of Bengal. The Siang river is the biggest river in AP which is originated in Tibet where it is known as Tsangpo. The major rivers traversing the state are Lohit, Dibang, Siang, Siyom, Subansiri and Kameng. The Lohit river or (Parnne and Rongtochu in Arunachal Pradesh) originates in the upper reaches of the Himalayas. It is one of the largest rivers of the Arunachal Himalayas, in which

[1] State Fisheries Department, Government of Arunachal Pradesh, Itanagar.

environmental conditions vary considerably along the elevational gradient; therefore, fish assemblages reflect more equitable structuring by biotic and abiotic factors. Ichthyofaunal diversity of Lohit river comprises of 46 species under 28 genera (Nath and Dey, 2000). The fish assemblages in lower stretch show close resemblance with the fishes of the other parts of the Himalayan region (e.g. *Barilius barna, B. bendelisis, Crossocheilus latius latius, Glyptothorax pectinopterus, Schizothorax progastus* etc) and Assam, Manipur and Nagaland (e.g. *Euchiloglanis kamengensis, Aborichthys kempi, Amblyceps arunachalensis, Amblyceps apangi, Bhavania arunachalensis, Sisor rhabdophorus, Olyra longicaudata* etc.).

Material and Methods

Twenty sites were randomly selected for the purpose of sampling of the probable fish breeding areas of river stretch of Lohit from Tidding confluence to its confluence with Brahmaputra river covering a stretch of 130 km. At each randomly selected sampling site the maximum possible length of the stream/ river was covered and fish breeding grounds were onserved. A review of existing literature on fish populations in the proposed area of study was undertaken. Both published and unpublished sources were studied.

Primary survey

Primary survey was conducted to locate the fish breeding areas in the stretch of Lohit river from Tidding confluence to its confluence with Brahmaputra river. Study was carried out for a period of six month. Local fishermen were also interviewed to gather information regarding the fish breeding sites in Lohit river. Hydrological study was carried out in the stretch of Lohit river from Tidding confluence to its confluence with river Brahmaputra. The water current was measured with the help of float method. A 20 m stretch of the river was measured and marked at both ends. A float was thrown at upper end and the time taken by the float to travel the marked distance, was recorded by a stop watch. The water temperature at surface, column and bottom of the river were recorded with the help of graduated mercury thermometer. Average value of these readings was considered. The pH was recorded with the help of pH Scan (Eutech) and pH meter (EI – 132 E) in the field. The turbidity was recorded with the help of Nephelometer or turbidometer (EI – 331 E).Dissolved oxygen was measured by using digital DO meters (Eutech ECDO 602K). Total hardness, were analyzed following APHA (1992) and Adoni (1985).

Results and Discussion

The temperature of the Lohit river water ranged from 12 to 18.5 ^{0}C in the stretch of Lohit from Tidding confluence to its confluence with Brahmaputra river. The dissolved oxygen showed a negative correlation with the water temperature in Lohit river. The dissolved oxygen concentration varied from 9.09 – 10.73 mg/l. In terms of Hardness, the water of Lohit river was soft and optimum ranges were recorded (36 – 52 mg/l). Free carbon dioxide (CO_2) was low in the river, which ranges between 0.5-0.8 mg l^{-1} (Nath and Dey, 2000). The velocity of Lohit river water in the stretch under consideration varied from 1.45-2.048 m/sec

Three categories of fish have been recorded from the stretch of river Lohit under consideration. They were stenothermic, eurythermic and exclusively warm water fish species.

In respect of the stenothermic species, they are located in the upper and middle stretches of the Lohit river where the substratum consists of rocks and boulders with sand and silt patches and some water pools having optimum temperature.

For eurythermic species inhabits substratum having stony boulders with pebbles and gravel, water pools etc. The favourable conditions required by the fishes are furnished below:

(i). Water should be warmer than that of the usual glacier streams

(ii) Water should be shallow and sheltered

(iii) Many snow-fed rivers have cold water even in the plains (same as in the case of Lohit river). In such cases, the fishes migrate up in small streams/ tributaries which contain warm water coming from low hills. The fishes, during floods apparently move up in the streams and after spawning there, they return back to the main river after the spate.

The warm water fish groups found in the lower stretches (upto elevation of 200m) and even in the Brahamputra river. They comes upstream during monsoon months and usually breeds in the ox-bow lakes, beels, swamps, etc. adjacent to the Lohit river. After spawning, these fishes return back to their original places.

The present study revealed that the breeding ground for stenothermic species were located at upper reaches, where the temperature of water is comparatively low. A good number of potential streams/ tributaries are forming the ideal habitats for breeding along with few water pools. The brood fishes and juvenile forms were found in these riffles and pools in almost dry river bed during monsoon and post monsoon season.

Similarly, for the eurythermic fish species, the breeding grounds have been observed in the tributaries located in the downstream at Digaru, Kamlang, Lai nala, Nao-dihing, etc. Besides along the bank of Lohit river, hiding places below the big boulders have also been noticed. Further, in respect of plain water forms (warm water fish species) have breeding grounds located in the downstreams reaches of Lohit river at several beels/ ox-bow lakes as well as its tributaries.

In downstream reaches (3 km and beyond) of Parsuramkund, there exist quite a good number of small water pools which are feeding and breeding grounds of various fish species, however, they are presently under threat due to non regulated fishing measures.

Conservation Steps

Some important conservation measures for fishes of Lohit river and their breeding grounds are furnished below:

1. Cage and Pen Culture methods may be adopted.

2. Supply of fish seeds for pond culture, paddy cum fish culture, integrated fish farming, ornamental fish culture, etc. may be proposed.
3. Strict ban on fishing the species from the wild during its peak breeding season of June, July and October may be implemented.
4. The use of gill and cast nets for catching the fish may be regulated.
5. The fishing with the use of mechanical devices such as bows and arrows, and spears, or firearms may be prohibited.
6. *In-situ* and *ex-situ* technologies to conserve and protect the fish stocks in our river systems may be developed for ensuring, natural breeding, habitat restoration and unstressed food-chain.
7. Alternative strategies have to be formulated to compensate the fish yield decline, from river systems, as far as possible, through ecological management of other manageable open-water ecosystems.
8. Efforts should be made to document the status, distribution pattern, habitant requirements and conservation strategy for the fish fauna falling under the RET schedule.

REFERENCES

Adoni, A.D., Ghosh, G., Chourasia, S.K., Vaishya, A.K., Yadav, M. and Verma, H.G. 1985. *Workbook on Limnology*. Pratibha Publishers, 216 pp.

APHA. 1992. *Standard Methods for the Examination of Water and Wastewater* (18th ed.). American Public Health Association, Washington D.C., 1198 pp.

Bagra K. , Kadu, K., Nebeshwar-Sharma, K., Laskar, B.A., Sarkar, U.K. and Das, D.N. 2009. Ichthyological survey and review of the checklist of fish fauna of Arunachal Pradesh, India. *Check List* 5(2): 330–350.

Bhakta, J., Bandyopadhyay, N. and Parasitology, P. K. 2008. *Fish Diversity in Freshwater Perennial Water Bodies in East Midnapore District of West Bengal*. India Laboratory, Department of Zoology, University of Kalyani, Kalyani-741 235 West Bengal, India

Colin, P., James S., and Peterson, T. 2007. An Evaluation of the Relative Influence of Habitat Complexity and Habitat Stability on Fish Assemblage Structure in Unregulated and Regulated Reaches of a Large Southeastern Warm water Stream. 136(4): 943-958.

Nath P. and Dey, S.C. 2000. *Fish and Fisheries of North Eastern India (Arunachal Pradesh)*, 159 pp.

Microbiological Studies on Effluent Points of Nambul River, Manipur

—R.K. Rajeshwari Devi[1] & Haobijam Dhamendra[2]

[Microbial fauna especially the coliform bacteria were surveyed and study in different seasons and at three effluents points of Nambul river. Highest contents of fecal coliform (6500/100ml) and total coliform bacteria (15200/ml) were recorded at Maharanithong, where Naga Nalla effluent meets the river. The presence of coliform bacteria in the river throughout the year and the ratio of FC/FS exceeding 4.0 suggest that the river is contaminated by domestic waste and human and animal excreta. The river water is considered not suitable for domestic used without proper treatment.

Key words: Coliform bacteria, Fecal streptococci, Nambul river, Manipur]

Introduction

The coliform group of bacteria gram negative nonperforming bacilli, which are commonly found in the intestinal tract of both human being and animals, are used as indicator of water pollution. They are mainly originated from the excreta of both human and animals. Their concentration is very high in the effluents sewages. Higher count of coliform group of bacteria indicates organic contamination in water (Royee and Prakasan, 2000). These coliform groups of organisms present in water causes various water born diseases like diarrhea, cholera and dysentery. These coliform bacteria include the genus *Aerobactor*, *Salmonella*, *Staphylococcus*, *Shingella*, *Streptococci*, *Enterobacter*, *Klebsiella*, *Proteus* etc. The *Escherichia coli* are entirely of human origin but exclusive estimation is difficult and hence the entire coliforms are used as indicator. The actual coliforms are not reported but they are reported as an approximate counts i.e., Most probable Number (MPN).

Bacterial population in water is highly variable. They show consistently higher during summer and rainy seasons than winter months. The seasonal fluctuation is mainly due to variation in water temperature and nutrient compositions (Shrama and Bhardwaj, 2000). Bacteria can survive and reproduce more successfully under warm environmental condition (Hendricks and Morrison, 1967). Seasonal variation in the densities of bacteria are also due to high degree of faecal waste run off from

[1] Department of Zoology, Oriental College, Imphal 795 001, Manipur
[2] Lake Monitoring and Research Lab., Loktak Development Authority, Ningthoukhong 795 126, Manipur

the disturbed as well as non-disturbed catchments areas (Sharma and Rajput, 1996). Moreover, physio-chemical parameters are also influenced on the seasonal variation of bacterial population. Some of the parameters like BOD, COD and chloride Ion concentration had found some influence on the seasonal variation of enteropathogenic bacterial population (WHO and UNEP, 1991). The nutrient availability and temperature are decisive factors in the selection of the types of bacterial genera in water bodies (Copper and Knight, 1989). The bacteriological analysis of water samples generally considered suitable for monitoring the safety aspect of treated water samples and for human consumption.

In Manipur, the Nambul river received maximum number of effluents and sewages, since it drains the thickly populated area of the Imphal city (Figure 21.1-21.7). In the present investigation, an attempt was made to assess the population and seasonal variability of the microbial organisms viz., the total coliform, fecal coliform and fecal streptococci in the Nambul river.

Materials and Methods

Water sample were collected from three different stations along the river on monthly basis from February 2005 to March 2006. Estimations of different bacterial (total coliform, fecal coliform, and fecal streptococci) were performed following standard method of SPC and membrane filter technique with selective media (Nutrient Agar, Endo-Agar, M-Fc Agar and M-Enterococcus Agar). Pour plate technique was employed for the enumeration of total bacterial count. Isolation and verification of bacteria were performed by using the methods of Trivedy and Goel (1984), Buchanan and Gilbon (1987) and APHA (1989). In the present study March to May is consider as summer season, June to October as rainy season and November to February as winter season.

Result and Discussion

The concentration of micro-organisms in Nambul river varies seasonally and at different locations of the river depending on the effluents and sewage discharge into the river (Figure 21.8-21.10). In the present study the fecal coliform ranged from 600 to 6500/100ml with an average value of 3550/100 ml. (Figure 21.11). The maximum value of fecal coliform was observed in Maharanithong area as this area receives the main effluent from Naga Nalha (containing Imphal market run-off). The minimum value of fecal coliform was observed in Naoremthong area. The total coliform (TC) was fluctuated from 1350 - 5200/100 ml. with the mean value of 8275/100 ml (Table 21.1). The mean value of total coliform upto 5000/100 ml. (MPN) in effluent water is categories as drinking water source with conventional treatment followed by disinfection (Effluent Standard 152296/1982). However, in the present study the average value of total coliform is more than the permissible limit. Therefore, the river water is not suitable for domestic use without proper treatment. The fecal streptococci (FS) ranges from 160 - 430/100 ml. with an average value of 295/100 ml (Table 21.1). The different in population might be due to survival rate to different environmental factors as well as the sources of pollution from where they have originated and accounted for survival. Hill and Webb (1958) also reported that the variability of bacterial count at different sites which they correlated with the sources of pollution.

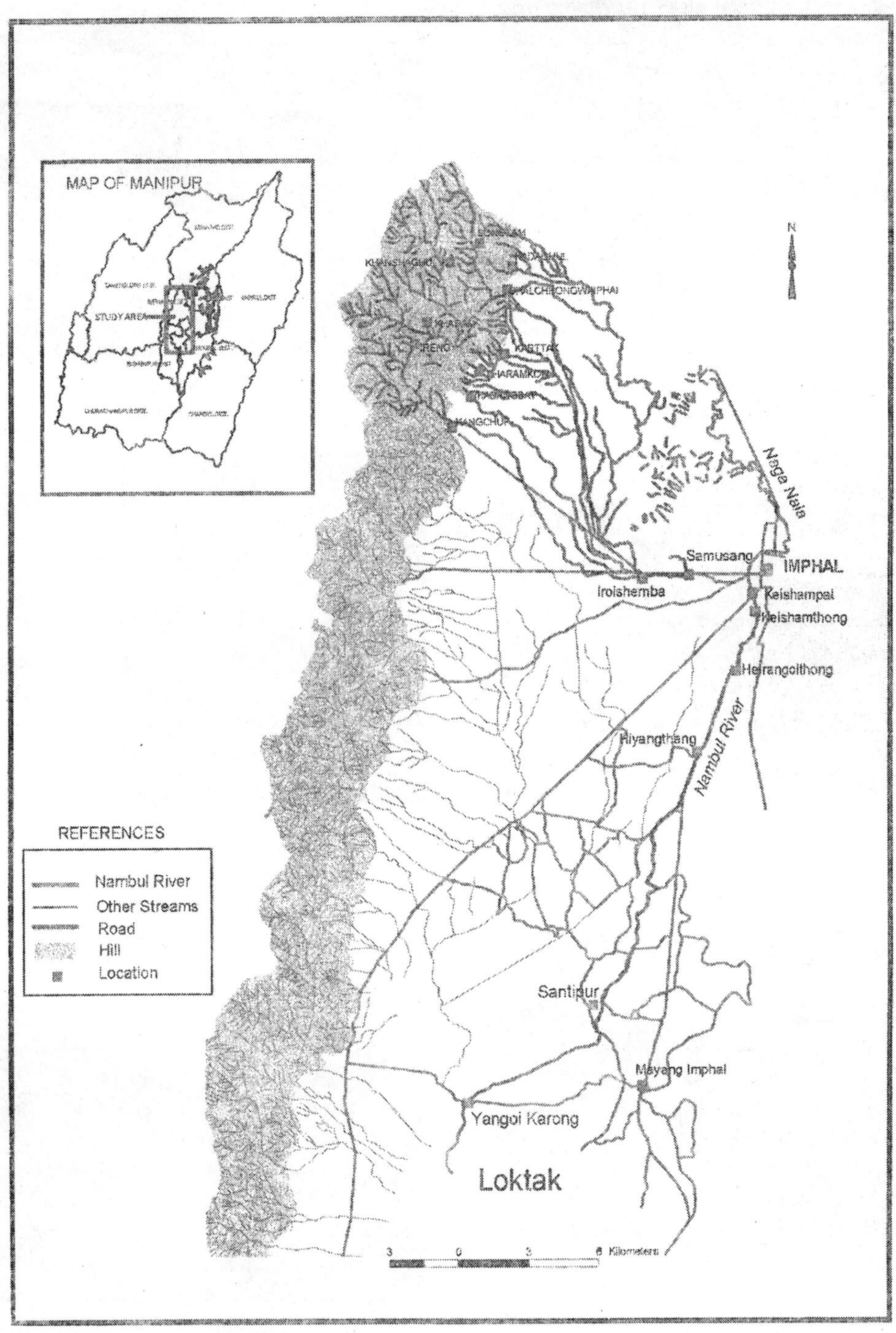

Figure 21.1. Location map of Nambul river, Manipur.

Figure 21.2. Samusang Effluent at Samusang.

Figure 21.3. Nambul river at Naoremthong.

Figure 21.4. Naga Nallha/Effluent at Maharanithong.

Figure 21.5. Nambul river at Wahengbam Leikai

Figure 21.6. A drain discharging into Nambul river.

Figure 21.7. Nambul river at Heirangoithong.

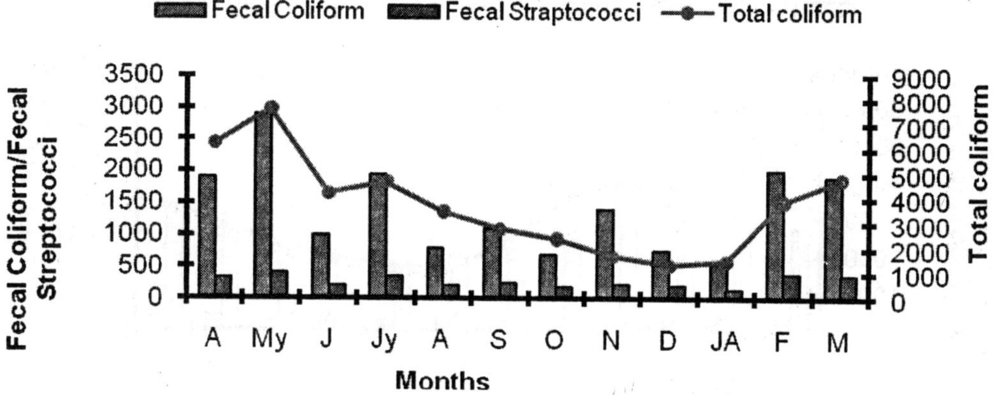

Figure 21.8. Seasonal variations in Fecal coliform, Fecal Straptococci and Total Coliform per 100 ml in Nambul river at Naoramthong for the period of 2005-06.

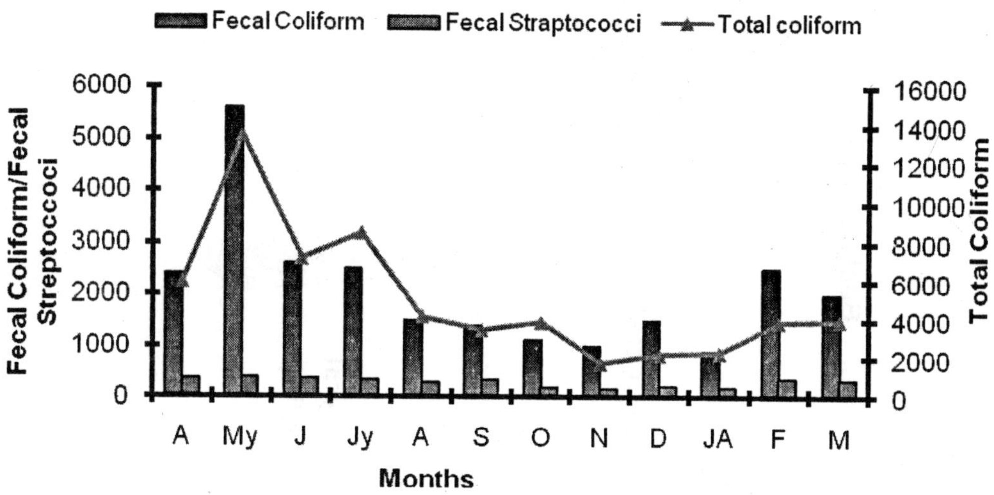

Figure 21.9. Seasonal variations in Fecal coliform, Fecal Straptococci and Total Coliform per 100 ml in Nambul river at Maharanithong for the period of 2005-06.

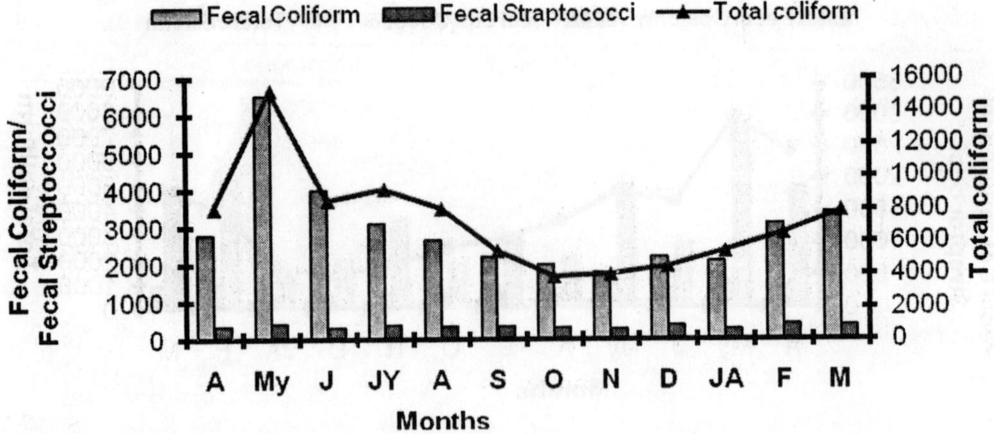

Figure 21.10. Seasonal variations in Fecal coliform, Fecal Straptococci and Total Coliform per 100 ml in Nambul river at Heirangoithong for the period of 2005-06.

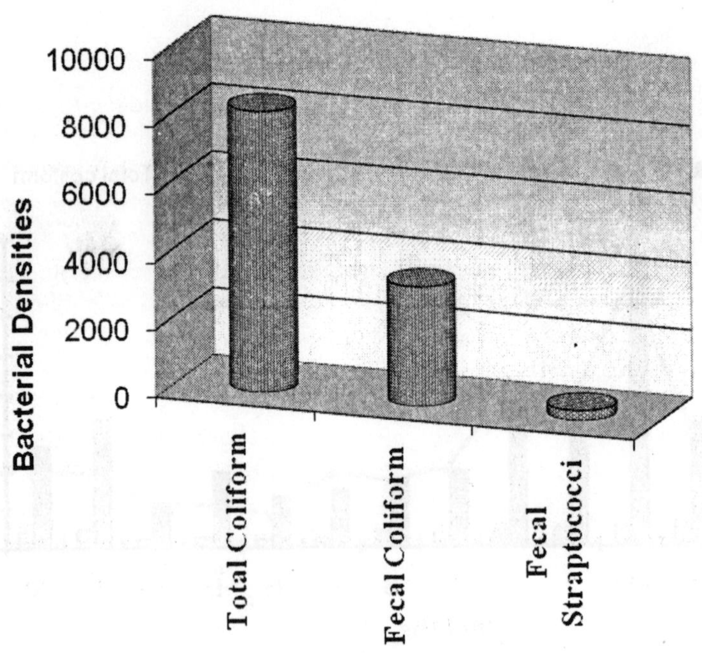

Figure 21.11. Mean bacterial densities of Nambul river for the period of 2005-06.

Table 21.1
Mean bacterial densities of Nambul river (2005-2006)

Sl.No.	*Parameter*	*Minimum*	*Maximum*	*Mean*
1.	Total Coliform	1350	15200	8275
2.	Fecal Coliform	600	6500	3550
3.	Fecal Straptococci	160	430	295

Note: Number of counts per 100 ml.

In the present study well marked seasonal variations of microbiological characteristics were also observed. The coliform group showed declining trend from the summer seasons towards rainy and winter seasons. It was observed that the total coliform, fecal coliform and fecal streptococci were consistently higher during summer and rainy season. This may be due to fluctuation of temperature and difference in input of waste run-off (effluent and sewage) during rainfall. During these seasons all the domestic waste, which are deposited along the river bank, animal waste, sanitary waste, market waste and municiapal waste are carried down into the Nambul river. It was reported by Sharma and Bharadwaj (2000) that during rainfall significant quantity of organic matter entered the water body from land with run-off, which favour the growth of bacteria. Several other workers in different water bodies also reported similar observations (Hill and Webb, 1958; Boxter-Pothe and Gilliland, 1988).

Fecal coliform/fecal streptococci (FC/FS) ratio was calculated for every month. FC/FS ratio less than 1.0 indicates warm blooded animal pollution while ratio of 4.0 and above has suggested domestic waste (Geldreich, 1976, Boxter-Potter and Gilliland, 1988 and Cooper and Knight, 1989). In the present study the ratio of FC/FS is higher than 4.0, so Nambul river is contaminated by domestic wastes. WHO and UNEP (1991) suggested that higher value of fecal coliform and fecal streptococii were indicator of fecal pollution in water body. The frequent and continuous occurrence of coliform in the river water may be due to continuous inflow of fecal discharge of human and animals either directly or indirectly through run off (Rajput, 1994). The presence of coliform group in the Nambul river throughout the year suggests that the river is contaminated by human and animal excreta.

REFERENCES

APHA. 1989. *Standard Methods for the examination of water and waste water analysis*, (17th edn.). American Public Health Association Inc., New York.

Boxter-Potter, W.R, and Gilliland, M.W. 1988. Bacterial pollution in runoff from agricultural land. *J. Environ. Qual, 17.27-34.*

Buchanan, R.E. and Gibbon, N.E. 1987. *Bergrey's Manual of Determinative Bacteriology* (9th Edn.). The Williams and Wikkins Co. Battimore, M.D.

Cooper, C.M. and Knight, S.S. 1989. Bacterial contamination in two agrarian Hill land streams in northern Mississippi,(USA). *J. Ecol. Environ. Sci. 15:85-96.*

Geldreich, E.E. 1976. Fecal coliform and fecal streptococci density relationships in waste discharge and receiving water. *CRC Crit. Rev. Environ. Control, 6;349.*

Hendricks, C.W. and Morrison, S.M. 1967. Multiplication and growth of selected entric bacteria in clear mountain stream water. *Water Res.* 1.567 - 596.

Hill, M.B. and Webb, J.E. 1958. *The Ecology of Logos Lagoon II. The topography and physical feature of Logos harbor and lagoon.* Philosophical Transaction of Royal Society of London, 214 (B): 314 - 333.

Rajput, S. 1994. Heterotrophic bacterial flora in the lake polluted by the effluents of standard coliform enumeration technique. *Am.Nat. Works. Assoc., 75: 538-542.*

Royee, M.K.P. and Prakasan, V.R. 2000, Water Quality analysis of natural ponds of Kollam municipality, Kerala. *Indian J. Environ. & Ecoplan. 6(3) :463-466.*

Sharma, A. and Bharadwaj, R. 2000. Bacteriological investigation of the water of Kolayatzi village pond of Bikaner (Rajasthan). *Uttar Pradesh J. Zool., 20(1): 79-82.*

Sharma, A. and Rajput, S. 1996. Microbial quality and persistence of entero-pathogenics in fresh water environment. *Eco. Env & Cons.,* 2:29-36.

Trivedy, R.K. and Goel, P.K. 1984. *Chemical and Biological Methods for water pollution studies.* Environmental Publication, Karad, 215 pp.

WHO. and UNEP. 1991. Pathogens in fresh waters. In Global freshwater quality. A first assessment (Ed. Miched, M.) *Env. Prod. Agency. Geneva.*

22

River Pollution Due to Unauthorised Dumping of Urban Solid Wastes: A Case Study of Two Important Rivers in Imphal City, Manipur

—*W. Chaoton Meetei[1], Koijam K.K. Manibhushan Singh[2] & N. Ibotombi Singh[3]*

[This paper deals with the pollution of two important rivers viz. Nambul river and Imphal river caused due to unauthorized dumping of urban solid wastes in Imphal city, Manipur. Water samples collected from two disposal sites and two non-disposal sites (one each from each river) were analyzed for physico-chemical characteristics like water temperature, pH, electrical conductivity, dissolved oxygen, hardness, alkalinity, chloride etc. The present study showed that there is a wide difference between the water quality of urban solid waste disposal area and non-disposal area of the two rivers in Imphal City. Lower values of dissolved oxygen and higher concentrations of free CO_2 and nutrients were observed in the disposal sites than the non-disposal sites.

Keywords: Urban solid waste, disposal sites, non-disposal sites, water quality].

Introduction

Water is the priceless gift of nature. Without water, there is no life on earth. None can deny that water is a friend to human race but it also acts as a foe by way of harboring disease producing micro-organisms and containing some substances that may lead to ill health (Manivasakam, 2001). Sinnarkar *et al.,* 1987 had also reported that Ganga River has suffered a lot and is becoming chemically and biologically polluted due to discharge of domestic, municipal and industrial wastes and several other factors.

The physico-chemical changes on river water due to illegal dumping of urban solid wastes were extensively documented (Devkota *et al.*, 2005; Zafar *et al.*, 2005; Ololade *et al.*, 2009 etc.).

Nambul and Imphal are the two important rivers that passes through Imphal City. These rivers are the main sources of water supply in the State. However, the water of these rivers are becoming unsafe for human consumption, agricultural and recreational uses. Haphazard dumping of urban solid wastes on the banks of

[1,2,3]D.M. College of Science, Imphal 795 001, Manipur

these rivers have become very common. Therefore, it is very much essential to know the pollution level of these two important rivers. The aim of the present study was to find the variation in water quality between urban solid waste disposal area and non-disposal area of these two important rivers regarding their physico-chemical characteristics.

Materials and Methods

To study the effects of urban solid waste disposal on river water and the level of contamination, water samples were collected in 1 litre clean plastic bottle from two non-disposal sites (NDS) and two disposal sites(DS) each from the Nambul and Imphal rivers of Manipur (Figure 22.1) from the month of July, 2007 to June, 2008 (Table 22.1). Before sampling bottles were cleaned thoroughly with double distilled water to remove the impurities.

Table 22.1
Sampling points, site code and type of water (NDS = Non-disposal sites and DS = Disposal sites)

Sl. No.	*Sampling Point*	*Site Code*	*Type of Water*
1	Naoremthong	NDS-1	Nambul river water
2	Singjamei, Pebiya Pandit Leikai	NDS-2	Imphal river water
3	Kakhulong	DS-1	Nambul river water
4	Minuthong Hatta	DS-2	Imphal river water

From each sample physico-chemical characteristics like water temperature, pH, electrical conductivity, dissolved oxygen, hardness, alkalinity, free carbon dioxide, chloride, phosphate, potassium and sodium were analyzed using standard methods as given in APHA (1995) and Trivedy, Goel and Trisal (1998).

Results and Discussion

Seasonal variations in physico-chemical characteristics of Nambul and Imphal rivers at water quality at non-disposal sites (NDS-1 and NDS-2) and disposal sites (DS-1 and DS-2) are shown in Figures 22.2-22.12.

Water Temperature

Non-Disposal Sites: The surface water temperature varied from 17°C (January, 2008) to 28°C (May, 2008) in NDS-1 and 14 °C (January, 2008) to 29 °C (August, 2007) in NDS-2.

Disposal Sites: The water temperature were ranged from 14°C (January, 2008) to 30°C (April, 2008) in DS-1 and 16°C (December, 2007) to 28°C (April, May, 2008) in DS-2.

It was observed that the water temperature was found to be higher in summer than winter in all the NDS and DS. The increase and decrease in water temperature may be due to low water level, atmospheric condition and greater solar radiation. Higher values of water temperature in the disposal sites may be due to the mixing of urban run-off and waste from washing activities etc. The higher water temperature during summer was because of higher air temperature and greater solar radiation and low water level. At all the study sites, water temperature was

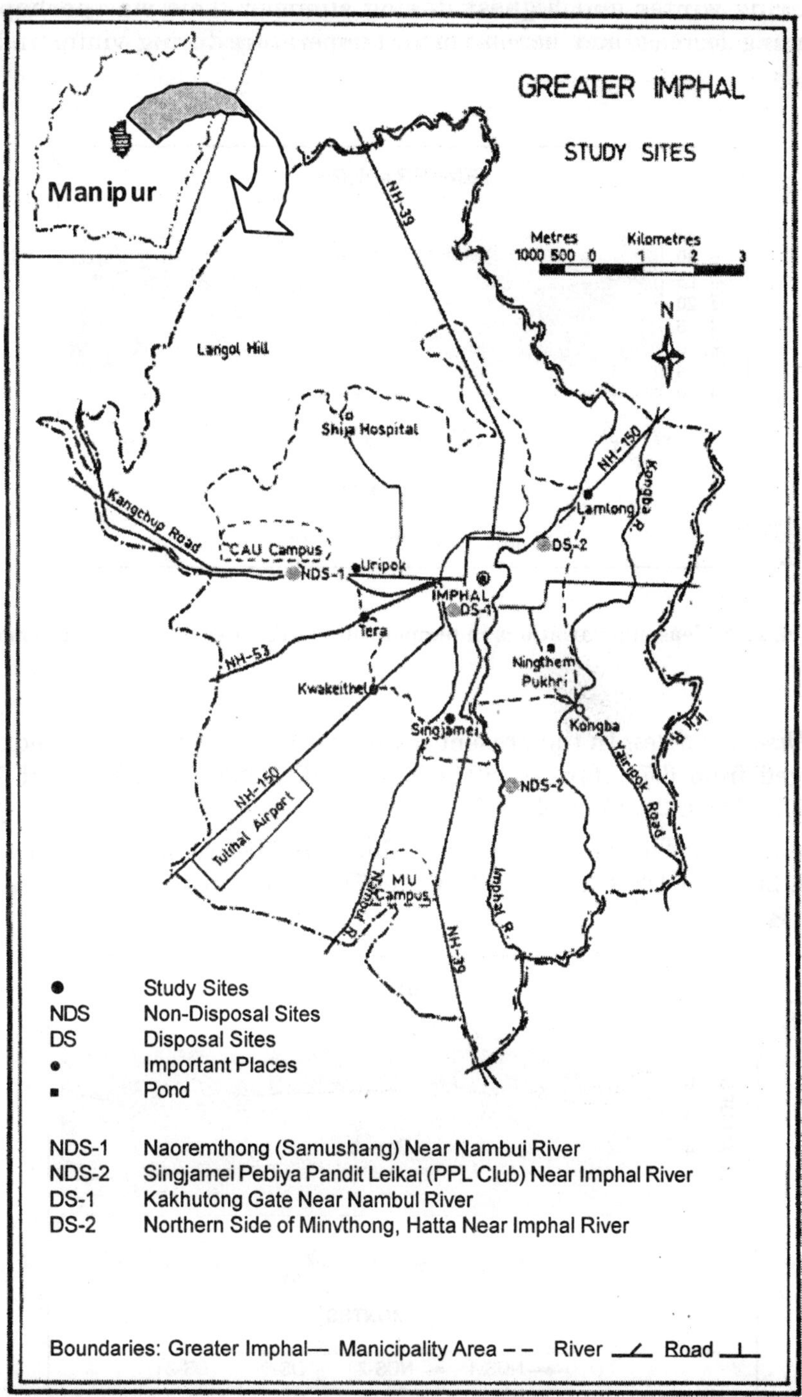

Figure 22.1 Map of the Greater Imphal showing locations of study sites.

lowest during winter and highest during summer. This was perhaps due to corresponding decrease and increase in air temperature during winter and summer respectively.

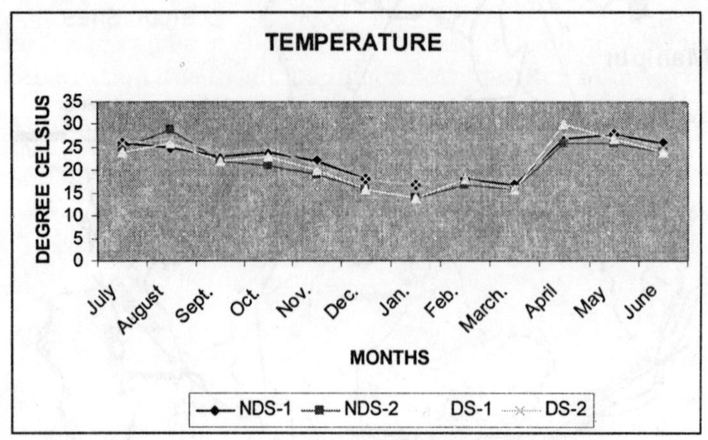

Figure 22.2. Seasonal variations in temperature at the disposal and non-disposal sites.

pH

Non-Disposal Sites: In the present investigation, pH value at the non-disposal sites ranged from 6.7 (May, 2008) to 8.3 (April, 2008) at NDS-1 and 7.1 (July, 2007) to 8.2 (April, 2008) at NDS-2.

Disposal Sites: The value of pH were fluctuated between 6.6 (August, 2007) to 8.1 (April, 2008) at DS-1 and 7.1 (July, 2007) to 8.5 (April, 2008) at DS-2.

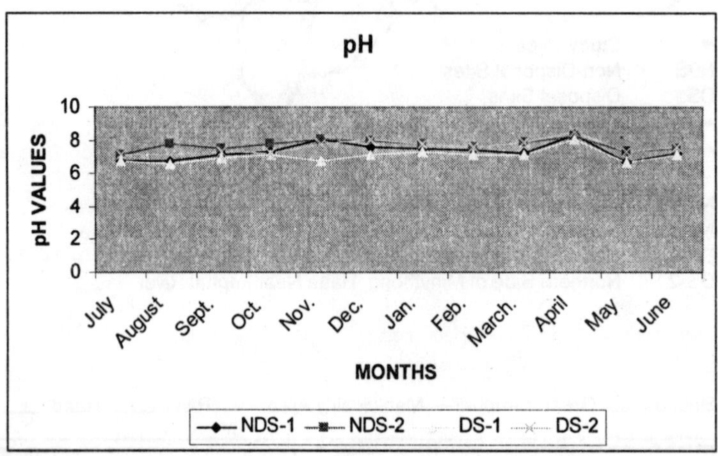

Figure 22.3. Seasonal variations in pH at the disposal and non-disposal sites.

It was observed that pH values were found to be maximum during summer and minimum during rainy months. This may be due to removal of sufficient amount of carbon dioxide by photosynthetic processes of the aquatic plants during summer while during rainy months, there were active decomposition of organic matters and rain water drained into the rivers resulting in the formation of carbonic acid which gives acidic characters. The lower values of pH in the disposal sites may be due to dumping of wastes which is found to be in agreement with the observation made by Muduli *et al.*, 2006.

Electrical conductivity

Non-Disposal Sites: The values of electrical conductivity in the non-disposal sites ranged from 101.92 µmho/cm (July, 2007) to 266.76 µmho/cm (February, 2008) in NDS-1 and 84.50 µmho/cm (July, 2007) to 270.72 µmho/cm (March, 2008) in NDS-2.

Disposal Sites: In the present study, the values of electrical conductivity varied from 119.34 µmho/cm (July, 2007) to 693.55 µmho/cm (May, 2008) in DS-1 and 71.50 µmho/cm (October, 2007) to 203.77 µmho/cm (April, 2008) in DS-2.

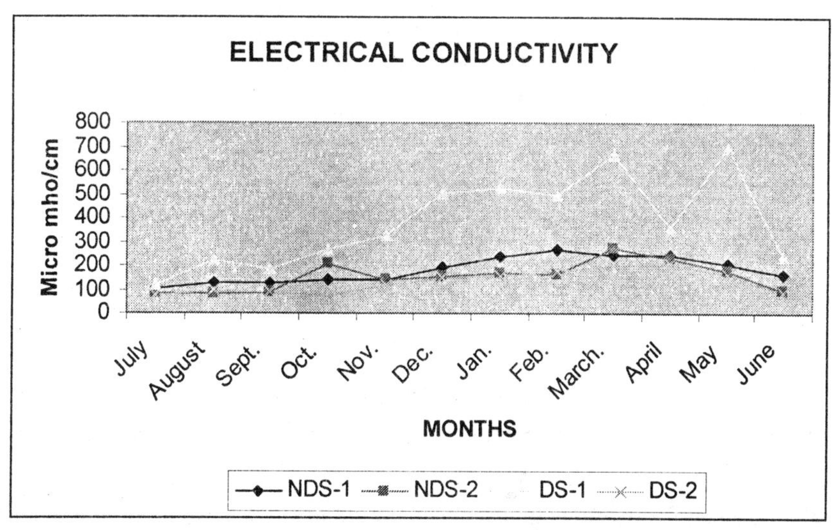

Figure 22.4. Seasonal variations in electrical conductivity at the disposal and non-disposal sites.

The values of conductivity were higher in the disposal sites than non-disposal sites. This may be attributed to the dissolved ions liberated from decomposed matter and input of organic and inorganic wastes from dumping of urban solid wastes.

Dissolved Oxygen

Non-Disposal Sites: The dissolved oxygen (DO)values of the non-disposal sites fluctuated between 2.03 mg/l (June, 2008) and 7.70 mg/l (December, 2007) in NDS-1 and 2.61 mg/l (April, 2008) to 8.10 mg/l(October, December, 2007) in NDS-2.

Disposal Sites: Dissolved oxygen were ranged between 0.202 mg/l (March, 2008) to 4.25 mg/l (August, 2007) at DS-1 and 2.43 mg/l (May, 2008) to 8.91 mg/l (June, 2008) at DS-2.

The lower values of DO at NDS in rainy season may be attributed to decreased light penetration due to silt loading inflow of water from surrounding making more turbid and reduced photosynthesis. The higher values of DO during winter might be associated with low temperature, more light penetration and more photosynthesis. Disposal sites have higher values of DO during rainy season and low in winter. This may be due to greater flow of water and turbulence. In winter,the value of DO drops down which was due to the fall of water level and more volume of oxygen demanding wastes being dumped at the disposal sites.

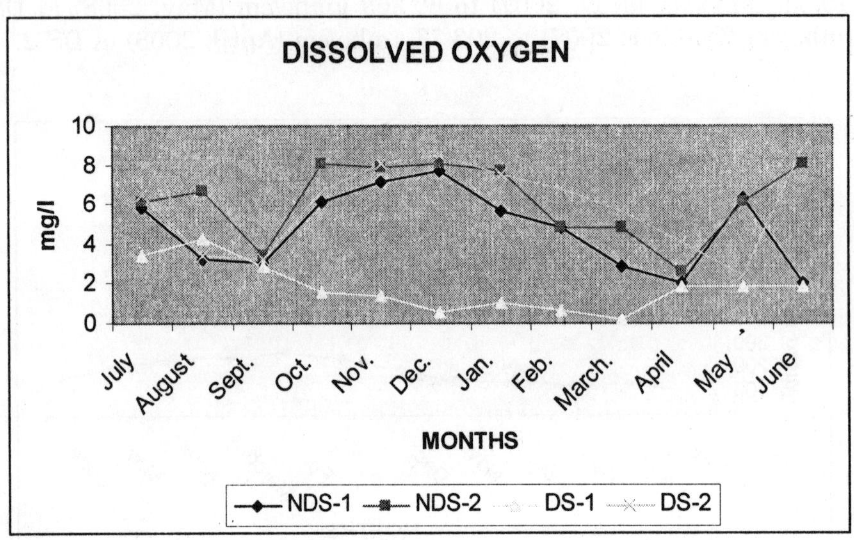

Figure 22.5. Seasonal variations in dissolved oxygen at the disposal and non-disposal sites.

factors in the non-disposal sites while different in quality and quantity of solid wastes dumped might be the reason for the disposal sites.

Hardness

Non-Disposal Sites: The hardness values of the non-disposal sites ranged from 28 mg/l (July, 2007) to 72 mg/l (April, 2008) in NDS-1 and 30 mg/l (July 2007) to 70 mg/l (April, 2008) in NDS-2.

Disposal Sites: The values of hardness varied from 32 mg/l (July, 2007) to 94 mg/l (March, 2008) in DS-1 and 26 mg/l (July, 2007) to 76 mg/l (April, 2008) in DS-2.

It was observed that the values of hardness were maximum in March and April and minimum in July. The lower values of total hardness during monsoon may be attributed to more dilution by rain.

Alkalinity

Non-Disposal Sites: The values of alkalinity were in the range of 10 mg/l (November, 2007) to 65 mg/l (April, 2008) in NDS-1 and 20 mg/l (December, 2007 and June, 2008) to 60 mg/l (April, 2008) in NDS-2.

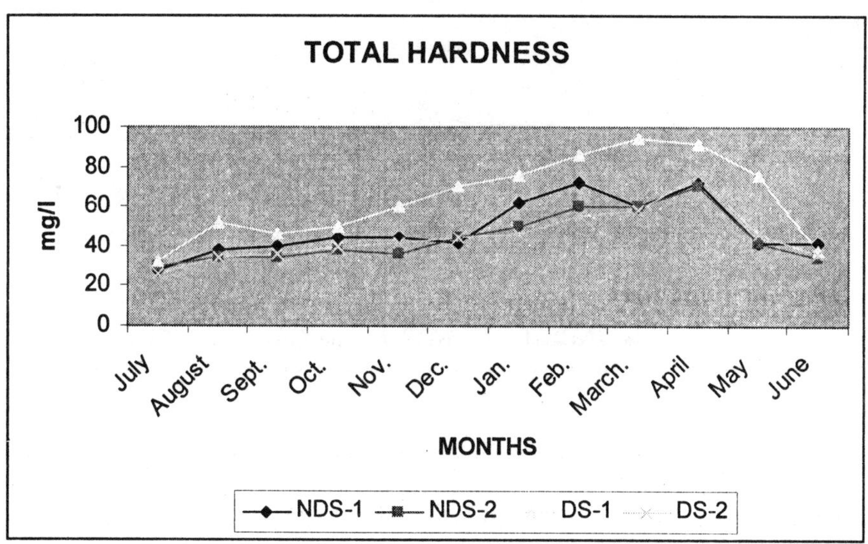

Figure 22.6. Seasonal variations in total hardness at the disposal and non-disposal sites.

Disposal Sites: The values of alkalinity of the present study were fluctuated from 20 mg/l (November, 2007 and June, 2008) to 170 mg/l (May, 2008) in DS-1 and 20 mg/l (October, 2007 and June, 2008) to 60 mg/l(April, 2008) in DS-2.

The total alkalinity values were found to be maximum during summer and minimum during winter .This finding is in agreement with the reported values of Chavan et al., 2005. Higher values observed in the non-disposal sites might be due to dissolution of calcium carbonate from the sediment and use of detergent and soap by local inhabitants in these sites. Higher values in the disposal sites might be due to the effect of urban solid waste disposal.

Free Carbon Dioxide

Non-Disposal Sites: The values of free carbon dioxide in the non-disposal sites varied from 4.4 mg/l (November, 2007) to 44 mg/l (July, 2007) in NDS-1 and from 4.4 mg/l (November, 2007) to 22 mg/l (April, 2008) in NDS-2.

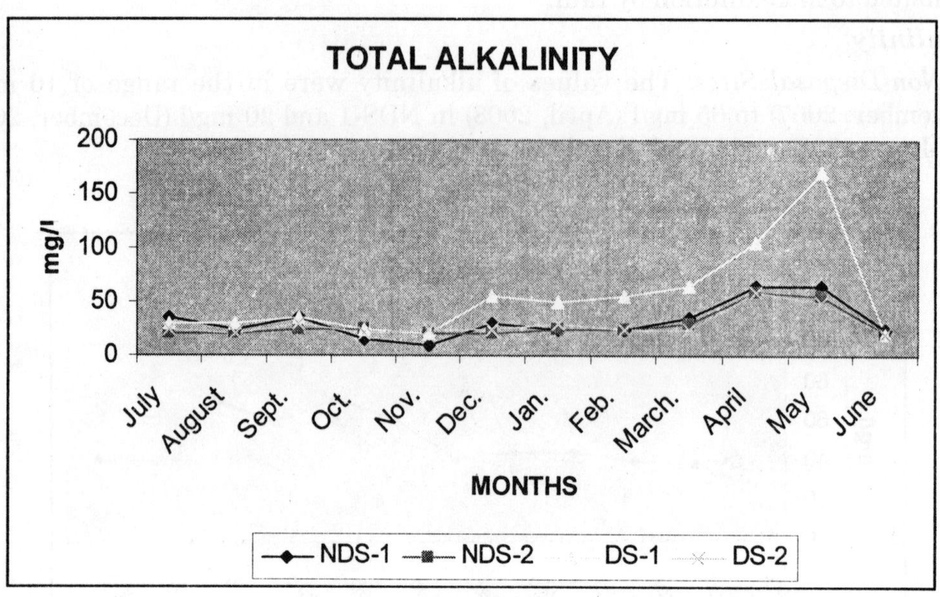

Figure 22.7. Seasonal variations in total alkalinity at the disposal and non-disposal sites.

Disposal Sites: In the present study, the values of free carbon dioxide fluctuated from 8.8 mg/l (November, 2007) to 92.4 mg/l (May, 2008) in DS-1 and 6.6 mg/l (December, 2007) to 30.8 mg/l (August, 2007) in DS-2.

High temperatures during summer and rainy months accelerates the process of the decay of organic matters and large amount of CO_2 is added in water. Higher values in the disposal sites might be due to higher organic load and subsequent faster decomposition rate.

Chloride

Non-Disposal Sites: The values of chloride in the non-disposal sites ranged from 22.72 mg/l (November, 2007) to 73.84 mg/l (May, 2008) in NDS-1 and 19.88 mg/l (November, 2007) to 51.12 mg/l (May, 2008) in NDS-2.

Disposal Sites: The chloride values in these sites varied from 19.88 mg/l (November, 2007) to 144.84 mg/l (May, 2008) in DS-1 and 18.43 mg/l (November, 2007) to 62. 48 mg/l (May, 2008) in DS-2.

Chloride values were found to be higher in pre-monsoon and lower in post-monsoon seasons. Similar observations were also made by Rani *et al.*, 2004 and Sanap *et al.*, 2006.

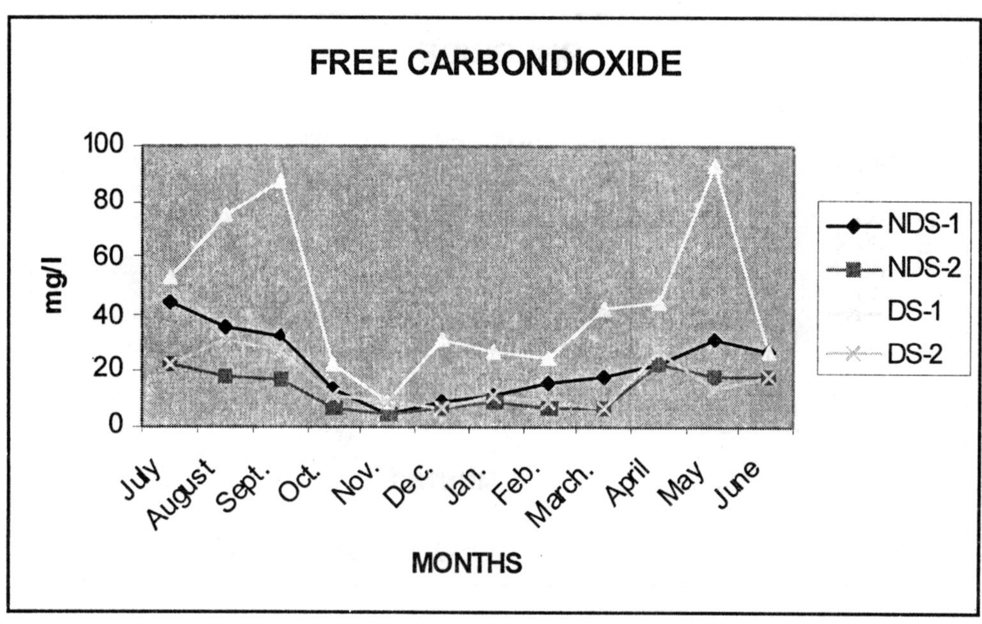

Figure 22.8. Seasonal variations in free carbondioxide at the disposal and non-disposal sites.

Generally, higher values were observed in the disposal sites which might be related to addition of wastes in river water. Such observation was also made by Sanap *et al.*, 2006.

Phosphate

Non-Disposal Sites: In the present study, the values of phosphate in the non-disposal sites varied from 0.14 mg/l (March, 2008) to 0.44 mg/l (June, 2008) in NDS-1 and 0.115 mg/l (March, 2008) to 0.345 mg/l (November, 2007) in NDS-2.

Disposal Sites: In the present investigation, phosphate values ranged between 0.115 mg/l (December, 2007) to 0.95 mg/l (March, 2008) in DS-1 and 0.115 mg/l (November, 2007) to 0.405 mg/l (June, 2008) in DS-2.

The higher values of phosphate in summer and rainy season might be due to mixed up with agricultural and urban run-off with river water and higher rate of evaporation of water.

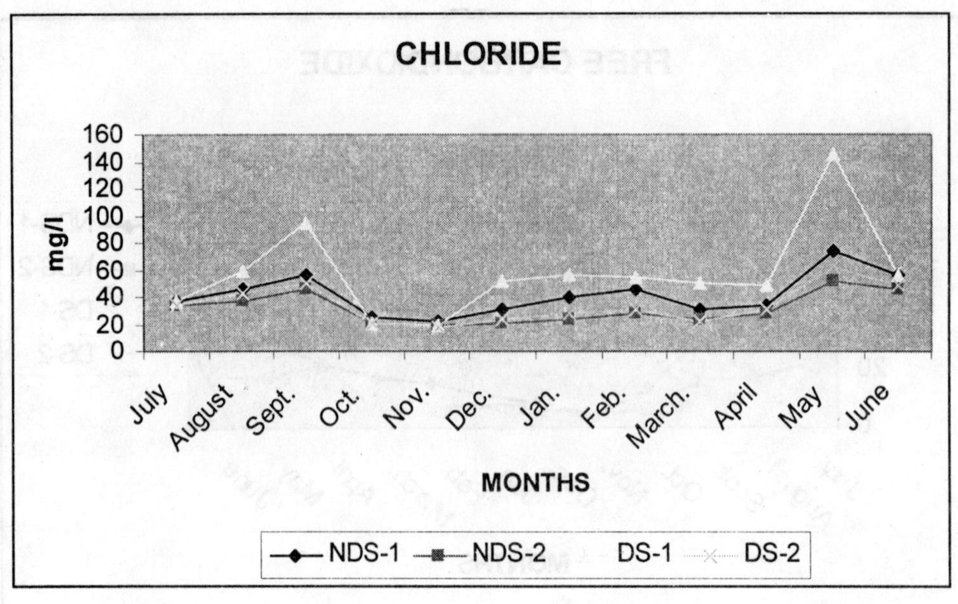

Figure 22.9. Seasonal variations in chloride at the disposal and non-disposal sites.

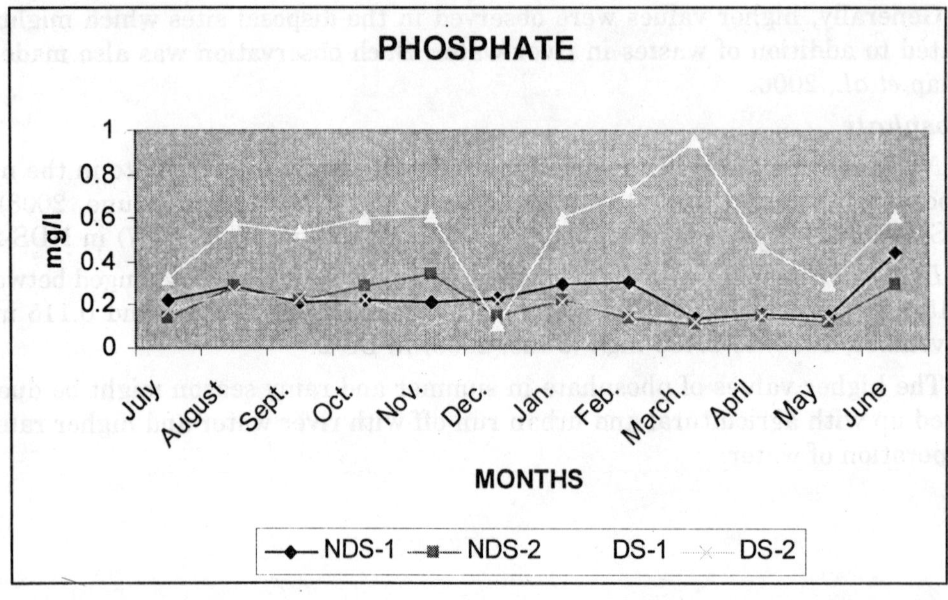

Figure 22.10. Seasonal variations in phosphate at the disposal and non-disposal sites.

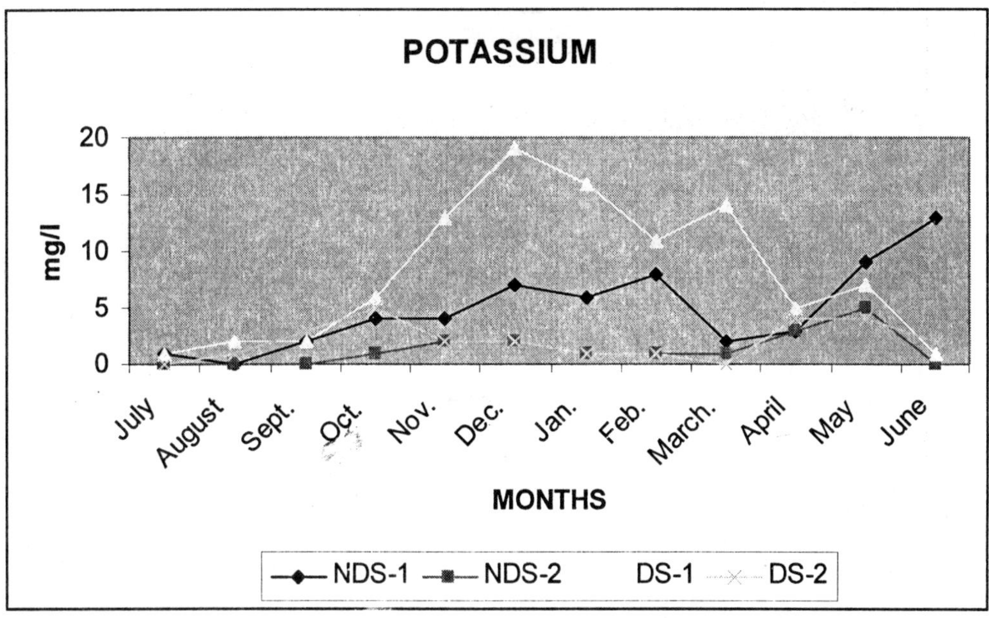

Figure 22.11. Seasonal variations in potassium at the disposal and non-disposal sites.

Potassium

Non-Disposal Sites: The values of potassium in the non-disposal sites varied from 0 mg/l (August, 2007) to 13 mg/l (June, 2008) in NDS-1 and 0 mg/l (June, 2008) to 5 mg/l (May, 2008) in NDS-2.

Disposal Sites: The values of potassium in the disposal sites ranged from 1 mg/l (July, 2007 and June, 2008) to 19 mg/l (December, 2007) in DS-1 and 0 mg/l (July, 2007 and March, 2008) to 5 mg/l (October, 2007) in DS-2.High values of potassium were due to more pollution by disposal of solid wastes. This is indicated in the previous work of Nambul River by Vidyarani *et al.*, 2006.

Sodium

Non-Disposal Sites: The values of sodium in the non-disposal sites varied from 5 mg/l (April, 2008) to 32 mg/l (June, 2008) in NDS-1 and 1 mg/l (February, 2008) to 24 mg/l (May, 2008) in NDS-2.

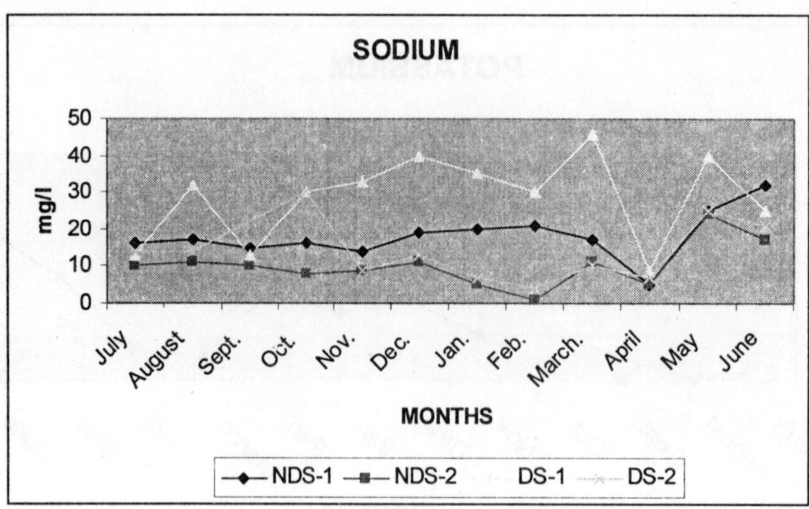

Figure 22.12. Seasonal variations in sodium at the disposal and non-disposal sites.

Disposal Sites: The disposal sites values of Sodium in the present study ranged from 9 mg/l (April, 2008) to 46 mg/l (March, 2008) in DS-1 and 4 mg/l (February, 2008) to 30 mg/l (October, 2007) in DS-2. The maximum or minimum values of sodium was not recorded in a particular month or season.

Conclusion

In the present study, it is observed that unauthorized dumping of urban solid wastes causes deterioration in the water quality of the two important rivers in the city. There is wide difference in the quality of river water between solid waste disposal sites and non-disposal sites. Analysis of physico-chemical properties of water did support the above inference. Variations in the degree of pollution in the disposal sites were perhaps due to factors like local land use pattern, quality and quantity of the solid wastes dumped, time and pattern of dumping etc. So, overall, these two rivers are under socio- ecological threat and require holistic measures to conserve them. Therefore, this study clearly warrants the need to control and manage allochthonous pollutants and to develop a long term strategy to protect these important rivers in the north-east India, particularly in Manipur.

REFERENCES

APHA, AWWA, WPCA. 1995. *Standard methods for examination of water and waste water.* 19th edition. American Public Health Association, Washington, DC.

Chavan, R.J., Mohekar, A.D., Sawant, R.J. and Tat, M.B. 2005. Seasonal variations of abiotic factors of Manjara Project Water Reservoir in Dist. Beed Maharashtra, *India. Poll. Res.* 24(3): 705-708.

Devkota, C.D. and Watanabe, K. 2005. Impact of solid waste on water quality of Bishnumati River and surrounding areas in Kathmandu, Nepal. *J. Nepal Geological Society*, 31: 19-24.

Mudule, S.D., G.D., Bhuyan, N.K. and Dhal, N.K. 2006. Physico-chemical characteristic assessment of Brahmani River, Orissa, India. *Poll. Res.* 25(4): 763-766.

Manivasakam, M. 2001. Chemical and Microbiological Analysis of Mineral Water and Packaged Drinking Water. Sakthi Book Services, Kovaipudur, Combatore (T.N.).

Ololade, I.A., Adewunmi, A., Ologundudu. A. and Adeleye, A. 2009. Effects of household wastes on surface and underground waters. *Int. J. Phy. Sci.* 4(1): 022-029.

Rani, R., Gupta, R.K. and Shrivastava, K.B.L. 2004. Studies on water quality assessment in Satna City (M.P.): Seasonal Parametric Variations. Nature, Environment and Pollution Technology © Techno-Science Publications, 3(4): 563-565.

Sanap, R.R., Mahite, A.K., Pingle, S.D. and Gunale, V.R. 2006. Evaluation of water qualities of Godawari River with reference to Physico-chemical parameters. Dist. Nasik, (M.S.), India. *Poll Res.*, 25(4): 775-778.

Sinnarkar, S.N., Kesarwani, S.K. and Bhat, S.H. 1987. *River Ganga: An overview of Environmental Research*, NEERI, Nagpur.

Trivedy, R.K., Goel, P.K. and Trisal, C.L. 1998. *Practical methods in Ecology and Environmental Science*. Reprint 1998. Enviromedia, Karad, India.

Vidyarani, W., Singh, H.B. and Sharma, M.G. 2006. Determination of Physico-chemcial and microbiological analysis of Nambul River at Imphal City, Manipur. *Indian J. Environ. & Ecoplan.* 12(3): 811-814.

Zafar, M. and Alappat, B.J. 2005. Landfill Surface Runoff and its effect on water quality on river Yamuna. *J. Environmental Science and Health*, Part-A, 39(2): 375-384.

23

Molecular Characterization of Cyanobacteria and their Diversity in Wetlands including the Paddy Fields

—*Ashish Kumar Srivastava*[1]

[Cyanobacteria are indispensible component of the wetland ecosystem, where they appreciably contribute in carbon and nitrogen dynamics. Cyanobacteria have a particular significance in rice fields due to their role in nitrogen economy. The classical methods of cyanobacterial characterization and classification are based on morphological attributes, which are liable to culture and environmental conditions. In this scenario, characterization of these oxygen evolving photosynthetic microorganisms using molecular tools seems promising. The molecular depiction generally involves DNA based approaches. Of the different strategies, which are discussed in this review and used for documentation of cyanobacterial diversity, 16S rRNA gene-PCR followed by DGGE analysis is most widely acclaimed. Although molecular methods are authentic and robust to environmental fluctuations, they suffer from PCR based biases. Therefore, identification based on morphological and molecular attributes is the best approach to assess the cyanobacterial diversity and their role in ecological processes of wetlands including rice fields.

Keywords: 16S rRNA gene, cyanobacteria, DGGE, PCR, rice field, wetland.]

Introduction

Wetland is one of the most diverse and fascinating ecosystems on the globe. Since wetland supports both terrestrial and aquatic ecosystems, it can sustain a vast variety of diversity. Total area occupied by wetlands of India is about 58.2 million hectares. About 38% of the inland wetlands have vanished in last decade; thus, fresh water ecosystems are most threatened. Therefore, it is of prime importance to conserve fresh water wetlands. The first and most significant step towards this goal is the systematic scientific documentation of its diversity and processes (Sarovar Saurabh, ENVIS Newsletter on Wetland Ecosystem). Among the different life forms present in wetland ecosystem, algae and cyanobacteria are of great significance

[1] Department of Botany, School of Life Sciences, Mizoram University, Aizwal–796009, Mizorm
Present Address: Institute of Enviornment and Sustainable Development, Banaras Hindu University, Varanasi, 221 005, India
Email: aksrivastava@mzu.edu.in

because they contribute about 50% in the ecosystem productivity. The luxuriant growth of these organisms suggests the stable wetland.

Cyanobacteria evolved almost 2.7 billion years ago and one of the oldest living prokaryotic organisms on the earth (Buick, 1992). After their evolution, the major transition occurred in the history of life. The concentration of oxygen increased in atmosphere due to oxygenic photosynthesis of cyanobacteria and the birth of oxygenic life took place. Cyanobacteria have an edge over the algae as certain cyanobacteria can fix atmospheric nitrogen. Thus cyanobacteria are only group of organisms, which are equally efficient to do oxygenic photosynthesis as well as nitrogen fixation. However, both the processes are spatially separated. The specialized heterocyst cells are the site of nitrogen fixation, whereas photosynthesis is confined to vegetative cells. Thus specialized cells for nitrogen fixation and two photosystems for oxygenic photosynthesis appeared first in cyanobacteria in due course of evolution. Thus cyanobacteria are very much close to bacteria in their cellular organization but to plants as far as their metabolism is concern (Kumari *et al.*, 2009). Microbial including cyanobacteria-mediated nitrogen dynamics are vital to biogeochemical functions in aquatic ecosystems such as wetlands, where many primary producers (like phytoplankton and plants) are nitrogen-limited. Further, enormous ability of cyanobacteria to adapt in any environmental condition makes them potential candidates of choice in all ecological niches. Though they are dominant component of wetlands in general and rice fields in particular, they are also present in extreme environment like Antarctica, hot springs (Srivastava *et al.*, 2009). Despite of the fact that prokaryotes play a crucial role in environmental processes and have a noble array of bioactive compounds, only 0.5 to 10 % of their actual diversity are known (Cases and de Lorenzo, 2002; Dorigo *et al.*, 2005). Microscopic size, simple morphology and specific nutritional requirement are some of the prominent reason behind their less known diversity (Rejmánková *et al.*, 2004).

Paddy fields are the temporary wetlands and dominated by the genera belonging to orders Nostocales and Stigonematales of cyanobacteria. The orders are exclusively consist of nitrogen fixing filamentous cyanobacteria having no branching to pseudo- and true- branching. The classification systems of cyanobacteria proposed to date are based on certain morphological attributes (Desikachary, 1959; Geitler, 1932; Rippika *et al.*, 1979). These attributes include cellular organization, mode of cell division, presence of heterocyst, position of akinetes to heterocysts, type of perenneting bodies, colour of cyanobacteria, presence of sheath, colonial or solitary, and branched or unbranched filaments. However, these characters largely depend on the culture conditions and have been criticized by several workers (Anagnostidis and Komárek, 1989; Rajaniemi *et al.*, 2005). The identification of filamentous multicellular cyanobacteria is not a serious problem, but those belong to orders Chroococcales and Pleurocapsales are very difficult (Srivastava *et al.*, 2009). Not only this, identification of certain cyanobacteria present as perennating bodies, detached heterocyst and akinetes can not be possible using these keys. Thus the authenticity of these classifications depends on the cyanobacteria in question.

In view of the above, molecular methods based on DNA and protein sequences proved more promising from last one decade. Morphological assessment supplemented with molecular identification is more precise method for identification, classification and diversity analysis of cyanobacteria. It is the molecular method, which proved the relationship of cyanobacteria with bacteria and plastids of higher plants. These methods involve DNA isolation, specific gene amplification, sequencing and then identification. Molecular methods to assess cyanobacteria include many flaws or biases related with DNA isolation and purification, primer selection, PCR conditions, cloning and sequencing (Farrelly *et al.*, 1995; Tebbe and Vahjen, 1993, Weller *et al.*, 1991). Notwithstanding this, these are most suitable method for this purpose to date.

Lots of report are available pertaining the cyanobacterial diversity assessment using molecular tools in different ecosystems like Colorado plateau (Garcia-Pichel *et al.*, 2001; Redfield *et al.*, 2002), exposed dolomite in central Switzerland (Sigler *et al.*, 2003), hot spring (Papke *et al.*, 2003), McMurdo Ice Self (Jungblut *et al.*, 2005), Southern Baltic Sea (Geiâ *et al.*, 2004) and rice fields of China (Song *et al.*, 2005). However, studies on cyanobacterial diversity in India are largely done at morphological level (Nayak and Prasanna, 2007; Tiwari and Singh 2005; Srivastava and Gupta, 2004; Rath and Adhikary, 2008). The cyanobacterial diversity in rice fields is studied in detail by Singh (1961), Gupta (1996) and Tiwari (1975). However, among northeast India, the cyanobacterial diversity has been studied in Arunachal Pradesh (Singh *et al.*, 1997a), Nagaland (Singh *et al.*, 1997b), Manipur (Tiwari and Singh 2005) and Assam (Saha *et al.*, 2007). Report of Srivastava *et al.* (2009) is the first to study the cyanobacterial diversity of rice fields of Eastern Uttar Pradesh at molecular level. Critical review of literature suggests a complete gap of knowledge regarding cyanobacterial diversity in Mizoram. However, work has already been started to document cyanobacterial diversity of rice fields of Mizoram.

Though reports are available regarding cyanobacterial diversity in particular ecosystem but very few studies have been reported so far dealing with cyanobacterial diversity in relation with there ecological functions. Thus, this mini-review discusses the methodologies used for documentation of cyanobacterial diversity using molecular tools and assessment of their ecophysiologies.

DNA Isolation

This is the first but most crucial step of any molecular method. The quality of isolated DNA decides the ultimate result of the whole protocol. For biodiversity assessment of cyanobacteria, DNA isolation protocol should be equally applicable to diverse samples like cyanobacterial culture, natural mats and soils. Unfortunately most methods used for cyanobacterial DNA isolation are specific for either culture or natural samples. Majority of the methods employed use either phenol (an irritant) or lysozyme (inefficient in exopolysaccharide containing cyanobacteria) or both. Taking recourse to the above, Srivastava *et al.* (2007) developed a rapid, phenol and lysozyme free DNA isolation protocol, which is equally applicable to diverse cyanobacterial samples. The details of this protocol are given in Srivastava *et al.*

(2007) and not discussed in detail here. This method includes physical breaking of cells using glass beads and ultrasonication rather than the chemical methods using lysozyme. A higher concentration of sodium chloride is used to stabilize DNA in *in-vitro* condition. Extra step of sapharose 4B purification has been included in the protocol for DNA isolation from natural samples to separate humic acid, which is a potent PCR inhibitor. Thus the protocol is rapid, cost effective (does not involve costly chemicals) and health friendly (does not include phenol extraction).

Gene of Interest for Cyanobacterial Biodiversity Assessment

Several genes are used to assess the cyanobacterial diversity. However, the gene of interest should (i) be present in all organisms to be studied, (ii) have some conserve and diverse region for primer binding and species identification respectively, (iii) be involved in certain metabolic processes, and (iv) not prone to mutation. In addition to this, gene sequence should be conserved within closely related organisms.

Based on above criteria different genes such as *nif* gene, phycocyanin locus (*cpc*), RNA polymerase (*rpo*) gene and 16S rRNA (*rrna*) gene are used to assess the cyanobacterial diversity.

16S rRNA gene

16S rRNA is most widely used molecular tool to investigate the diversity of cyanobacteria because of its conservness and ubiquity in entire group (Neilan, 2002). However, genome size, genome number and *rrna* gene organization vary significantly among cyanobacteria. The 16S rRNA library is extensively available library on the database as compared to other genes used for taxonomic purposes. Srivastava *et al.* (2007; 2009) reported the cyanobacterial diversity in rice fields of Eastern Uttar Pradesh based on 16S rRNA gene sequence. 16S rRNA gene based phylogeny of *Microcystis* strains producing hepatotoxin was found more satisfactory than those of phycocyanin gene. Iteman *et al.* (2002) have reported 16S rDNA based heterocystous cyanobacterial diversity. The shortcomings of 16S rRNA gene based biodiversity assessment have been examined by Farrelly *et al.* (1995). One of the significant limitations of 16S rRNA gene is presence of multiple copies in the genome with different evolution of each copy (Dahllöf *et al.,* 2000). This generally causes artifacts on DGGE gels as separate bands and thus overestimates the diversity and community structure (**Martinez-Murcia *et al.,* 1999; Nübel *et al.,* 1996; Clayton *et al.,* 1995).** This problem is addressed in many papers but not solved to a significant level. Not only this, the 16S rRNA gene sometime shows high degree of conservation among the closely related strains and misleads the interpretation of the result (Fox *et al.,* 1992). Recently Wang and Qian (2009) have reported the inefficiency of PCR primers to amplify the entire rRNA gene in a sample. They reported that most of the isolates are left and can not be detected and the results underestimate the actual diversity. Despite of all these limitations, 16S rRNA gene is the most popular choice for the assessment of cyanobacterial diversity.

RNA polymerase (rpo) gene

To overcome the problem of multiple copies of 16S rRNA gene within a genome, taxonomy based on gene present in single copy has been started. Of the different genes, *rpo* gene encoding RNA polymerase is used (Palenik, 1992). The *rpo*C1 gene has quite conserve regions which can be used for primer designing. The significant works have been done by Palenik (1994) and Palenik and Swift (1996), who reported diversity among strains of *Prochlorococcus-* and *Synechococcus-* like cyanobacteria found in Sargasso Sea using this gene. The comparison of bacterial identification based on *rpo*B gene with that of 16S rRNA gene has been done by Dahllöf *et al.* (2000). Wilson *et al.* (2000) reported the diversity of bloom farming toxic cyanobacterium *Cylindrospermopsis raciborskii* based on *rpo*C1 gene. Diversity among different *Synechococcus* strains inhabiting California stream has been assessed by Toledo and Palenik (1997) using *rpoC*1 gene sequence.

Phycocyanin (cpc) locus

Neilan *et al.* (1995) and Haande *et al.* (2007) studied the diversity of *Microcystis aeruginosa* isolates using intergenic spacer (IGS) between the bilin subunit genes *cpc*B and *cpc*A. This genus is studied most extensively due to its cosmopolitan distribution across the water bodies of the world. Ivanikova *et al.* (2007) reported the diversity of freshwater autotrophic picoplankton present in the Lake Superior using the 16S rRNA and *cpc*BA genes. They found that the phycocyanin and phycoerythrin dominating community in winter changes into phycoerythrin dominating community in summer. They also reported that community showed close similarity with freshwater and marine *Synechocystis* and *Prochlorococcus*. Further, Robertson *et al.* (2001) investigated phylogeny of *Synechococcus* based on *cpc* gene.

nif gene

Since *nif* gene is highly conserve (Young, 1992), it is a very potential tool to assess the cyanobacterial diversity (Ben-Porath *et al.*, 1993; Ben-Porath and Zehr, 1994). The gene presents a promising model to separate non-nitrogen- and nitrogen-fixers as later form a coherent cluster in phylogenetic tree (Zehr *et al.*, 1997). Among the different genes of *nif* operon, *nifH*, encoding the two identical subunits of Fe protein in nitrogenase complex, has been used widely to ascertain cyanobacterial phylogeny (Zehr *et al.*, 1997; 2003). They further reported that since the *nif* gene is not transferred laterally but descended vertically from the ancient progenitor, it may be promising to identify uncultured nitrogen-fixing cyanobacteria from various geographic locations using *nif* gene. Jackman and Mulligan (1995) have characterized a 6 kb region containing the *nifB* operon and the *nifH* gene from cyanobacterium *Anabaena azollae* 1a and compared it with that of *Anabaena* PCC7120.

Molecular Methods Used for Cyanobacterial Characterization

RFLP and RAPD

Restriction fragment length polymorphism (RFLP) is one of the traditional molecular techniques used for identification of cyanobacteria and development of genetic fingerprints. The RFLP technique is used in two ways: either digestion of whole genome with restriction enzymes and visualization of specific fragment using designed radioactive or florescent probes, or amplify the specific gene and then restriction digestion of the gene. The most commonly used gene is small ribosomal DNA and the technique is known as ARDRA (Amplified ribosomal DNA restriction analysis). Van Coppenolle *et al.* (1995) assessed the genetic diversity and phylogeny of *Anabaena azollae* using RFLP and *nif* gene probe. In contrast to this, amplified fragment length polymorphism (AFLP) involves restriction digestion of genomic DNA followed by PCR amplification of selected restriction fragments (Chial, 2008). Satish *et al.* (2001) and Dvornyk and Nevo (2003) have reported the genetic diversity and evolution of *Nostoc linckia* isolated from Evolution Canyon, Israel using AFLP. Lyra *et al.* (1997) have characterized the cyanobacteria using RFLP.

Generally each band obtained using these techniques are known as restriction DNA markers and based on these markers the species is identified. Terminal-restriction fragment length polymorphism (T-RFLP) is another technique used to assess cyanobacteria in a community. The technique constitutes restriction digestion of double stranded PCR products with fluorescently labeled end (Dorigo *et al.,* 2005; Liu *et al.,* 1997, Widmer *et al.,* 2006). The labeling is basically done by using 5′ florescent labeled primers. In the community fingerprint, each terminal fragment of specific size represents a species, though many species have same size bands. The technique can detect even the sparse member of the community. Ren *et al.* (2006) have assessed the cyanobacterial diversity from Jiaozhao Bay using the T-RFLP. The cyanobacterial community in the polluted reservoirs of Korea was analyzed using the same technique (Kim *et al.,* 2004).

Randomly amplified DNA polymorphism (RAPD) is PCR based genetic fingerprinting method. The technique involves use of random primers, which are complimentary to repetitive DNA sequences and capable of amplifying the different regions of the genome. The primers should be 10 nucleotides long. Since the repetitive DNA sequences are uniquely distributed in entire genome of a species, the resulted genetic fingerprint specifically represents that species. The technique is less tedious and time conscious as compared to RFLP. However, the resulting fingerprint is prone to DNA contamination and PCR related biases (Srivastava *et al.,* 2007). To overcome the problem of PCR conditions, Neilan (1995) proposed the multiplex RAPD-PCR, which involves use of two random primers per reaction instead of one used in normal RAPD-PCR. Srivastava *et al.* (2007) reported the genetic fingerprint of nitrogen-fixing cyanobacteria isolated from rice fields using both RAPD-PCR and multiplex RAPD-PCR. The genetic diversity of marine cyanobacterium *Leptolyngbya valderiana* has also been done by Premanandh *et al.* (2009) using RAPD and 16S rDNA-PCR. They reported intra-specific genetic diversity in the genus.

DGGE and TGGE

RFLP and RAPD are generally used for identification of pure cyanobacteria and not applicable for study of natural samples. Therefore, to assess microbial diversity in natural samples, denaturing gradient gel electrophoresis (DGGE) and temperature gradient gel electrophoresis (TGGE) are commonly used (Muyzer and Smalla, 1998). These are culture independent techniques involve PCR amplification of targeted genes and separation on denaturing gel based on their sizes and sequences. The gradient may be of denaturing chemicals (usually deionized formamide and urea) or temperature in DGGE or TGGE respectively. Since the methods are free of tedious isolation and culturing of cyanobacteria, these are rapid and less labour requiring.

Originally DGGE was developed to detect point mutation in biomedical research. However, its sensitivity to detect even one nucleotide change made it extensively useful in microbial ecology (Srivastava *et al.,* 2009). This technique is based on the concept of melting temperature (Tm). The DNA molecules of different sizes and sequences have different Tm and thus separate at specific concentration of denaturant or temperature (Muyzer, 1999). Each band in DGGE gels corresponds to individual organism. For example, five different bands in a DGGE gel of natural sample represent five cyanobacterial strains. Since metagenome of soil or water samples consists of many organisms, specificity of analysis depends on the particular primers. For example, cyanospecific primers only amplify the cyanobacterial genes and leave DNA of other organisms.

It is pertinent to mention that DGGE is not always an absolute method to quantify the microbial community (Jackson *et al.,* 2000). The efficiency of DGGE actually dependents on the amount and quality of metagenome to be analyzed, PCR associated biases and health of community to be examined. Despite of the limitation of DGGE, it is one of the best methods for the rapid characterization of the cyanobacterial communities in different ecosystems as well as to correlate their ecophysiologies with different environmental factors (Nübel *et al.,* 2000; Srivastava *et al.,* 2009).

Further, DGGE analysis based on total mRNA associated with reverse transcription gives a better picture of total functional community. Thus a comparison of DGGE gels based on DNA and cDNA-PCR provides one of the best ways to analyze the total community and its functional part (Nocker *et al.,* 2007; Wartiainen *et al.,* 2008). This analysis is very useful in all ecosystems especially the rice fields because a better understanding of microbial population and its processes may enhance the rice productivity in sustainable manner (Fernández-Valiente and Quesada, 2004). Further, in different wetlands and water bodies, characterization of cyanobacterial population especially toxin producing strains are very essential to help our surroundings for utilizing the water for domestic purposes. To achieve this target, DGGE/ TGGE may play a crucial role.

For better understanding of the procedure of molecular characterization of cyanobacteria, an example is given below with a flow chart (Figure 23.1) which depicts the major steps involved in molecular characterization.

16S rRNA PCR

Total genomic DNA from natural samples (cyanobacterial mat and soil) is isolated as mentioned in Srivastava *et al.* (2007). The primers CYA106F (CGC ACG GGT GAG TAA CGC GTG A) and CYA359F (GGG GAA TYT TCC GCA ATG GG) with a 40 nucleotide GC clamp (5'-CGC CCG CCG CGC CCC GCG CCG GTC CCG CCG

CCC CCG CCC G-3') on the 5' end (forward primer) and CYA781R (equimolar mixture of CYA781Ra (GAC TAC TGG GGT ATC TAA TCC CAT T) and CYA781Rb (GAC TAC AGG GGT ATC TAA TCC CTT T)) (reverse primers) is used for amplification of a segment of cyanobacterial *16S rRNA* gene (Nübel *et al.,* 1997). The GC clamp is used for better separation of amplified products in the gel. Amplification of selected gene is achieved in a semi-nested PCR reaction with the first reaction using primers CYA106F and CYA781R followed by a reaction with primers CYA359F and CYA781R. The composition of PCR mix per sample should be: 100 ng of DNA, 2.5 μl of 10X PCR buffer, 15 mM $MgCl_2$, 200 μM dNTPs, 10 pmol of each primer, 200 μg bovine serum albumin (nuclease free) and 0.2 U *Taq* DNA polymerase. The thermal cycling profile is as follows: initial denaturation for 3 min at 94°C, followed by 35 amplification cycles, each consisting of 1.5 min denaturation at 94°C, 1 min annealing at 59°C, and a 2 min elongation at 72°C, with a final 5 min elongation at 72°C.

DGGE Analysis

The PCR products of mat and soil samples obtained after the second PCR reaction are subjected to DGGE analysis. An aliquot of 25 μl of PCR product should be mixed with 5 μl of 10X gel loading solution (100% glycerol, 0.25% bromophenol blue and 0.25% xylenecyanole) and applied directly onto a 6% polyacryamide gel (acrylamide/bis 38.93/1.07) (w/v) in 1X Tris-acetate-EDTA (TAE) buffer with a linear 35-55% denaturant gradient. The gradient concentration may vary according to composition of community (Muyzer, 1999). A gradient dye solution (0.5% bromophenol blue, 0.5% xylenecyanole and 1X TAE buffer) can be used to examine the gradient formation. DGGE should be carried out at 60ÚC (constant temperature) for 16 h at 100 V (35 mA). The gel thus obtained is stained for 15 min in ethidium bromide (1 μg ml^{-1} in 1X TAE buffer) and visualized by UV transillumination and photographed. The PCR (16S rRNA gene) products of some pure cyanobacteria should be used as genetic markers on each gel alongside the environmental samples to compare the different gels.

Sequencing of Genes and Homology Search

The amplified PCR products or specific DGGE bands are subjected to sequencing. The PCR products having single amplified gene can be used for DNA sequencing after the cleaning with 70 % ethanol. Whereas DGGE bands should be excised from gels and the DNA is eluted in MilliQ water (Boutte *et al.* 2005). This DNA is used as template for next PCR reaction (all conditions should be same as for DGGE analysis) and the amplified product is subjected to sequencing. The resulting sequences are used for homology search in BLAST program provided by NCBI (http://blast.ncbi.nlm.nih.gov/Blast.cgi). By general convention the organism to which the sequence in question shows maximum similarity identifies the unknown organism. However, the *16S rRNA* gene sequences obtained from DGGE of environmental samples do not share absolute identity to the sequences obtained from cultured cyanobacteria (Giovannoni *et al.,* 1990; Rajaneimi *et al.,* 2005; Srivastava *et al.,* 2009). Therefore, it is important to identify the cyanobacteria

using both morphological and molecular attributes. The sequences obtained should be submitted in appropriate database under particular accession numbers.

Phylogenetic Analysis

Several bioinformatics tools can be used to document genetic diversity and assign the correct lineage of different cyanobacteria. First step is to perform multiple sequence alignment using CLUSTAL_X software. The alignment is manually corrected by identifying the gaps using JalView. This is followed by construction of phylogenetic tree. There are two main algorithm maximum-parsimony and maximum-likelihood used for this purpose (for detail see Yang, 1992). These methods are provided by different free (PHYLIP, MEGA4) or paid (PAUP) software. Other methods involve UPGMA (Unweighted Pair Group Method with Arithmetic mean) and NJ (Neighbor-Joining algorithm) (Saitou and Nei, 1987). Generally one thousand bootstrap replicates (random jumbling) of the alignment data should also be performed and the consensus tree is constructed.

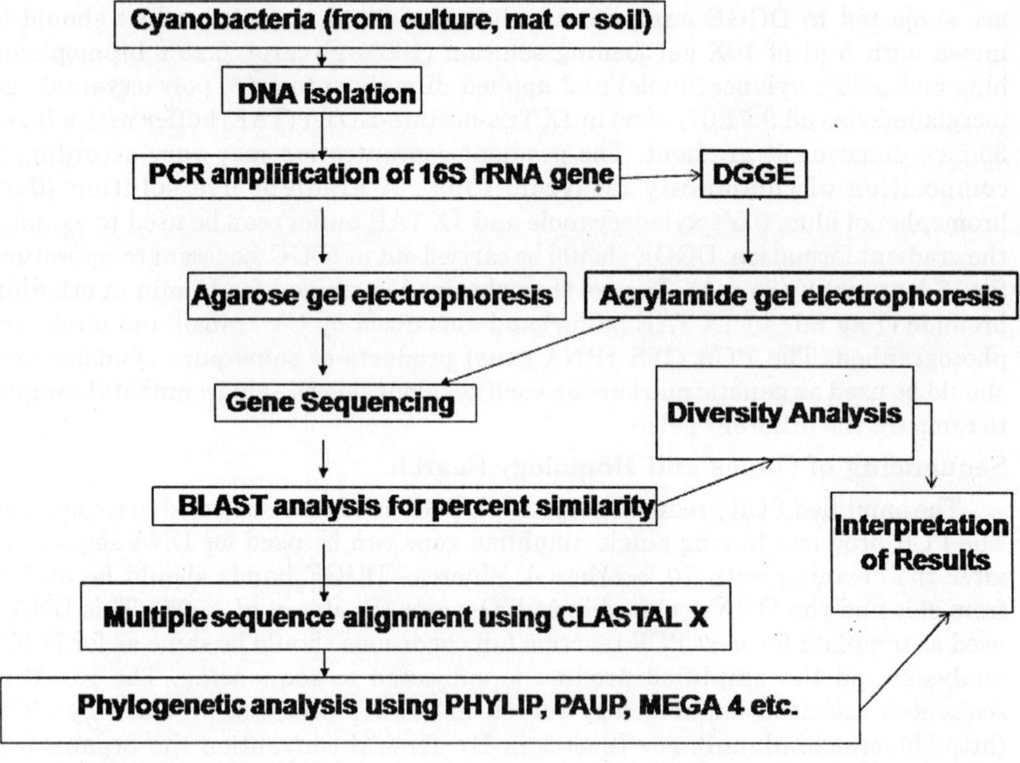

Figure 23.1. Various steps involved in characterization of cyanobacteria using 16S rRNA gene and DGGE. For pure cyanobacteria of culture, direct sequencing of amplified PCR product can be done. However, for cyanobacteria of natural sample, PCR products should be subjected to DGGE followed by sequencing.

Cyanobacterial Diversity in Wetlands Including Rice Fields

In the world

Characterization of cyanobacteria from Antarctic wetland and their ability to tolerate desiccation stress has been done by Šabacka′ and Elster (2006). They reported that cyanobacteria from these wetland are more desiccation tolerant than other wetland. Foster *et al.* (2009) have studied the cyanobacterial diversity in the stromatolites of Highborne Cay, Bahamas. They found *Apahanothece, Hyella, Leptolyngbya* and *Phormidium* and concluded that these cyanobacteria play a key role in initiation and stability of stromatolites and whole ecosystem. Molecular chacterization of bloom forming toxic cyanobacterium *Cylindrospermopsis raciborskii* isolated from different parts of Australia have also been done (Wilson *et al.*, 2000). This cyanobacterium is infamous for its role in secreting hepatotoxin namely cylindrospermopsin associated with human poisoning in Palm Island in Australia. Hamed *et al.* (2007) analyzed cyanobacterial community in saline lakes of Egypt using remote sensing technology. They confirmed the remote sensing data of *Spirulina platensis* with the ground verification. Further, the ecological genomics of marine picocyanobacteria has been presented by Scanlan *et al.* (2009). Documentation of *Microcystis* from several water bodies including those used for drinking water, of Uganda and Kenya has been done (Haande *et al.*, 2007). They reported four isolates producing microcsystin, however, most of them reported to produce aeruginosin and cyanopeptolin. Acinas *et al.* (2009) reported the phenotypic and genotypic diversity of *Pseudanabaena* strains isolated from Baltic Sea. This genus is involved in bloom formation in both brackish and freshwater bodies (Vasconcelos and Pereira, 2001; Stal *et al.*, 2003; Gkelis *et al.*, 2005; Zwart *et al.*, 2005; Willame *et al.*, 2006).

Quantitative real time PCR assessment method of bloom forming toxic cyanobacterium *Nodularia* in Baltic Sea has also been developed (Koskenniemi *et al.*, 2007). They reported that the Baltic Sea is highly exposed to hepatotoxin nodularin. Cyanobacterial diversity in Salar de Huasco, a high altitude saline wetland of Chile has been assessed by Dorador *et al.* (2008). The community was dominated by members of Chroococcales, Pleurocapsales and Oscillatoriales. In contrast to this, Vargas and Novelo (2007) reported the nitrogen-fixing cyanobacterial diversity in protected tropical Caribbean wetland. Microphytobenthos including cyanobacteria in southern California marine wetland were studied by Janousek (2009). The molecular characterization of cyanobacteria present in James Ross Island of Antarctica has also been done (Komárek *et al.*, 2008). Cyanobacteria in a Prairie Pothole wetland soil was investigated using 16S rDNA-PCR and denaturing gradient gel electrophoresis (Bleakley and Kruse, 2002). Implication of bacterial population on health of wetland of Utah, USA was examined by Merkley *et al.* (2004). They reported the greater metabolic diversity of aerobic bacteria including cyanobacteria is a very promising indication of degradation of wetland ecosystem.

The cyanobacterial communities and their dynamics have been widely studied in rice fields of different part of the world (Chunleuchanon *et al.*, 2003; Fernández-Valiente and Quesada, 2004; Song *et al.*, 2005; Teaumroong *et al.*, 2002). Fernández-Valiente and Quesada (2004) discussed in detail the role of cyanobacteria in rice ecosystem. They reported the succession of rice field microflora. Rice fields are first dominated by phytoplankton especially chlorophyceans and diatoms followed by presence of filamentous green algae and non-nitrogen fixing cyanobacteria at the time of initiation of panicle. After the development of panicle to harvest the nitrogen-fixing cyanobacteria dominate in rice fields and contribute in nitrogen economy of soil. Among the different cyanobacteria, *Gloeotrichia* and *Cylindrospermum* are the pioneer followed by *Scytonema, Microcheate, Nostoc* and *Anabaena* and finally *Nostoc* in Spanish rice fields. Total coverage area of cyanobacteria may be as much as 45 % after 120 days of rice plantation (Fernández-Valiente *et al.*, 2000). The growth and nitrogen fixation of cyanobacteria in rice fields depend on rice canopy vis-à-vis light intensity, addition of chemicals, drying and rewetting of rice fields, physicochemical properties of soil, agronomic practices and predation.

Song *et al.* (2005) reported the diversity of cyanobacteria in rice fields of Fujian, China and changes in community due to seasonal variation. Cyanobacteria in rice fields of Uruguay, their ability to fix nitrogen and tolerance to herbicides have been done by Irisarri *et al.* (2001). Kim and Lee (2006) reported diversity of nitrogen-fixing cyanobacteria in the Korean rice fields. They also compared their susceptibility to ten fungicides used in rice fields. They reported presence of *Anabaena, Calothrix, Cylindrospermum, Nodularia, Nostoc, Scytonema* and *Tolypothrix*. They further accounted that *Nostoc* isolates are more tolerant to fungicides than *Anabaena* isolates, however, use of fungicide benomyl should be avoided due to its toxicity to most nitrogen-fixing cyanobacteria. Teaumroong *et al.* (2002) and Chunleuchanon *et al.* (2003) have reported the diversity of nitrogen-fixing cyanobacteria in different ecosystems including the wetlands and rice fields of Thailand. It is worthwhile to mention that a significant increase in rice production has been achieved by cyanobacterial inocula along with other agriculturally important microorganisms in rice fields of Vietnam (Cong and Dung, 2008).

In India

Rice is staple crop of India and therefore its productivity and good growth is important. Cyanobacteria are dominant inhabitant and contribute significantly in the nitrogen economy of rice agro-ecosystem. Thus cyanobacterial diversity is a matter of great interest in Indian rice fields. The classical work of Late Professor R. N. Singh in 1961 provides the most magnificent document on role of nitrogen fixing cyanobacteria in Indian rice fields. The report also deals with the role of cyanobacteria in reclamation of saline soils commonly known as Usar soil. Cyanobacteria not only contribute to nitrogen economy but also improve the soil structure through their exo-polysaccharides and lipid. The dominance of nitrogen fixing cyanobacteria in rice fields can be assumed by the fact that their biomass can range from few kilograms to as much as 500 kg dry weight per hectare. It is worthwhile to mention that these strains can contribute as much as 30 kg nitrogen / ha/ season (Nayak and Prasanna, 2007). Not only this, these microorganisms also play a significant role in decreasing ammonia volatilization, reduction of methane from rice fields by making rice microclimate oxygenic, degradation of chemical compounds and weed control (Prasanna *et al.*, 2008). Further, Prasanna *et al.* (2009) reported the role of *Nostoc* and *Anabaena* in promoting germination of rice by producing growth promoting compounds especially IAA.

Prasanna and Nayak (2007) reported the cyanobacterial diversity in rice fields across the India including the rice fields of Shillong. They compared the soil ecologies (pH and electrical conductivity) with the cyanobacterial distribution based on morphological data. They found that *Anabaena* and *Nostoc* are the dominant genera of selected rice fields. Further, Nayak *et al.* (2008) characterized 70 different strains of *Anabaena* isolated from rice fields using molecular tool. Chaodhary (2009) reported the colonization of Chrooccocalean genera during rice cultivation in North Bihar. Cyanobacterial population in the soils of Orissa and West Bengal was investigated by Saxena *et al.* (2007).

Molecular characterization of cyanobacteria inhabiting rice fields of Eastern Uttar Pradesh was for the first time done by Srivastava *et al.* (2009). They used 16S rRNA-PCR and DGGE for this investigation. Soil and cyanobacterial samples were collected from rice fields of twenty villages situated in six districts of Eastern UP. After DGGE analysis, 51 different bands were selected for sequencing. Of these bands only 31 sequences showed significant differences. After BLAST analysis provided by NCBI, these sequences showed similarity with cyanobacteria and plastids. They further correlated the distribution of cyanobacteria with soil salinity measured in terms of electrical conductivity. Based on the results it was concluded that salinity has a significant role in distribution of cyanobacteria across the paddy fields. Non-heterocystous cyanobacteria dominated when salinity was high; however, low salinity (normal condition) favoured growth of heterocystous cyanobacteria (Figure 23.2). They further narrated that increasing salinity is posing a negative pressure on fertility of soil and leading to decline in biological nitrogen fixation and increased demand of chemical fertilizers in the paddy fields.

Muthukumar *et al.* (2007) reported cyanobacterial diversity from the different fresh water ponds of Thanjavur, Tamilnadu. They compared their presence with the physicochemical properties of water. They reported the dominance of cyanobacteria belonging to order Chrooccocales and Oscillatoriales with frequent presence of *Aphanothece microscopica, Synechocystis aquatilis, Merismopedia glauca, Oscillatoria limnetica* and *O. subbrevis*. Further, Thajuddin and Subramanian (2005) presented a good piece of survey on cyanobacterial diversity from different environments. Saha *et al.* (2007) reported the biodiversity of epilithic cyanobacteria from the fresh water streams of Kakoijana reserve forest, Assam and neighbouring paddy fields. They reported different species of *Aphanocapsa, Aphanothece, Chrocooccus, Calothrix, Leptolyngbya, Lyngbya, Nostoc, Hapalosiphon* and *Scytonema*. Cyanobacterial diversity in Loktak Lake and acidic soil of Manipur was documented by Tiwari and Singh (2005).

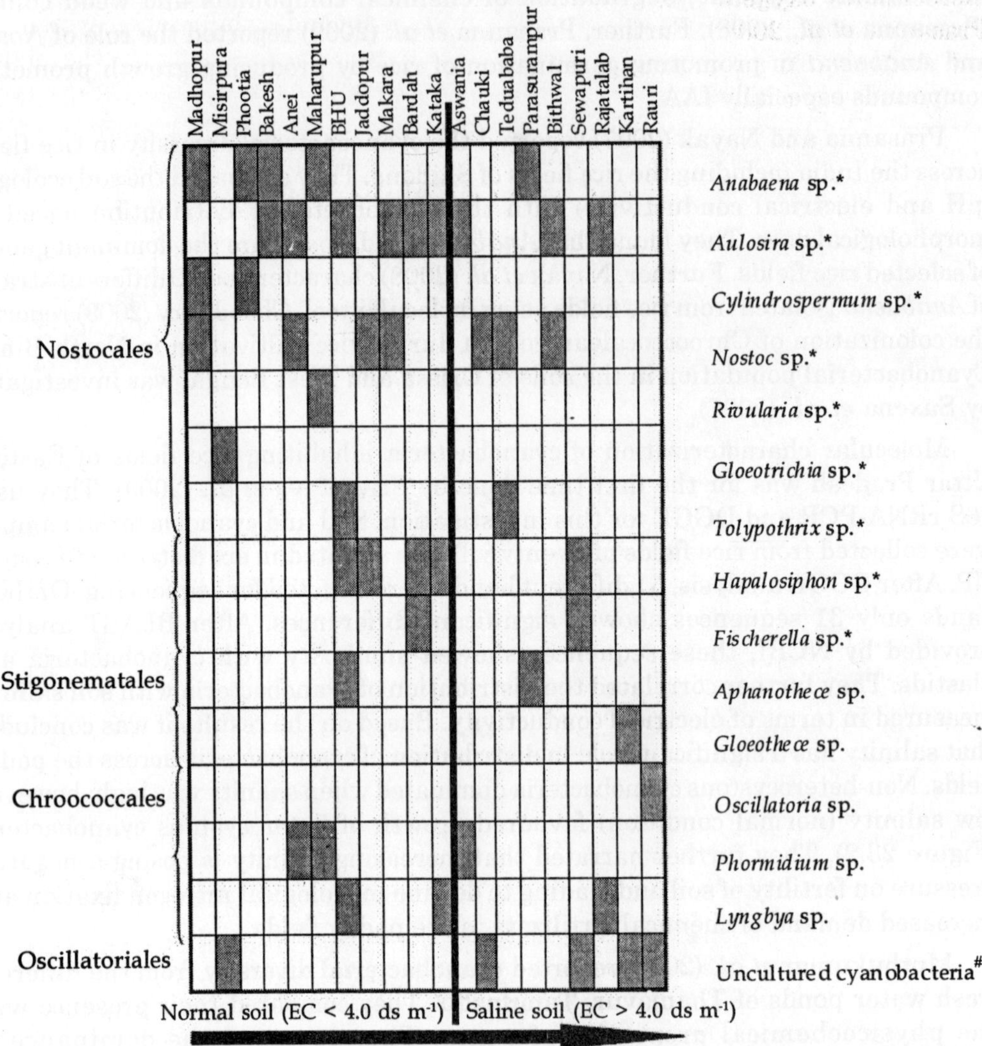

Figure 23.2. Distribution of different cyanobacteria in the selected rice fields of Eastern Uttar Pradesh (India). The names mentioned on top of the figure are name of experimental sites. Different orders of cyanobacteria are mentioned left hand side. The arrow denotes the increasing level of salinity measured in terms of electrical conductivity. The cyanobacterial genus name (mentioned at right hand site) includes all the DGGE bands showed similarity with the corresponding organism. Sings * and # represent the heterocystous cyanobacteria and unknown cyanobacteria respectively. Figure depicts dominance of heterocystous cyanobacteria in normal soils while saline soils were dominated by non-heterocystous forms. For detail see Srivastava *et al.* (2009). The figure is reproduced after the consultation with the journal "Saline Systems".

Future Perspective

The fundamental role of cyanobacteria in evolution of aerobic life and their cosmopolitan presence on modern earth make them potential candidate of scientific researches. Cyanobacteria are also the backbone of aquatic ecosystems including the wetlands due to their role in primary productivity and stability of ecosystem. It is worthwhile to conserve the diverse wetlands because they support the major component of vast biodiversity of India. In order to save wetlands it is necessary to document their enormous biodiversity. Therefore, it is of utmost importance to catalogue cyanobacterial diversity due to their role in nitrogen and carbon dynamics thus stability of this ecosystem. This will help us in sustainable management of wetlands. Assessment of toxic cyanobacteria in wetlands and water bodies used for agricultural and domestic purposes is another exciting area of future research. Research on seasonal changes in diversity pattern of cyanobacteria in wetlands may depict the bloom formation and thus the water quality. This can be supplemented with remote sensing for assessment bloom formation in large water bodies, wetlands, rivers as well as sea.

Further, cyanobacterial role in rice agro-ecosystem has been acknowledged by many scientists. Cyanobacteria are the major source of nitrogen pool of paddy fields and are being used as biofertilizer. Formulation of location specific biofertilizers, as cyanobacterial growth largely depends on soil type, is another future perspective. This will be a major step towards the organic farming and help to achieve evergreen revolution of India.

[**Acknowledgements:** The author is thankful to DST, New Delhi for financial support. Facilities provided by Head, Department of Botany, MZU is gratefully acknowledged. I am thankful to Prof. L.C. Rai, Department of Botany, BHU for his suggestions.]

REFERENCES

Acinas, S.G., Haverkamp, T.H.A., Huisman, J. and Stal, L.J. 2009. Phenotypic and genetic diversification of *Pseudanabaena* spp. (cyanobacteria). *ISME J.* 3: 31-46.

Anagnostidis, K. and Komárek, J. 1989. Modern approach to the classification system of cyanophytes. 3. Oscillatoriales. *Arch Hydrobiol.,* 80(Suppl): 327-472.

Ben-Porath, J. and Zehr, J.P. 1994. Detection and characterization of cyanobacterial *nif* H genes. *Appl. Environ. Microbiol.,* 60: 880-887.

Ben-Porath, J., Carpenter, E.J. and Zehr, J.P. 1993. Genotypic relationship in *Trichodesmium* (cyanophyceae) based on *nif*H sequence comparison. *J. Phycol.,* 29: 806-810.

Bleakley, B.H. and Kruse, J. 2002. Investigating cyanobacteria in a Prairie Pothole Wetland soil using 16S rDNA sequences revealed by PCR and denaturing gradient gel electrophoresis. *Soil/Water Research, 2002 Progress Report,* South Dakota State University, Brookings, SD.

Boutte, C., Grubisic, S., Balthasart, P. and Wilmotte, A. 2005. Testing of primers for the study of cyanobacterial molecular diversity by DGGE. *J. Microbiol. Methods,* 65: 542-550.

Chaodhary, K.K. 2009. Occurrence of chroococcaceae during rice cultivation of North Bihar, India. *Bangladesh J. Plant Taxon.,* 16: 57-63.

Chial, H. 2008. DNA fingerprinting using amplified fragment length polymorphisms (AFLP): No genome sequence required. *Nature Edu.,* 1(1).

Chunleuchanon, S., Sooksawang, A., Teaumroong, N. and Boonkerd, N. 2003. Diversity of nitrogen-fixing cyanobacteria under various ecosystems of Thailand: population dynamics as affected by environmental factors. *World J. Microbiol. Biotechnol.*, 19: 167–173.

Clayton R.A., Sutton, G., Hinkle, P.S., Bult, C. and Fields, C. 1995. Intraspecific variation in small-subunit rRNA sequences in GenBank: Why single sequences may not adequately represent prokaryotic taxa? *Int. J. Syst. Bacteriol.*, 45: 595-599.

Cong, P.T. and Dung, T.D. 2008. Interaction effects of BioGro with nitrogen and phosphorus on grain yield and nutrient uptake of rice in light-textured soil of southern Vietnam. pp. 24-31. In: *Efficient Nutrient Use in Rice Production in Vietnam Achieved Using Inoculant Biofertilizers.* (eds. I.R. Kennedy, A.T.M.A. Choudhury, M.L. Kecskes and M.T. Rose) ACIAR Proceedings Series, Australia.

Dahllöf, I., Baillie, F.H. and Kjelleberg, S. 2000. *rpoB*-based microbial community analysis avoids limitations inherent in 16S rRNA gene intraspecies heterogeneity. *Appl. Environ. Microbiol.*, 8: 3376–3380.

Desikachary, T.V. 1959. *Cyanophyta.* ICAR: New Delhi.

Dorador, C, Vila, I., Imhoff, J.F. and Witzel, K.P. 2008. Cyanobacterial diversity in Salar de Huasco, a high altitude saline wetland in northern Chile: an example of geographical dispersion. *FEMS Microbiol. Ecol.*, 64: 419 – 432.

Dorigo, U., Volatier, L. and Humbert, J.F. 2005. Molecular approaches to the assessment of biodiversity in aquatic microbial communities. *Water Res.*, 39: 2207-2218.

Dvornyk V. and Nevo, E. 2003. Genetic polymorphism of cyanobacteria under permanent natural stress: A lesson from the "Evolution Canyons". *Res. Microbiol.*, 154: 79-84.

Farrelly, V., Rainey, F.A. and Stackebrandt, E. 1995. Effect of genome size and rrn gene copy number on PCR amplification of 16S rRNA genes from a mixture of bacterial species. *Appl. Environ. Microbiol.*, 61: 2798-2801.

Fernández-Valiente, E. and Quesada, A. 2004. A shallow water ecosystem: rice-fields. The relevance of cyanobacteria in the ecosystem. *Limnetica,* 23: 95-108.

Fernández-Valiente, E., Ucha, A., Quesada, A., Leganés, F. and Carreres, R. 2000. Contribution of N_2 fixing cyanobacteria to rice production: availability of nitrogen from ^{15}N-labelled cyanobacteria and ammonium sulphate to rice. *Plant Soil,* 221: 107-112.

Foster, J.S., Green, S.J., Ahrendt, S.R., Golubic, S., Reid, R.P., Hetherington, K.L. and Bebout, L. 2009. Molecular and morphological characterization of cyanobacterial diversity in the stromatolites of Highborne Cay, Bahamas. *ISME J.*, 1–15.

Fox, G. E., Wisotzkey, J. D. and Jurtshuk Jr., P. 1992. How close is close: 16S rRNA sequence identity may not be sufficient to guarantee species identity. *Int. J. Syst. Bacteriol.*, 42: 166-170.

Garcia-Pichel, F., Cortes, A.L. and Nübel, U. 2001. Phylogenetic and morphological diversity of cyanobacteria in soil desert crusts from the Colorado plateau. *Appl. Environ. Microbiol.*, 67: 1902-1910.

Geiß, U., Selig[1]University of Rostock, Department Biological Sciences, Aquatic Ecology, Albert-Einstein-Str. 3a, D-18051 Rostock, Germany. , U., Schumann[3]University of Rostock, Applied Ecology, Albert-Einstein-Str. 3a, D-18051 Rostock, Germany., R., Steinbruch[2]University of Rostock, Plant Physiology, Albert-Einstein-Str. 3a, D-18051 Rostock, Germany. , R., Bastrop, R., Hagemann, M.[2]University of Rostock, Plant Physiology, Albert-Einstein-Str. 3a, D-18051 Rostock, Germany. and Schoor, A. [1]University of Rostock, Department Biological Sciences, Aquatic Ecology, Albert-Einstein-Str. 3a, D-18051 Rostock, Germany. 2004. Investigations on cyanobacterial diversity in a shallow estuary (Southern Baltic Sea) including genes relevant to salinity resistance and iron starvation acclimation. *Environ. Microbiol.*, 6: 377-387.

Gietler, L. 1932. Cyanophyceae 14. In *Kryptogamenflora von Deutschland Österreich und der Schweiz by Rabenhorst L.* Leipzig. Akademische. Verlagsgesellschaft.

Giovannoni, S.J., Britschgi, T.B., Moyer, C.L. and Field, K.G. 1990. Genetic diversity in Sargasso Sea bacterioplankton. *Nature,* 345: 60-63.

Gkelis, S., Rajaniemi, P. Vardaka, E., Moustaka-Gouni, M., Lanaras, T. and Sivonen, K. 2005. *Limnothrix redekei* (Van Goor) Meffert (Cyanobacteria) strains from Lake Kastoria, Greece form a separate phylogenetic group. *Microb. Ecol.,* 49: 176–182.

Gupta, A.B. 1996. Algal flora and its importance in the economy of rice fields. *Hydrobiologia,* 28: 213-222.

Haande, S., Ballot, A., Rohrlack, T., Fastner, J., Wiedner, C. and Edvardsen, B. 2007. Diversity of *Microcystis aeruginosa* isolates (Chroococcales, Cyanobacteria) from East-African water bodies. *Arch. Microbiol.,* 188: 15 – 25.

Hamed, A.F., Salem, B.B. and Abd El-Fatah, H.M. 2007. Floristic survey of blue-green algae / cyanobacteria in saline-slkaline lake of Wadi El- Natrun (Egypt) by remote sensing application. *J. Appl. Sci. Res.,* 3: 495-506.

Irisarri, P., Gonnet, S. and Monza, J. 2001. Cyanobacteria in Uruguayan rice fields: diversity, nitrogen fixing ability and tolerance to herbicides and combined nitrogen. *J. Biotechnol.,* 91: 95-103.

Iteman, I., Rippka, R., de Marsac, N.T. and Herdman, M. 2002. rDNA analyses of planktonic heterocystous cyanobacteria, including members of the genera *Anabaenopsis* and *Cyanospira. Microbiology.,* 148: 481-496.

Ivanikova, N.V., Popels, L.C., Michael, R., McKay, L. and Bullerjahn, G.S. 2007. Lake Superior supports novel clusters of cyanobacterial picoplankton. *Appl. Environ. Microbiol.,* 73: 4055-4065.

Jackman, D.M. and Mulligan, M.E. 1995. Characterization of a nitrogen-fixation (*nif*) gene cluster from *Anabaena azollae* 1a shows that closely related cyanobacteria have highly variable but structured intergenic regions. *Microbiology.,* 141: 2235-2244.

Jackson, C.R., Roden, E.E. and Churchill, P.F. 2000. Denaturing gradient gel electrophoresis can fail to separate 16S rDNA fragments with multiple base differences. *Mol. Biol. Today,* 1: 49-51.

Janousek, C.N. 2009. Taxonomic composition and diversity of microphytobenthos in southern California marine wetland habitats. *Wetlands.* 29:163-175.

Jungblut, A.D., Hawes, I., Mountfort, D., Hitzfeld, B., Dietrich, D.R., Burns, B.P. and Neilan, B.A. 2005. Diversity within cyanobacterial mat communities in variable salinity meltwater ponds of McMurdo ice shelf, Antarctica. *Environ. Microbiol.,* 7: 519–529.

Kim, J.B., Moon, M.S., Lee, D.H., Lee, S.T., Bazzicalupo, M. and Kim, C.K. 2004. Comparative analysis of cyanobacterial communities from polluted reservoirs in Korea. *J. Microbiol.,* 42: 181-187.

Kim, J.D. and Lee, C.G. 2006. Diversity of heterocystous filamentous cyanobacteria (blue-green algae) from rice paddy fields and their differential susceptibility to ten fungicides used in Korea. *J. Microbiol. Biotechnol.,* 16: 240-246.

Komárek, J., Elster, J. and Komárek, O. 2008. Diversity of the cyanobacterial microflora of the northern part of James Ross Island, NW Weddell Sea, Antarctica. *Polar Biol.,* 31: 853-865.

Koskenniemi, K., Lyra, C., Rajaniemi-Wacklin, P., Jokela, J. and Sivonen, K. 2007. Quantitative real-time PCR detection of toxic *Nodularia* cyanobacteria in the Baltic Sea. *Appl. Environ. Microbiol.,* 73: 2173-2179.

Kumari, N., Srivastava, A.K., Bhargava, P. and Rai, L.C. 2009. Molecular approaches towards assessment of cyanobacterial biodiversity. *African J. Biotechnol.*, 8: 4284-4298.

Liu, W.T., Marsh, T.L., Cheng, H. and Forney, L.J. 1997. Characterization of microbial diversity by determining terminal restriction fragment length polymorphisms of genes encoding 16S rRNA. *Appl. Environ. Microbiol.*, 63: 4516-4522.

Lyra, C., Hantula, J., Vainio, E., Rapala, J., Rouhiainen, L. and Sivonen, K. 1997. Characterization of cyanobacteria by SDS-PAGE of whole-cell proteins and PCR/RFLP of the 16S rRNA gene. *Arch. Microbiol.*, 168: 176-184.

Martinez-Murcia, A.J., Anton, A.I. and Rodriguez-Valera, F. 1999. Patterns of sequence variation in two regions of the 16S rRNA multigene family of *Escherichia coli*. *Int. J. Syst. Bacteriol.*, 49: 601 -610.

Merkley, M., Rader, R.B., McArthur, J.V. and Eggett, D. 2004. Bacteria as bioindicators in wetlands: Bioassessment in the Bonneville Basin of Utah, USA. *Wetlands.*, 24: 600-607.

Muthukumar, C., Muralitharan, G., Vijaykumar R., Panneerselvan A. and Thajuddin, N. 2007. Cyanobacterial diversity from different freshwater ponds of Thanjavur, Tamilnadu, India. *Acta Bot. Malac.*, 32: 17-25.

Muyzer, G. 1999. DGGE/TGGE a method for identifying genes from natural ecosystems *Curr. Opin. Microbiol.*, 2: 317-322.

Muyzer, G. and Smalla, K. 1998. Application of denaturing gradient gel electrophoresis (DGGE) and temperature gradient gel electrophoresis (TGGE) in microbial ecology. *Antonie van Leeuwenhoek.*, 73: 127-141.

Nayak, S. and Prasanna, R. 2007. Soil pH and its role in cyanobacteria abundance and diversity in rice field soils. *Appl. Ecol. Environ. Res.* 5: 103-113.

Nayak, S., Prasanna, R. Prasanna, B.M. and Sahoo, D. 2008. Genotypic and phenotypic diversity of *Anabaena* isolates from diverse rice agro-ecologies of India. *J. Basic Microbiol,.* 49: 165-177.

Neilan, B. A., Jacobs, D. and Goodman, A. E. 1995. Genetic diversity and phylogeny of toxic cyanobacteria determined by DNA polymorphisms within the phycocyanin locus. *Appl. Environ. Microbiol.*, 61: 3875-3883.

Neilan, B.A. 1995. Identification and phylogenetic analysis of toxigenic cyanobacteria by multiplex randomly amplified polymorphic DNA PCR. *Appl. Environ. Microbiol.*, 61: 2286-2291.

Neilan, BA. 2002. The molecular evolution and DNA profiling of toxic cyanobacteria. *Curr. Issues Mol. Biol.*, 1:1-11.

Nocker, A., Burr, M. and Camper, A.K. 2007. Genotypic microbial community profiling: A critical technical review. *Microb. Ecol.*, 54: 276-289.

Nübel, U., Engelen, B., Felske, A., Snaidr, J., Wieshuber, A., Amann, R. I., Ludwig, W. and Backhaus, H. 1996. Sequence heterogeneities of genes encoding 16S rRNAs in *Paenibacillus polymyxa* detected by temperature gradient gel electrophoresis. *J. Bacteriol.*, 178: 5636-5643.

Nübel, U., Garcia-Pichel, F., Clavero, E. and Muyzer, G. 2000. Matching molecular diversity and ecophysiology of benthic cyanobacteria and diatoms in communities along a salinity gradient. *Environ. Microbiol.*, 2: 217-226.

Nübel, U., Garcia-Pichel, F. and Muyzer, G. 1997. PCR primers to amplify 16S rRNA genes from cyanobacteria. *Appl. Environ. Microbiol.* 63: 3327-3332.

Palenik B. 1992. Polymerase evolution and organism evolution. *Curr. Opin. Genet. Dev.* 2: 931-936.

Palenik, B. 1994. Cyanobacterial community structure as seen from RNA polymerase gene sequence analysis. *Appl. Environ. Microbiol.* 60: 3212-3219.

Palenik, B. and Swift, H. 1996. Cyanobacterial evolution and prochlorophyte diversity as seen in DNA-dependent RNA polymerase gene sequences. *J. Phycol.* 32: 638 – 646.

Papke, R.T., Ramsing, N.B., Bateson, M.M. and Ward, D.M. 2003. Geographical isolation in hot spring cyanobacteria. *Environ. Microbiol.* 5: 650–659.

Prasanna, R., Jaiswal, P. and Kaushik, B.D. 2008. Cyanobacteria as potential options for environmental sustainability — promises and challenges. *Indian J. Microbiol.* 48: 89-94.

Prasanna, R., Jaiswal, P., Nayak, S., Sood, A. and Kaushik, B.D. 2009. Cyanobacterial diversity in the rhizosphere of rice and its ecological significance. *Indian J. Microbiol.* 49: 89-97.

Premanandh, J., Priya, B., Prabaharan, D. and Uma, L. 2009. Genetic heterogeneity of the marine cyanobacterium *Leptolyngbya valderiana* (Pseudanabaenaceae) evidenced by **RAPD** molecular markers and **16S rDNA** sequence data. *J. Plankton Res.* 31:1141-1150.

Rajaniemi, P., Hrouzek, P., Kastovšká, K., Williame, R., Ranata, A., Hoffmann, L., Komárek, J. and Sivonen, K. 2005. Phylogenetic and morphological evaluvation of the genera *Anabaena, Aphanizomenon, Trichormus* and *Nostoc* (Nostocales, Cyanobacteria). *Int. J. Syst. Evol. Microbiol.* 55: 11-26.

Rath, J. and Adhikary, S.P. 2008. Biodiversity assessment of algae in Chilka lake, East coast of India. pp. 22-33 In: *Monitoring and Modelling Lakes and Coastal Environments.* (Ed. P.K. Mohanty) Springer, The Netherlands.

Redfield, E., Barns, S.M., Belnap, J., Daane, L.L. and Kuske, C.R. 2002. Comparative diversity of composition of cyanobacteria in three predominant soil crusts of the Colorado Plateau. *FEMS Microbiol. Ecol.* **40**: 55–63.

Rejmánková, E., Komárek, J. and Komárková, J. 2004. Cyanobacteria – a neglected component of biodiversity: patterns of species diversity in inland marshes of northern Belize (Central America). *Diversity Distrib.* 10: 189-199.

Ren J., Dang, Song, L.S., Sun, S. and An, L.G. 2006. Bacterial and cyanobacterial diversities determined by T-RFLP analyses in Jiaozhou Bay. *Acta. Oceanol. Sin.* 25:113–123.

Rippka, R., Deruelles, J., Waterbury, J.B., Herdman, M. and Stenier, R.Y. 1979. Generic assignments, strain histories and properties of pure cultures of cyanobacteria. *J. Gen. Microbiol.* 111: 1-61.

Robertson, B.R., Tezuka, N. and Watanabe, M.M. 2001. Phylogenetic analyses of *Synechococcus* strains (cyanobacteria) using sequences of 16S rDNA and part of the phycocyanin operon reveal multiple evolutionary lines and reflect phycobilin content. *Int. J. Syst. Evol. Microbiol.* **51**: 861–871.

Sabacka, M. and Elster, J. 2006. Response of cyanobacteria and algae from Antarctic wetland habitats to freezing and desiccation stress. *Polar Biol.* 30: 31-37.

Saha, S.K., Das, R., Bora, K.N. and Uma, L. 2007. Biodiversity of epilithic cyanobacteria from freshwater streams of Kakoijana reserve forest, Assam, India. *Indian J. Microbiol.* 47: 219-232.

Saitou, N. and Nei, M. 1987. The neighbor-joining method: A new method for reconstructing phylogenetic trees. *Mol. Biol. Evol.* 4: 406-425.

Satish, N., Krugman, T., Vinogradova, O.N., Nevo, E. and Kashi, Y. 2001. Genome evolution of the cyanobacterium *Nostoc linckia* under sharp microclimatic divergence at "Evolution Canyon," Israel. *Microb. Ecol.* 42: 306- 316.

Saxena, S., Singh, B.V., Tiwari, S. and Dhar, D.W. 2007. Physiological characterization of cyanobacterial isolates from Orissa and West Bengal. *Indian J. Plant Physiol.* 12.

Scanlan, D. J., Ostrowski, M., Mazard, S., Dufresne, A., Garczarek, L., Hess, W.R., Post, A.F., Hagemann, M., Paulsen, I. and Partensky, F. 2009. Ecological genomics of marine picocyanobacteria. *Microbiol. Mol. Biol. Rev.* 73: 249-299.

Sigler, W.B., Bachofen, R. and Zeyer, J. 2003. Molecular characterization of endolithic cyanobacteria inhabiting exposed dolomite in central Switzerland. *Environ. Microbiol.* 5: 618-627.

Singh, N.I., Dorycanta, H., Devi, G.A., Singh, N.S. and Singh, S.M. 1997b. Blue-green algae from rice field soils of Nagaland. *Phykos.* 36:115–120.

Singh, N.I., Singh, N.S., Devi, G.A. and Singh, S.M. 1997a. Blue-green algae from rice growing areas of Arunachal Pradesh. *Phykos.* 36:21–26.

Singh, R.N. 1961. *Role of blue green algae in the nitrogen economy of Indian agriculture.* ICAR, New Delhi.

Song, T., Mårtensson, L., Ericsson, T., Zheng, W. and Rasmussen, U. 2005. Biodiversity and seasonal variation of the cyanobacterial assemblage in a rice paddy in Fujian, China. *FEMS Microbiol. Ecol.* 54: 131-140.

Srivastava, A.K., Ara, A., Bhargava, P., Mishra Y., Rai, S.P. and Rai, L.C. 2007. A rapid and cost effective method of genomic DNA isolation from cyanobacterial mat and soil suitable for genomic fingerprinting and community analysis. *J. Appl. Phycol.* 19: 373-382.

Srivastava, A.K., Bhargava, P., Kumar, A., Rai, L.C. and Neilan, B.A. 2009. Molecular characterization and effect of salinity on cyanobacterial diversity in the rice fields of Eastern Uttar Pradesh, India. *Saline Sys.* 5: 4.

Srivastava, S.K. and Gupta, R.K. 2004. Studies on the algal plant diversity of Pong Dam wetland. *Indian J. Forestry.* 27: 103-112.

Stal, L.J. 2007. Cyanobacteria: Diversity and versatility, clues to life in extreme environment. pp. 659-680. In: *Algae and Cyanobacteria in Extreme Environment.* (Ed. J. Seckbach) Springer, The Netherlands.

Stal, L.J., Albertano, P., Bergman, B., von Brockel, K., Gallon, J.R., Hayes, P.K., Sivonen, K. and Walsby, A. E. 2003. BASIC: Baltic Sea cyanobacteria. An investigation of the structure and dynamics of water blooms of cyanobacteria in the Baltic Sea responses to a changing environment. *Cont. Shelf Res.* 23: 1695–1714.

Teaumroong, N., Innok, S., Chunleuchanon, S. and Boonkerd, N. 2002. Diversity of nitrogen-fixing cyanobacteria under various ecosystems of Thailand: I. Morphology, physiology and genetic diversity. *World J. Microbiol. Biotechnol.* 18: 673-682.

Tebbe, C.C. and Vahjen, W. 1993. Interference of humic acids and DNA extracted directly from soil in detection and transformation of recombinant DNA from bacteria and a yeast. *Appl. Environ. Microbiol.* 59: 2657-2665.

Thajuddin, N. and Subramanian, G. 2005. Cyanobacterial biodiversity and potential applications in biotechnology. *Curr. Sci.* 89: 47-57.

Tiwari, G.L. 1975. A study of blue green algae from paddy field soil of India II. Taxonomic consideration of non heterocystous blue green alage. *Nova Hedwigia* 25: 765-798.

Tiwari, O.N. and Singh, H.T. 2005. Biodiversity of cyanobacteria in Loktak lake and rice fields of Manipur, India having acidic properties. *Proc. Natl. Acad. Sci. India B Bio. Sci.* 75: 209-213.

Toledo, G. and Palenik, B. 1997. *Synechococcus* diversity in the California current as seen by RNA polymerase (*rpo*C1) gene sequences of isolated strains. *Appl. Environ. Microbiol.* 63: 4298-4303.

Van Coppenolle B., McCouch, S. R., Watanabe, I., Huang, N. and Van Hove, C. 1995. Genetic diversity and phylogeny analysis of *Anabaena azollae* based on RFLPs detected in *Azolla-Anabaena azollae* DNA complexes using nif gene probes. *Theor. Appl. Genet.* 91: 589- 597.

Vargas, R. and Novelo, E. 2007. Seasonal changes in periphyton nitrogen fixation in a protected tropical wetland. *J. Biol. Fert. Soils.* 43: 367-372.

Vasconcelos, V.M. and Pereira, E. 2001. Cyanobacteria diversity and toxicity in a wastewater treatment plant (Portugal). *Water Res.* 35: 1354–1357.

Wang, Y. and Qian, P.Y. 2009. Conservative fragments in bacterial 16S rRNA genes and primer design for 16S ribosomal DNA amplicons in metagenomic studies. *PLoS One.* 4: e7401.

Wartiainen, I., Eriksson, T., Zheng, W. and Rasmussen, U. 2008. Variation in the active diazotrophic community in rice paddy—*nifH* PCR-DGGE analysis of rhizosphere and bulk *soil. Appl. Soil Ecol.* 39: 65-75.

Weller, R., Weller, J.W. and Ward, D.M. 1991. 16S rRNA sequences of uncultivated hot spring cyanobacterial mat inhabitants retrieved as randomly primed cDNA. *Appl. Environ. Microbiol.* 57: 1146- 1151.

Widmer, F., Hartmann, M., Frey, B. and Kölliker, R. 2006. A novel strategy to extract specific phylogenetic sequence information from community T-RFLP *J. Microbiol. Methods.* 66: 512-520.

Willame, R., Boutte, C., Grubisic, S., Wilmotte, A., Komarek, J. and Hoffmann, L. 2006. Morphological and molecular characterization of planktonic cyanobacteria from Belgium and Luxembourg. *J. Phycol.* 42: 1312–1332.

Wilson, K. M., Schembri, M. A., Baker, P.D. and Saint, C.P. 2000. Molecular characterization of the toxic cyanobacterium *Cylindrospermopsis raciborskii* and design of a species-specific PCR. *Appl. Environ. Microbiol.* 66: 332-338.

Yang, Z. 1992. Phylogenetic analysis using parsimony and likelihood methods. *J. Mol. Evol.* 42: 294-307.

Young, J.P.W. 1992. Phylogenetic classification of nitrogen fixing organisms. pp. 43-86. In: *Biological Nitrogen Fixation.* (Eds. G. Stacey, H.J. Evans, and R.H. Burris). Chapman and Hall, New York.

Zehr, J.P., Jenkins, B.D., Short, S.M. and Steward, G.F. 2003. Nitrogenase gene diversity and microbial community structure: a cross-system comparison. *Environ. Microbiol.* 5: 539–554.

Zehr, J.P., Mellon, M.T. and Hiorns, W.D. 1997. Phylogeny of cyanobacterial nifH genes: evolutionary implications and potential applications to natural assemblages. *Microbiology* 143: 1443-1450.

Zwart, G., Kamst-van Agterveld, M.P., van der Werff-Staverman, I., Hagen, F., Hoogveld, H.L. and Gons, H.J. 2005. Molecular characterization of cyanobacterial diversity in a shallow eutrophic lake. *Environ. Microbiol.* 7: 365–377.

24

Water Resources in Arunachal Pradesh: An Overview

—*J.S. Rawat*[1] & *Boni Amin Laskar*[2]

[Water resources, both ground and surface, of the Arunachal Pradesh in the Eastern Himalaya region is very rich and most of the runoff drains in to the river Brahmaputra. Out of the total basin area of Brahmaputra, 41.9% lies in the State which cascades through very steep gorges by the name Siang and finally debouches to Assam plain. Bountiful tributaries of this mighty river winds in the hills of Arunachal Pradesh; which were studied in the light of potentialities and bottlenecks of water resource. Besides other characteristics like lengths, discharge volumes etc. were assessed using the Digital Elevation Model of the Shuttle Radar Topographic Mission (SRTM). Nearly 50% river length is under low and gentle stream gradient while about 64% of the river length falls under the very low and low stream elevation. Therefore, the paper suggests for installation of mini hydel projects instead of mega hydel project for the judicious utilization of vast water resources. The paper also implores the catchments treatment plans and land use planning in the watershed basins of the state.

Key Words: River, discharge, river length, gradient, basin, hydropower.]

Introduction

Water is vital component of the biosphere and a base for all life system on the earth. It is closely related to biotic population, demand for food, production of non–agricultural and industrial items, production of energy and improvement in the quality of life and preserving the ecology. The National Water Policy 1987 stresses that 'Water is a prime natural resource, a basic human need and a precious national asset'. The country's per capita availability of water at the beginning of last century (in terms of annual water renewable sources) was 8192 m^3 $year^{-1}$, and 5694 m^3 $year^{-1}$ at the time of Independence. Due to rapid population growth, the per capita availability is decreasing. It was 1953 m^3 $year^{-1}$ in 1991 which is estimated to go down to 1235 m^3 $year^{-1}$ by 2050. The optimum utilization of this vital resource, therefore, needs sincerity. Indeed, there is a need of the proper inventory and

[1] Department of Geography, Rajiv Gandhi University, Rono Hills, Doimukh, Itanagar, 791111, Arunachal Pradesh, Email:jsr_06@rediffmail.com

[2] Department of Zoology, Rajiv Gandhi University, Rono Hills, Doimukh, Itanagar, 791111 Arunachal Pradesh.

assessment of water resource as well as identification of potentialities and bottlenecks. For this the application of remote sensing and Geographical Information System (GIS) is essential to predict of water quality parameters with accuracy (Rajitha *et al.*, 2007). The satellite images are also used to derive information on hydrogeomorphology of the terrain (Dwevedi *et al.*, 2006). In recent past, development of stereo sensor in space platforms, capable of acquiring terrain information in 3-dimensional form has added its unique potential for topographic analysis in relatively large area (Kumar *et al.*, 2008). The stereo pair and Interferometric Synthetic Aperture Randar (InSAR) have emerged as powerful tools for extraction of terrain parameters (Jayaprasad *et al.*, 2008). This paper aims to provide an overview of the water resource in the state Arunachal Pradesh (AP). It is largest state in northeast in terms of geographical extends covering 83, 743 km^2 area and is bestowed with the plentiful water resources. The State receives heavy rainfall and the mountain peaks remain snow-capped which enriches the rivers with runoff and perennial flow respectively.

Methodology

The information about the water resources, river lengths, volume, flow characteristics were consulted from various secondary sources. Rainfall data of 23 meteorological stations of RWD, Government of Arunachal Pradesh, for the period of 15 years (1991 to 2005) are collected. A point layer created by the average rainfall of station is interpolated by moving average method in GIS platform. Finally, an isohyets map is developed showing rainfall distribution. The InSAR Digital Elevation Model (DEM) of Shuttle Radar Topographic Mission (SRTM) is downloaded from the United States Geological Survey (USGS) website (http://seamless.usgs/gov). The tiles Z_55_7 and Z_56_7 containing 3 arc second DEM with a default pixel size of 87.413 metre are imported in the ILWIS 3.4 GIS (Figure 24.1). Using 'DEM hydro-processing' operation drainage network is extracted from the DEM. In order to reduce the computational robustness, drainages are extracted with a stream threshold value of 500 pixels. It also eliminates ephemeral and other insignificant streams. The slope and elevation along the rivers are derived from DEM and grouped into categories with suitable scale. The total length of the river in each category of slope and elevation is obtained ˆ ᵛ the analysis.

Results and Discussion

Rainfall

The rainfall data of 15 years of Arunachal Pradesh shows that the state receives an average, 3015 mm rainfall (Table 24.1). The annual average rainfall ranges from 2477 mm to 3997 mm. On the other hand, the station-wise average rainfall of the state varies between 1205 mm in Ziro to 4506 mm in Tuting. The places receiving more than 4000 mm annual rainfall are Khonsa, Tezu and Tuting. Whereas, Namsai, Doimukh, Likabali, Roing, Itanagar, Senwa, Hayuliang and Pasighat receive annual rainfall between 3000 to 4000 mm. Ziro, Bomdila, and Seppa receive annual rainfall below 2000 mm. Thus, the lower reaches (foothills) receive more rainfall than the areas lying in higher altitude as shown by isohyets in Figure 24.2 with some exceptions of Tuting and Hayuliang.

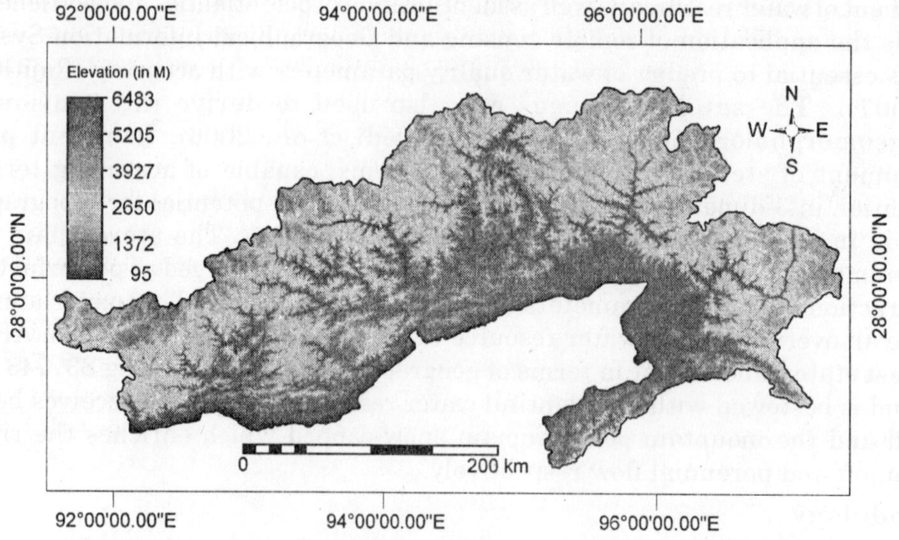

Figure 24.1. Digital Elevation Model (DEM) of Arunachal Pradesh.

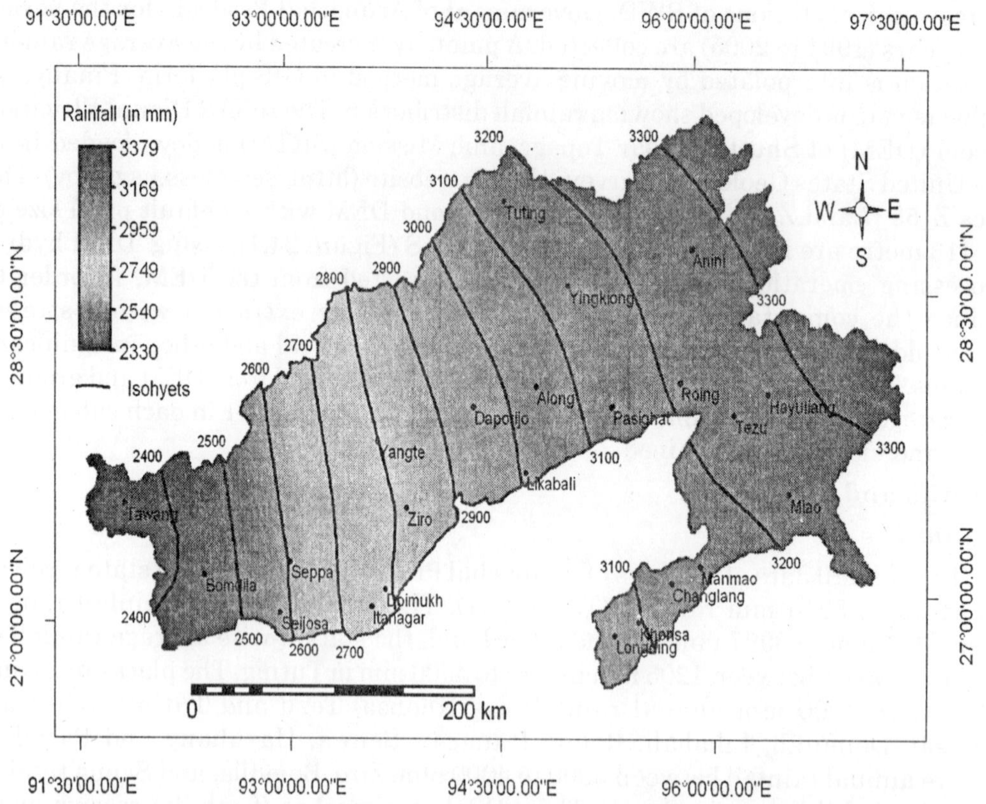

Figure 24.2. Rainfall distribution in Arunachal Pradesh.

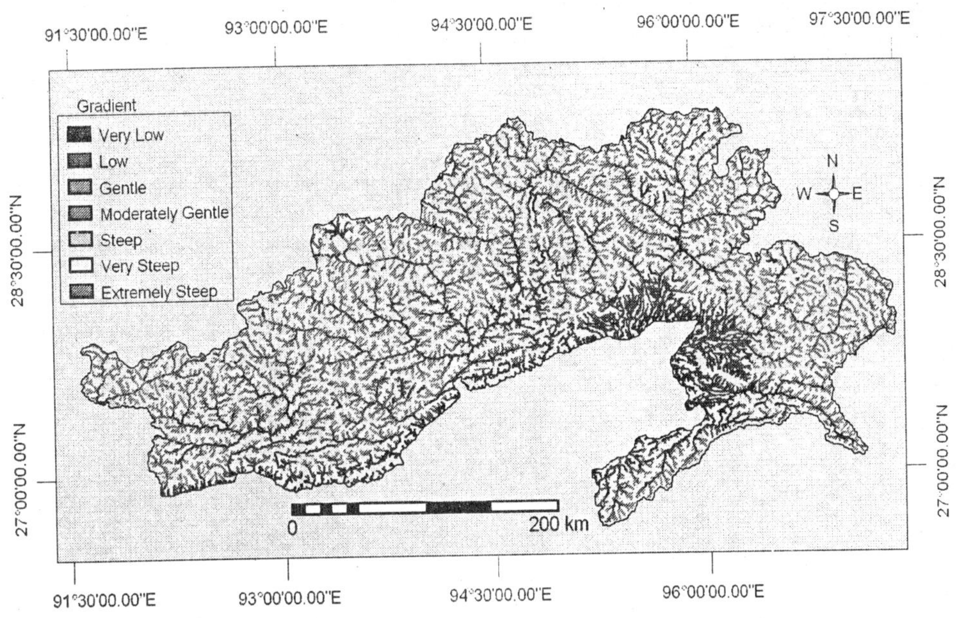

Figure 24.3. Stream gradient vs length of Arunachal Pradesh.

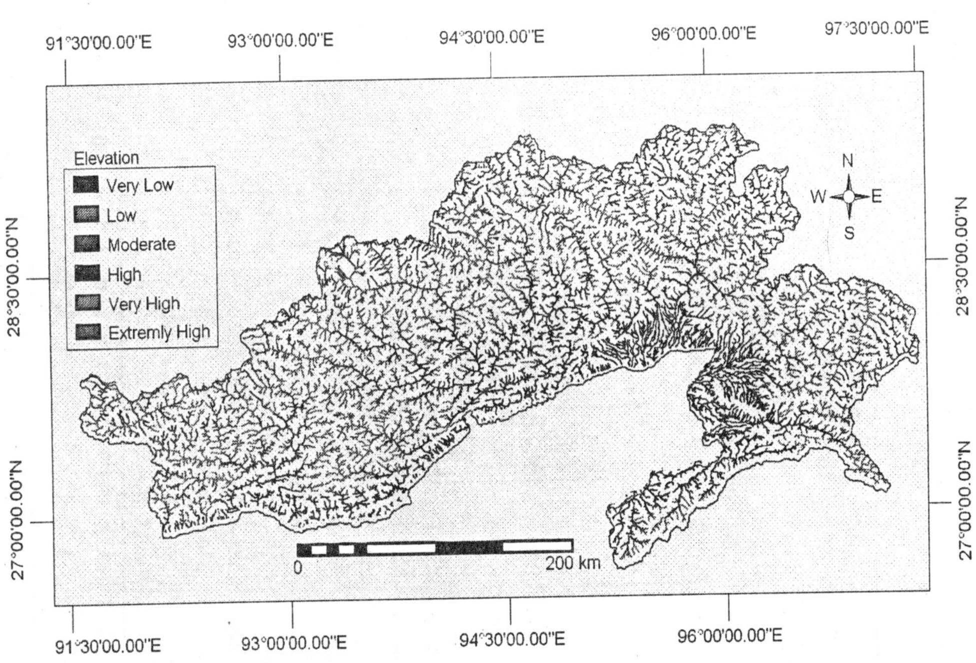

Figure 24.4. Stream elevation vs length of Arunachal Pradesh.

Table 24.1
Rainfall data (in mm), Arunachal Pradesh (1991 to 2005)

Stations	*1991*	*1992*	*1993*	*1994*	*1995*	*1996*	*1997*	*1998*	*1999*	*2000*	*2001*	*2002*	*2003*	*2004*	*2005*	*Average Rainfall*
Tawang	3377.6	2764.2	2816.6	2696.5	3413.6	3004.6	2580.9	2760.1	1755	1857.7	1291.4	1353.1	1543.3	1874.6	1737	2321.75
Bomdila	1680.1	1699	1616.6	1413.6	1904.9	1660.7	1885.1	1620.7	1536.6	1433.3	1384.6	2363.5	2049.1	2024.1	1474.6	1716.43
Seppa	2628.6	1605	2193	1407	1952	1864	1771	2596	1769	1742	1512.4	1703	1951	2143	2152	1932.60
Seijosa	2535	2417.3	3768.2	2807.7	2520.5	3021.2	2546	3655.9	2848.3	2753.5	2590.7	2575	3121.5	N/A	N/A	2858.52
Yingkiong	3056.4	2887.6	2786.2	3119.8	3107.6	3010	3324.8	3307.5	NA	2365.5	2878.5	2538.5	2911.8	1811	2588.7	2835.28
Tuting	N/A	N/A	N/A	N/A	N/A	N/A	4385.5	5388.5	4513.7	4933.6	4475.1	4203.8	N/A	3648	NA	4506.89
Pasighat	3234.6	2842.7	4752.6	2553.8	4037.9	4673.9	3035.8	5570.6	3419.2	3888.4	N/A	N/A	N/A	4696.8	4612	3943.19
Along	2322.8	2108.8	2573.4	1939.5	2433.5	2167.4	2270.6	2608.1	2048.5	2376.4	1676.5	2376.8	2427	2135.9	1933.1	2226.55
Likabali	N/A	N/A	N/A	N/A	N/A	N/A	3123.9	4360.2	3497.4	3180.7	2809.6	2455.6	3754.6	3756.6	3027.2	3329.53
Daporijo	2611.1	3226	2401.8	1604.3	1223.6	N/A	N/A	N/A	N/A	N/A	N/A	N/A	1645.2	1500.5	NA	2030.36
Ziro	3163.6	1069.4	1344.5	1144.4	1252.8	889.8	N/A	N/A	843	1131.6	1341.6	923.7	667.5	962.6	940.6	1205.78
Itanagar	3854.1	2682	6784.6	5763.6	2920.9	3338.3	3459.4	5171.6	3738.2	3094	2289.2	2864.6	3300.8	4250.2	2864.8	3758.42
Doimukh	3911.7	2937.4	3802.4	3163.3	3324.9	3496.4	3303.2	3914.8	3575.4	2707.2	2274.9	2954.2	3083.6	4226.7	2880.4	3303.77
Changlang	2708.8	2355.6	2768.4	1577	3575.5	1953.1	2716.9	2083	2126	1783.5	1792	1628.4	2435.3	3050	2469.4	2334.86
Manmao	3321.1	2904.2	2806.2	2437.4	3362.4	2552.5	2411.1	4128.1	2992.1	2691.6	2403.6	2616.6	2643.8	2512	2688.5	2831.41
Miao	2594.6	1757.4	2038.9	N/A	N/A	N/A	N/A	3970.4	3192	2823	2088.8	2159.2	2844	3163.5	2689.3	2665.55
Longding	3553.9	2884.2	3073.1	2377	2557.6	3068.1	2687.8	3321.2	2560.2	2577.2	2455.1	2430.9	3109.5	2494.9	2398.6	2769.95
Senuwa	2489.4	2491.7	1964.5	1173.4	4123.8	3006.2	1694.9	5754.8	3582	4334.3	5429.2	4867	6432	4774.2	5136.1	3816.90
Khonsa	4214.6	4379.8	3558	3168.5	2988.5	3307.3	3685.2	6208.7	4712.6	N/A	N/A	3807.9	4082.6	3904.1	NA	4001.48
Hayuliang	4323.6	3516.7	3838.5	2895.2	5147.5	4435	N/A	N/A	N/A	3393.4	N/A	N/A	N/A	N/A	N/A	3935.70
Tezu	3566.8	2871.8	5442.4	5588.4	5265.5	5816.8	2685.9	4258.5	4162.7	3965.3	N/A	3954.2	4002.1	5179	4823.1	4398.75
Namsai	3161.1	2351.4	4810.4	2298.1	N/A	N/A	N/A	N/A	N/A	N/A	N/A	N/A	N/A	N/A	N/A	3155.25
Roing	3616.2	2502.6	2844.6	2558.4	2893.4	3028.4	1239.4	5277.5	3825.6	4723.7	3418.8	3484	3889.8	4883.2	3977.6	3477.55
State's Annual Average	**3139.3**	**2583.6**	**3237.4**	**2584.3**	**3053.0**	**3016.3**	**2711.5**	**3997.7**	**2984.1**	**2887.8**	**2477.2**	**2697.9**	**2941.8**	**3149.5**	**2846.6**	**3015.50**

Source: Rural Work Department, Itanagar, Government of Arunachal Pradesh
Note: N/A means data not available

Table 24.2
Water resource of Arunachal Pradesh (after Sarmah 2008)

RIVERS	*Source Altitude (metre)*	*Length (Km)*	*Total basin (km^2)*	*Basin in AP (km^2)*	*Av. annual discharge ($m^3 s^{-1}$)*	*Sediment Yield (ton $km^{-2}y^{-1}$*	*Suspended Load (ha. m)*
Tawang	1577	111	1611	321	66		0.7004*
Kameng	4520	247	11,716	1476.2	349487	4721	2013
Ranganadi		160	2941	2206	74309	1598	186
Subansiri	5059	450	27400	16440	755771	959	992
Jia Dhal	1200	46	752	368	217	-	-
Dibang	5355	190	12270	-	-	6150	15500
Lohit	6614	360	-	23400	-	1960	14700
Buri Dihing		128	6000	1800	141539	1129	210

Note: *Discharge in millions tons, AP means Arunachal Pradesh, Dash (-) means data not available and Av. stands for average.

Table 24.3
Distribution of various types of wetlands in Arunachal Pradesh

Type of Inland Wetland	*No. of Wetland*	*Area of Wetland (in km^2)*
Lakes/ponds	380	30.99
Oxbow lakes	6	2.45
Waterlogged (seasonal)	396	451.87
Swamps/marshes	24	77.93
Total	806	563.24

Source: SAC (1998)

Table 24.4
Major wetlands in Arunachal Pradesh

Sl. No.	*Name of the Lake*	*Place/ District*
1	Mehao Lake	Roing (Dibang Valley)
2	Glao Lake	Kamlang Sanctuary, Anjaw
3	Dungpin tso	Tawang
4	Lamchuer tso	Tawang
5	Dhongchang tso	Tawang
6	Japang beel	D'Ering Memorial Sanctuary W/Siang
7	Mandela Phudung	West Kameng
8	Sangti marshes	West Kameng
9	Gorjo tso	Tawang
10	Sela Lake	Sela (Tawang)
11	Moti jheel-Rani jheel Raja jheel	Namdapha (Changlang)
12	Ganga Lake	Itanagar (Papum Pare)
13	PT tso	Tawang
14	Sally Lake	Lower Diband Valley
15	Sangetsar Lake	Tawang
16	Near Tawang Gompa (unnamed)	Tawang
17	Complex in Gashela and Y-Jn	Tawang
18	Near Y-Jn	Tawang

Source: Prasad *et al.* (2004).

Table 24.5
Power projects under implementation in Arunachal Pradesh

Name of projects	*Prime mover*	*Capacity (MW)*	*Agency*	*Status*
Ranganadi	Hydro	405	NEEPCO	Commissioned (2002)
Subansiri Lower	Hydro	2,000	NHPC	Under Construction
Kameng	Hydro	600	NEEPCO	Under construction
Ranganadi Stage-II	Hydro	130	NEEPCO	Investigation
Papum Pam	Hydro	100	NEEPCO	-do-
Pakke	Hydro	105	NEEPCO	-do-
Dikrong	Hydro	10	NEEPCO	Survey & Investigation
Siang Upper	Hydro	11,000	NHPC	-do-
Siang Middle	Hydro	700	NHPC	-do-
Siang Lower	Hydro	1,700	NHPC	-do-
Subansiri Upper	Hydro	2,500	NHPC	-do-
Subansiri Middle	Hydro	2,000	NHPC	-do-

Note: NEEPCO means North Eastern Electrical Power Corporation, and NHPC means National Hydel Power Corporation.
Source: NEDFI Data Bank Quarterly, October, 2002

Table 24.6
Hydroelectric projects for which MoA is signed

	Installed Capacity (MW)	*Annual Energy Generation (in MU)*	*Location (District)*	*Agency*	*Basis of Funding*
Tawang I Hydroelectric Project	750	3156.10	Tawang	NHPC	BOO
Tawang II Hydroelectric Project	750	3480.40	Tawang	NHPC	BOO
Dibang Multipurpose Project	3000	12270	Lower Dibang Valley	NHPC	GoAP and NHPC Joint venture
Kameng – I Hydroelectric Project	1120	4112.40	East Kameng	NEEPCO	BOO
Pare Hydroelectric Project	110	491.74	Papum Pare	NEEPCO	BOO
Etalin Hydro Electric Project	4000	16071.60	Dibang Valley	NTPC	BOO
Atunli Hydro Electric Project	500	2247.32	Dibang Valley	NTPC	BOO

Note: BOO stands for Build, Own and Operate and GoAP stands for Government of Arunachal Pradesh

Table 24.7
Stream gradient vs stream length

Gradient (degree)	*Category*	*Length (in km)*	*Length (%)*	*Cumulative (%)*
<2	Very Low	9897.07	34.12	34.12
2-5	Low	6393.94	22.05	56.17
5-10	Gentle	8118.82	27.99	84.16
10-15	Moderately Gentle	3592.97	12.39	96.55
15-20	Steep	787.33	2.71	99.26
20-30	Very Steep	210.85	0.73	99.99
>30	Extremely Steep	2.26	0.01	100
Total		29003.23	100	

Table 24.8
Stream elevation vs stream Length

Elevation (metre)	*Category*	*Stream Length (km)*	*Length (%)*	*Cumulative (%)*
<1000	Very Low	11293.88	38.94	38.94
1000-2000	Low	7668.34	26.44	65.38
2000-3000	Moderate	4954.20	17.08	82.46
3000-4000	High	4000.75	13.79	96.25
4000-5000	Very High	922.18	3.18	99.43
>5000	Extremely High	163.87	0.57	100
Total		29003.23	100	

The Rivers of Arunachal Pradesh

The Siang river, known as Tsangpo in Tibet and Brahmaputra in Assam, originates at an elevation of 5150 metre from Chema-Yung-Dung glacier, south-east of Manasarovar Lake in Tibet. It is a major international river covering 5,80,000 km^2, 50.50% of which is lying in China, 33.60% in India, 8.10% in Bangladesh and 7.80% in Bhutan. Out of its total basin in India, 41.88% is shared by Arunachal Pradesh alone, 36.33% by Assam, 5.57% by Nagaland, 6.10%by Mehalaya, 3.75% by Sikkim and 6.47% by West Bengal (Goswami, 2008). It enters Arunachal Pradesh at an elevation of 3500 metre and come across Greater, Middle and Sub-Himalaya finally to debouch Assam Plain near Pasighat at 155 m altitude and travelling 278 km distance. These different geo-ecological zones have distinctive topographical, geological, climatological and floral characteristics (Goswami, 2008). The gradient of the river near gorge ranges from 4.3 to 7 metre/km and 0.27 metre/km near Pasighat. It cascades through one of the deepest gorges and suddenly debouches the wide plain unloading large amount of sediments. The river is characterized by marked seasonality and high variability of daily flow. The highest recorded daily discharge at Pandu was 72, 726 m^{3}s^{-1} in August 1962 while the lowest was 1,757 m^3 s^{-1} in February 1968 (Goswami and Das, 2003). The Dihing, Dibang, Lohit and Subansiri are estimated to contribute more than 60% of the average flow of the Brahmaputra (Bora, 2008).

The Tawang river is characterized by high magnitude of water and sediment. The water discharge during monsoon and lean season are 111 m^3 s^{-1} and 19 m^3 s^{-1} respectively. The maximum discharge and average flood discharge of the river is recorded to be 1388 m^3 s^{-1} and 590 m^3 s^{-1} respectively. The Kameng originates at 4520 metre and traverses 247 km before joining the Barahmaputra. Its length in Arunachal Pradesh is 190 km and total basin area of 11,716 km^2, out of which 1476.2 km^2 falls in Arunachal Pradesh. The average annual water discharge is 74, 309 m^3s^{-1} (Table 24.2). Ranganadi covers 2941 km^2 basin area out of which 2206 km^2 is in Arunachal Pradesh and rest in Assam. The annual average discharge is 74, 309 m^3s^{-1}.

The Subansiri river originates at 5059 m altitude and traverses a length of 450 km. Out of the total 27,400 km^2 basin area, 16,440 km^2 lies in Arunachal Pradesh. Its average annual discharge is 755,771 m^3s^{-1}. The Jia Dhal originates at 1200 m altitude and covers 46 km of total length. The basin area in Arunachal Pradesh is 368 km^2 with an annual average discharge of 217 m^3 s^{-1}. Dibang originates at an altitude of 5355 m and traverse 190 km to join Brahmaputra. The total basin area is 12,270 km^2. Lohit originates in China at an elevation of 6614 m and enters Arunachal Pradesh near Du pass at 5233 m elevation. Its total basin area and length in Arunachal Pradesh are 23, 400 km^2, and 190 km respectively (Singh, 1999). The Buri Dihing, in the southern bank of Brahmaputra, traverses a flow distance of 128 km. The total basin area of the river is approximately 6000 km^2, out of which 1800 km^2 is in Arunachal Pradesh. It has an average annual discharge of 141,539 m^3s^{-1}. However, these rivers also discharge huge amount of sediments and suspended loads (Table 24.2). The Brahmaputra, as a whole, is considered as second most heavily sediment-charged large rivers in the world after the Yellow (Hwang Ho) river in China. Its sediment transported per unit drainage area is 1,128 tons km^{-2} $year^{-1}$ at Bahadurabad in Bangladesh. The river carries an average annual suspended load of 400 million metric tonnes at Pandu with an average daily rate of nearly two million metric tonnes in the rainy season (Goswami and Das, 2003).

Wetlands in Arunachal Pradesh

According to SAC (1998) Arunachal Pradesh has 806 wetlands covering an area of about 563 sq km which forr · 0.67% of the total area of the state. These wetlands are divided into four types – lakes/ponds, oxbow lakes, waterlogged ans swamps/marshes (Table 24.3). However, Sahariah and Bora (2008) describes three types of wetlands in Arunachal Pradesh – freshwater marsh, riparian wetlands and Alpine wetlands. The oxbow lakes are not found in all parts of Arunachal Pradesh but mostly along the rivers Noa-Dihing, Lohit and Siang. Most of them are being silted up because of the reclamation of hills in the catchments. There mehao lake in Dibang Valley is of tectonic/ landslide origin, which was created during Earthquake of 1950. However, there are innumerable glacial lakes in the Great Himalaya and a few in the Mishmi hills. These high altitude lakes are less studied and there is dearth of detail information. There is one artificial reservoir near Yazali in Paum Pare district due to the construction of dam on Ranga Nadi. The Table 24.4 shows the major wetlands of Arunachal Pradesh.

Ground Water Resource

Ground water potential in the State is equally impressive. The quality of ground water in the State is good for domestic, agricultural and industrial use. The ground water has pH value ranging from 6.5 – 8.5, and its electrical conductivity varies from 46 to 316 micro mhos cm^{-1} at 25° C. All other cations and anions are within the permissible limit. The following suggestions are given by the Central Ground Water Board (2004) to augment water resources: to construct deep tube wells, and to construct bunds/check dams, etc., on perennial streams in the valleys and to develop perennial springs and to provide storage for better supply by applying scientific methods, such as arresting rain water for drinking and creating artificial recharge through construction of suitable roof top rain water harvesting structures in the hilly areas.

Hydropower Resources

The vast resource of Arunachal Pradesh can effectively be converted into hydro-power resource, besides drinking, irrigation and other industrial purposes. Large river system along with the structure of the landforms provides ideal conditions to generate large amount of electricity. According to various estimates, the state has one third of the country's total hydropower potentials. The Central Electricity Authority (CEA) estimates the capacity of State to produce 50,238 MW of hydro power from the 89 major identified projects.

The Department of Power, Government of Arunachal Pradesh has commissioned 35 mini hydel projects with an installed capacity totaling to 463.95 MW. North Eastern Electric Power Corporation (NEEPCO) and Brahmaputra Board have taken several steps for setting mega projects in Arunachal Pradesh. Ranganadi project with installed capacity of 405 MW is already commissioned in March 2002 by NEEPCO. Some other projects like Kameng project (600MW), Subansiri project (4500 MW), Siang project (2000MW) and Damwe project (600MW) identified by NEEPCO, Brahmaputra Board and Central Water Commission are in initial stage of survey and investigation. Besides, there are number of projects, as presented in Table 24.5, under different stages of consideration and investigation with capacity ranging from 10 MW to 11000 MW: Since 94% of State's hydro power potential is yet to be tapped, the 10th and 11th Plans envisaged the generation of one lakh MW of power in the state. In a landmark move, the State Government in September 2006 entered a Memorandum of Agreement (MoA) with National Hydro Power Corporation Ltd (NHPC), National Thermal Power Corporation Ltd (NTPC) and NEEPCO for harnessing 10,230 MW of hydroelectricity out of the CEA estimated potentiality of 50,238 MW. The Table 24.6 shows the projects for which MoA has been entered into.

Gradient-wise Stream Lengths

With the help of slope, elevation and drainage network derived from the SRTM DEM, the stream lengths under different gradients and elevations have been analysed. The gradient of the river indicates the flow velocity and the erosion capacity. In the areas with low gradient, velocity decreases which reduce both erosion and transportation capacities. The river starts precipitating sediments resulting river braiding, riparian condition, delta formation, sand bar formation, etc. The steep gradient on the other hand results high velocity, rapids, cascades, waterfalls and turbidity. The total length of the rivers with a threshold value of 500 pixels is 29003.23 km distributed along the gradients varying from 0Ú to 39.90Ú. This range of gradients has been further classified into seven categories and length of river in each category has been obtained and presented in Table 24.7.

The stream lengths in the different categories of the stream gradients are shown Figure 24.3. The maximum stream lengths i.e. 9897 km (32.12%) is found in the gradient below 2Ú which is put under very low category. It is followed by the low gradient (2-5Ú) which accounts 6393 km (22.05%) stream lengths. The gentle and moderately gentle categories with gradient from 5 to 10Ú and 10 to 15Ú have stream length of 8119 km (27.99%) and 35.92 km (12.39%) respectively. The steep gradient varying from 15 to 20Ú and very Steep river having gradient between 20 to 30Ú have the stream length of 787.33 km (2.71%) and 210.85 km (0.73%) respectively. The extremely steep gradient with gradient of bed more than 30Ú comprises only 2.26 km of the river lengths which account 0.01% of the total.

Elevation-wise Stream Lengths

The elevation directly influence the climate specially the distribution of temperature and orographic rainfall thereby other physical and chemical properties of the river water. The elevation also influences the control on vegetational adaptation of the area. The vegetation on the other hand influences runoff, infiltration and sediment supply. The elevation along the stream line in Arunachal Pradesh ranges from 107 metre in the southern part to 5699 metre in the northern margins. This elevation have been divided into very low, low, moderate, high, very high and extremely high elevations categories. The length of the river in each elevation category has been computed as in Table 24.8 and Figure 24.4.

The elevations below 1000 m have been kept under the very low stream elevation category which accounts 11293.88 km (38.94%) river length. The low elevation category ranging between 1000 to 2000 m has the river length of 7668.34 km which accounts 26.44% of the total. The moderate stream elevation (2000 to 3000 m) and high stream elevation (3000 to 4000 m) constitutes 4954.20 km (17.08%) and 4000.75 km (13.79%) length of river respectively. The stream length in the very high elevation (4000 to 5000 m) and extremely high elevation (>5000m) progressively decreases to 922.18 km (3.18%) and 163.87 km (0.57%) respectively.

Conclusion

Arunachal Pradesh has plentiful of perennial surface water and good quality ground water resources. The 41% of the basin of river Brahmaputra lies in this state. About 50% river length in Arunachal Pradesh is under the stream gradient of low and gentle stream gradient. The river lengths pass through wide range of altitude ranging from 107 to 5699 metre. The maximum stream lengths, however is in very low and low altitudes which together accounts about 64% of the total river length. Due to the hilly and mountainous nature of the area and very high monsoon rainfall, the discharges of the rivers are very high. This makes the state the potential power house of the country with about 50% of the power production capacity. However, the fragile geological structure and the rich biodiversity of the State stood in the way to tap the vast hydro-potentiality. Therefore, much precision is required before taking any step to harness the power potentiality. At the same time the sparsely distributed population over the geographical area of Arunachal Pradesh makes it difficult to ensure power distribution. Considering all these, a stand-alone mini hydel project would be more ideal to meet the energy requirement of the people. On the other hand, these rivers also transport huge quantity of sediments and suspended loads. Proper measures of catchments treatment and land use planning in the watershed basins are imperative to check the sediments delivery.

REFERENCES

Bora, A.K. 2008. Water Resources of North-East India: Potentials, Development and Management, pp. 173-197. In: *North-East India Geo-Environmental Issues* (Ed: S. Deka,). EBH Publication, Guwahati.

Central Ground Water Board. 2004. *Ground Water Resources of Arunachal Pradesh,* NE Region. Central Ground Water Board, Guwahati. 10 pp.

Dwivedi, R.S., Sreenivas, K., Ramana, K.V., Reddy, P.R. and Sankar, G.R. 2006. Sustainable Development of Land and Water Resources Using Geographic Information System and Remote Sensing. *Journal of the Indian Society of Remote Sensing,* 34(4): 351-367

Goswami, D.C. 2008 Fluvial Geomorphology of the Brahmaputra River, Assam: An Overview, pp. 31-61. In: *North-East India Geo-Environmental Issues* (ed: S. Deka,). EBH Publication, Guwahati.

Goswami, D.C. and Das, P.J. 2003. The Brahmaputra River, India : The eco-hydrological Context of Water use in One of World's Most unique River Systems, Ecologists Asia. *Special Issue on large dams in north east India-rivers, forests, people and power,* 11 (1): 9-14

Jayaprasad, P., Narender, B., Pathan, S.K. and Ajai. 2008. Generation and Validation of DEM using SAR Interferometry and Differential GPS Supported by Multispectral Optical Data. *Journal of the Indian Society of Remote Sensing,* 36(4): 313-322

Kumar, S., Kumar, A., Saha S.K. and Kumar, A. 2008. Stereo Cartosat-1 Satellite Remote Sensing Data in Assessing Topographic Potential of Soil Erosion. *Journal of the Indian Society of Remote Sensing,* 36: 159-165

Prasad, S.N., Jaggi, A.K., Tiwari, A.K., Kaushik P., Vijayan, L., Muralidharan, S. and Vijayan, V.S. 2004. *Inland Wetlands of India: conservation Atlas.* SACON Publication, Coimbatore, 222 pp.

Rajitha, K., Mukherjee, C.K. and Chandran, V. 2007. Applications of remote sensing and GIS for sustainable management of shrimp culture in India. *Aquaculture engineering,* 36 (1): 1-17

Sahariah, D. and Bora, A.K. 2008. *Wetlands of North-East India,* pp. 249-272. In *North-East India Geo-Environmental Issues* (ed. S. Deka,). EBH Publication, Guwahati.

Sarmah, R. 2008. Glimpse of the River Basins of North-East India, pp. 63-96. In: *North-East India Geo-Environmental Issues* (Ed: S. Deka,). EBH Publication, Guwahati.

Singh, S. 1999. *A Resource Atlas of Arunachal Pradesh.* Department of Planning, Government of Arunachal Pradesh. 161 pp.

25

Ecology and Biodiversity of Ningthoukhong River, Manipur, India

—*S. Shyamjai Singh*[1], *Haobijam Dhamendra*[2] & *Laishram Kosygin*[3]

[An attempt has been made to evaluate physico-chemical characteristics and aquatic biodiversity of the Ningthoukhong river. The river water is considered eutophic, as it has high concentrations of nutrients. Seventeen species of fishes which belongs to 14 genera and 11 families and 30 species of macrophytes which belongs to 27 genera and 14 families were recoreded from the river. The floral and faunal compositions of the river were mostly widely distributed forms that can survive in polluted environments. The river may be considered as a threat to the Loktak lake as its nutrient concentration is exceeding the tolerance limit of any tributaries discharging into a lake.

Key words: Ningthoukhong river, physico-chemical characteristics, biodiversity, eutrophic.]

Introduction

The Ningthoukhng river originates from Churchanpur district of Manipur. It is a small river formed by the combination of two rivulets arising one from the Sadu hill and another from the Tukhul (kom) villages of the Laimaton ranges of Churachanpur District, Manipur. The two tributaries are met at the downstream near Turel Maru. This river flows further downstream and after passing the Ningthoukhong munical area joins the Loktak lake in Bishnupur district. The Loktak Lift Irrigation canal (LLI) also discharged into the river which provides water to the villagers for their domestic and irrigation uses in the dry season. The river is located at about 35 km south from Imphal, the capital city of state. It has deposited heavy soil load at the mouth of the stream in the lake.

Assessments of river water quality have been done by various workers in India (Sangu and Sharma, 1985; Das and Sinha, 1993; Jameson and Rana, 1996; Sharma and Agarwal, 1999; Mahanta and Patra, 2000; Baruah and Baruah, 2003; Kumar, 2003; Kulshrestha and Sharma, 2006). In Manipur there are numerous rivers and streams, which are directly or indirectly feeding the important lakes of the state and many of them, enhancing the process of eutrophication and pollution in the

[1,2] Lake Monitoring and Research Laboratory, Loktak Development Authority, Ningthoukhong 795126, Manipur, India

[3] Zoological Survey of India, Hilltop, Gopalpur-on-Sea 761 002, Orissa

lake ecosystem as they discharge huge amounts of nutrients and polluted water (Tombi Singh and Shyamanda Singh, 1994; Trisal and Manihar, 2004). However, there are few reports on the status of riverine ecosystem of the state (Kosygin and Dhamendra, 2005; Rajeshwari Devi *et al.*, 2005; Dhamendra and Kosygin, 2005; Kosygin *et al.*, 2007 and Kosygin *et al.*, 2009). In the present study an attempt has been made to study the water quality and biodiversity of the Ningthoukhong river.

Materials and Methods

Surface water samples were collected from the downstream of Ningthoukhong river at Ningthoukhong on quarterly basis during March 2004 to March 2005. For each season, five replicates of water samples were collected and physico-chemical and biological parameters were analyzed following APHA (1989) and Trivedi and Goyal (1986). Surveys on the fish fauna and macrophytes of the river were conducted in the downstream of the river upto its mounth in the Loktak lake. Fishes were identified following Jayaram (1981) and Talwar and Jhingran (1991). Aquatic Macrophytes were identified following Adoni *et al.* (1985) and by comparing with the preserved herbaria available in the Loktak Development Authority, Manipur.

Results and Discussion

Physico-chemical characteristics

The result of physico-chemical characteristics of the Ningthoukhong river are summarised in Table 25.1. It indicates water temperature varies from 20.0 °C – 24.6°C with an average value of 22.5°C. The maximum value is observed in the month of September. The pH value ranged from 7 to 7.3 with an average value of 7.1. Therefore, the river water may be considered slightly alkaline in nature. It is negatively correlated with water temperature (r= -0.275). Dissolved oxygen ranges from 4.2 to 6.4 mg/l with an average value of 5.3 mg/l. The free CO_2 concentration was usually high, which ranged from 7.6 to 17.7 mg/l with an average value of 12.5 mg/l. The pH and DO levels of the river were within the tolerance limit of ISI (1982) whereas CO_2 exceeded the tolerance limit. High concentration of free CO_2 may be attributed to the heavy discharged of organic waste into the river from the Ninthoukhong Market and surrounding human settlements etc. Concentration of Cl ranged from 5.2mg/ to 18.5 mg/l with average value of 9.04 mg/l during studies period. The concentrations of sodium ranged from 15.0mg/l to 29.0 mg/l with an average value of 20.4 mg/l. Potassium concentration ranged from 2.0 – 7.0 mg/l with an average value 3.8 mg/l. Concentration of calcium ranged from 4.5mg/l during September to 11.5 mg/l during March with an average value of 7.02mg/l. All the values of sodium, calcium and potassium occurred within the tolerance limit of ISI (1982). The concentration of nutrients in the river water was high. The concentration of dissolved inorganic phosphate phosphorus (DIPP) in the present investigation was observed to vary from 0.069 mg/l to 0.152 mg/l with an average value of 0.097mg/l. The average concentration of inorganic phosphorus (0.097mg/l) in the present investigation was found to be higher than the standard permissible limit recommended by USEPA (1976), for any tributary discharging to a lake (Trivedy and Goyal, 1986). The high concentration of nutrients in Ningthoukhong river may be attributed to the fact that the river receives huge amount of domestic and municipal waste from the Ninthoukhong Market, apart from the surface runoff and effluents discharging from the catchment area and surrounding agricultural fields.

Biodiversity

Fishes and macrophytes identified during the present study are shown in Table 25.2 and 25.3 respectively. The collection included 17 species of fishes which belongs to 14 genera and 11 families and 30 species of macrophytes which belongs to 27 genera and 14 families. Most of the flora and fauna were widely distributed forms, which can thrive well in the polluted waters. Heavy infestation of the macrophytes near the Loktak lake shows that the river is rich in nutrients and these macrophytes may be considered as nutrient sink.

The river water may be considered suitable for fish culture and wildlife propagation as the pH and DO values are within the limit of Class D water (ISI, 1982). However, the value of free CO_2 exceeds the tolerance limit (6 mg/l). Therefore, water quality of the river needs to be improved, if it is to be used for fishery purposes.

From the above discussion, it is clear that down stream of the Ningthoukhong river is eutophic as it has high concentration of nutrients, free CO_2 and infestation of macrophytes at the mouth area. The river water also seems to be polluted as it has occurrence of flora and fauna that can thrive well in the polluted environment. It may be considered as a threat to the Loktak lake ecosystem as its nutrient concentration is exceeding the tolerance limit of any tributaries discharging into a lake. Proper conservation measures may be taken up in order to save the river and the Loktak lake from further deterioration.

Table 25.1
Physico-chemical characteristics of Ningthoukhong rivers during March, 2004 to March 2005

Parameters	*Range*		*Mean*	*sd*
	Min.	*Max.*		
Water Temperature (°C)	19.0	27.6	22.9	2.3
pH	7.0	7.3	7.12	0.13
Dissolved oxygen (mg/l)	4.2	6.4	5.3	0.64
Free CO_2 (mg/l)	7.6	17.7	12.5	3.1
TDS (mg/l)	158	226	176.8	28.1
Chloride (mg/l)	5.2	18.5	9.04	5.4
Calcium (mg/l)	4.5	11.5	7.02	2.7
Magnesium (mg/l)	3.2	8.2	5.24	1.9
Sodium (mg/l)	15.0	29.0	20.4	5.6
Potassium (mg/l)	2.0	7.0	3.8	1.9
Nitrate-nitrogen (mg/l)	0.428	0.942	0.67	0.2
Inorganic phosphorus (mg/l)	0.059	0.098	0.078	0.03

Note: All the values are in mg/l, except temperature and pH.

Table 25.2
Fish fauna of Ningthoukhong river, Manipur

	Scientific Name	*Local Name*	*Family*	*Fishery Potentials*
1.	*Puntius sophore* (Hamilton)	Phabounga	Cyprinidae	Medium
2.	*Puntius manipurensis* (Menon, Rema and Vishwanath)	Ngakha meingangbi	Cyprinidae	Low
3.	*Amblypharyngodon mola* (Hamilton)	Mukanga	Cyprinidae	Low
4.	*Esomus danricus* (Hamilton)	Ngasang	Cyprinidae	Low
5.	*Aplocheilus panchax* (Hamilton)	Lameithanbi	Aplocheilidae	Low
6.	*Clarias batrachus* (Linnaeus)	Ngakra	Clariidae	Medium
7.	*Heteropneustes fossilis* (Bloch)	Ngachik	Heteropneustidae	Low
8.	*Monopterus albus* (Zuiew)	Ngaprum	Symbranchidae	Medium
9.	*Chanda nama* (Hamilton)	Ngamhai	Chandidae	Low
10.	*Anabas testudineus* (Bloch)	Ukabi	Anabantidae	Medium
11.	*Colisa* sp.	Ngabemma	Belontiidae	Low
12.	*Channa gachua* (Hamilton)	Meitei ngamu	Channidae	Medium
13.	*Channa punctatus* (Hamilton)	Porom	Channidae	Medium
14.	*Channa striatus* (Bloch)	Ngamu bogra	Channidae	Medium
15.	*Mystus* sp.	Ngasep	Bagridae	Medium
16.	*Glossogobius giuris* (Hamilton)	Nailonngamu	Gobiidae	Low

Table 25.3
Floral diversity of Ningthoukhong river along with their economic utilization

	Botanical name	*Local name*	*Family*	*Economic utilization*
1.	*Ageratum conizoides* Linn	Khongjainapi	Asteraceae	*NA*
2.	*Polygonum glabrum* Willd	Chouhaisabi	Polygonaceae	*NA*
3.	*Polygonum hydropiper* Linn	Lilhar	Polygonaceae	WE- the tender leave taken as vegetable.
4.	*Polygonum barbatum* Linn	Yellang	Polygonaceae	WE- tender shoot used as vegetable.
5.	*Polygonum barbatum* Linn	Moroksabi	Polygonaceae	*NA*
6.	*Equisetum sp.*	Horse tail plant	*Equisitaceae*	*NA*
7.	*Dryopteris marginata* (Wall) Christ	Laichangkhrang	*Polypodiaceae*	*NA*
8.	*Echinocloa stagnina* Retz	Hoop	*Poaceae*	MED, CUL- the twigs.
9.	*Colocasia esculenta* Linn Schott	Lambal	*Araceae*	WE-the green leaves used for making curry and feed for pig.

(Contd...)

Botanical name	*Local name*	*Family*	*Economic utilization.*
10. *Alternanthera pheloxeroides* (Mart) Griseb	Kabonapi	*Ameranthaceae*	FOD- whole plant used as fodder.
11. *Selaginella monosrerme* Spring	Eesang	*Selaginallaceae*	O- fish feed.
12. *Paspalum species*	Napakpi	*Poaceae*	FOD- fodder for cattle.
13. *Monochoria hastaefolia* Presl.	Kakla	*Pontederiaceae*	MED-plant tonic, cooling, insanity, juice in boils.
14. *Saccharum munja* Roxb.	Khoimom	*Poaceae*	MED- stem refrigerant, aphrodisiac; O- the floral product used in making pillow, mattress etc.
15. *Mikania cordata* Burm.	Uri-hingchabi	*Asteraceae*	WE- tribal people eat the tender leaves; O- fish poisoning.
16. *Gynura angulosa* Var. *G. cusimba* (D. Don) Moore	Terapaibi	*Asteraceae*	MED-cuts and wound, the leaves used for stomach ulcer.
17. *Rubus ellipticus* Sm.	Heijampet	*Rosaceae*	WE- the red fruit
18. *Spilanthus acmella* Murr.	Manjreng	*Asteraceae*	MED- the inflorescence used in blood pressure control.
19. *Osbeckia nepalensis* Hk.f.	Yachubi/ Tejpatmanbi	*Melastomaceae*	NA
20. *Primula euosma* Craib	Hidakma-namanbi	*Primulaceae*	NA
21. *Jussiaea suffroticosa* (L)	Cheiteklei	*Onagraceae*	NA
22. *Gnaphalium luteoalbum* (L)	Phunil	*Asteraceae*	WE- tender shoot eaten as vegetable.
23. *Alocasia indica* (Roxb.) Schott	Hongu	*Araceae*	CUL, WE-stem added to a traditional fish fermentation- <u>*hentak*.</u>
24. *Rumex maritimus* Linn	Torong khongchak	*Polygonaceae*	WE- tender leaves.
25. *Brachiaria mutica* (Forssk) Stapf	Paragrass	*Poaceae*	FOD- tender

(Contd...)

Botanical name	*Local name*	*Family*	*Economic utilization.*
26. *Salvinia cucullata* Roxb.	Salvinia	*Salviniaceae*	O- fertilizer
27. *Eichhornia crassipes* (Mart.) Solms.	Kabokang	*Pontederiaceae*	WE-shoot; MED-flower for skin of horse.
28. *Cynodon dactylon* Linn	Tingthou	*Poaceae*	CUL, MED- cuts and wound, dropsy, hysteria, epilepsy, tonic.
29. *Phragmites karka* (Retz) Trin	Tou	*Poaceae*	O- stem used for fencing and firewood.
30. *Arundo donax* Linn	Yengthou	*Poaceae*	MED- rhizome useful in emollient, dieuretic, stimulant, retard lactation; O-firewood, stem making mats and baskets.

Note:-

WE - Wild edible	CUL - Cultural
MED - Medicinal	HM – House making
FOD - Fodder	O - Others

[Acknowledgement: We are grateful to Shri Th. Manihar, Former Project Director, Loktak Development Authority, Manipur for providing laboratory facilities. We are also thankful to Shri P. Shyamkumar Singh, Shri M. Birabal Singh, Shri Deban Singh, Shri Kh. Gandhi Singh, and Shri Chotelal Sharma of Loktak Development Authority, Manipur for their valuable supports and help.]

REFERENCES

Adoni, A. D. 1985. *Work Book on Limnology*. Pratibha Publishers, Sagar. 216 pp.

APHA-AWWA and WPCF. 1989. *Standard methods for the examination of waters and waste waters.* 17th Edn. American Public Health Association, Inc. New York.

Baruah, B. K. and Baruah, D. 2003. Study on water quality of Subansiri river in Asssam – An EIA approach for a proposed hydroelectric power project. *Ind. J. Environ. Ecoplan.*, 7, 381-384.

Das, N.K. and Sinha, R.K. 1993. Assessment of water quality of Ganga river, Patna, *India. Environ. Ecol.*, 11, 829-832.

Dhamendra, H. and Kosygin, L. 2005. Water quality and aquatic bio-diversity of Merakhong River, Manipur, pp. 193-200. In *Fish Management and Aquatic Environment* (Kumar A. and Dubey, P., Eds.). Daya Publishing House, Delhi.

ISI. 1982. Indian standard - Tolerance limit for inland surface waters subject to pollution (second revision). IS: 2296. Indian Standard Institution Publication, New Delhi, 18 pp.

Jameson, J. and Rana, B.C. 1996. Pollution status of the river complex Sabarmati at Kheda region of Gujrat 1. Physico-chemical characteristics. *Pollut. Res.*, 15, 53-55.

Jayaram. K.C. 1981. *Freshwater fishes of India, Pakistan, Bangladesh, Burma, and Shri Lanka.* Handbook , Zoological Survey of India, Calcutta, 475pp.

Kosygin, L. and Dhamendra, H. 2005. Limnological studies of Potsangbam river, Manipur. Pp. 16-24. In *Fundamentals of Limnology* (ed. Kumar A.). APH Publication, New Delhi.

Kosygin, L., Dhamendra, H. and Gyaneshwari, R.K. 2007. Pollution status and conservation strategies of Moirang river, Manipur with a note on its aquatic bio-resources. *J. Environmental Biology*, 28 (3): 669-673.

Kosygin, L., Premananda, N. and Dhamendra, H. 2009. Water quality and fish fauna of the Irumbi River, Manipur with a note on prospects of fish and shellfish culture. Pp 293-300. In *Wetlands of North East India: Ecology, Aquatic Bioresources and Conservation* (ed. Kosygin L.). Akansha Publishing House, New Delhi.

Kosygin, L. and Dhamendra, H. 2009. Physico-Chemical and Biological Characteristics of Awang Khujairok River, Manipur, India. *Environment and Ecology*, 27(1): 82-85.

Kumar, A. 2003. Ecology of river Mayurakshi in Jharkhand, India. Pp. 01-38. In: *Ecology and conservation of lakes, reservoirs and rivers* (ed. A. Kumar). APH Publishing Corporation, New Delhi.

Kulshrestha, H. and Sharma, S. 2006. Impact of mass bathing during Ardhkumbh on water quality status of river Ganga. *J. Environ. Biol.*, 27, 437-440.

Mahanta, B.K. and Patra, A.K. 2000. Studies on the water quality index of river Sanamachhakandana at Keonjhar Garh, Orissa, India. *Pollut. Res.*, 19, 371-385.

Rajeshwari Devi, R.K., Dhamendra, H., Gyaneshwari. R.K. and Kosygin, L. 2005. Limnological studies of Nambol river, Manipur with a note on its aquatic bio-resources. *Indian J. Environ & Ecoplan.*, 10(3): 831-834.

Sand-Jensen K. 2001. Freshwater ecosystems, human impact on. Pp. 89 – 108. In *Encyclopedia of Biodiversity* (ed. S. A. Levin), Volume 3. Academic Press, New York.

Sangu, R. P.S. and Sharma, K.D. 1985. Studies of water pollution of Yamuna river at Agra. *Ind. J. Environ. Hlth.*, 27, 257-261.

Sharma, B.S. and Agarwal, A. 1999. Assessment of water quality of river Yamuna at Agra. *Pollut. Res.*, 18, 109-110.

Talwar, P. K. and Jhinhran, A. G. 1991. *Inland fishes of India and adjacent countries*, I, Oxford and IBH Publ. Co. Pvt. Ltd. New Delhi, Bombay, Calcutta, 541 pp.

Tombi Singh, H. and Shyamanda Singh, R.K. 1994. *Ramsar sites of India Loktak lake Manipur.* World Wide Fund for Nature, India.

Trivedy, R.K. and Goyal, P. K. 1986. *Chemical and biological methods for water pollution studies.* Environmental Publication, Karad, 248 pp.

U.S.E.P.A. 1976. *Quality Criteria for water.* EPA-440/9-76-023, Washington D.C.

Index

V

W

Z

■■■